A Level
Advancing
Physics
for OCR

Third Edition

B

Series Editor
Lawrence Herklots

Authors
Lawrence Herklots
John Miller
Helen Reynolds

Great Clarendon Street, Oxford, OX2 6DP, United Kingdom

Oxford University Press is a department of the University of Oxford. It furthers the University's objective of excellence in research, scholarship, and education by publishing worldwide. Oxford is a registered trade mark of Oxford University Press in the UK and in certain other countries

British Library Cataloguing in Publication Data
Data available

978-0-19-834094-2

10 9 8 7 6 5 4

Paper used in the production of this book is a natural, recyclable product made from wood grown in sustainable forests. The manufacturing process conforms to the environmental regulations of the country of origin.

Printed and bound by CPI Group (UK) Ltd, Croydon, CR0 4YY

This resource is endorsed by OCR for use with the specification H157 AS Level Physics B (Advancing Physics) and with the specification H557 A Level Physics B (Advancing Physics). In order to gain endorsement this resource has undergone an independent quality check. OCR has not paid for the production of this resource, nor does OCR receive any royalties from its sale. For more information about the endorsement process please visit the OCR website www.ocr.org.uk

Index compiled by INDEXING SPECIALISTS (UK) Ltd., Indexing House, 306A Portland Road, Hove, East Sussex BN3 5LP United Kingdom

This book has been written to support students studying for OCR A Level Physics B. It covers all A Level modules from the specification. The content covered is shown in the contents list, which also shows you the page numbers for the main topics within each chapter. There is also an index at the back to help you find what you are looking for. If you are studying for OCR AS Physics B, you will only need to know the content in the blue box.

AS exam

A level exam

Year 1 content

1   Development of practical skills
2   Fundamental data analysis
3.1 Imaging, signalling, and sensing
3.2 Mechanical properties of materials
4.1 Waves and quantum behaviour
4.2 Space, time, and motion

Year 2 content

5.1 Creating models
5.2 Matter
6.1 Fields
6.2 Fundamental particles

A Level exams will cover content from Year 1 and Year 2 and will be at a higher demand. You will also carry out practical activities throughout your course.

This book contains many different features. Each feature is designed to support and develop the skills you will need for your examinations, as well as foster and stimulate your interest in physics.

Terms that you will need to be able to define and understand are highlighted by **bold text**.

Practical features

These features support further development of your practical skills, and cover the required practicals for this course.

Extension features

These features contain material that is beyond the specification. They are designed to stretch and provide you with a broader knowledge and understanding and lead the way into the types of thinking and areas you might study in further education. As such, neither the detail nor the depth of questioning will be required for the examinations. But this book is about more than getting through the examinations.

1   Extension features also contain questions that link the off-specification material back to your course.

Summary Questions

1   These are short questions at the end of each topic.

2   They test your understanding of the topic and allow you to apply the knowledge and skills you have acquired.

3   The questions are ramped in order of difficulty. Lower-demand questions have a paler background, with the higher-demand questions having a darker background. Try to attempt every question you can, to help you achieve your best in the exams.

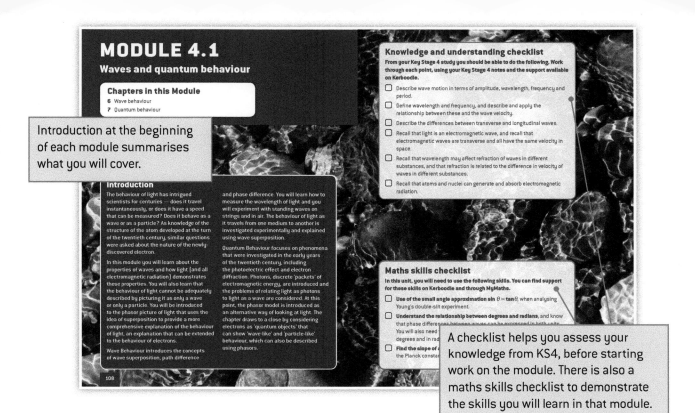

Introduction at the beginning of each module summarises what you will cover.

A checklist helps you assess your knowledge from KS4, before starting work on the module. There is also a maths skills checklist to demonstrate the skills you will learn in that module.

Module summaries highlight the key concepts of each module. They break down the important facts and formulas by chapter and topic from across the module.

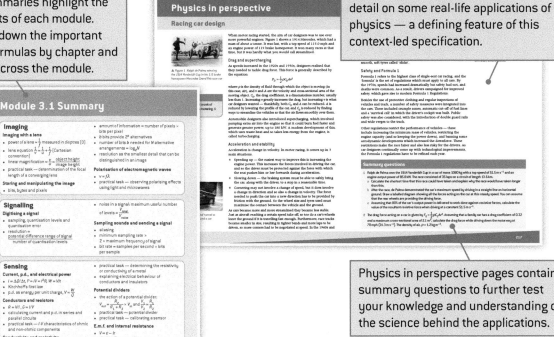

Physics in perspective pages go into detail on some real-life applications of physics — a defining feature of this context-led specification.

Physics in perspective pages contain summary questions to further test your knowledge and understanding of the science behind the applications.

Practice questions

1 An electron travelling at $1 \times 10^6\,\mathrm{ms^{-1}}$ has a

Practice questions at the end of each chapter including questions that cover practical and maths skills.

... emitted per second by the laser. *(4 marks)*

3 An electron has kinetic energy = 1.6 keV.
Use the equation kinetic energy = $\frac{momentum^2}{2m}$ to calculate the momentum of the electron and use your value to calculate its de Broglie wavelength. *(4 marks)*

4 The energy required to release an electron from the surface of magnesium is 3.7 eV. Radiation of wavelength 170 nm is incident on the surface of the metal. Calculate the maximum kinetic energy of the photoelectrons released. *(3 marks)*

5 Calculate the velocity of an electron with a de Broglie wavelength of $6.6 \times 10^{-10}\,\mathrm{m}$. *(3 marks)*

6 This question is about photons and phasors. Figure 1 shows two paths of photons from slits S_1 and S_2 to a point on the screen P_1. The path difference $S_2P_1 - S_1P_1$ is $\frac{\lambda}{3}$ where λ is the wavelength of the light.

▲ Figure 1

a The angles of the phasor arrows arriving at P_1 from slits are shown on the right hand side of the diagram. State why the phase difference is $\frac{2\pi}{3}$ radians. *(1 mark)*

b Draw a scale diagram to find the length of the resultant phasor arrow. *(2 marks)*

c Phasors meet at P2 with no phase difference. Calculate the ratio $\frac{\text{length of the resultant arrow at } P_2}{\text{length of resultant arrow at } P_1}$. *(2 marks)*

d Calculate the ratio $\frac{\text{probability of photon arriving at } P_1}{\text{probability of photon arriving at } P_2}$. *(2 marks)*

e A third slit is added, as shown in Figure 2.

▲ Figure 2

not to scale

Explain the effect this has on the probability of a photon arriving at P1. You may include a diagram of the phasor arrows at P1 in your answer. *(3 marks)*

7 This question is about the photoelectric effect. When light above the threshold frequency f_0 is incident on a metal surface, photoelectrons are emitted. The maximum kinetic energy of the photoelectrons is given by Einstein's equation:
$E_{kmax} = hf - \phi$
where ϕ is the minimum energy required to remove an electron from the surface of the metal and f is the frequency of the light incident on the surface.
Red light of frequency 4.5×10^{14} Hz is incident on a metal surface. The maximum energy of the ejected photoelectrons is 0.2 eV.

a State how the number and energy of the emitted photoelectrons will change when the intensity of the light incident on the surface is doubled. Explain your answer in terms of photons interacting with electrons in the surface of the metal. *(4 marks)*

b Violet light of frequency 7.5×10^{14} Hz ejects electrons with maximum energy of 1.4 eV. Use the data for red and violet light to calculate a value for the Planck constant h. *(4 marks)*

c Calculate the minimum frequency for the release of photoelectrons from this surface. *(2 marks)*

8 In a simple wave model to explain the diffraction of waves at a gap, the gap of width b is divided into three equal parts as shown in Figure 3a.
The centre of each part is treated as a source of waves.

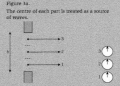

▲ Figure 3a   ▲ Figure 3b

a The phasors for the waves from each of the three parts of the gap reaching a distant screen in the straight-on direction are shown in Figure 3b.

(i) The paths taken by the waves in Figure 3a are all equal in length. Explain how the phasors in Figure 3b confirm this. *(1 mark)*

(ii) Each phasor has an amplitude A. Write down the amplitude of the resultant phasor at the distant screen. *(1 mark)*

b At an angle θ to the straight-on direction, the path difference between neighbouring paths is Δx, as show in Figure 4a. For one particular value of θ, the resultant intensity is **zero**.

▲ Figure 4a   ▲ Figure 4b

(i) Explain why the phasor for path 1 has rotated 120° more than the phasor for path 2, when $\Delta x = \frac{1}{3}\lambda$, where λ is the wavelength of the waves. *(2 marks)*

(ii) Draw arrows on a copy of Figure 4b to represent the phasors for waves 2 and 3. Explain, using a diagram, why the three phasors have a zero resultant. Label your phasors in the diagram 1, 2, and 3. *(2 marks)*

(iii) Use Figure 4a and the fact that $\Delta x = \frac{1}{3}\lambda$ to show that $\lambda = b\sin\theta$ where b is the total width of the gap. Show your working clearly. *(2 marks)*

c Use the equation $\lambda = b\sin\theta$ to calculate the angle θ at which a minimum signal occurs when microwaves of wavelength 2.4 cm are incident on a gap of width 6.0 cm. *(2 marks)*

OCR Physics B Paper G492 Jan 2011

Paper 1 Practice questions (A Level)

Section A — Multiple Choice

1 Here is a list of units:
A Js^{-1} B Jkg^{-1}
C Jm^{-1} D JC^{-1}

Practice questions at the end of the book, with multiple choice questions and synoptic style questions, also covering the practical and math skills.

3 An initially uncharged capacitor is charged by a constant current of 1.6 mA over a period of 2.0 s. At the end of this period the potential difference across the capacitor is 8.0 V. What is the capacitance of the capacitor?
A 0.2 μF   B 0.4 μF
C 200 μF   D 400 μF   *(1 mark)*

4 The graph shows how variable y changes with x.

▲ Figure 1

Which pair(s) of variables will produce a graph of a similar shape?
Pair 1: activity of a radioisotope (y) against time (x)
Pair 2: p.d. across a capacitor during discharge (y) against time (x)

Pair 3: Electric field strength of a point charge (y) against distance from charge (x)
A 1, 2 and 3 are correct
B Only 1 and 2 are correct
C Only 2 and 3 are correct
D Only 1 is correct   *(1 mark)*

5 Figure 2 shows four versions of the same circuit with different component values.

▲ Figure 2

In all four circuits the capacitor is uncharged before the switch is closed.
a Which circuit (A, B, C, or D) has the greatest final charge on the capacitor when the switch is closed? *(1 mark)*
b Which circuit (A, B, C, or D) takes the least time to charge up the capacitor when the switch is closed? *(1 mark)*

OCR Physics B Paper G494 January 2012

6 The half-life of a source is 7.0×10^2 s. The activity of the source at time $t = 0$ s is 1.7 kBq. How many radionuclides are present in the source at this time?
A 1.1×10^3 B 1.7×10^3
C 1.1×10^6 D 1.7×10^6 *(1 mark)*

7 Here are three statements about a mass oscillating in undamped simple harmonic motion between two springs.
1 The total energy of the oscillator remains constant throughout the oscillation.
2 The maximum kinetic energy of the oscillator is equal to the total energy of the oscillator.

3 The total energy of the oscillator will double if the amplitude is doubled.
Which of these statements is/are correct?
A 1, 2 and 3 are correct
B Only 1 and 2 are correct
C Only 2 and 3 are correct
D Only 1 is correct   *(1 mark)*

8 A simple pendulum oscillates in simple harmonic motion with a time period of 1.2 s. What is the angular frequency, ω, of the oscillation?
A 0.19 rad s^{-1} B 1.7 rad s^{-1}
C 2.6 rad s^{-1} D 5.2 rad s^{-1} *(1 mark)*

9 An object of mass m hangs from a spring and oscillates with time period T. The object is replaced with one of half the mass. What is the new time period of the oscillation?
A $\frac{T}{2}$ B $\frac{T}{\sqrt{2}}$
C $\sqrt{2}\,T$ D $2T$ *(1 mark)*

10 The acceleration due to gravity near the surface of the Moon is about 1.7 m s^{-2}. The radius of the Moon is about 1.7×10^6 m. What is the mass of the Moon?
A 2.5×10^{22} kg   B 7.3×10^{22} kg
C 1.0×10^{24} kg   D 1.2×10^{29} kg   *(1 mark)*

11 The planet Jupiter orbits the Sun in 3.8×10^8 s. The mean distance between Jupiter and the Sun is 7.8×10^{11} m. What is the centripetal acceleration of Jupiter?
A 3.4×10^{-5} m s^{-2} B 6.8×10^{-5} m s^{-2}
C 1.0×10^{-4} m s^{-2} D 2.1×10^{-4} m s^{-2} *(1 mark)*

12 An electron is accelerated in a laboratory to a relativistic factor of 3.0. What is the velocity of the electron relative to the laboratory?
A 2.0×10^8 m s^{-1} B 2.4×10^8 m s^{-1}
C 2.7×10^8 m s^{-1} D 2.8×10^8 m s^{-1} *(1 mark)*

13 Here is a list of particles emitted in radioactive decays.
A alpha particle   B beta particle
C positron   D neutrino

a Which particle is a positively charged lepton? *(1 mark)*
b Which particle does not cause ionisation as it passes through matter? *(1 mark)*
c Which particle has a neutral anti-particle? *(1 mark)*

14 The graph in Figure 3 shows how binding energy per nucleon varies with nucleon number.

▲ Figure 3

Which of the following statements about the graph is/are correct?
1 Energy is released when nuclei in region C split, producing lighter nuclei.
2 The most stable nuclei are at B.
3 The process in which nuclei in region A fuse to produce more massive nuclei is the source of the Sun's energy.
A 1, 2, and 3 are correct
B Only 1 and 2 are correct
C Only 2 and 3 are correct
D Only 1 is correct   *(1 mark)*

15 A proton is accelerated to a relativistic factor of 3.0. What is the ratio $\frac{\text{kinetic energy of proton}}{\text{rest energy of proton}}$?
A 1.5   B 2.0
C 3.0   D 4.0   *(1 mark)*

16 Here is an equation representing the beta decay of a neutron.
$^1_0n \rightarrow {}^1_1p + {}^0_{-1}e + {}^0_0\bar{\nu}$
Which of these statements about the decay is/are correct?
1 Mass is conserved.
2 Lepton number is conserved.
3 Momentum is conserved.
A 1, 2, and 3 are correct.
B Only 1 and 2 are correct.
C Only 2 and 3 are correct.
D Only 1 is correct.

This book is supported by next generation Kerboodle, offering unrivalled digital support for independent study, context, differentiation, assessment, and the new practical endorsement.

If your school subscribes to Kerboodle, you will also find a wealth of additional resources to help you with your studies and with revision.

- Study guides
- Maths skills boosters and calculation worksheets
- Practicals activities to support the practical endorsement
- Interactive progress quizzes that give question-by-question feedback
- Animations and revision podcasts
- Self-assessment checklists

Test your knowledge with the progress quizzes, and learn from your mistakes with the detailed explanations given for each answer.

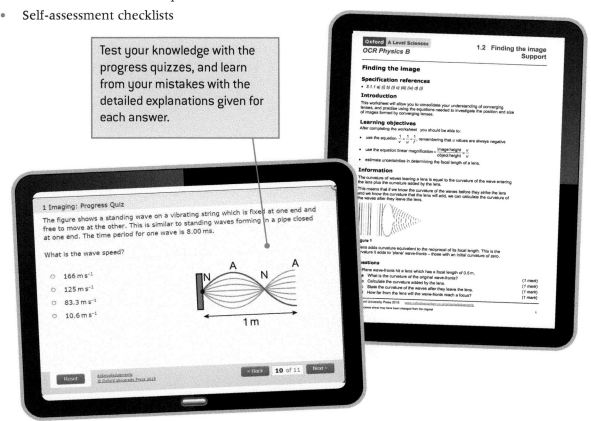

For teachers, Kerboodle also has plenty of further support, including answers to the questions in the book and a digital markbook. There are also full teacher notes for the activities and worksheets, which include suggestions on how to support students and engage them in their own learning. All of the resources are pulled together into teacher guides that suggest a route through each chapter.

Studying Advancing Physics

▲ **Figure 1** *Rutherford, although critical of stamp collectors, actually appeared on stamps. This stamp shows a representation of the alpha-particle scattering experiment that led to the nuclear model of the atom*

Advancing Physics is about *understanding* the world around us, and appreciating how it is revealed by the experiments we perform and the explanations we study. Ernest Rutherford, one of the finest experimental physicists of all time, is often quoted as saying

All science is either physics or stamp-collecting.

Whether or not he actually said this, the statement can be interpreted as suggesting that physics is more than just collecting facts. That is not to say that facts aren't important — it is still useful to *know* fundamental facts when studying physics, such as the fact that an atom is around 1×10^{-10} m in diameter, or that the Universe is about 14 billion years old.

Remembering facts is one thing, but understanding ideas is quite another. To use the example of Rutherford, whilst it is useful to remember the diameter of an atom, it is far more interesting to understand how we measured this value. As you progress through the Advancing Physics course you will be encouraged to seek explanations for the statements you are given and to use your understanding of basic physics in a range of different contexts.

Experiments

Facts do not begin in books: they begin in the world around us. The study of what we now call science really got going when people began to perform controlled experiments and make carefully recorded observations. Experimental work remains a very important part of physics and you will be given many opportunities to develop your practical skills.

Experiments come in a variety of forms — some demonstrate effects or processes, whilst others require careful measurements to investigate relationships. Some experiments aim to find values for fundamental characteristics, such as the refractive index of a transparent material.

Careful experimentation is extremely important. Many advances in science have stemmed from experimental results not quite matching with the expected theoretical value. You should always try to make your practical work as good as it can be — it's vital to be self-critical and estimate the uncertainty in your measurements. Don't think to yourself that *this is good enough*, but rather think *how can this experiment be improved?*

Technology

We are in a time of rapid technological change. This is perhaps most clearly seen in the fields of electronics and communication, but technological developments are having an impact on different areas of life on a much wider scale than just mobile phones and tablet computers. Automotive design and digital imaging in the health sciences are just two more examples of the positive impact of technology on our lives.

▲ **Figure 2** *The advances in communication technology since the last century are such that the old phone is barely even recognisable anymore*

Much of this development relies on fundamental physics, and this course will highlight some of the links between physics and developing technology.

Mathematics

Physics is a science that relies upon mathematical reasoning, so much so that many explanations in physics are actually mathematical relationships written in word form. You will be introduced to any new mathematical ideas you need as you meet them. The technique of modelling physical processes such as accelerating bodies, or the decay of radioisotopes, using repetitive or 'iterative' calculations is particularly powerful.

Graphical representations of relationships are extremely useful. Choosing the correct graph to draw and techniques such as estimating the area beneath a graph line or calculating its gradient are skills that will be used and developed throughout the course.

Asking Questions

Perhaps a more suitable heading would be asking the *right* questions. Simple questions can lead to great developments in our understanding of the world around us. For example, the question 'why does light travel in a straight line?' appears to be so simple that it is hardly worth asking, but answering this question requires a theory (or model) of how light behaves — something that physicists debated for centuries.

▲ **Figure 3** *Light beams travelling in straight lines*

To take another example, Albert Einstein asked himself the question, 'if I jump off a roof why do I feel no force, even though I am accelerating towards the ground?' He later called this the happiest thought of his life and the answer led to his greatest work — the General Theory of Relativity.

Excitement

Physics is an exciting and challenging subject. Some of the excitement comes from new discoveries or technological developments arising from fundamental physics. We cannot tell how important the development of graphene and carbon-nanotube technologies will be in the future economy, just as we cannot predict what future space missions will reveal about our local Solar System or the greater reaches of space. Then there is another form of excitement — the enjoyment of finding things out yourself, of deepening your understanding of the world around you. The more you understand, the more you will appreciate the new discoveries, and studying physics is a very good place to start.

▲ **Figure 4** *Photo of comet Churyumov-Gerasimenko taken in 2014 by Philae, the first human-built spacecraft to successfully land on a comet*

MODULE 1
Development of practical skills in physics

Physics is a practical subject and experimental work provides you with important practical skills, as well as enhancing your understanding of physical theory. You will be developing practical skills by carrying out practical and investigative work in the laboratory throughout both the AS and the A Level Physics course. You will be assessed on your practical skills in two different ways:

- written examinations (AS and A level)
- practical endorsement (A level only)

Practical coverage throughout this book

Practical skills are a fundamental part of a complete education in science, and you are advised to keep a record of your practical work from the start of your A level course that you can later use as part of your practical endorsement. You can find more details of the practical endorsement from your teacher or from the specification.

In this book and its supporting materials practical skills are covered in a number of ways. By studying Application boxes and practice questions in this student book, and by using the Practical activities and Skills sheets in Kerboodle, you will have many opportunities to learn about the scientific method and carry out practical activities.

1.1 Practical skills assessed in written examinations

The practical skills of planning, implementing, analysis, and evaluation will be assessed in all the written papers at both AS and A Level. The A level examination paper 3, *Practical skills in physics*, will include longer questions focusing on practical skills and analysis. This section summarises the skills that you need to develop when answering questions in the written papers.

1.1.1 Planning

- Designing experiments
- Identifying variables to be controlled
- Evaluating the experimental method

Skills checklist

- ☐ Selecting apparatus and equipment
- ☐ Selecting appropriate techniques
- ☐ Selecting appropriate quantities of materials and substances and scale of working
- ☐ Solving physical problems in a practical context
- ☐ Applying physics concepts to practical problems
- ☐ Identifying and controlling variables (where appropriate)

1.1.2 Implementing

- Using a range of practical apparatus
- Carrying out a range of techniques
- Using appropriate units for measurements
- Recording data and observations in an appropriate format

Skills checklist

- ☐ Understanding practical techniques and processes
- ☐ Identifying hazards and safe procedures
- ☐ Using SI units
- ☐ Recording qualitative observations accurately
- ☐ Recording a range of quantitative measurements
- ☐ Using the appropriate precision for apparatus

1.1.3 Analysis

- Processing, analysing, and interpreting results
- Analysing data using appropriate mathematical skills
- Using significant figures appropriately
- Plotting and interpreting graphs

Skills checklist

- ☐ Analysing qualitative observations
- ☐ Analysing quantitative experimental data, including
  - calculation of means
  - amount of substance and equations
- ☐ For graphs,
  - selecting and labelling axes with appropriate scales, quantities, and units
  - drawing tangents and measuring gradients
  - representing uncertainty in graphs correctly

1.1.4 Evaluation

- Evaluating results to draw conclusions
- Identify anomalies
- Explain limitations in method
- Identifying uncertainties and errors
- Suggesting improvements

Skills checklist

- ☐ Reaching conclusions from qualitative observations
- ☐ Identifying uncertainties and calculating percentage errors
- ☐ Identifying procedural and measurement errors
- ☐ Refining procedures and measurements to suggest improvements
- ☐ Understanding accuracy and precision

1.2 Practical skills assessed in practical endorsement

You will also be assessed on how well you carry out a wide range of practical work and how to record the results of this work. These hands-on skills are divided into 12 categories and form the practical endorsement. This is assessed for the A Level Physics qualification only.

The endorsement requires a range of practical skills from both years of your course. If you are taking only AS Physics, you will not be assessed through the practical endorsement but the written AS examinations will include questions that relate to the skills that naturally form part of the AS common content to the A level course.

The practicals you do as part of the endorsement will not contribute to your final grade awarded to you. However, these practicals must be covered and your teacher will go through how this is to be done in class. It is important that you are actively involved in practical work because it will help you with understanding the theory and also how to effectively answer some of the questions in the written papers.

The practical activities you will carry out in class are divided into Practical Activity Group (PAGs). PAG1 to PAG 6 will be undertaken in Year 1, PAG7 to PAG 10 in Year 2, and PAG11 to PAG 12 throughout the two-year course.

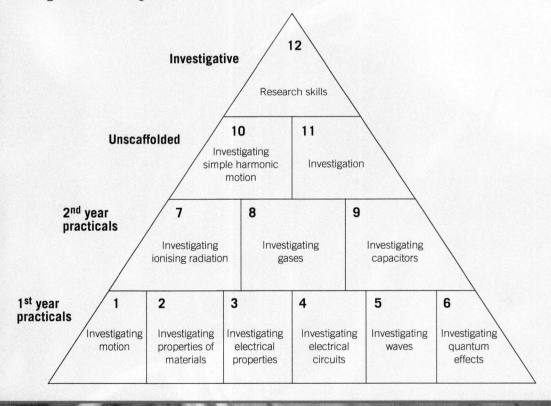

The PAGs are summarised below, together with the topic references in the book that relate to the specific PAG.

PAG	Topic reference
1 Investigating motion	8.1
2 Investigating properties of materials	4.3
3 Investigating electrical properties	3.2, 3.3
4 Investigating electrical circuits	3.5
5 Investigating waves	6.1, 6.3, 6.4
6 Investigating quantum effects	7.1
7 Investigating ionising radiation	10.1
8 Investigating gases	14.1
9 Investigating capacitors	10.4
10 Investigating simple harmonic motion	11.3, 11.4
11 Investigation	Throughout
12 Research skills	Throughout

Maths skills and How Science Works across Module 1

In order to develop your knowledge and understanding in A Level Physics, it is important to have specific skills in mathematics. All the mathematical skills you will need during your physics course have been embedded into the individual topics for you to learn as you meet them. An overview is available in each of the module openers and these skills are further supported by the worked examples, summary questions, practice questions and the Maths appendix.

How Science Works (HSW) is another area required for success in A Level Physics, and helps you to put science in a wider context, helping you to develop your critical and creative thinking skills in order to solve problems in a variety of contexts. Once again, this has been embedded into the individual topics covered in the books, particularly in application boxes and examination-style questions. The application and extension boxes cover some of the HSW elements.

You can find further support for maths skills and HSW on Kerboodle.

Base units

When you make a measurement, or calculate a physical quantity, you need both a number *and* a unit. In the **SI system** of units (SI comes from *Le Système International d'Unités*), there are seven **base units** from which all the others are derived. The base units are the metre (m), kilogram (kg), second (s), ampere (A), kelvin (K), candela (cd), and mole (mol). They are the units of distance, mass, time, electric current, temperature, luminosity, and amount of substance, respectively.

The units of other quantities can all be expressed in base units, though we tend to use **derived units**.

▼ Table 1 *Equivalent and derived units*

Quantity	Can be calculated with	Equivalent to	Derived unit
force	mass × acceleration	$kg\,m\,s^{-2}$	newton N
energy	force × distance	$N\,m$	joule J
power	energy per second	$J\,s^{-1}$	watt W
charge	current × time	$A\,s$	coulomb C
potential difference	energy per coulomb	$J\,C^{-1}$	volt V

Note that m/s and $m\,s^{-1}$ are equivalent, but it is better to write $m\,s^{-1}$ at A Level.

Checking equations with units

You can use units to check that equations make physical sense. The units on both sides of an equation must match. If they do not, the equation is wrong.

 Worked example: Power = *IV*

Power is measured in watts and $1\,W = 1\,J\,s^{-1}$. Show that the units of *IV* are also $J\,s^{-1}$.

Step 1: Lay out the units for $I \times V$.

   Units of $IV = A\,J\,C^{-1}$

Step 2: Since the coulomb C = A s, this can be substituted into the equation above.

   Units of $IV = A\,J\,A^{-1}\,s^{-1}$

Step 3: Cancel out terms where possible.

   Units of $IV = J\,s^{-1}$ This is the same as the unit of power.

Radian measure

Although you can define an angle using degrees (°), where there are 360° in a circle, you can also define an angle by the ratio of the arc length to the radius. If the arc length is equal to the radius then the angle at the centre is equal to one **radian** (^{c} or rad). When the arc length equals the circumference then the angle at the centre is 360°, or $2\pi^c$. So $180° = \pi$ radians, and $90° = \frac{\pi}{2}$ radians. This means that 1 radian $= \frac{180}{\pi}$.

$$\text{angle (radians)} = \frac{\text{arc length (m)}}{\text{radius (m)}}$$

> **Synoptic link**
>
> The use of radians is important when discussing phasors, which you will meet in Topic 6.1, Superposition of waves.

 Worked example: Radians and degrees

A circle has a radius of 0.5 m. Calculate the angle at the centre if the arc length is 1.0 m. Convert the angle to degrees.

Step 1: Use the equation to calculate the angle.

$$\text{angle (radians)} = \frac{\text{arc length (m)}}{\text{radius (m)}}$$

$$\text{angle} = \frac{1.0\,\text{m}}{0.5\,\text{m}} = 2\ \text{radians}$$

Step 2: Convert from radians to degrees using 1 radian $= \frac{180}{\pi}$.

$$2\ \text{radians} = 2 \times \frac{180}{\pi} = 110° \ (2\ \text{s.f.})$$

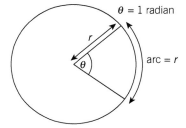

▲ **Figure 1** *An arc length equal to r means the angle is one radian*

Standard form

In physics, numbers can be very large, like the distance from the Earth to the Sun, or very small, like the size of an atom. In **standard form** (also called scientific notation) you can write numbers with one digit in front of the decimal place, multiplied by the appropriate power of ten, positive or negative.

For example, 1000 m can be written 1.0×10^3 m, $0.1\,\text{s} = 1.0 \times 10^{-1}$ s, $20\,000\,\text{Hz} = 2 \times 10^4$ Hz, and $0.0005\,\text{kg} = 5 \times 10^{-4}$ kg. Note that 1.0×10^3 m is the same as 10^3 m. This means that:

- the distance from the Earth to the Sun = 150 000 000 000 m = 1.5×10^{11} m
- the diameter of an atom is 0.000 000 000 1 m = 1×10^{-10} m.

You will see many physical constants written in standard form, for example, the speed of light $= 3 \times 10^8\,\text{m s}^{-1}$.

▲ **Figure 2** *The Sun is 1.5×10^{11} m from the Earth*

Multiplying numbers in standard form

When you multiply two numbers in standard form together you *add* the powers of ten. When you divide two numbers you *subtract* the powers of ten of the second number from that of the first.

- If the powers *add* when you multiply then $10^2 \times 10^3 = 10^5$ or 100 000.
- If the powers *subtract* when you divide then $10^2 \div 10^4 = 10^{-2}$.

▲ **Figure 3** *You need a scientific calculator to do calculations involving standard form*

Hint

When you enter a number in standard form into your calculator, such as 3×10^{-3}, you just need to press 3, EXP, +/−, 3. You do not need to press the 1 or 0 buttons.

Scientific calculators have a button that you use when you are calculating with numbers in standard form. Work out which button you need to use (it could be EE, EXP, 10^x, $\times 10^x$).

 Worked example: How many?

A full library contains 200 000 books. Each book contains 400 pages. Calculate the total number of pages. The total length of the shelves in the library = 4800 m. Calculate the thickness of a page.

Step 1: Convert the numbers to standard form.

$$200\,000 = 2 \times 10^5, \quad 400 = 4 \times 10^2$$

Step 2: Multiply the numbers and work out the answer by adding the powers.

Number of pages = number of books × pages per book

$$= 2 \times 10^5 \times 4 \times 10^2 = (2 \times 4) \times (10^5 \times 10^2)$$
$$= 8 \times 10^7 \text{ pages}$$

Step 3: Divide the numbers and work out the answer by subtracting the powers.

$$\text{Width of a page} = \frac{\text{width of all the books}}{\text{number of pages in all the books}} = \frac{4.8 \times 10^3}{8 \times 10^7}$$

$$= \frac{4.8}{8} \times \frac{10^3}{10^7} = 0.6 \times 10^{-4}\,\text{m because you subtract}$$

the powers: $3 - 7 = -4$. This can be written as $= 6 \times 10^{-5}\,\text{m}$.

Metric prefixes

You can use various **metric prefixes** to show large or small multiples of a particular unit. Here the unit is the metre. For example, 3 nanometres $= 3 \times 10^{-9}\,\text{m}$. Most of the prefixes that you will use in physics involve multiples of 10^3.

▼ **Table 2** *The most common prefixes*

T	G	M	k	c	m	μ	n	p	f
tera	giga	mega	kilo	centi	milli	micro	nano	pico	femto
10^{12}	10^9	10^6	10^3	10^{-2}	10^{-3}	10^{-6}	10^{-9}	10^{-12}	10^{-15}

Converting between units

It is helpful to use standard form when you are converting between units. You need to think about how many of the smaller units are in the bigger one.

- There are 1000 mm in 1 m. So $1\,\text{mm} = \dfrac{1}{1000}\,\text{m} = 10^{-3}\,\text{m}$.

- There are 1000 mm × 1000 mm $= 10^6\,\text{mm}^2$ in $1\,\text{m}^2$. So $1\,\text{mm}^2 = 10^{-6}\,\text{m}^2$.

- There are 100 cm × 100 cm × 100 cm $= 10^6\,\text{cm}^3$ in $1\,\text{m}^3$. So $1\,\text{cm}^3 = 10^{-6}\,\text{m}^3$.

Worked example: Cross-sectional area of a wire

A wire has a diameter of 0.5 mm. Calculate the cross-sectional area using $A = \pi r^2$.

Step 1: Convert mm to m using standard form.

$$\text{Diameter } d = 0.5 \times 10^{-3}\,\text{m} = 5 \times 10^{-4}\,\text{m}$$

Radius $r = 2.5 \times 10^{-4}\,\text{m}$, or you could simply use $0.25 \times 10^{-3}\,\text{m}$

Step 2: Substitute the number in standard form into the equation.

$$A = \pi r^2 = \pi \times (2.5 \times 10^{-4}\,\text{m})^2 = 2 \times 10^{-7}\,\text{m}^2 \text{ (1 s.f.)}$$

Hint

When carrying out multiplications or divisions using standard form, add or subtract the powers of ten to work out roughly what you expect the answer to be. This will help you to avoid mistakes.

Uncertainty

There are two main ways to estimate the **uncertainty** of a measurement:

- repeat it many times and make an estimate from the variation you get
- look at the process of measurement used, and inspect and test the instruments used.

Usually, you should focus mainly on the second way, which is the process of measuring, and on the qualities of the instruments you have. The main reason for being interested in the quality of a measurement is to try to do better.

Properties of instruments

The essential qualities and limitations of measuring instruments are:

- **resolution** – the smallest detectable change in input, for example, 1 mm on a standard ruler
- **sensitivity** – the ratio of output to input, for example, the change in potential difference across a thermistor when the temperature changes by 1 °C
- **stability (repeatability, reproducibility)** – the extent to which repeated measurements give the same result, including gradual change with time (drift)
- **response time** – the time interval between a change in input and the corresponding change in output, for example, how long it takes a temperature sensor to respond when you put it in hot water
- **zero error** – the output for zero input, for example, a newtonmeter that reads 0.1 N when there is no force acting
- **noise** – variations, which may be random, superimposed on a signal, for example, changes to a reading on a temperature sensor as someone opens a door
- **calibration** – determining the relation between output and true input value, including linearity of the relationship, for example, the relationship between the resistance of a thermistor (in Ω) and the temperature (in °C).

▲ **Figure 4** *The resolution of this multimeter is ±0.01 V*

Estimating uncertainty

Uncertainty can be estimated in several ways:

- from the resolution of the instrument concerned. For example, the readout of a digital instrument ought not to be trusted to better than ±1 in the last digit
- from the stability of the instrument, or by making deliberate small changes in conditions (a tap on the bench, maybe) that might anyway occur, to see what difference they make
- by trying another instrument, even if supposedly identical, to see how the values they give compare
- from the range of some repeated measurements.

When comparing uncertainties in different quantities, it is the **percentage uncertainties** that need to be compared, to identify the largest. You can make the biggest improvement to your measurements by trying to reduce the largest uncertainty.

 Worked example: Calculating percentage uncertainty

The meter is reading 12.51 V, and the uncertainty in the reading is ±0.01 V (±1 in the last digit). Calculate the percentage uncertainty of this reading.

Step 1: Identify the equation for percentage uncertainty.

$$\text{The percentage uncertainty} = \frac{\text{uncertainty in reading} \times 100\%}{\text{actual reading}}$$

Step 2: Substitute values.

$$\text{Percentage uncertainty} = \frac{0.01\,\text{V} \times 100\%}{12.51\,\text{V}} = 0.08\% \ (1 \text{ s.f.})$$

Analysing uncertainties

The final uncertainty in an answer depends on how quantities are combined. Here are three important rules about the way uncertainties propagate.

1 **Adding or subtracting quantities**

When you add or subtract quantities in an equation, you add the absolute uncertainties for each value.

 What is the extension?

The original length of a spring is 2.5 ± 0.1 cm and the final length is 15.0 ± 0.2 cm. Calculate the extension of the spring and the absolute uncertainty.

Step 1: Calculate the extension by subtracting the lengths.

$$\text{extension} = 15.0 - 2.5 = 12.5\,\text{cm}$$

Step 2: Add the absolute uncertainties.

absolute uncertainty = 0.1 + 0.2 = 0.3

Step 3: Write the answer in the normal convention.

extension = 12.5 ± 0.3 mm

2 Multiplying or dividing quantities

When you multiply or divide quantities, you add the percentage uncertainties for each value.

 What is the resistance?

The current I in a resistor is 1.60 ± 0.02 A and the potential difference V across the resistor is 6.00 ± 0.20 V. Calculate the resistance and the absolute uncertainty.

Step 1: Calculate the resistance R of the resistor.

$$R = \frac{V}{I} = \frac{6.00}{1.60} = 3.75\,\Omega$$

Step 2: Calculate the percentage uncertainty in each measurement.

% uncertainty in $I = \frac{0.02}{1.60} \times 100 = 1.25\%$

% uncertainty in $V = \frac{0.20}{6.00} \times 100 = 3.33\%$

Step 3: Add the percentage uncertainties.

% uncertainty in $R = 1.25 + 3.33 = 4.58\%$

Step 4: Calculate the absolute uncertainty in R.

absolute uncertainty in $R = 0.0458 \times 3.75 = 0.2\,\Omega$ (1 s.f.)

Step 5: The uncertainty in R limits the precision with which the final answer for resistance can be given. The final answer should be rounded to the same precision, which is one decimal place.

$R = 3.8 \pm 0.2\,\Omega$ (1 d.p.)

3 Raising a quantity to a power

When a measurement in a calculation is raised to a power n, your percentage uncertainty is increased n times. The power n can be an integer or a fraction.

 Cross-sectional area of a wire

The diameter of a wire is recorded as 0.51 ± 0.02 mm. Calculate the cross-sectional area of the wire and the absolute uncertainty.

Step 1: Calculate the cross-sectional area A of the wire.

$$A = \frac{\pi d^2}{4} = \frac{\pi \times (0.51 \times 10^{-3})^2}{4} = 2.04 \times 10^{-7}\,m^2$$

Step 2: The percentage uncertainty in A is equal to 2 times the percentage uncertainty in d.

(The π and the 4 are numbers and therefore have no uncertainty associated with them.)

% uncertainty in $A = 2 \times \left(\dfrac{0.02}{0.51} \times 100\right) = 7.84\%$

Step 3: Calculate the absolute uncertainty in A.

absolute uncertainty in $A = 0.0784 \times 2.04 \times 10^{-7} = 0.16 \times 10^{-7}\,\text{m}^2$

Step 4: The diameter of the wire is quoted to 2 significant figures, therefore the final answer for the cross-sectional area must also be written to 2 significant figures.

$A = (2.0 \pm 0.2) \times 10^{-7}\,\text{m}^2$ (2 s.f.)

Why results vary

There are different kinds of *variation*, *uncertainty*, or *error*:

- inherent *variation* in the measured quantity, for example, variation in the value amongst a set of 'identical' commercial resistors
- small (maybe random) uncontrollable variations in conditions, including noise, leading to *uncertainty*
- simple mistakes, for example, misreading a scale, or a *one-off* accidental *error*, which needs to be detected and removed – **outliers** often turn out to be due to such mistakes
- systematic *error* or bias, for example, reading a scale at an angle – this may show up as an intercept on a suitable graph
- a genuine outlying value, which may be an *error*, or may be significant for another reason.

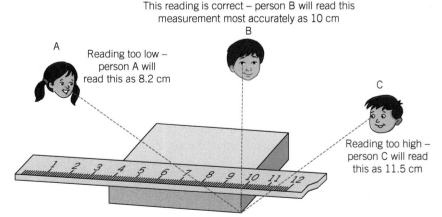

▲ **Figure 5** *If you always read a scale at an angle you can introduce a systematic error to your measurements*

Dot plots

Suppose you measure the resistance of resistors that all come out of a packet labelled $47\,\Omega$. There will be a variation in the value of resistance that you measure. You can display the values on a simple **dot plot**, such as the one in Figure 6.

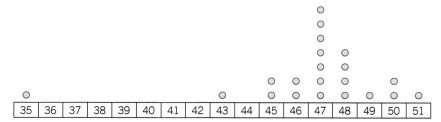

▲ **Figure 6** *A dot plot can show the spread of results*

You can estimate the uncertainty of a measurement in terms of the **range** of typical values, excluding outliers. Here the value of $35\,\Omega$ is probably an outlier. The **spread** is then

$$\text{spread} = \pm\frac{1}{2}\text{range}$$

A good rule of thumb is that a value is likely to be an outlier if it lies more than 2 × spread from the **mean**, but this is only a rule of thumb. There may be a reason that you have that reading.

To calculate the mean of a set of numbers you need to add up all the values and divide by the number of values.

 Worked example: Estimating uncertainty from the range

Use the dot plot in Figure 6 to find the mean resistance, the spread, and the percentage uncertainty.

Step 1: Identify any outliers.

$35\,\Omega$ is an outlier.

Step 2: Calculate the mean.

Mean = $(43 + 2 \times 45 + 2 \times 46 + 7 \times 47 + 4 \times 48 + 49 + 2 \times 50$

$+ \, 51)\,\Omega \div 20 = \dfrac{946}{20}\,\Omega = 47\,\Omega$ (2 s.f.)

Step 3: Calculate the spread.

Spread = $\pm\dfrac{1}{2}(51\,\Omega - 43\,\Omega) = \pm\,4\,\Omega$

Step 4: Calculate the percentage uncertainty.

Measurement of resistance = $47\,\Omega \pm 4\,\Omega$

Percentage uncertainty = $\dfrac{\text{uncertainty} \times 100\%}{\text{reading}} = \dfrac{4\,\Omega \times 100\%}{47\,\Omega}$

$= 9\%$ (1 s.f)

Hint

You usually use the mean of a set of results, but there may be occasions when the mean is not representative – for example, when there are results that are far apart from the rest but not clearly outliers.

Hint

Don't just completely discard outliers – you should investigate them carefully and see if there's a reason for them.

Graphs and uncertainty

The uncertainty in a measurement can be used to give a small range or **uncertainty bar** for each measurement. Instead of plotting just the points on a graph, you can plot an uncertainty bar for all of your measurements.

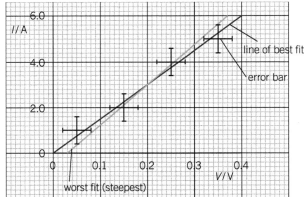

▲ **Figure 7** *Uncertainty bars are useful when you draw the line of best fit and the worst line for your measurements*

Your straight best fit line must pass through all the uncertainty bars (Figure 7). You would use this line to determine the value of the gradient. How can you determine an approximate value for the uncertainty in the gradient? You would draw the **line of worst fit** – the least acceptable straight line through the data points – this can either be the steepest or the shallowest line.

The absolute uncertainty in the gradient is the positive difference between the gradient of the line of best fit and the gradient of the line of worst fit.

The percentage uncertainty in the gradient can be calculated as

$$\% \text{ uncertainty in gradient} = \frac{\text{absolute uncertainty}}{\text{gradient best fit line}} \times 100\%$$

Precision and accuracy

In everyday life, people often use the words *precise* and *accurate* to mean the same thing – this is not the case in physics.

- **Accuracy** is to do with how close a measurement result is to the true value – the closer it is, the more **accurate** it is.
- **Precision** is to do with how close repeated measurements are to each other – the closer they are to each other, the more **precise** the measurement is.

Figure 8 is a visual way of appreciating the terms accuracy and precision using a dartboard as an example. You are aiming for bullseye – it represents the true value.

Estimations

Sometimes an exact answer is absolutely necessary, such as when calculating the speed of a spacecraft at various stages on a trip to the

not accurate
not precise

accurate
not precise

not accurate
precise

accurate
precise

▲ **Figure 8** *Accuracy and precision*

Moon. Sometimes a rough estimate is helpful. It can help you to work out if a calculation is correct, or where you expect an answer to be. A simple estimate is an **order of magnitude** estimate, which is an estimate to the nearest power of 10. For example, to the nearest power of 10 you are 1 m tall, and can run $10\,\mathrm{m\,s^{-1}}$. You may also simplify the question by imagining objects as squares or cubes, for example, when estimating how many atoms there are in a given volume.

You, your desk, and your chair are all of the order of 1 m tall.

▲ Figure 9 *Order of magnitude of 1 m*

 Worked example: Uncertainties and significant figures

The radius of a circle $r = 1.3 \pm 0.1$ m. Estimate the area of the circle. Calculate the area of the circle and estimate the uncertainty in the area. Justify the number of significant figures in your answer.

Step 1: Estimate the area by using approximations.

The area = πr^2. Assume that π approximately equals 3. So the area lies between $3 \times (1\,\mathrm{m})^2 = 3\,\mathrm{m}^2$ and $3 \times (2\,\mathrm{m})^2 = 12\,\mathrm{m}^2$. 1.3 m is closer to 1 than to 2, so the area is closer to $3\,\mathrm{m}^2$ than $12\,\mathrm{m}^2$. Therefore, $5\,\mathrm{m}^2$ is a reasonable estimate.

Step 2: Calculate the area of the circle using the measurement.

The area of the circle = $\pi r^2 = \pi(1.3\,\mathrm{m})^2 = 5.30929218...\,\mathrm{m}^2$

Step 3: Use the uncertainty to identify how big or small the measurement could be.

The radius could be as big as $1.3\,\mathrm{m} + 0.1\,\mathrm{m} = 1.4\,\mathrm{m}$, or as small as $1.3\,\mathrm{m} - 0.1\,\mathrm{m} = 1.2\,\mathrm{m}$.

Step 4: Calculate the area using these numbers.

The area of the circle = $\pi r^2 = \pi(1.4\,\mathrm{m})^2 = 6.157521601...\,\mathrm{m}^2$. This is bigger by about $0.8\,\mathrm{m}^2$. (Note that you would not write down a number with this number of significant figures for any answer.)

The area of the circle = $\pi r^2 = \pi(1.2\,\mathrm{m})^2 = 4.523893421...\,\mathrm{m}^2$. This is smaller by just under $0.8\,\mathrm{m}^2$.

Step 5: Use the largest and smallest values to estimate the uncertainty.

The area lies somewhere between $4.5\,\mathrm{m}^2$ and $6.1\,\mathrm{m}^2$, so the uncertainly = $\pm0.8\,\mathrm{m}^2$ and area = $5.3 \pm 0.8\,\mathrm{m}^2$.

Step 6: Justify the number of significant figures in your answer.

We are effectively using two significant figures for the answer, which is what we were given in the question. All the numbers after the 3 are not significant because of the value of the uncertainty.

Hint

You should only write your final answer using a justifiable number of significant figures. If you need to use one part of a question to calculate something in another part of the question, you should use the unrounded number in your calculator, and then round the answer to the same number of significant figures as before.

MODULE 3.1
Imaging, signalling and sensing

Introduction

This module introduces you to some fundamental ideas in physics and shows how they are used in a variety of situations. It gives you the first taste of a very useful rule — learn the fundamental ideas in detail and you will be able to apply them to all kinds of situations or contexts. For example, the physics of the simple lens can lead to a better understanding of imaging by satellite. Similarly, understanding the humble potential divider can help you appreciate how all kinds of sensor systems work. You will also be introduced to some of the mathematical and experimental skills you will use throughout the course.

Imaging begins by looking at a simple converging lens, the sort that has been used for hundreds of years in reading glasses and magnifying glasses. We then consider how the image is captured as an array of numbers. This is a very important area of technology to understand as nearly all images are now stored digitally as an array of numbers. This allows manipulation and editing of the images, a feature useful in the word of medical imaging as well as simple photo manipulation and editing.

Signalling looks at how information is sampled and represented by a chain of digital data. This means information from images, text, music, and more can be transmitted quickly across the room, across the country or across space. This will give you an appreciation of the basic principles used in video and music streaming, digital radio, and digital television. Nearly all modern communication technology depends on digital transmission — imagine the world without the Internet.

Sensing introduces you to electrical circuits and some basic ideas such as internal resistance. You will learn to use potential dividers to make simple sensors for detecting light level, temperature, and other variables. You will investigate the sensitivity and resolution of sensors, which are important features when choosing a sensor for a particular task. These basic sensors are at the core of many, more complex circuits, and often produce data that is stored and transmitted digitally.

Knowledge and understanding checklist

From your Key Stage 4 study you should be able to do the following. Work through each point, using your Key Stage 4 notes and the support available on Kerboodle.

☐ Describe the difference between a transverse and a longitudinal wave and recall that electromagnetic waves are transverse.

☐ Recall that different substances may refract waves.

☐ Recall that current is a rate of flow of charge.

☐ Recall the units for current, resistance and potential difference and apply the relationship $V = I R$.

☐ Describe the difference between series and parallel circuits.

☐ Calculate the current, potential difference and resistance in d.c. series circuits.

☐ Recognise and use common circuit symbols including diodes, LDRs and thermistors.

Maths skills checklist

Maths is a vitally important aspect of Physics. In this unit, you will need to use the following maths skills. You can find support for these skills on Kerboodle and through MyMaths.

☐ **Change the subject of non-linear equations** when using the lens equation in this chapter.

☐ **Use a calculator to find logs** in calculations involving the number of bits of data required to code images and signals.

☐ **Sketching graphs of relationships** in studying the relationship between resistance and temperature for components such as thermistors.

☐ **Calculate cross-sectional areas** of wires when investigating conductivity and resistivity.

☐ **Use appropriate numbers of significant figures** such as in experimental work in Sensing — you will need to think carefully about how many significant figures to use in your calculations.

☐ **Make calculations using the appropriate units** such as in calculations involving quantities like coulombs, newtons or metres. Using units correctly is an important part of physics.

☐ **Rearranging and solving algebraic equations** in many situations throughout the course

MyMaths.co.uk
Bringing Maths Alive

1 IMAGING

1.1 Bending light with lenses

Specification references: 3.1.1a(i), 3.1.1b(i), 3.1.1b(ii), 3.1.1c(ii)

▲ Figure 1 *The curvature of ripples and wave-fronts becomes less as the distance from the source increases*

What do you notice when you throw a stone into the calm water of a pond? Circular ripples spread out from the point where the stone enters the water. These ripples are examples of **wave-fronts.** A wave-front can be thought of as a line of disturbance moving through a material or through space. In the case of the stone in a pond, the wave-front is a ripple moving through the water. The distance between two consecutive wave-fronts is the wavelength.

Wave-fronts in water give us a useful way of thinking about light, as light spreads out from a small source in a similar manner to the ripples spreading out on a pond (Figure 1) – although light wave-fronts will spread out as three-dimensional spheres.

Wave-fronts and curvature

The curvature of the circumference of a circle is the reciprocal of its radius. For a circle of radius r the curvature is $\dfrac{1}{r}$.

As $r \rightarrow \infty$, $\dfrac{1}{r} \rightarrow 0$. This means that, as the distance from the source increases, the curvature of the wave-fronts decreases. Wave-fronts from a very distant source will appear to not curve at all. These are often called **plane wave-fronts**.

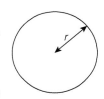

▲ Figure 2 *The curvature of the circumference of this circle is $\frac{1}{r}$*

 Worked example: Curvature

Calculate the curvature of a wave emitted from a point 1.5 m away.

Step 1: Select the appropriate equation.

$$\text{curvature of wave} = \frac{1}{r}$$

Step 2: Substitute values and evaluate.

$$\text{curvature of wave} = \frac{1}{1.5\,\text{m}} = 0.67\,\text{m}^{-1}\ (2\ \text{s.f.})$$

Light as rays and wave-fronts

Light can be thought of as wave-fronts or rays (Figure 3). These ways of thinking about light are connected. A ray of light points along the direction of motion of the wave-front and is always at right angles to the wave-front.

Ray point of view

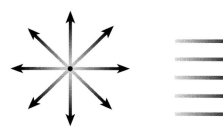

Light travels in straight lines from a small source.

Light in a parallel beam, or from a very distant light source, has rays (approximately) parallel to one another.

Wave point of view

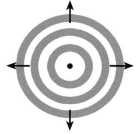

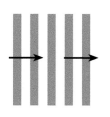

Light spreads out in spherical wave-fronts from a small source.

Wave-fronts in a parallel beam, or from a very distant light source, are straight (not curved) and parallel to one another.

▲ **Figure 3** *Rays and waves – two ways of picturing light*

Light through a converging lens

A converging lens is a lens that focuses light to a point behind it. It can be used to concentrate light rays as a *burning glass*. This concentrates sunlight striking the surface of the lens into a tiny image of the Sun. The intensity of the light at this point, the **focus** of the lens, can be great enough to start a fire.

In terms of rays, the burning glass works by bending the parallel rays from the Sun and focusing them together at the focus. In wave terms, the lens works by altering the curvature of the waves, changing the plane waves that strike it from the Sun into spherical ripples that converge on a focus (see Figure 4).

The distance from the lens to the focus is the radius of the wave-fronts just after passing through the lens. This is the focal length, *f*. Powerful lenses have small focal lengths, for example, lenses in smartphone cameras have focal lengths of about 4 mm.

A converging lens adds a curvature of $\frac{1}{f}$ to wave-fronts passing through the lens.

$\frac{1}{f}$ is the **power** of the lens. This is measured in dioptres (D).

$$\text{Lens power (D)} = \frac{1}{\text{focal length } f \text{ (m)}}$$

Wave point of view:
The lens adds curvature to the waves, centering them on the focus.

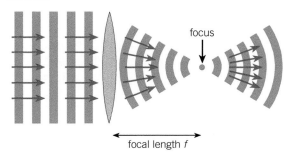

focal length *f*

Ray point of view:
The lens bends the rays, bringing them to a focus.

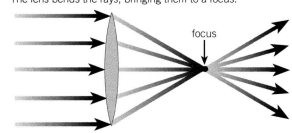

focus

▲ **Figure 4** *Rays and waves – two ways of representing the burning glass*

🖩 Worked example: Calculating lens power

Calculate the power of a lens with a focal length of 4.0 mm.

Step 1: Identify the equation to use.

$$\text{Lens power} = \frac{1}{f}$$

Step 2: Convert focal length into metres.

$$4.0\,\text{mm} = 4.0 \times 10^{-3}\,\text{m}$$

Step 3: Substitute value into the equation in Step 1, and evaluate.

$$\text{Lens power} = \frac{1}{4.0 \times 10^{-3}} = 250\,\text{D}$$

Study tip

1 D is equivalent to 1 m^{-1}

Study tip

Remember to always check the units before performing a calculation. The formula for curvature uses radius in metres.

Summary questions

1 A wave-front has a radius of 32 cm. Calculate the curvature of the wave-front. *(1 mark)*

2 A camera lens has a focal length of 50 mm.
   a Calculate the lens power in dioptres. *(2 marks)*
   b State the additional curvature the lens adds to a wave-front as it passes through the lens. *(1 mark)*

3 A lens has a power of 7.4 D. Calculate its focal length in millimetres. *(2 marks)*

4 a Draw a diagram of plane wave-fronts passing through a converging lens towards the focal point. *(2 marks)*
   b Draw a diagram showing plane wave-fronts of the same wavelength (spacing between wave-fronts) passing through a lens of half the power of the lens in **a**. *(2 marks)*

1.2 Finding the image

Specification references: 3.1.1a(i), 3.1.1b(i), 3.1.1c(iii), 3.1.1c(iv), 3.1.1d(i)

A converging lens makes an image of the source of light (the object) because light from every part of the object goes through the lens. The cornea and the lens of the eye focus light from an object onto the retina. The image on the retina is upside down as rays from the top of the object are focused at the bottom of the image and vice versa (Figure 1). The brain flips the image so you *see* the world the right way up.

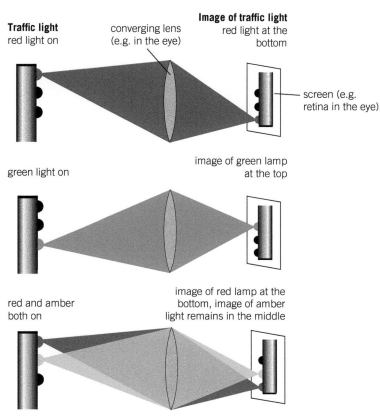

Traffic light
red light on

converging lens
(e.g. in the eye)

Image of traffic light
red light at the bottom

screen (e.g. retina in the eye)

green light on

image of green lamp at the top

red and amber both on

image of red lamp at the bottom, image of amber light remains in the middle

▲ Figure 1 *How a lens makes an image*

Making images

The rule for how a lens shapes light is simple.

| **Curvature of waves leaving the lens** | = | **curvature of waves entering the lens** | + | **curvature added by the lens** |

The distance from the lens to the image of the source (the image distance) is represented by v. The distance from the lens to the source (the object distance) is represented by u.

After passing through the lens, the waves form part of spheres centred on a point at distance v. So the curvature of the waves leaving the lens $= \frac{1}{v}$. The curvature of the waves entering the lens $= \frac{1}{u}$.

21

We can translate the rule for shaping light into a rule for the distances of the object and image from the lens:

$$\frac{1}{v} = \frac{1}{u} + \frac{1}{f}$$

 Worked example: Using the lens equation

a   Calculate the curvature of the wave-fronts before and after passing through the lens, and use your results to calculate the power of the lens.

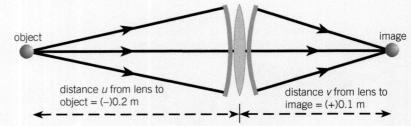

object

image

distance u from lens to object = (−)0.2 m

distance v from lens to image = (+)0.1 m

▲ **Figure 2**

Step 1: Calculate the curvature of the wave-fronts after passing through the lens:

$$\text{curvature} = \frac{1}{v} = \frac{1}{0.1\,\text{m}} = 10\,\text{D}$$

Step 2: Calculate the curvature of the wave-fronts reaching the lens:

$$\text{curvature} = \frac{1}{u} = \frac{1}{-0.2\,\text{m}} = -5\,\text{D}$$

Step 3: Curvature added by lens = power of lens = curvature after − curvature before

$$= 10\,\text{D} - (-5)\,\text{D} = 15\,\text{D}$$

b   Calculate the focal length of the lens.

Step 1: Select the appropriate equation: $\text{focal length} = \dfrac{1}{\text{power}}$

Step 2: Substitute and evaluate: $\text{focal length} = \dfrac{1}{15} = 0.067\,\text{m}$ or 67 mm (2 s.f.)

The image distance v is different from the focal length f, because the focal point is the point at which parallel waves from a very distant object are brought to focus. The focal length is the distance from the centre of the lens to the focal point, and is a constant for a particular lens with a fixed shape. Waves from a nearer object are brought together beyond the focal point. The image distance v is greater than the focal length f, except for very distant objects.

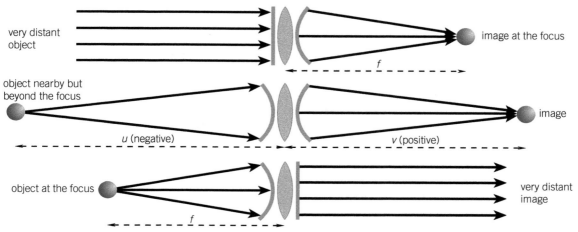

▲ **Figure 3** *Effect of a converging lens on light from a distant object, a nearby object, and an object at the focus*

Practical 3.1.1d(i): Determining power and focal length of a converging lens

A small filament lamp is used as an object and a lens is used to project its image onto an opaque screen. The object distance u is changed several times, and the screen is moved each time so that the image is in focus. At each object distance, the values of u and v (the image distance) are measured.

The values of $\frac{1}{v}$ (y-axis) and $\frac{1}{u}$ (x-axis) are plotted on a graph. This will give a straight line. The intercepts on the axes both give $\frac{1}{f}$, the power of the lens. The reciprocal of this value gives the focal length of the lens.

Magnification

Lenses can also be used to **magnify** the image of an object. This means that the size of the image appears larger than the original object. The linear magnification of an image is defined as:

$$\text{linear magnification } m = \frac{\text{image height (m)}}{\text{object height (m)}}$$

Linear magnification is also related to the object distance and the image distance by:

$$\text{linear magnification } m = \frac{\text{image distance } v \text{ (m)}}{\text{object distance } u \text{ (m)}}$$

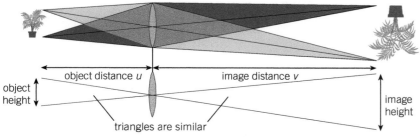

▲ **Figure 4** *Magnification by a lens*

Worked example: Magnification using the lens equation

a A lens produces an image of a lamp. The distance from the lamp to the lens is −0.40 m. The distance from the lens to the focused image is 0.95 m. Calculate the magnification of the image.

Step 1: Choose the appropriate equation: $m = \dfrac{v}{u}$

Step 2: Substitute and evaluate. $m = \dfrac{0.95\,\text{m}}{-0.40\,\text{m}} = (-)2.4$ (2 s.f.).

b Calculate the focal length of the lens.

Step 1: Choose the appropriate equation: $\dfrac{1}{v} = \dfrac{1}{u} + \dfrac{1}{f}$

Step 2: Substitute and evaluate to find $\dfrac{1}{f}$:

$$\frac{1}{0.95\,\text{m}} = \frac{1}{-0.40\,\text{m}} + \frac{1}{f}$$

$$\therefore\ \frac{1}{f} = \frac{1}{0.95\,\text{m}} + \frac{1}{0.40\,\text{m}}$$

$$\frac{1}{f} = 3.553\,\text{m}^{-1}$$

Step 3: Take the reciprocal to find f:

$$\therefore\ f = 0.28\,\text{m}\ (2\ \text{s.f.})$$

▲ **Figure 5** *Painting from 1466 AD, showing the use of reading glasses*

 Reading glasses

As we get older, the muscles in our eyes weaken. Close objects can no longer be focused. From about the year 1300, those who needed to work with close objects (e.g. reading a book) began to wear reading glasses, often with quartz lenses. The lenses added curvature to the light waves from the object, helping the eye focus near objects. The same principle was employed in the development of magnifying glasses around the same time.

1 The invention of reading glasses has been called one of the greatest inventions of the Medieval period. Why do you think this is? What effect might the development of reading glasses and magnifying glasses have had on the spread of learning and knowledge? Explain your answer.

Summary questions

1 A converging camera lens with a focal length of 50 mm produces a focused image of a face. The distance between the lens and the face is 1 m. Calculate the image distance. *(2 marks)*

2 An object is placed 1.5 m in front of a converging lens. A focused image is formed on a screen 2.5 m from the lens. Calculate the magnification of the image. *(2 marks)*

3 You tell a colleague that the magnification of a particular system is 0.27. You are told that this must be wrong because images can't be magnified *and* be smaller than the object. Explain what is meant by 'magnification' in this context. *(2 marks)*

4 A lamp 400 mm from a lens is in focus on a screen 400 mm from the other side of the lens.
  a Calculate the power of the lens. *(2 marks)*
  b Use the lens equation to explain why the lamp must be moved nearer to the lens to project its image further away. *(3 marks)*

1.3 Storing and manipulating the image

Specification reference: 3.1.1a(ii), 3.1.1b(i), 3.1.1c(i), 3.1.1c(vi)

Digital cameras are getting ever more compact. Figure 1 shows a tiny camera that can be swallowed. It transmits video images as a stream of binary numbers as it passes through the digestive system. White LEDs (light-emitting diodes) provide the light source for this camera.

CCDs and pixels

In a digital camera, the image from the lens is focused onto a light-sensitive microchip called a charge-coupled device or CCD. This is a screen covered by millions of tiny 'picture elements' or **pixels** (Figures 2 and 3). Each pixel stores electric charge when light falls on it – the brighter the light falling on the pixel, the greater the charge stored on it. The image becomes an array of numbers, which can then be manipulated to edit the image.

Bits and bytes

Numbers in computers, such as the numbers representing the image in a digital camera, are stored as *on* or *off* values. 'On' may be a high potential difference, 'off' may be low. This is a **binary** system. The two values can be thought of as two digits, 1 and 0.

If one pixel simply records 'high' or 'low', then only one memory location storing a 1 or 0 is needed. This is called one **bit** (binary digit) of information. In practice, pixels do not simply record dark = black and bright = white. Pixels can often have 256 shades of grey, with each shade recorded as a number from 0 to 255.

> ## Learning outcomes
>
> Describe, explain, and apply:
>
> → the storing of images in a computer as an array of numbers that may be manipulated to enhance the image
>
> → the terms: pixel, bit, byte, resolution, noise
>
> → the equation: amount of information in an image = no. of pixels × bits per pixel
>
> → the equation for the number of bits b, using $b = \log_2 N$.

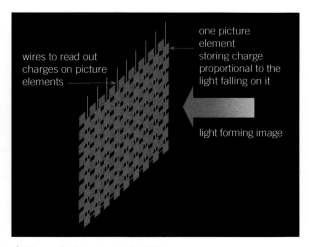

▲ Figure 1 A capsule endoscope

▲ Figure 2 A typical CCD. In modern cameras a CCD can have many millions of pixels

▲ Figure 3 Structure of a CCD

one picture element storing charge proportional to the light falling on it

wires to read out charges on picture elements

light forming image

Surprisingly, the number of bits required to store 256 alternatives is only eight. A group of eight bits is called a **byte**, so one byte can store 256 alternatives.

Table 1 shows you the basic idea. Consider the decimal number 5. This is represented by a 1 in the 1s column and a 1 in the 4s column, that is, $5 = 1 + 4$, represented in binary as 101.

How about the decimal number 13? That is $8 + 4 + 1$, or 1101 in binary.

▼ Table 1 *Bit representations of decimal numbers*

Bit D (8s)	Bit C (4s)	Bit B (2s)	Bit A (1s)	Decimal value
0	0	0	0	0
0	0	0	1	1
0	0	1	0	2
0	0	1	1	3
0	1	0	0	4
0	1	0	1	5
0	1	1	0	6
0	1	1	1	7

The number of arrangements of bits N can also be calculated using the equation

$$N = 2^b \text{ or } b = \log_2 N$$

where b is the number of bits available.

 Worked example: Calculations using bits

a Show that 8 bits can have 256 arrangements.

Step 1: Identify the equation.

$$N = 2^b$$

Step 2: Substitute values and evaluate.

$$N = 2^8 = 256$$

b Calculate the number of bits required to give 65 536 possible arrangements.

Step 1: Identify the equation required.

$$b = \log_2 N$$

Step 2: Substitute values and evaluate.

$$b = \log_2 65\,536 = 16$$

Study tip

To use the equation $b = \log_2 N$, you will need the button on your calculator that looks like $\boxed{\log_\bullet\square}$. You might need to change your calculator to maths mode.

Image resolution

The **resolution** of a digital image is the scale of the smallest detail that can be distinguished. If an image is about 20 mm across, with approximately 40 pixels across, this means that the resolution of the image is 0.5 mm per pixel (0.5 mm pixel⁻¹). To calculate the resolution, divide the distance represented in the image by the number of pixels in that distance.

 Worked example: Resolution of an image

Figure 4 shows the surface of Europa, a satellite of Jupiter. It represents an area of 30 × 70 km. There are 1300 pixels across the image. Calculate the resolution of the image.

Step 1: Identify the correct equation to use.

$$\text{resolution} = \frac{\text{width of object in image}}{\text{number of pixels across object}}$$

Step 2: Substitute and evaluate.

$$\text{resolution} = \frac{70\,000\,\text{m}}{1300} = 54\,\text{m pixel}^{-1}\ (2\ \text{s.f.})$$

▲ **Figure 4** *Surface of Europa imaged by the Galileo spacecraft*

Information in an image

The amount of information stored in an image can be calculated using the equation

amount of information in image = no. of pixels × bits per pixel

 Worked example: Information in an image

Calculate the storage required for a six megapixel camera that uses three bytes to encode colour information for each pixel.

Step 1: Identify the correct equation required.

the amount of information in an image
= no. of pixels × bits per pixel

Step 2: Convert values given so that they may be substituted into the equation in Step 1.

6 megapixels = 6 × 10⁶ pixels

3 bytes = 3 × 8 = 24 bits

Step 3: Substitute values into the equation in Step 1 and evaluate.

amount of storage required = 6 × 10⁶ × 24
= 144 Mbits (18 Mbytes)

Image processing

In a digital image each pixel is coded by a number. An 8 bit (1 byte) pixel will usually be coded so that the darkest value is represented by 0 and the brightest by 255. Changing the number will change the appearance of the pixel in the image. This means that mathematical operations can be used to edit or process the images. Examples of how digital images can be processed are shown in Table 2.

▼ Table 2 *Image processing*

Changing brightness	A dim image can be brightened by increasing the value on each pixel by the same amount until the brightest pixel in the image is coded at 255.
Removing noise	Noise in images refers to the random speckles across the image. This can be reduced by smoothing, where the value of each pixel is replaced with the median or mean of its value and those around it.
Edge detection	To enhance edges in an image the average value of the pixel's neighbours is subtracted from each pixel. This removes uniform areas of brightness and picks out the places where the gradient of the brightness changes abruptly – at the edges.
Changing contrast	An image with little contrast will not use the full range of pixel values. An image may only use the values between 75 and 150. To improve the contrast this range is stretched across the 256 possible values so the value 75 becomes 0 and 150 becomes 255.

Study tip

Remember that the median is the middle value in a series of values arranged in an ordered list, whereas the mean is the sum of all the values divided by the number of values.

Summary questions

1 Replace the pixel in the middle of Table 3 with
   a the mean of all the values, *(2 marks)*
   b the median of all the values. *(2 marks)*
   c Compare the processes above for eliminating possible noise. Suggest which is the better choice. Explain your answer. *(3 marks)*

▼ Table 3

100	100	100
100	200	100
100	100	100

2 Calculate the number of bits required to code for 4096 alternative values. Express this as a number of bytes. *(3 marks)*

3 A satellite system to image the Earth's surface is designed to have a resolution of 10 m pixel^{-1} and to cover an area of 100 km^2 in each image.
   a State what the term 'resolution' means in this context. *(1 mark)*
   b Calculate the number of pixels required to achieve this resolution. *(2 marks)*
   c Each pixel requires 3 bytes. Calculate the amount of memory in Mbytes that each image requires. *(2 marks)*

1.4 Polarisation of electromagnetic waves

Specification references: 3.1.1a(iv), 3.1.1b(i), 3.1.1b(iii), 3.1.1c(v), 3.1.1d(ii)

Some snow goggles and sunglasses have polarising lenses or filters in them to dramatically cut down the glare in bright environments. These are particularly useful in the mountains on sunny days. For this reason polarising lenses (or filters) are used by sportspeople and photographers. They are also used in microscopy.

Frequency, speed, and wavelength

Light can be thought of as a wave. Three important characteristics of waves are: speed v, frequency f, and wavelength λ. These are related by the equation

$$\text{speed } v \text{ (m s}^{-1}) = \text{frequency } f \text{ (Hz)} \times \text{wavelength } \lambda \text{ (m)}$$

Frequency is the inverse of the time period T of the wave – the time for one complete oscillation of the wave. This is expressed as $f = \dfrac{1}{T}$.

Time picture

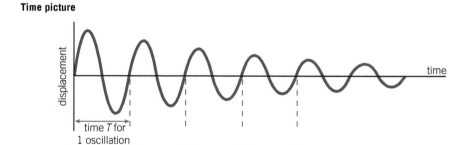

time T for 1 oscillation

▲ Figure 1 *Polarising filter in ski goggles*

Position picture: two different wave speeds compared (same frequency)

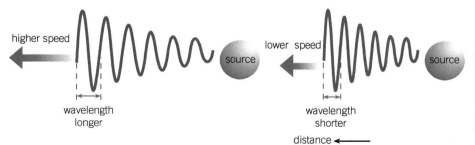

higher speed

lower speed

wavelength longer

wavelength shorter

distance ←

▲ Figure 2 *Frequency, speed, and wavelength*

Synoptic link

Waves will be discussed in more detail in Chapter 6, Wave behaviour.

Study tip

The speed of light c is sometimes used in equations instead of v, so for example the equation $v = f\lambda$ becomes $c = f\lambda$.

Study tip

It is a good idea to memorise approximate values of wavelengths on the electromagnetic spectrum. This means that you can check you have obtained the correct order of magnitude in a calculation.

 Worked example: Frequency and wavelength of ultrasound

The time period of an ultrasound signal is $0.1\,\mu s$. It travels at $2000\,m\,s^{-1}$.

a   Calculate the frequency and wavelength of this signal.

Step 1: Identify the correct equation to use for frequency.

$$\text{frequency} = \frac{1}{T}$$

Step 2: Substitute values into the equation and evaluate.

$$\text{frequency} = \frac{1}{1 \times 10^{-7}\,s} = 1 \times 10^7\,Hz \text{ or } 10\,MHz$$

Step 3: Identify an equation that links frequency to wavelength.

$$v = f\lambda$$

Step 4: Rearrange $v = f\lambda$ to make λ the subject.

$$\lambda = \frac{v}{f}$$

Step 5: Substitue values and evaluate.

$$\lambda = \frac{2000\,m\,s^{-1}}{1 \times 10^7\,Hz} = 2 \times 10^{-4}\,m \text{ (or 0.2 mm)}$$

The Electromagnetic spectrum

Visible light is part of the electromagnetic spectrum. All electromagnetic waves are transverse waves and have the same speed of propagation c in empty space. $c = 3.00 \times 10^8\,m\,s^{-1}$.

You can see from Figure 3 that visible light is a narrow band of wavelengths. The range of wavelengths of visible light is roughly $4 \times 10^{-7}\,m$ (violet) to $7 \times 10^{-7}\,m$ (red).

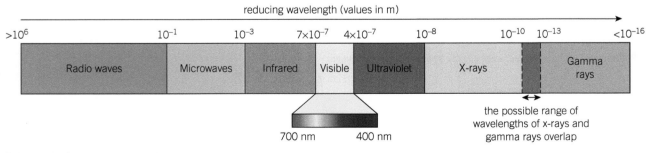

▲ **Figure 3** *The electromagnetic spectrum*

Polarisation

Electromagnetic waves can be **polarised**. This is a property of transverse waves. Transverse waves are polarised if they vibrate in one plane only. Unpolarised transverse waves vibrate in a randomly changing plane.

Electromagnetic waves are waves of oscillating magnetic and electric fields. The two fields are at right angles to each other and to the direction of travel of the wave, as shown in Figure 4. You can see that the electric field is oscillating in the vertical plane (up and down) and the magnetic field is oscillating in the horizontal plane. All electromagnetic waves can be polarised.

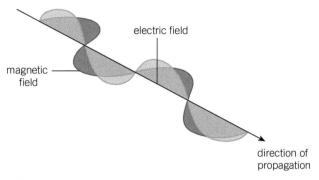

▲ Figure 4 *An electromagnetic wave*

If a wave is plane-polarised, the direction of oscillation remains fixed. In our example, the electric field strength will always be oscillating in the vertical plane. If the wave is unpolarised the direction of the oscillation will not be fixed. This is represented in Figure 5.

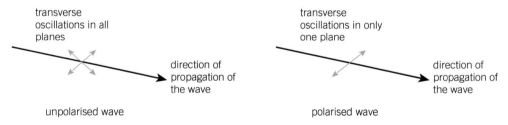

▲ Figure 5 *Polarised and unpolarised waves*

Demonstrating polarisation

Unpolarised light is polarised when it passes through a polarising filter. Figure 6 represents unpolarised light passing through a pair of filters. The first filter only allows through the vertical component of the wave, and the second filter only allows through the horizontal component of the wave. No light is transmitted through the second (horizontal) filter because the light reaching it after passing through the first filter is fully polarised in the vertical direction.

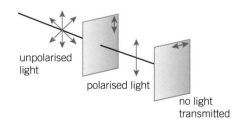

▲ Figure 6 *The effect of filters on polarised light*

You can detect whether light is polarised by observing it through a single polarising filter and rotating the filter. If the intensity of the light remains constant, the light source is emitting unpolarised light. If the intensity varies as the filter rotates, the source is emitting polarised light. The polarisation of radio waves can be detected by rotating the receiving aerial. The aerial will pick up the strongest signal when it is set up parallel to the plane of polarisation of the radio waves.

Practical 3.1.1 d(ii): Observing the polarisation of light and microwaves

Light can be polarised by shining a narrow beam of light through a tank of water that contains a few drops of milk (Figure 7).

By looking through a polarising filter at light scattered vertically upwards (looking down at the tank from above), and rotating the filter slowly, the light observed will vary in intensity, showing that the light has been polarised. A similar effect is seen when you view the tank side on.

If light is completely cut out at one angle you will know that the polarised light at that point is vibrating perpendicular to the filter.

A similar experiment can be performed with microwaves (Figure 8). If a metal grille is placed between a transmitter of polarised waves and a receiver, the strength of the signal at the detector can be changed by rotating the grille in the same way as rotating a polarising filter for the light experiment above.

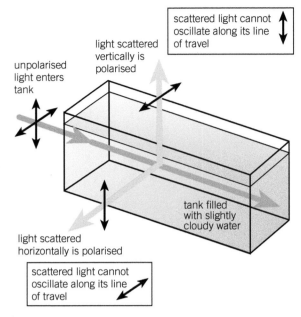

▲ **Figure 7** *Polarisation of light by scattering*

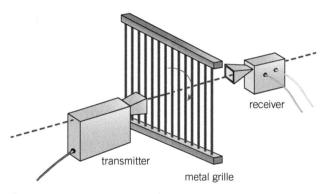

▲ **Figure 8** *An experiment to show the scattering of microwaves*

Hint

Radio waves and microwaves are both electromagnetic waves. The same principles apply to both.

Summary questions

1 A radio station broadcasts at a frequency of 91 MHz. Calculate the wavelength of the waves assuming the speed of the waves is $c = 3.0 \times 10^8 \, \text{m s}^{-1}$. Explain how you decided how many significant figures to use in your answer. *(2 marks)*

2 a Draw a displacement–time graph showing a wave with time period of 0.1 s. Draw three complete waveforms. *(3 marks)*
  b Add to your graph a waveform with a frequency of 20 Hz. *(2 marks)*

3 The radio emission from some galaxies is polarised. Suggest how this fact could be confirmed. *(2 marks)*

Practice questions

1 A digital camera has a 6 megapixel lens. Each pixel is coded by 24 bits. The memory card on the camera can store 160 images. What is the minimum memory required? *(1 mark)*

   A 23 kbyte   **B** 5 Mbyte

   C 2.9 Gbyte   **D** 23 Gbyte

2 An electronic display uses a grid of 16×16 pixels. Each pixel is coded by 4 bits.

   a State the number of bits the display memory stores for one image. *(1 mark)*

   b Give your answer to **a** in bytes.

                                     (1 mark)

3 Here is some data about ultrasound used in medical imaging:

   wavelength in soft tissue $= 2.9 \times 10^{-4}\,\text{m}$

   speed in soft tissue $= 1450\,\text{m s}^{-1}$

   a Calculate the frequency of the ultrasound.

                                     (1 mark)

   b The speed of ultrasound in air is about $340\,\text{m s}^{-1}$. State what happens to the frequency and wavelength of ultrasound as it travels from soft tissue into air.

                                    (2 marks)

4 Radio waves from a transmitter are vertically polarised. A teacher demonstrates this by positioning a radio with its aerial vertical. The signal is at a maximum with the radio in this orientation. She then rotates the radio to position B and then position C.

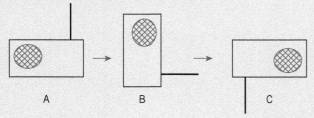

A   **B**   **C**

▲ Figure 1

   a Explain what is meant by polarisation. You may include diagrams in your answer. *(2 marks)*

   b Describe how the signal will vary as the radio is rotated to position B and then C. *(2 marks)*

5 This question is about finding the focal length of a converging lens.

A student varies the distance between a point source of light and the lens *u* and measures the distance between the lens and the focused image, *v*.

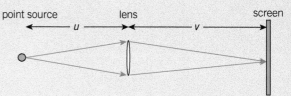

▲ Figure 2

Here is one pair of readings:
object distance $= -0.300\,\text{m}$
image distance $= 0.145\,\text{m}$

The student suggested that the uncertainty in the image distance is $\pm 0.01\,\text{m}$

   a Suggest why there is uncertainty in the image distance measurements, and why the student chooses to ignore the uncertainty in the object distance when calculating the focal length of the lens.

                                    (2 marks)

   b Calculate the focal length of the lens, including the uncertainty in the value.

                                    (4 marks)

The graph shows more data gained from the experiment, plotted as a graph of $\frac{1}{v}$ against $\frac{1}{u}$. The uncertainty bars have not been included.

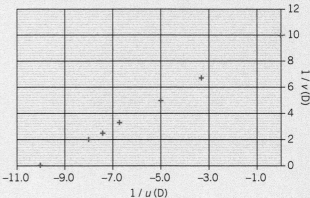

▲ Figure 3

   c On a copy of the graph, add the uncertainty on the data point $(-6.7, 3.3)$. *(1 mark)*

   d Use the equation $\frac{1}{v} = \frac{1}{u} + \frac{1}{f}$ to explain why the value of the $\frac{1}{u}$ intercept is equal to the negative of the $\frac{1}{v}$ intercept. *(2 marks)*

SIGNALLING
2.1 Digitising a signal

Specification references: 3.1.1a(iii), 3.1.1b(i), 3.1.1c(ix), 3.1.1c(x)

Signals come in many different forms. A signal transfers information from one location to another – it can be conveyed through sound or light. A red traffic light is a signal. So is a fire alarm, or even a wink. A signal can be a variation of current in a telephone cable or a stream of binary highs and lows (1s and 0s).

Digitising signals

We live in a digital world. Information is digitised and sent around the globe at speeds that could not have been imagined 50 years ago. Digitised signals are possible because information can be coded into a string of binary digits that is transmitted (through the air, through space, through wires, or through optical fibres) from the sender to the receiver. For example, emails use digital signals. Each character is sent as a 1 byte number code.

 Worked example: Encoding characters into digital signals

A book contains 1000 pages. Each page of text contains 500 words that equates to 3000 characters including spaces. Each character is encoded with a 1 byte number.

Calculate the number of bits required to encode the book.

Step 1: Convert bytes per character to bits per character.

1 byte = 8 bits

Step 2: Multiply the number of bits per character by the number of characters in the book.

Number of bits required = number of bits per character × number of characters per page × number of pages = 8 × 3000 × 1000 = 2.4×10^7 bits or 24 Mbits

▲ Figure 1 *All aspects of communication technology are rapidly changing*

Advantages of digital signals

Analogue signals vary continuously from one value to the next, without fixed values. For most of the 20th century, long-distance communication systems such as telephone, radio, and television were analogue systems. For example, in an analogue telephone a sound vibration is changed into matching oscillations of potential difference.

One problem with analogue systems is the need for amplification as the signal becomes weaker. If the signal becomes distorted or 'noisy',

Hint

Do not confuse Mbit with Mbyte. 1 Mbit = 1×10^6 bit, whilst 1 Mbyte = 1×10^6 byte.

the amplification boosts the signal *and* the noise. In this context, noise is the random variation on the signal. For example, holding a conversation in a crowded room is difficult because what your friend is saying (the signal) has to be picked out from the background chatter (the noise). Simply amplifying the sound won't help, because that will also amplify the effect of the background chatter. The hiss on a badly tuned radio is another example of noise – turning the radio up increases the hiss and the signal.

Noise can be filtered out, but this causes a loss of detail in the signal. This is not a problem in a digital signal. It is easy to detect binary 'on/off' signals even when they are weak and noisy. A perfect copy of a message can be regenerated and sent on. Digital signals can also transmit information much faster than analogue signals.

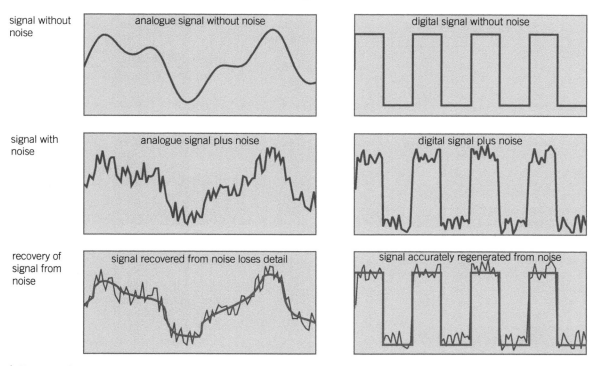

signal without noise
analogue signal without noise
digital signal without noise

signal with noise
analogue signal plus noise
digital signal plus noise

recovery of signal from noise
signal recovered from noise loses detail
signal accurately regenerated from noise

▲ Figure 2 *Signals and noise*

Sampling

When musicians refer to sampling, they often mean short extracts of music that are edited and mixed with other recordings. In physics, **sampling** is the process in which the displacement of a continuous (analogue) signal is measured at small time intervals and turned into a digital string of binary numbers (samples). This is shown in Figure 3.

Analogue to digital conversion

Figure 4 (next page) shows the principle of converting a varying analogue signal into a stream of numbers. In this example, each sample is coded with three bits.

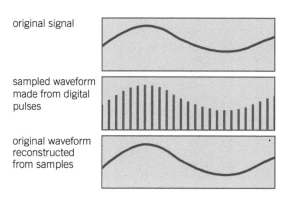

original signal

sampled waveform made from digital pulses

original waveform reconstructed from samples

▲ Figure 3 *Sampling a signal*

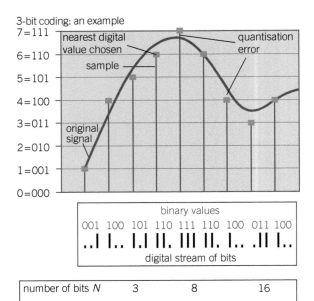

▲ **Figure 4** *Digitising a signal*

number of bits N	3	8	16
number of levels 2^N	$2^3 = 8$	$2^8 = 256$	$2^{16} = 65536$

This means that there are eight (2^3) levels to represent the signal value for any sample. These eight levels, known as quantisation levels, cover the range of signal values from the lowest (000) to the highest (111). You can see that the signal does not always match exactly with a quantisation level. In such cases the system records the nearest level to the signal. The difference between the signal value and the quantisation level is the quantisation error. Increasing the number of quantisation levels produces a better match to the original signal.

Resolution of a sample

The **resolution** of a sample is the smallest change in potential difference that can be determined. This is given by the equation

$$\text{resolution} = \frac{\text{potential difference range of signal}}{\text{number of quantisation levels}}$$

Strictly speaking, the p.d. should be divided by the number of steps between levels, which is 1 less than the number of levels. For real systems, 2^b and $2^b - 1$ are very close so we can use the number of levels.

▦ Worked example: Calculating resolution of a sample

Consider a signal that is detected over a range 0.0 V to 12.0 V. An 8-bit sample of this signal is produced. Calculate the resolution of this sample.

Step 1: Identify the equation required.

$$\text{resolution} = \frac{\text{potential difference range of signal}}{\text{number of quantisation levels}}$$

Step 2: Calculate the number of quantisation levels possible.

number of possible arrangements N for 8 bits = $2^b = 2^8 = 256$

Step 3: Substitute values into the equation in Step 1 and evaluate.

$$\text{resolution} = \frac{\text{potential difference range of signal}}{\text{number of quantisation levels}}$$

$$= \frac{12.0 \, \text{V}}{256} = 0.047 \, \text{V (2 s.f.)}$$

Synoptic link

You have met the number of arrangements of bits, N, and the available number of bits, b, in Topic 1.3, Storing and manipulating the image.

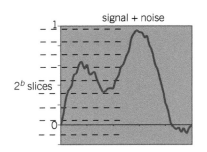

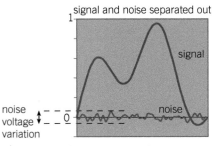

noise voltage variation

▲ **Figure 5** *Signal, noise and quantisation levels*

Number of useful quantisation levels

Increasing the number of quantisation levels increases demands on data storage and transmission. Noise also plays a part in the choice of the number of levels to use.

Look at Figure 5. The diagram above shows a noisy signal. The lower diagram shows the signal and noise separated out. It is pointless to have a smaller gap between quantisation levels than the size of the noise variation, as this would mean sampling the noise to great detail, rather than ignoring it.

The maximum number of useful quantisation levels is given by the equation

$$\text{Maximum useful number of levels} = \frac{\text{total noisy signal variation}}{\text{noise variation}}$$

$$= \frac{V_{total}}{V_{noise}}$$

Since number of possible arrangements (or levels) $N = 2^b$, this equation becomes

$$2^b = \frac{V_{total}}{V_{noise}}$$

Taking $\log_2$ on both sides

$$b = \log_2 \frac{V_{total}}{V_{noise}}$$

 ### Worked example: Bits and quantisation levels

A signal has a maximum total variation of 200 mV. The noise variation is 5 mV. Calculate the largest number of bits per sample worth using to encode the variation.

Step 1: Identify the equation required. $b = \log_2 \frac{V_{total}}{V_{noise}}$

Step 2: Substitute values and evaluate.

$$b = \log_2 \frac{V_{total}}{V_{noise}} = \log_2\left(\frac{200}{5}\right) = 5.3, \text{ which rounds } up \text{ to 6 bits.}$$

We round up, because 5 bits gives 32 levels and 40 is the maximum useful number of levels.

Summary questions

1 Calculate the number of bits required to encode this sentence in a signal if each character and space is represented by 8 bits. The previous sentence is 127 characters in length. *(2 marks)*

2 A signal variation is sampled with 6 bits per sample.
  a Calculate how many quantisation levels are possible in each sample. *(2 marks)*
  b The potential difference is sampled on a scale from 0.00 V to 12.00 V. Calculate the resolution of the sample. *(2 marks)*

3 Give two advantages of digital recording over analogue recording. Explain your answers. *(4 marks)*

4 The ratio $\frac{V_{total}}{V_{noise}}$ in a signal is 75. Calculate the largest number of bits per sample worth using to encode the variation. *(3 marks)*

5 a Draw a diagram showing a waveform sampled with 8 quantisation levels. You should include 6 sampling points at regular intervals on your diagram. *(4 marks)*
  b Use your diagram to explain what 'quantisation error' means. *(2 marks)*

2.2 Sampling sounds and sending a signal

Specification references: 3.1.1a(iii), 3.1.1c(vii), 3.1.1c(viii)

Learning outcomes

Describe, explain, and apply:

→ disadvantages of digital signals

→ the equation: minimum rate of sampling > 2 × maximum frequency of signal

→ the equation: rate of transmission of digital information = samples per second × bits per sample.

Have you ever watched old films where the wheels on a speeding car appear to turn slowly or even backwards? This happens because the time interval between individual images (frames) captured by the camera is too long compared to the speed of the rotating wheel. This causes the wheels to appear as though they move backwards as the wheels make nearly a full rotation between each frame. This highlights the problem of not taking enough samples each second. The sampling rate (sampling frequency) is the number of samples taken each second. For a varying signal to be sampled accurately the time interval between samples must be shorter than the time in which important changes in the signal occur. If the time interval is too large the reconstructed signal will lose detail. This is illustrated in Figure 1.

In fact, to ensure that the original signal can be reconstructed accurately two conditions have to be met:

- the signal cannot contain frequencies above a certain maximum
- minimum sampling rate > 2 × highest frequency component.

If the signal contains frequencies above the maximum component the signal will not be reconstructed accurately. Lower frequency signals called *aliases*, which are not present in the original signal, will be generated. The backward-rotating wheels in the film described above are an example of this.

Sampling too slowly misses high frequency detail in the original signal

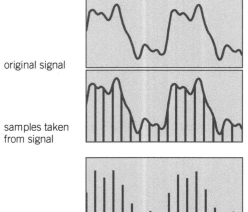

original signal

samples taken from signal

samples alone

signal reconstructed from samples

Sampling too slowly creates spurious low frequencies (aliases)

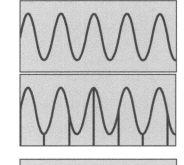

original signal

samples taken from signal

samples alone

signal reconstructed from samples

▲ Figure 1 *Losing high frequency detail*

▲ Figure 2 *Aliasing*

The human ear cannot detect sound above frequencies of about 20 kHz. Therefore, for music to be sampled accurately, the sampling frequency should be greater than 40 kHz. The standard frequency used is 44.1 kHz. Filters remove frequencies above 20 kHz in the original signal so aliasing (the formation of aliases) will not be a problem.

Bit rate

The **bit rate** is the rate of transmission of digital information. It can be calculated using the equation

$$\text{bit rate (bit s}^{-1} \text{ or Hz)} = \text{samples per second} \times \text{bits per sample}$$

 Worked example: Calculating bit rate

A CD quality sound uses 16 bits per sample at a sample rate of 44.1 kHz for each of the two stereo channels. Calculate the combined bit rate for the two stereo channels.

Step 1: Identify the equation required.

$$\text{bit rate (bit s}^{-1} \text{ or Hz)} = \text{samples per second} \times \text{bits per sample}$$

Step 2: Substitute values and evaluate.

bit rate for one stereo channel (bit s^{-1} or Hz) = $44.1 \times 10^3 \times 16$
= 7.1×10^5 bit s^{-1} or 7.1×10^5 Hz

bit rate for both stereo channels = $2 \times 7.1 \times 10^5 = 1.4 \times 10^6$ bit s^{-1}
= 1.4 Mbit s^{-1} or 1.4 MHz

sampling stereo sound for a compact disc

sampling rate
minimum sampling rate = 2 × maximum frequency
maximum frequency = 20 kHz
actual sampling rate = 44.1 kHz
time between samples = 22.7 µs

rate to send bits
32 bits every 22.7 µs
rate of sending bits = 32 bit/22.7 µs
rate of sending bits = 1.4 million bits per second
1.4 MHz

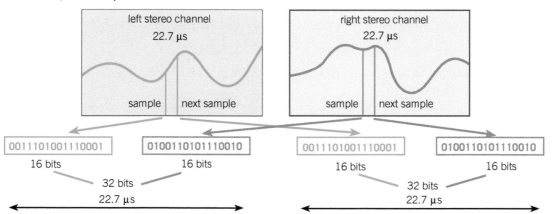

▲ Figure 3 *Samples and strings of bits*

We can also use the idea of bit rate to calculate the playing time of a recorded sound signal of known size.

$$\text{duration of signal (s)} = \frac{\text{number of bits in signal}}{\text{bit rate (bit s}^{-1} \text{ or Hz)}}$$

Study tip

Watch out for data given in bytes when the question asks for bits, or the other way round. Remember that 1 byte = 8 bits.

 Worked example: Signal duration from bit rate

A song occupies 28 Mbytes of memory on a mobile phone. If the bit rate of the system is 1.4 MHz, calculate the duration of the song.

Step 1: Select the appropriate equation:

$$\text{duration of signal (s)} = \frac{\text{number of bits in signal}}{\text{bit rate (bit s}^{-1}\text{ or Hz)}}$$

Step 2: Convert bytes to bits, and MHz to Hz:

$$28 \text{ Mbytes} = 8 \times 28 \text{ Mbits} = 228 \text{ Mbits} = 224 \times 10^6 \text{ bits}$$
$$1.4 \text{ MHz} = 1.4 \times 10^6 \text{ Hz}$$

Step 3: Substitute and evaluate.

$$\text{duration of signal (s)} = \frac{224 \times 10^6 \text{ bits}}{1.4 \times 10^6 \text{ Hz}}$$
$$= 160 \text{ s} = 2 \text{ minutes } 40 \text{ seconds}$$

Disadvantages to digital signals

How will digital signals affect society in the twenty-first century? Does seeing tragic world events almost as they happen, but being powerless to do anything about them, affect people's sense of responsibility?

Digital signals are numbers and so can be changed or scrambled. Who should have the right to do this? Are online banking details completely secure? How can or should the web be policed? Will cyber-terrorism become an increasing problem? What about cyber-bullying? Images of actors we see on television screens and in cinemas can be digitally enhanced and give the impression of a level of perfection that cannot be met in reality – should this be a cause of concern?

The digital revolution of the twentieth- and twenty-first centuries has produced great benefits, from medicine to entertainment and academic study, but it also raises difficult questions.

Summary questions

1   A recording app on a mobile phone samples at a rate of 32.0 kHz. There are 8 bits per sample. Calculate the bit rate of the system. *(2 marks)*

2   This topic contains approximately 4500 characters including spaces. If each character is encoded with 8 bits, calculate the time duration of the encoded topic when transmitted at a bit rate of 10 Mbit s^{-1}. *(2 marks)*

3   Explain why the sampling rate needs to be greater than 40 000 Hz to encode an audio recording accurately when the highest frequency present is 20 000 Hz. *(3 marks)*

4   If the human ear cannot detect frequencies above 20 000 Hz, explain why it is necessary to filter out frequencies above this maximum before encoding takes place at a sampling rate of 40 000 Hz. *(2 marks)*

5   a   Give two disadvantages to society of digital recording of sounds and images. Explain your answers. *(4 marks)*

    b   Suggest an aspect of digital technology that can be considered as both an advantage and disadvantage to society. Explain your choice. *(3 marks)*

Practice questions

1 A signal is sampled at a rate of 44.1 kHz. Each sample is coded with 2 bytes. *(1 mark)*

Which figure gives the bit rate of the sampling?

A 88 bits s^{-1} B 706 bits s^{-1}

C 88 200 bits s^{-1} D 705 600 bits s^{-1}

2 Here are three statements about hearing and sampling.

1 The limit of human hearing is about 20 kHz.

2 Recordings can use a filter so that no variations above 20 kHz are recorded.

3 Recordings are often sampled at 44.1 kHz, over double the highest frequency that has been recorded.

Explain why it is necessary to limit the frequency range and sample at a rate over double the highest frequency of human hearing to obtain a high-quality recording. *(3 marks)*

3 An analogue signal has a total voltage variation (signal + noise) of 600 mV. The noise variation is 9 mV.

a Calculate the maximum useful number of bits to code the signal. *(2 marks)*

b Explain why it is not worth using more bits to code the signal. *(2 marks)*

4 An analogue signal has a voltage variation of 380 mV. It is sampled with 8 bits per sample. Calculate the resolution of each sample. *(2 marks)*

5 A digital photograph has a file size of 28 Mbyte. Calculate the time it will take the image to download at a download rate of is 1 Mbit s^{-1}. *(2 marks)*

6 Music is streamed to a mobile phone at a rate of 96 kilobits per second.

How many kilobytes per second does this represent? *(1 mark)*

7 The graph shows a signal coded with three bits per sample. The blue curved line represents the original waveform.

a How can you tell that there are three bits per sample? *(1 mark)*

b How does the graph show that the original signal cannot be perfectly reconstructed from the digital sample? *(1 mark)*

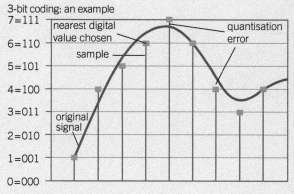

▲ Figure 1

c State two changes to the sampling system that would reduce the error in the reconstructed signal. *(2 marks)*

8 This question is about a sampling system. The graph shows a section of a waveform that has been sampled. Each small triangle represents a sample point.

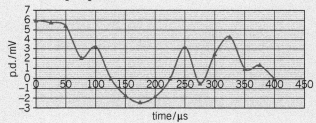

▲ Figure 2

a Show that the sample rate is 40 000 Hz. *(1 mark)*

b State an estimate for the highest frequency sound that can be accurately sampled by this system. Why might a higher frequency not be sampled accurately? *(2 marks)*

c The range of voltage variation is −3 mV to + 7 mV. There are 16 bits per sample. Calculate the resolution of the sample. *(2 marks)*

d The system is used to store a music track lasting $3\frac{1}{2}$ minutes. Calculate the number of Mbytes of memory that will be required. *(3 marks)*

e A student suggests that a digital recording can never be as accurate as an analogue recording. Suggest an argument against this point of view. State and explain an advantage of digitally recording music. *(3 marks)*

3 SENSING

3.1 Current, p.d., and electrical power

Specification references: 3.1.2a(i), 3.1.2a(ii), 3.1.2a(v), 3.1.2a(ix), 3.1.2b(i), 3.1.2c(i), 3.1.2c(ii)

Learning outcomes

Describe, explain, and apply:

→ the concept of current as a flow of charged particles

→ the need for charge and current to be conserved in any circuit loop

→ the equation $I = \dfrac{\Delta Q}{\Delta t}$

→ the term potential difference (p.d.) as the energy change per unit charge moved between two circuit points

→ the equation $V = \dfrac{W}{Q}$

→ the dissipation of power in electric circuits and the equations $P = IV = I^2R$ and $W = VIt$.

▲ Figure 1 *The aurora borealis*

▲ Figure 2 *The ion drive of the SMART-1 probe was driven by a stream of xenon ions*

Electrically charged particles are around us in abundance. You may have heard of them moving at high speed in the Large Hadron Collider at CERN, in the aurora borealis (Northern Lights), and in ion drives for propelling space probes. Ions are electrically charged atoms or molecules in which the number of electrons is different from the number of protons. They form the structure in crystals of ionic compounds, such as salt, and in metals. It is the presence of positive ions and the movement of negative electrons around them that gives metals their electrical properties.

Electric current

A flow of charged particles produces an electric current – this could be electrons in a wire (Figure 3), or positive ions leaving a space probe's ion drive – each of which has the same electrical charge q, measured in **coulombs** (C). As they flow, they pass through the section of the stream shown coloured in Figure 3.

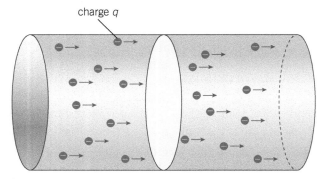

▲ Figure 3 *Flow of charges*

If the number of charged particles passing through the pale blue section in a time Δt is N, the total charge ΔQ flowing across the section is Nq and the **current** I, defined as the rate of flow of charge, is given by

$$I = \frac{\Delta Q}{\Delta t}$$

and is measured in **amperes**, or amps (A).

The symbol Q originally stood for 'quantity of electrical charge'. The French physicist André-Marie Ampère originally gave the name 'intensity of current', together with the symbol I, to what we now call current. The SI unit of current is named after him (but without the accent).

Worked example: Electrons in a wire

In Figure 3, a 1.0 m length of copper wire contains 1.7×10^{22} free electrons, each of charge 1.6×10^{-19} C. These move along the wire with a mean speed $v = 7.4 \times 10^{-4}$ m s^{-1}. Calculate the current I in the wire.

Step 1: Find the time Δt for all 1.7×10^{22} electrons in the 1.0 m wire to move past the end point.

$$\text{speed} = \frac{\text{distance}}{\text{time}} \text{ so time } \Delta t = \frac{\text{distance}}{\text{speed}} = \frac{1.0 \,\text{m}}{7.4 \times 10^{-4}\,\text{m s}^{-1}}$$
$$= 1351.3...\,\text{s}$$

Step 2: Find the total charge of these electrons.

Total charge passing the end point in this time =
$$\Delta Q = Nq = 1.7 \times 10^{22} \times 1.6 \times 10^{-19}\,\text{C} = 2720\,\text{C}$$

Step 3: Calculate the current, which is the charge flowing per unit time.

$$\text{current } I = \frac{\Delta Q}{\Delta t} = \frac{2720\,\text{C}}{1351.3...} = 2.0128 = 2.0\,\text{A (2 s.f.)}$$

Note how the combination of a tiny quantity (the electronic charge) and a huge quantity (the number of free electrons) gives a current of everyday proportions.

In Figure 3, the pale blue section could have been drawn anywhere along the stream of charges. They are all moving the same way with the same mean speed, so *the current is the same at all points* along the stream of charged particles.

If some of the charges are diverted, as in the junction in Figure 4, then the current I must divide in such a way that the two divided currents I_1 and I_2 add to give the original current I: all of the moving charges must go one way or the other.

This rule, that the currents at a junction must add up, is called **Kirchhoff's first law**. It is a consequence of the conservation of charge – all of the charges in the stream of charges must go one way or the other.

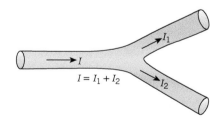

▲ **Figure 4** *Current dividing at a junction*

Potential difference

Charges will move when attracted by charges of the opposite sign or when repelled by charges of the same sign. Inside a battery cell, chemical reactions produce an electrical potential energy difference between the two terminals, resulting in a positively charged terminal and a negatively charged terminal. This can be compared with the gravitational potential energy difference down a hill, with the positive pole being the higher point and the negative the lower.

If the poles of the cell are joined by a conducting path, charges will flow as shown in Figure 6 (next page).

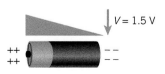

▲ **Figure 5** *Potential difference produced by a battery cell*

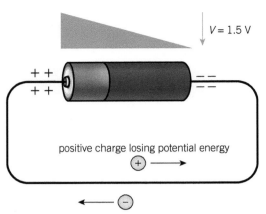

$V = 1.5\,V$

positive charge losing potential energy

negative charge losing potential energy

▲ **Figure 6** *Potential energy changes for charges in a circuit*

The direction in which charges flow depends on the sign of the charges: positive charges flow from the positive pole to the negative pole, and negative charges move the other way. As metals contain free electrons, which are negatively charged, the movement of negative charges is more usual in a circuit.

Potential difference and potential energy

When charges move between two points in a circuit, their electrical potential energy changes by an amount ΔE. By the principle of conservation of energy, the work done W in this movement is given by $W = \Delta E$.

Potential difference (p.d.), or 'voltage', V is the potential energy difference ΔE per unit charge moving between two points:

$$V = \frac{\Delta E}{Q} = \frac{W}{Q}$$

so the potential energy change $\Delta E = W = VQ$. For negative charges moving 'uphill' from $V = 0$ to $V = 1.5\,V$ the p.d. = $+1.5\,V$, but as Q is negative, the potential energy change, $\Delta E = VQ$, is negative – the charges have lost potential energy, just like positive charges moving 'downhill' through a p.d. of $-1.5\,V$.

Power dissipated in electrical circuits

A p.d. between the ends of a wire will accelerate the free electrons in it, but their movement down the wire is obstructed by their interactions with the positive ion cores of the metal atoms, so they do not gain kinetic energy. The potential energy lost does work on the wire, heating it. This 'wasted' energy is called **dissipation** and the process is often called 'Joule heating'.

$$V = \frac{W}{Q} \Rightarrow W = VQ$$

and

$$I = \frac{Q}{t} \Rightarrow Q = It$$

so

$$W = V(It) = VIt$$

Power P is the rate at which energy is transferred, and it is measured in $J\,s^{-1}$, which is the **watt** (W).

$$P = \frac{W}{t} = \frac{VIt}{t} = IV$$

Figure 7 shows a component called a resistor, with p.d. V between its ends and a current I through it. As you will see in Topic 3.2, the ratio $\frac{V}{I}$ for any conductor is called its **resistance** R, measured in **ohms** (Ω). The equation $R = \frac{V}{I}$ rearranges to give $V = IR$.

This means that there is an alternative equation to use for the dissipative heating in any component of resistance R – the rate of heating or power, P is given by:

$$P = IV = (IR)I = I^2R$$

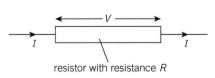

I V I

resistor with resistance R

▲ **Figure 7** *The p.d. across, and current through, a resistor*

Worked example: The maximum current taken by a resistor

A resistor is labelled $10\,k\Omega$ $0.60\,W$. The power rating is the maximum safe power output. What is the maximum current it can take without overheating?

Step 1: Write down the relationship between power, resistance, and current.

$P = I^2R$ so $P_{max} = I_{max}^2 R$

Step 2: Rearrange to give

$$I_{max}^2 = \frac{P}{R} = \frac{0.60\,W}{10 \times 10^3\,\Omega} = 6.0 \times 10^{-5}\,A^2$$

Step 3: Take the square root.

$$I = \sqrt{(6.0 \times 10^{-5}\,A^2)} = 0.00774...\,A$$
$$= 7.7\,mA \text{ (2 s.f.)}$$

Summary questions

1 A charge of $10\,C$ passes through a conductor in $2\,s$. The potential difference between the ends of the conductor is $12\,V$.
   a Calculate the current. (*1 mark*)
   b Calculate the power dissipated in the conductor. (*1 mark*)

2 Copy the circuit of Figure 8, replacing V_1, V_2, and I_1 with their values.

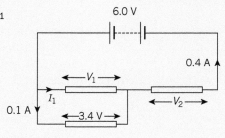

▲ Figure 8 (*3 marks*)

3 The current in a torch bulb is $0.24\,A$, which is a flow of electrons, each of charge $-1.6 \times 10^{-19}\,C$. Calculate the number of electrons entering (and leaving) the filament of the bulb each second. (*2 marks*)

4 In the electrolytic cell shown in Figure 9, copper ions of charge $+3.2 \times 10^{-19}\,C$ move from the positive plate to the negative plate, and chloride ions of charge $-1.6 \times 10^{-19}\,C$ travel in the opposite direction. A current of $0.35\,A$ flows for 2 minutes. Calculate the number of ions of each type which reach the plates.

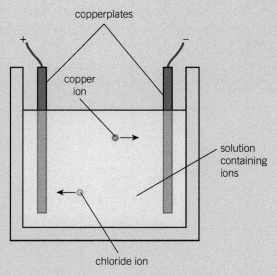

▲ Figure 9 (*3 marks*)

3.2 Conductors and resistors

Specification references: 3.1.2a(iii), 3.1.2a(vi), 3.1.2b(i), 3.1.2b(ii), 3.1.2b(iii), 3.1.2c(i), 3.1.2c(ii), 3.1.2d(i)

Learning outcomes

Describe, explain, and apply:

→ the terms resistance (R) and conductance (G)

→ the equations $R = \dfrac{V}{I}$ and $G = \dfrac{I}{V}$

→ the equations $R = R_1 + R_2 + ...$ or $\dfrac{1}{G} = \dfrac{1}{G_1} + \dfrac{1}{G_2} + ...$ for a series combination of conductors

→ the equations $G = G_1 + G_2 + ...$ or $\dfrac{1}{R} = \dfrac{1}{R_1} + \dfrac{1}{R_2} + ...$ for a parallel combination of conductors

→ graphs of current against potential difference, and of resistance or conductance against temperature

→ an experiment to obtain the I–V characteristic of an ohmic and a non-ohmic conductor.

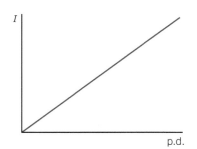

▲ **Figure 1** *Resistance at work on a moving vehicle*

▲ **Figure 2** *Current against p.d. for an ohmic conductor*

When a car is moving at its top, constant, speed along a straight, level, road, the forwards push from the engine balances friction and drag forces from the air it is travelling through. In a similar way, when there is a constant current in a conductor, the forwards push on the moving charges from the p.d. between the ends of the conductor is balanced by the obstructing effect of the atoms in the conductor interacting with the charges. In each case we can use the word *resistance* – air resistance stops the car accelerating, and electrical resistance stops the charges accelerating.

Resistance and Ohm's law

The ratio of the p.d. V across a component to the current I which it produces for a component is defined as its **resistance**, R.

$$\text{resistance } R = \frac{V}{I}$$

The unit of resistance is $V\,A^{-1}$, which is the **ohm** (Ω). Resistance is a measure of how difficult it is to get current through the component – you can think of it as the number of volts of p.d. you need to get a current of 1 ampere.

For most metals at a constant temperature, the resistance of a wire is constant, so $I \propto V$ – this is **Ohm's law**. A conductor that obeys Ohm's law is described as **ohmic**.

Conductance

Although the concept of resistance was developed first, metals are generally such good conductors it is more natural to define them by the opposite of resistance, that is, **conductance**. This is given the symbol G and is defined as:

$$\text{conductance } G = \frac{I}{V}$$

The unit of conductance is $A\,V^{-1}$, which is the **siemen** (S). It is the inverse of resistance.

$$G = \frac{1}{R}$$

You can think of conductance as how many amperes you get from a p.d. of 1 V.

🖩 Worked example: Conductance and resistance of a car headlamp

A standard car headlamp bulb is rated 55 W 12 V. What are the resistance and conductance of the filament?

Step 1: To find the current through the bulb, write down the relationship between power, p.d., and current.

$P = IV$ so current $I = \dfrac{P}{V} = \dfrac{55\,\text{W}}{12\,\text{V}} = 4.58...\,\text{A}$

Step 2: Now calculate the resistance, using the definition of resistance.

$R = \dfrac{V}{I} = \dfrac{12\,\text{V}}{4.58...\,\text{A}} = 2.61...\,\Omega = 2.6\,\Omega$ (2 s.f.)

Step 3: Then calculate the conductance, using the definition of conductance.

$G = \dfrac{I}{V} = \dfrac{4.58...\,\text{A}}{12\,\text{V}} = 0.381...\,\text{S} = 0.38\,\text{S}$ (2 s.f.)

Or you can use $G = \dfrac{1}{R} = \dfrac{1}{2.61...\,\Omega}$ to get the same answer.

Series and parallel combinations of conductors

Parallel circuits

Figure 3 shows two ohmic conductors **1** and **2** connected **in parallel** with a battery of voltage V.

The charges moving from the battery divide at one junction, pass along two different routes, and then recombine, so $I = I_1 + I_2$. The potential 'hill' down which each charge falls as it moves around the circuit is the same, whichever path it takes, so the p.d. across each conductor is V, assuming the connecting wires are of very high conductance (low resistance).

The conductance of the parallel combination is

$$G = \frac{I}{V} = \frac{I_1 + I_2}{V} = \frac{I_1}{V} + \frac{I_2}{V} = G_1 + G_2$$

This shows that if you have components in parallel, the total conductance is obtained by adding the conductance of each component. For more than two conductors in parallel,

$$G = G_1 + G_2 + ...$$

Because $G = \dfrac{1}{R}$, this expression $G = G_1 + G_2 + ...$ can be written as

$$\frac{1}{R} = \frac{1}{R_1} + \frac{1}{R_2} + ...$$

where R is the resistance of the combination and R_1, R_2, etc. are the resistances of the different components.

Study tip: Decoding circuits

Before analysing a circuit diagram, first be sure that you know the circuit symbols that you met at GCSE. Then trace circuit loops around from the positive battery terminal to the negative battery terminal, looking out for intersections where there is a choice of routes.

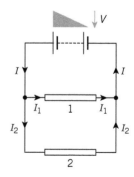

▲ **Figure 3** *Parallel conductors*

 Worked example: Adding parallel resistances

Two resistors are connected in parallel as shown in Figure 4. What is the resistance of the combination?

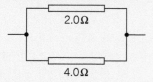

▲ **Figure 4** *Resistors in parallel*

Step 1: Even though resistances are given here, it is easier for a parallel circuit to work in terms of conductances. Express each resistance as a conductance.

$$\frac{1}{2.0\ \Omega} = 0.5\ \text{S and } \frac{1}{4.0\ \Omega} = 0.25\ \text{S}$$

Step 2: Add the conductances, because they are in parallel.

Total conductance = 0.5 S + 0.25 S = 0.75 S

Step 3: Find the total resistance by calculating the inverse of the conductance.

$$R = \frac{1}{G} = \frac{1}{0.75\ \text{S}} = 1.3\ \Omega \ (2\ \text{s.f.})$$

Study tip

Note that adding resistors in parallel always results in an overall resistance that is smaller than the smallest of the values being added. This is because the overall conductance is the sum of the individual conductances, and so must be greater than the highest conductance in the combination.

Series circuits

Figure 5 shows two ohmic conductors **1** and **2** connected **in series** with a battery of voltage V.

The charges moving from the battery have no choice in where to go, so the current I is the same throughout. As a charge Q moves around the circuit, potential energy is dissipated as heat in the two conductors, and the total energy must be that obtained from the battery, so $VQ = V_1Q + V_2Q$. Dividing by Q gives $V = V_1 + V_2$. The total p.d. is shared between the two conductors.

The resistance of the series combination is:

$$R = \frac{V}{I} = \frac{V_1 + V_2}{I} = \frac{V_1}{I} + \frac{V_2}{I} = R_1 + R_2$$

If you have components in series, the total resistance is obtained by adding the resistance of each component. For more than two conductors in series,

$$R = R_1 + R_2 + \ldots$$

Because $R = \dfrac{1}{G}$, $R_1 = \dfrac{1}{G_1}$, and $R_2 = \dfrac{1}{G_2}$, this expression $R = R_1 + R_2 + \ldots$ can be written as:

$$\frac{1}{G} = \frac{1}{G_1} + \frac{1}{G_2} + \ldots$$

Therefore, for combinations of components in series circuits it is easier to use resistance than conductance.

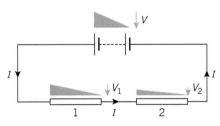

▲ **Figure 5** *Series conductors*

🖩 **Worked example: P.d. across series resistances**

A 2.0 Ω and a 4.0 Ω resistor are connected in series to a 4.5 V battery, in a circuit like that in Figure 5. What is the p.d. across the 4 Ω resistor?

Step 1: You need to find the current through the 4 Ω resistor, which is the current throughout the series circuit. So first calculate the total resistance.

$$R = R_1 + R_2 = 2.0\ \Omega + 4.0\ \Omega = 6.0\ \Omega$$

Step 2: Find the current.

$$R = \frac{V}{I} \text{ so } I = \frac{V}{R} = \frac{4.5\,\text{V}}{6.0\,\Omega} = 0.75\,\text{A}$$

Step 3: Now you can find the p.d. across the $4\,\Omega$ resistor.

$$R = \frac{V}{I} \text{ so } V = IR = 0.75\,\text{A} \times 4.0\,\Omega = 3.0\,\text{V}$$

Figure 6 gives a summary for dealing with parallel and series combinations of components.

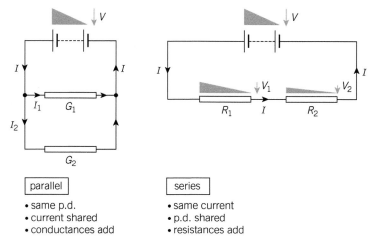

parallel
- same p.d.
- current shared
- conductances add

series
- same current
- p.d. shared
- resistances add

▲ **Figure 6** *Parallel and series circuits – a summary*

More complex networks

Real circuits are often networks with combinations of both series and parallel components. To analyse these, you need to look at the separate parallel and series parts one step at a time, converting from resistance to conductance and back with $G = \frac{1}{R}$ where necessary.

 Worked example: Step-by-step analysis of a network

What are the total conductance and resistance of the combination shown in Figure 7?

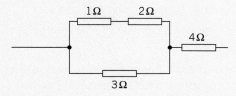

▲ **Figure 7** *The network to be analysed*

Figure 8 shows the sequence of steps in the solution.

Step 1: add series resistances and find the conductance of parallel components

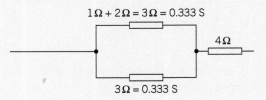

Step 2: add parallel conductances

$$0.333\,S + 0.333\,S = 0.666\,S = 1.5\,\Omega \quad 4\,\Omega$$

Step 3: add series resistances

$$1.5\,\Omega + 4\,\Omega = 5.5\,\Omega = 0.182\,S$$

▲ **Figure 8** *The solution*

So the circuit has an overall conductance of 0.2 S and resistance of 5 Ω.

Resistance and temperature

Metals are ohmic conductors if the temperature can be kept constant, but in reality a current in a wire dissipates energy through Joule heating (at a rate of $P = I^2R$), which will raise the temperature. This will increase the resistance of the wire.

As an example, if the current in a filament lamp is measured for a range of p.d.s up to the working voltage, when the filament glows white-hot it appears not to obey Ohm's law. Figure 9 shows the variation of current, resistance, and conductance for a filament lamp as the p.d. across it is increased. The y-axis scales have been adjusted to allow the graphs to be compared easily. Such a component, which does not obey Ohms' law under ordinary working conditions, is referred to as a **non-ohmic** conductor.

P.d. is plotted as the independent variable in Figure 9, because it is the variable you change to obtain the data. However, the real cause of variation of R and G is changing temperature, which is caused by the current and the I^2R heating it produces.

This increase in resistance (decrease in conductance) in metals as they get hotter has both negative and positive aspects. If precise, known, and constant resistance values are needed, it can be a nuisance, so alloys such as constantan have been developed where the change is very small. On the other hand, the known variation of the resistance of platinum with temperature allows measurement of its resistance to be used as a temperature sensor.

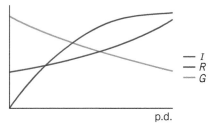

▲ **Figure 9** *Variation of current, resistance, and conductance with p.d. for a filament lamp*

When samples of semiconducting non-metals such as silicon are heated, the resistance does not increase, but decreases. The way that temperature changes affect resistance is discussed further in Topic 3.4, Conductance under the microscope.

Practical 3.1.2d(i): Investigating *I–V* characteristics for ohmic and non-ohmic components

Figure 10 shows a circuit which may be used to investigate how the current through a component X, shown here as a labelled box, varies with the p.d. across it.

In this investigation, it is important to bear in mind that the component may behave differently if reversed.

It is also important to consider:

- the range of p.d.s applied to the component – changing the range will probably affect the choice of ammeter used to measure the current

- the resolution of the ammeter and the voltmeter used

- the reproducibility of the readings obtained.

The source of variable p.d. may be a variable d.c. power supply, or a battery and rheostat (variable resistor).

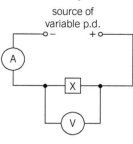

▲ Figure 10 *Investigating an electrical component, X*

Summary questions

1 A conductor has a current of 250 mA when there is a p.d. of 5.0 V across it. Assuming that it is ohmic, calculate the p.d. necessary to produce a current of 1.3 A. *(2 marks)*

2 a Explain why the *I–V* graph in Figure 9 is not linear, but curves downwards. *(2 marks)*
  b Explain how the shapes of the *R–V* and *G–V* graphs in Figure 9 are related to the *I–V* graph. *(2 marks)*

3 One type of platinum resistance thermometer has a resistance of 100.00 Ω at a temperature of exactly 0 °C. The variation of the resistance of platinum with temperature is given by the equation

$$R_\theta - R_{0°C} = \alpha R_{0°C}\theta$$

where θ is the temperature in °C and α is a constant for platinum with value $\alpha = 0.003\,926\,°C^{-1}$. At a certain temperature θ_x, the resistance, is measured to be 106.24 Ω. Calculate the temperature θ_x to 3 s.f. and explain why very accurate and precise measurement of resistance is needed in this case. *(3 marks)*

4 The graph of Figure 11 shows how the current through a metal wire varies with temperature. The p.d. between the ends of the conductor was constant throughout. Sketch a copy of this graph, and add lines to show how the resistance and the conductance are varying with temperature. Do not do any calculations.

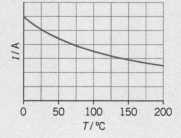

▲ Figure 11

(2 marks)

5 In the circuit shown in Figure 12, the current in the 3.0 Ω resistor is measured to be 0.45 A. Calculate the p.d. *V* of the battery.

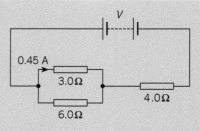

▲ Figure 12

(4 marks)

3.3 Conductivity and resistivity

Learning outcomes

Describe, explain, and apply:

→ the concepts of resistivity (ρ) and conductivity (σ)

→ the equations $R = \dfrac{\rho L}{A}$ and $G = \dfrac{\sigma A}{L}$

→ an experiment using one method of measuring resistivity or conductivity.

The resistance and the conductance of a metal wire are examples of *extensive* properties, because they depend on the length and thickness of the wire concerned, as well as the metal from which it is made. To describe and make appropriate use of the electrical properties of the metal concerned we need some measure which does not depend on the dimensions of the sample used. Such a measure is called an *intensive* or *bulk* property of the material. Other bulk properties that you will meet in the course are density, the Young modulus and refractive index.

Dependence of resistance and conductance of a wire on its dimensions

The way in which the conductance and resistance of a sample of wire depend on the dimensions of the wire can be shown by considering identical wires in series and in parallel, as in Figure 1.

Doubling the length L doubles the resistance, implying that $R \propto L$ and so $G \propto \dfrac{1}{L}$.

In a similar way, doubling the cross-sectional area A doubles the conductance, implying that $G \propto A$ and so $R \propto \dfrac{1}{A}$.

Combining these two relationships for R and for G we get:

$$R \propto \frac{L}{A} \text{ and } G \propto \frac{A}{L}$$

These two equivalent relationships apply for any given material. Each needs a constant of proportionality, or *bulk constant*, true for a particular material, to allow calculations for and comparisons between different materials.

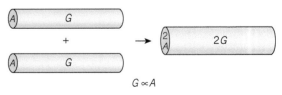

▲ **Figure 1** *Dependence of R and G on the dimensions of a wire*

Conductivity and resistivity

The bulk properties for describing electrical conduction are the constants that must replace the proportionalities in the relationships above.

$$G \propto \frac{A}{L} \Rightarrow G = \frac{\sigma A}{L}$$

where σ (sigma) is the electrical **conductivity** whose unit is $S\,m^{-1}$.

$$R \propto \frac{L}{A} \Rightarrow R = \frac{\rho L}{A}$$

where ρ (rho) is the electrical **resistivity** whose unit is $\Omega\,m$.

Just as $G = \dfrac{1}{R}$, so $\sigma = \dfrac{1}{\rho}$.

To get a feel for the meaning of resistivity and conductivity, think of the cube of material shown in Figure 2.

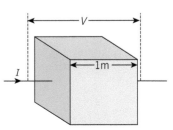

▲ **Figure 2** *Visualising conductivity and resistivity*

For this huge conductor

$$G = \frac{\sigma A}{L} = \frac{\sigma \times 1\,\text{m}^2}{1\,\text{m}}$$

so that conductivity σ, in $\text{S}\,\text{m}^{-1}$, is numerically the same as the conductance G in S, and the resistivity ρ, in $\Omega\,\text{m}$, is numerically the same as the resistance R in Ω.

If the conductor is a metal, the conductance G across a 1 m cube, being the current in A produced by a p.d. of 1 V, is huge, and its resistance R is correspondingly tiny. Some values are shown in Table 1.

▼ Table 1 *Values of conductivity and resistivity for some metals*

Metal	Conductivity, $\sigma / \text{S}\,\text{m}^{-1}$	Resistivity, $\rho / \Omega\,\text{m}$
copper	6.41×10^7	1.56×10^{-8}
iron	1.12×10^7	8.90×10^{-8}
silver	6.62×10^7	1.51×10^{-8}

 Worked example: Resistance of copper conductors

A 6 V 0.2 A bulb is connected to its battery via copper wires of diameter 0.5 mm and length 0.5 m. Do these wires add significantly to the resistance of the circuit? Use the data in Table 1.

Step 1: Find the resistance of the bulb.

$$R = \frac{V}{I} = \frac{6\,\text{V}}{0.2\,\text{A}} = 30\,\Omega$$

Step 2: You need to find the resistance of the wires. You can obtain the resistivity of copper from Table 1. From the given dimensions of the wires, calculate the cross-section area.

$$A = \pi r^2 = \pi \times (0.25 \times 10^{-3}\,\text{m})^2 = 1.96... \times 10^{-7}\,\text{m}^2$$

Step 3: Now calculate the resistance of the copper wires:

$$R = \rho \frac{L}{A} = \frac{1.56 \times 10^{-8}\,\Omega\,\text{m} \times 0.5\,\text{m}}{1.96... \times 10^{-7}\,\text{m}^2} = 0.040\,\Omega \ (2\ \text{s.f.})$$

This is negligible compared with the resistance of the bulb, $30\,\Omega$.

Conductors, insulators, and semiconductors

Conductivity and resistivity are measures for the electrical properties of materials that we can use to classify the materials as conductors, insulators or **semiconductors**. As Figure 3 (next page) shows, the range of values is immense. If a linear scale were used, with a resistivity of $10^{-8}\,\Omega\,\text{m}$ taking up 1 cm, then a book page over 100 000 light-years long would be needed to include polystyrene! A **logarithmic scale** must be used, with steps going up in factors of 100.

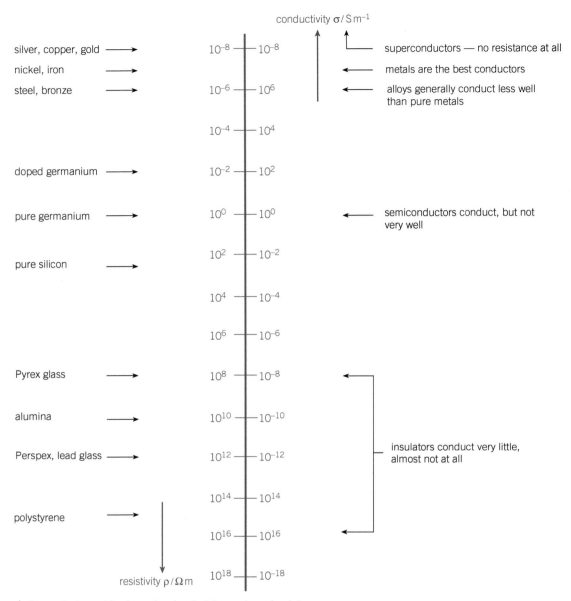

▲ **Figure 3** *Logarithmic scale of resistivity and conductivity*

Measuring resistivity or conductivity

The particular problems in measuring these electrical quantities depend on the material you are investigating. First it is necessary to make an order of magnitude calculation of the measurements you might expect to be able to make, using approximate values taken from Figure 3.

- For an insulator, you need a tiny value of L, a large value of A, a large p.d., and a very, very sensitive ammeter to have any chance of making a measurement at all.

- For a metal, the problems are quite different. You need a large L and a very small A, and you must be aware of complications caused by I^2R heating of the wire and systematic errors such as p.d. drops in places other than across the conductor in question.

Synoptic link

This experiment involves measurements on a long wire. Refer to the issues involved in measurement of the Young modulus of a wire in Chapter 4, Testing materials.

Practical 3.1.2d(ii): Determining conductivity or resistivity

The circuit used in Practical 3.1.2d(i) in Topic 3.2 can be used here, with the component **X** being a sample of the material being investigated, which could be a metal wire. It may be necessary to do a preliminary experiment to find the most suitable dimensions for this sample.
Practical considerations in this experiment include:

- appropriate ranges for the p.d. (and voltmeter) and ammeter used
- the uncertainty in the measurements, particularly of the most uncertain measurement
- how the wire is connected into the circuit, and how the p.d. between the ends of the measured length is measured.

Extension: Semiconductors

Metals have very large conductivities ($>10^6 \, \mathrm{S\,m^{-1}}$) and insulators very small ones ($<10^{-8} \, \mathrm{S\,m^{-1}}$). One class of materials has conductivities in the range $0.01-100 \, \mathrm{S\,m^{-1}}$. These are semiconductors, of which germanium and silicon have been most important historically, although compounds such as gallium arsenide are also widely used.

The electrical properties of semiconductors are not constant, as with metals, but can be modified. Some of this modification is through *doping* a pure semiconductor with other elements which greatly affect its conductivity, and by combining thin layers of differently doped semiconductors.

Changing the voltages applied to semiconductor devices also changes their electrical behaviour, so they can actively modify signals applied to them. Since the 1950s the electronics industry has developed this behaviour and created a range of devices from transistors and amplifiers to complex computers. Each year these devices

have become smaller, faster and cheaper as ever-tinier semiconductor and metal elements are 'grown' in situ on integrated circuits.

Questions
The integrated circuit in Figure 4 contains many tiny transistors about 22 nm in size. What is the conductance across opposite faces of a silicon cube of side length 22 nm (assume $\sigma = 1 \, \mathrm{S\,m^{-1}}$)? What would the conductance be if this cube could be made half as big in each dimension?

◀ **Figure 4** *Scanning electron micrograph of the surface of an integrated circuit from a computer's Arithmetic Logic Unit*

Summary questions

1. Explain why resistivity has the unit $\Omega\,\mathrm{m}$ and conductivity has the unit $\mathrm{S\,m^{-1}}$. *(3 marks)*

2. A 230 V 50 W lamp has a filament which is a coil of tungsten wire whose resistivity, at the working temperature, is $6.5 \times 10^{-7} \, \Omega\,\mathrm{m}$. The diameter of the filament is $3.2 \times 10^{-5} \, \mathrm{m}$. Calculate the length of the filament. *(5 marks)*

3. It is planned to measure the conductivity of polythene, thought to be about $10^{-12} \, \mathrm{S\,m^{-1}}$, using a sheet of polythene 0.1 mm thick sandwiched between aluminium foil electrodes, as shown in Figure 5, with an EHT supply of 5 kV and an ammeter capable of reading currents down to 0.1 µA.
Make an appropriate calculation to predict whether this experiment could be successful.

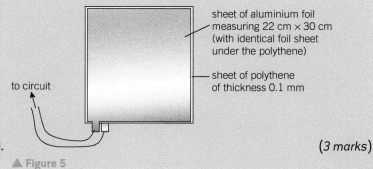

to circuit

sheet of aluminium foil measuring 22 cm × 30 cm (with identical foil sheet under the polythene)

sheet of polythene of thickness 0.1 mm

(3 marks)

▲ **Figure 5**

X-ray crystallography, pioneered by William and Lawrence Bragg in the early years of the 20th century, revealed the crystalline structure of a number of compounds and elements, including metals. The regular lattice structure of metallic crystals, together with the high melting point and stiffness of metals, shows that the attractive forces in metals are large.

A conduction model for metals

In a metal, each atom contributes one or more electrons to a 'soup' of free, mobile electrons which move about in the lattice of positive ion cores. It is this delocalised cloud of electrons which provides the strong metallic bond. In Figure 1 the electrons are represented as individual charged points moving with random velocities, very much like the molecules in a gas.

When a p.d. is applied across the lattice, the electrons each experience a force pulling them to the positive side (or pushing them from the negative side). This produces a slow general drift in one direction, superimposed on the rapid random motion – imagine a swarm of bees drifting slowly in the breeze. The rate of flow of charge due to this net movement is the current produced by the p.d.

Figure 1 shows the lattice of fixed positive ions. At room temperature, the electrons have kinetic energy and their motion is obstructed by the positive ions. The positive ions also have kinetic energy but are bound in their positions, and vibrate. These vibrations, which increase with temperature, mean that the electron paths in the lattice are obstructed more in a hotter metal than in a cooler one. This is illustrated in Figure 2, where, for clarity, only two electrons are shown in each lattice.

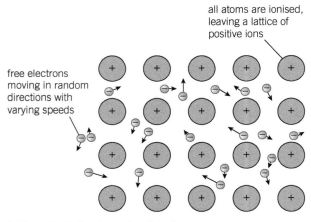

all atoms are ionised, leaving a lattice of positive ions

free electrons moving in random directions with varying speeds

▲ Figure 1 A simple model of the ions and electrons in a metal

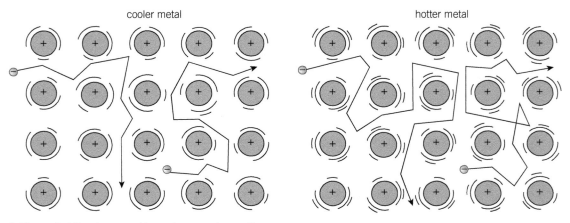

cooler metal   hotter metal

▲ Figure 2 Vibrating metal ions obstruct charge flow

Number density of charges in a metal

The simplified model in Figure 1 suggests that each atom donates one free electron to the delocalised cloud. If the **number density** n of these electrons (the number per m^3) is known, the mean **drift velocity** of the electrons can be calculated.

🖩 Worked example: Drift velocity of electrons

$1\,m^3$ of copper contains 8.5×10^{28} atoms, so the number density of free electrons $n = 8.5 \times 10^{28}\,m^{-3}$. What is the drift velocity of electrons in a copper wire of cross-sectional area $3.0 \times 10^{-6}\,m^2$ carrying a current of $10\,A$?

Step 1: Think about a length of copper wire of cross-sectional area A, as shown in Figure 3. You need to find the length L for which all the electrons in that length of wire will drift through the coloured end of wire in 1 second.

charge e moving with drift velocity v

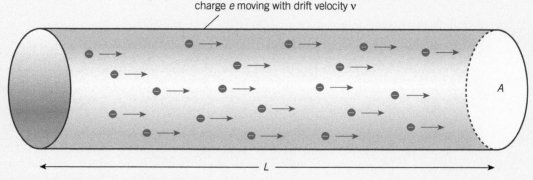

▲ **Figure 3** *Drift velocity of electrons*

Step 2: Find the number of electrons that drift through the coloured cross-section in 1 second.

Current $I = \dfrac{\Delta Q}{\Delta t} = 10\,A = \dfrac{\Delta Q}{1\,s}$, so $\Delta Q = 10\,C$.

Each electron has charge $e = 1.6 \times 10^{-19}\,C$, so the number of electrons is given by

$$N = \frac{\Delta Q}{e} = \frac{10\,C}{1.6 \times 10^{-19}\,C} = 6.25 \times 10^{19}$$

Step 4: Volume V is equal to $LA = L \times 3.0 \times 10^{-6}\,m^2$.

$$\text{So } L = \frac{7.4 \times 10^{-10}\,m^3}{3.0 \times 10^{-6}\,m^2} = 2.5 \times 10^{-4}\,m$$

As all the charges in the length L (including those at the very back) move through the coloured end in 1 second, their drift velocity $= 2.5 \times 10^{-4}\,m\,s^{-1}$.

Step 3: You are given the number density of electrons, so now you can calculate the volume of the length L. As $1\,m^3$ contains 8.5×10^{28} free electrons, the volume of the wire V is given by

$$\frac{V}{1\,m^3} = \frac{6.25 \times 10^{19}}{8.5 \times 10^{28}}$$

giving $V = 7.4 \times 10^{-10}\,m^3$

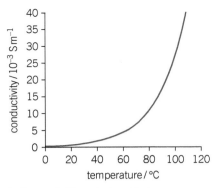

▲ **Figure 4** *The relationship between temperature and conductivity for a semiconductor*

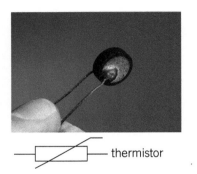

▲ **Figure 5** *Thermistor*

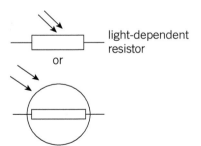

▲ **Figure 6** *Two ways of showing a light-dependent resistor (LDR)*

Insulators and semiconductors

Electrical **insulators** are materials that do not have mobile charges. There may be ions and electrons within them, but these are not free to move. An example is glass, which is a good insulator at normal temperatures, but will conduct if heated until it begins to soften, because the sodium ions in the glass can then move freely.

Semiconductors do have free charges, but very few – about 1 in 10^{12} atoms are ionised. If Figure 1 were to be drawn for a semiconductor, in order to show one mobile charge carrier the diagram would need to be 200 000 times larger in each dimension!

When semiconductors are heated, the proportion of atoms that are ionised increases greatly. This results in the rapid increase in number density of charge carriers, and therefore conductivity, as shown in Figure 4.

Thermistors and LDRs

Thermistors are cheap, useful sensors made from semiconducting material. An increase in temperature liberates electrons, causing the conductivity to increase, so the resistance of the component drops significantly.

Light-dependent resistors (LDRs) are also semiconductor devices, usually constructed from cadmium sulfide. Light falling on the exposed semiconductor liberates electrons, which increases the conductivity much like a temperature increase does in thermistors, resulting in a drop in resistance of the LDR.

Both thermistors and LDRs can be incorporated as sensors in potential divider circuits, as you will see in Topic 3.5.

Summary questions

1   Silver has an electrical conductivity nearly six times that of iron. With reference to Figure 1, suggest differences between the microscopic structure of the two which may account for this.   *(2 marks)*

2   Two conductors **A** and **B** of identical length and cross-sectional area are connected in series and carry a constant current of 0.5 A. The number density n of free electrons in **A** is ten times that in **B**. Without calculations, explain carefully how the drift velocities of the electrons in the two conductors compare.   *(2 marks)*

3   Use the approach in the worked example to show that the drift velocity v of electrons in a conductor is given by

$$v = \frac{I}{nAe}$$

where n, A and e have the same meaning as in that example.   *(2 marks)*

3.5 Potential dividers

Specification references 3.1.2a(vii), 3.1.2b(i), 3.1.2c(iv), 3.1.2d(iii), 3.1.2d(iv)

A prematurely born baby needs careful monitoring. The baby's temperature, pulse rate, and breathing need to be recorded, and the conditions in the incubator need careful control. These are all done with small electronic sensors, some attached to the baby and some in the hood of the incubator. For most of these sensors, a potential divider provides the output p.d. responsible for alerting medical staff, or for switching equipment on and off.

Sharing voltages

In Topic 3.2, Conductors and resistors, you saw that series resistors share the applied p.d. Consider the circuit in Figure 2.

From the definition of resistance,

$$R = \frac{V}{I} \text{ so } I = \frac{V}{R}$$

The current I is the same throughout a series circuit, so for the two resistors R_1 and R_2

$$I = \frac{V_1}{R_1} = \frac{V_2}{R_2} \text{ so } \frac{V_1}{V_2} = \frac{R_1}{R_2}$$

In a similar way,

$$I = \frac{V}{R_1 + R_2}$$

giving

$$\frac{V_1}{V} = \frac{R_1}{R_1 + R_2} \text{ and } \frac{V_2}{V} = \frac{R_2}{R_1 + R_2}$$

This type of circuit is called a **potential divider**, because the applied p.d. is divided according to the relative values of V_1 and V_2.

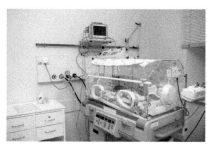

▲ **Figure 1** *Newborn premature baby in an incubator*

> ### 📖 Worked example: The potential divider equation
>
> Find the p.d. across each resistor in the circuit of Figure 3.
>
> 6.0V
>
> $R_1 = 18\,\Omega$ $R_2 = 12\,\Omega$
>
> ▲ Figure 3
>
> **Step 1:** Find one p.d. in terms of the other, from the ratio of their resistances.
>
> $$\frac{V_1}{V_2} = \frac{R_1}{R_2} = \frac{18\ \Omega}{12\ \Omega} = 1.5$$
>
> Therefore, $V_1 = 1.5V_2$

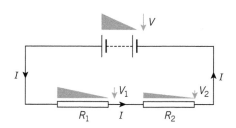

series

• same current
• p.d. shared
• resistances add

▲ **Figure 2** *The potential divider circuit*

→

Step 2: Calculate one of the p.d.s, knowing the supply p.d.

$$V_1 + V_2 = 6.0\,\text{V} \Rightarrow 1.5V_2 + V_2 = 2.5V_2 = 6.0\,\text{V} \Rightarrow V_2 = \frac{6.0\,\text{V}}{2.5} = 2.4\,\text{V}$$

Step 3: Calculate the other p.d.

$$V_1 = 6.0\,\text{V} - V_2 = 6.0\,\text{V} - 2.4\,\text{V} = 3.6\,\text{V}$$

Or you could use $V_1 = 1.5V_2 = 1.5 \times 2.4\,\text{V} = 3.6\,\text{V}$.

Changing the value of an output voltage

For the potential divider considered in Figure 2,

$$\frac{V_2}{V} = \frac{R_2}{R_1 + R_2}$$

So we can write:

$$V_2 = \frac{R_2}{R_1 + R_2} V$$

The p.d. V_2 across R_2 is the same fraction of the total voltage V that the resistance R_2 is of the total resistance. This is useful in **sensor circuits** where one resistor (say R_1) is a fixed resistor and the other (R_2) an intrinsically variable one, such as a thermistor (the resistance of which depends on temperature) or an LDR (the resistance of which depends on light intensity). In that case, V_2 is the **output voltage** of the sensing circuit, generally denoted by V_{out}, and the battery voltage V is the **input voltage**, V_{in}, as shown in Figure 4.

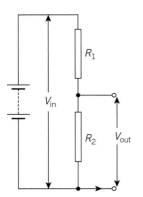

▲ **Figure 4** *Potential divider as used in a sensing circuit*

Choice of the fixed resistor in a sensor circuit

Consider the temperature-sensing circuit in Figure 5. A thermistor is used as R_1 in this case, because its resistance, and hence the p.d. across it, falls as the temperature increases. The output voltage across R_2 will therefore increase with increasing temperature. Careful choice of the value of R_2 will give a large range of output voltages over the temperature range to be used.

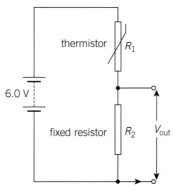

▲ **Figure 5** *Temperature-sensing circuit*

🖩 Worked example: Choosing a fixed resistor

A thermistor is chosen for a temperature-sensing circuit that needs to work between 0 °C and 100 °C. The data sheet for the thermistor chosen states that its resistance varies from 16 300 Ω at 0 °C to 340 Ω at 100 °C. What would be the best choice for the fixed resistor? Choose from 20 000 Ω, 2000 Ω, or 200 Ω.

Step 1: You need to calculate the expected output voltages at 0 °C and 100 °C for each possible fixed resistor. Starting with the 20 000 Ω resistor.

At 0 °C,

$$V_{\text{out}} = \frac{R_2}{R_1 + R_2} V = \frac{20\,000\ \Omega}{16\,300\ \Omega + 20\,000\ \Omega} \times 6.0\,\text{V} = 3.3\,\text{V}$$

At 100 °C,

$$V_{out} = \frac{20\,000\ \Omega}{340\ \Omega + 20\,000\ \Omega} \times 6.0\,V = 5.9\,V$$

Steps 2 and 3: Do the same calculations for the other two resistors. Table 1 shows the set of possible output voltages.

▼ Table 1

R_2 / Ω	20 000	2000	200
V_{out} at 0 °C /V	3.3	0.66	0.07
V_{out} at 100 °C /V	5.9	5.1	2.2

The intermediate value (2000 Ω) gives the best (greatest) range of output voltages.

Note that the resistance of a thermistor does not vary in a linear way with temperature, so the output voltage will not vary uniformly over the temperature range.

Practical 3.1.2d(iii): Using a potential divider including a sensor

Many sensor circuits use an intrinsically variable resistor, such as a thermistor or an LDR, to switch an electronic switch that operates at a fixed voltage. For some electronic switches, this is a value of 0.7 V, and for others it is half the voltage of the battery in the circuit.

Before setting up the circuit, it may be necessary to calibrate the potential divider (see Practical 3.1.2d(iv)).

When designing a sensor circuit, you need to decide whether you want the output p.d. from the sensor circuit to go from low to high as the switching point is reached, or the other way round. This will affect whether the output p.d. is across the variable component or across the fixed resistor. Figure 6 shows an example. The input to the electronic switch is the output V_{out} from the sensor circuit. The switch will switch *on* the alarm when V_{out} from the sensor circuit *drops*. This means that the potential divider in the sensor circuit must be set up so that V_{out} is normally high, but falls below the switching value when the environment reaches the condition required for the alarm to sound.

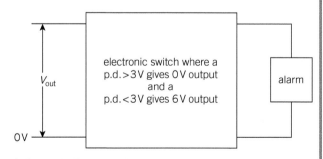

▲ **Figure 6** *Using an electronic switch*

A suitable approximate value of the fixed resistor can be obtained from knowledge of the resistance variation of the fixed resistor (see the worked example above) and from the calibration graph (see Practical 3.1.2d (iv)). It may be necessary to adjust the fixed resistor so that switching occurs at the right point. This would involve using a potentiometer (variable resistor) as the 'fixed' resistor in the potential divider.

Practical 3.1.2d(iv): Calibrating a sensor

A simple sensor measuring the change in an environmental variable (such as temperature or light intensity) will need an intrinsically variable resistor (such as a thermistor or an LDR) as well as an appropriate fixed resistor chosen to give an output p.d. with good variation. It will be necessary also to have some way of measuring that environmental variable, so that the output p.d. can be tabulated alongside the value of that variable over an appropriate range. A suitable calibration graph will then permit the output p.d. reading to be converted to the required values of the environmental variable.

Summary questions

1 Calculate the p.d. across the 100 Ω resistor in Figure 7. *(2 marks)*

←—2.0V—→   ←— V_{100} —→
40Ω   100Ω

▲ Figure 7

2 The resistance of an LDR varies between 5.4 kΩ in bright sunlight and 1.0 MΩ in darkness. Draw a sensor circuit to work with a 3.0 V battery, including a fixed resistor of suitable value, which will give an output voltage decreasing as the light gets brighter. *(3 marks)*

3 Two 10 000 Ω resistors **A** and **B** are connected in series to a 6.0 V battery. A voltmeter, also of resistance 10 000 Ω, is connected across each resistor in turn, and then across the battery, giving the results in Table 2.

▼ Table 2

p.d. across A / V	p.d. across B / V	p.d. across battery / V
2.0	2.0	6.0

Explain these results. *(5 marks)*

3.6 E.m.f. and internal resistance

Specification references: 3.1.2a(iv), 3.1.2b(i), 3.1.2c(i), 3.1.2d(v)

The temperature sensor in Figure 1 is a thermocouple. The probe inserted into the cooking food has different metals in contact, which produces a voltage by the thermoelectric effect (discovered by the Estonian physicist Thomas Seebeck in 1820). This effect is just one of the phenomena by which a p.d. can be generated by a physical or chemical change. In this topic we will consider battery cells, but the principles developed can be applied to any source of potential difference. Other sources include generators and photocells.

Electrical sources and e.m.f.

An electrical supply, or source, can be connected into a circuit to provide a p.d. which will make charges move. In this process, energy is given to the charges, and this is dissipated in the circuit. The potential difference (p.d.) V between the external terminals of the source is the potential energy difference per coulomb as the charges move 'downhill', releasing energy as they do so, as shown in Figure 2.

The question must arise, what pushes the charges uphill to start with? Within battery cells, chemical processes liberate electrons at the negative terminal (cathode), and consume them at the positive terminal (anode), effectively moving electrons from the anode to the cathode and providing the energy to 'lift them' up the potential hill.

The **e.m.f.**, symbol ε, is the energy that the source gives to the charges for every coulomb of charge flowing from the source. The abbreviation e.m.f. stands for 'electro-motive force', but it is not strictly a force, so it is better just to use the abbreviation.

Applying the principle of conservation of energy to the circuit in Figure 2, you can see that the energy supplied by the battery is equal to the energy dissipated by the circuit resistance. This is true for every coulomb passing, so as p.d. = energy transfer per coulomb, we can say that, for any complete circuit, e.m.f. = sum of all the p.d.s across the resistances of the circuit – this is known as **Kirchhoff's second law**.

Internal resistance

The e.m.f. produced by a battery results in a current through the entire circuit, including the body of the battery itself, and the chemicals in the battery provide resistance to the flow of charges. The battery has **internal resistance**, given the symbol r, so this means there is a p.d. drop inside the battery itself. Figure 2 has an 'ideal' battery with no internal resistance. For a real battery, a better representation is Figure 3 (next page). The dotted outline around the battery helps to identify where the battery terminals are.

▲ Figure 1 *A thermocouple in use in the kitchen*

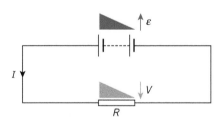

▲ Figure 2 *The p.d. (or V) and e.m.f. ε for a battery in a circuit*

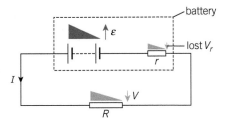

▲ **Figure 3** *A source of e.m.f. with internal resistance r*

In the circuit of Figure 3, the e.m.f. provides the total p.d. across both the external and internal resistances.

$$\varepsilon = V + V_r$$

where V_r is the p.d. drop, or 'lost volts', within the battery itself. V is the p.d. across the resistor R, and is also the p.d. across the battery terminals – the **terminal p.d.** – provided the resistance of the connecting cables is negligible.

The current I is the same throughout the circuit. Rearranging

$$r = \frac{V_r}{I} \text{ gives } V_r = Ir$$

We can use this and the equation above for ε to give

$$\varepsilon = V + Ir$$

This equation can be rearranged into

$$V = (-r)I + \varepsilon$$

As both ε and r are constants of the battery, this is in the form of the straight line equation $y = mx + c$ (gradient m and y-axis intercept c), as shown in Figure 4. The gradient is $-r$.

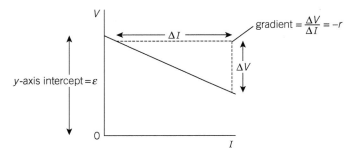

▲ **Figure 4** *Variation of terminal p.d. with current*

 Worked example: Calculating internal resistance

A battery of e.m.f. 6.0 V has a terminal p.d. of 5.6 V when delivering a current of 0.5 A. Calculate the internal resistance of the battery.

Step 1: Substitute into the equation $\varepsilon = V + Ir$
$$\Rightarrow 6.0\,\text{V} = 5.6\,\text{V} + 0.5\,\text{A} \times r$$

Step 2: Rearrange to find r.
$$r = \frac{6.0\ \text{V} - 5.6\ \text{V}}{0.5\ \text{A}} = \frac{0.4\ \text{V}}{0.5\ \text{A}} = 0.8\,\Omega$$

In any real circuit, the terminal p.d. V is less than the e.m.f. ε, but this difference is not constant. As the **load** resistance R (the resistance of the circuit attached to the battery terminals) decreases, the current I will increase. This happens because the total circuit resistance, $R + r$, has been reduced. This increase in I will increase the 'lost volts' Ir and so reduce the terminal p.d.

Worked example: Calculating e.m.f. and internal resistance

A battery has a terminal p.d. of 5.6 V when delivering a current of 0.5 A, but this drops to 5.2 V when delivering a current of 1.0 A. Calculate the e.m.f. and the internal resistance of the battery.

Step 1: Substitute the data into the equation.

For the 0.5 A current, $\varepsilon = V + Ir$ becomes
$\varepsilon = 5.6\,\text{V} + 0.5\,\text{A} \times r$ Equation 1

For the 1.0 A current, $\varepsilon = V + Ir$ becomes
$\varepsilon = 5.2\,\text{V} + 1.0\,\text{A} \times r$ Equation 2

Step 2: Subtract Equation 2 from Equation 1 to give:

$(\varepsilon - \varepsilon) = (5.6\,\text{V} - 5.2\,\text{V}) + (0.5\,\text{A} \times r - 1.0\,\text{A} \times r)$

$0 = 0.4\,\text{V} - (0.5\,\text{A} \times r)$

$r = \dfrac{0.4\,\text{V}}{0.5\,\text{A}} = 0.8\,\Omega$

Step 3: Substitute this value of r into Equation 1 or 2. Using Equation 1:

$\varepsilon = 5.6\,\text{V} + 0.5\,\text{A} \times 0.8\,\Omega = 5.6\,\text{V} + 0.4\,\text{V} = 6.0\,\text{V}$

Measuring internal resistance

Some battery cells, such as those in lead–acid car batteries, have such a low internal resistance that it is difficult to measure. However, some cells have a significant internal resistance, and for these cells the internal resistance r can be found.

 Practical 3.1.2d(v): Determining the internal resistance of a cell

If the circuit of Figure 3 is set up using a source of e.m.f. with high internal resistance (one example might be a piece of fruit or vegetable into which two dissimilar metals have been stuck), with a variable resistor in place of the load resistor R, it is possible to obtain a range of values of I and corresponding values of V, the p.d. across the load resistor. V and I are variables, and ε and r are constants for this battery, so a suitable graph can be drawn and the internal resistance found.

Appropriate voltmeter and ammeter ranges need to be chosen. The most suitable range of values of R may need to be found in a preliminary experiment.

▲ **Figure 5** *Two different metals in a conduction fluid form a battery — the fluid here is lemon juice*

Summary questions

1 Explain why the p.d. across the internal resistance of a cell is often referred to as 'lost volts'. *(1 mark)*

2 In an experiment to measure p.d. across the terminals of a battery for different values of current delivered, the data in Table 1 was obtained.

▼ Table 1

Current / A	0.5	1.0	1.5	2.0	3.0	4.0
P.d. / V	3.4	3.25	3.0	2.75	2.35	2.0

Plot an appropriate graph to find the e.m.f. and the internal resistance of the battery used. *(5 marks)*

3 All electrical appliances in a car are connected in parallel to the terminals of a 12 V battery. A car is parked with its lights on. The driver starts the engine, using the powerful starter motor. The car lights become dim while the starter motor is operating. Explain this observation in terms of internal resistance. *(3 marks)*

Module 3.1 Summary

Imaging

Imaging with a lens

- power of a lens $= \dfrac{1}{f}$ measured in dioptres (D)
- lens equation $\dfrac{1}{v} = \dfrac{1}{u} + \dfrac{1}{f}$ (Cartesian convention)
- linear magnification $= \dfrac{v}{u} = \dfrac{\text{object height}}{\text{image height}}$
- practical task — determination of the focal length of a converging lens

Storing and manipulating the image

- bits, bytes and pixels
- amount of information = number of pixels × bits per pixel
- b bits provide 2^b alternatives
- number of bits b needed for N alternative arrangements $= \log_2 N$
- resolution as the smallest detail that can be distinguished in an image

Polarisation of electromagnetic waves

- $v = f\lambda$
- practical task — observing polarising effects using light and microwaves

Signalling

Digitising a signal

- sampling, quantisation levels and quantisation error
- resolution $= \dfrac{\text{potential difference range of signal}}{\text{number of quantisation levels}}$
- noise in a signal: maximum useful number of levels $= \dfrac{V_{total}}{V_{noise}}$

Sampling sounds and sending a signal

- aliasing
- minimum sampling rate > 2 × maximum frequency of signal
- bit rate = samples per second × bits per sample

Sensing

Current, p.d., and electrical power

- $I = \Delta Q / \Delta t$, $P = IV = I^2 R$, $W = VIt$
- Kirchhoff's first law
- p.d. as energy per unit charge, $V = \dfrac{W}{Q}$

Conductors and resistors

- $R = V/I$, $G = I/V$
- calculating current and p.d. in series and parallel circuits
- practical task — I-V characteristics of ohmic and non-ohmic components

Conductivity and resistivity

- the equations $R = \dfrac{\rho L}{A}$ and $G = \dfrac{\sigma A}{L}$
- practical task — determining the resistivity or conductivity of a metal
- explaining electrical behaviour of conductors and insulators

Potential dividers

- the action of a potential divider:
$V_{out} = \dfrac{R_2}{R_1 + R_2} \times V_{in}$ and $\dfrac{V_1}{V_2} = \dfrac{R_1}{R_2}$
- practical task — potential divider
- practical task — calibrating a sensor

E.m.f. and internal resistance

- $V = \varepsilon - Ir$
- Kirchhoff's second law
- practical task — determining the internal resistance of a cell

Strain gauges

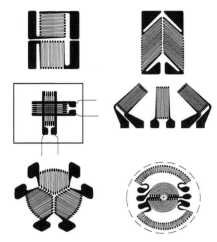

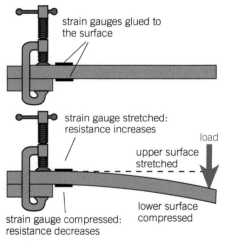

▲ **Figure 1** *These gauges are made to detect strains in several directions*

▲ **Figure 2** *The resistances of both strain gauges change when the beam bends*

Buildings in many parts of the world can crack or even collapse if the ground beneath them moves. It is important to monitor movement to provide advance warning of a possible collapse. This was once done by cementing microscope slides to key structural parts of buildings at risk, and inspecting them regularly to check if any had cracked. Strain gauges now perform this task, with the advantage that they can detect much smaller movements and produce a continuous recording of the movement, which can be logged by digital systems.

How does a strain gauge work?

The resistance of a metal increases when it is stretched, simply because it becomes longer and thinner. A strain gauge is a zigzag strip of metal foil glued to the surface of the component whose strain is to be monitored. The foil is stretched if the component is stretched. Stretching along the length of the strips in the foil is detected as an increase in resistance. The zigzag arrangement simply increases the length of conductor being stretched, which increases the change in resistance. Such strain gauges are cheap and convenient to use.

Suppose the bending of a beam of material has to be measured. Such measurements are needed in investigating the strengths of materials. A clever trick is to use two strain gauges, so that as the beam bends, one is stretched and the other is compressed. See Figure 2. Connected in a potential divider, the two gauges give twice the output that one would give alone.

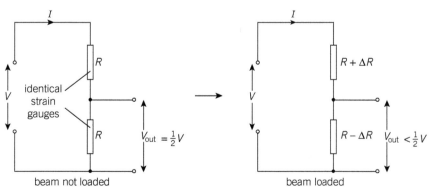

▲ **Figure 3** *Strain gauges in a potential divider*

With this setup there is a bonus — it is not so sensitive to changes in temperature. The resistance of a metal increases with temperature, so by using just one metal strain gauge in a potential divider, a rise in temperature might be registered as a strain. With two gauges, if the temperature goes up, the resistance of both gauges increases and so the output of the potential divider does not change.

This is an example of something that is important in scientific measurement. Either the apparatus is designed to eliminate the systematic effect of outside factors, or the effect is measured independently and accounted for.

A further enhancement of the circuit can be made by using not two but four identical strain gauges. Two extra strain gauges are kept unstrained and connected in parallel with the two fastened to the beam. This makes what is called a bridge circuit, shown in Figure 4.

This is done because the voltage at point **B** will vary from 0.50 times the battery voltage to about 0.49 times the battery voltage at the lowest, which is a small change.

If a meter is connected between points **A** and **B**, however, the change becomes much easier to detect. When the beam is not loaded, both **A** and **B** will be at the same p.d., so the p.d. between them will be zero. As the strain builds up, the p.d. between **A** and **B** gives the change in the p.d. at **B**, not its actual value. This can be read with a more sensitive meter.

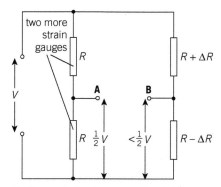

▲ **Figure 4** *Strain gauges in a bridge circuit*

Summary questions

1 State and explain the advantages of using strain gauges, instead of cemented microscope slides, to check for subsidence in ancient monuments such as cathedrals.

2 One type of strain gauge is shown in Figure 5. It consists of 14 strands, each 1.0 cm long, of foil 3.5 µm thick. The width of each strip is 55 µm. The wider strips of foil holding these strands have much lower resistance, and can be ignored. The strain gauge has a resistance of 360 Ω when unstrained. Calculate the resistivity of the metal used.

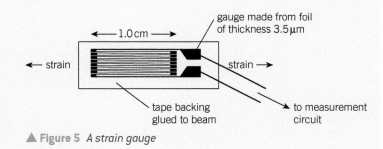

▲ **Figure 5** *A strain gauge*

3 When the strain gauge in Figure 5 is stretched, the fourteen 1.0 cm lengths get longer, but the volume of metal in them stays the same. Show that a 1% increase in length means that there is a 1% decrease in cross-sectional area, and a 2% increase in resistance.

4 The maximum strain that the gauge in Figure 5 can measure is 4% (that is, a length increase of 4%). Show that the change in resistance of a 360 Ω strain gauge under 4% strain is nearly 30 Ω.

5 If the bent beam in Figure 2 produces 4% strain in each gauge, use the circuit in Figure 3 to find the output voltage V_{out} from the potential divider. The circuit voltage $V = 6.0$ V.

6 If the bridge circuit in Figure 4 had been used with the two more 360 Ω strain gauges and a 6.0 V battery, find the p.d. between **A** and **B**.

7 Strains are usually much less than 4%. Use the answers to 5 and 6 to explain why the bridge circuit in Figure 4 is often used in practice.

8 When strain gauges are used in bridge circuits, the p.d. between **A** and **B** is usually connected to a strain gauge amplifier rather than just measured directly with a voltmeter. Suggest a reason for this.

Practice questions

1 Which combination of units is equivalent to J?

   a   A s

   b   A V

   c   V C

   d   W s^{-1} *(1 mark)*

2 Figure 1 shows an *R*-*V* graph for an electric circuit component.

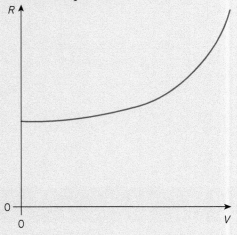

▲ Figure 1

Which electric circuit component would give this graph?

   A   a cell with internal resistance

   B   a filament lamp

   C   a fixed resistor

   D   a semiconductor   *(1 mark)*

3 Three identical resistors can be connected together in a network in the four different ways shown in Figure 2

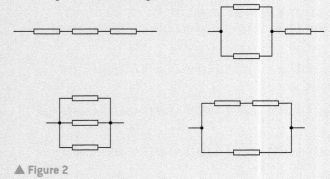

▲ Figure 2

Which of the following resistance values can be made with one of these networks of three 12 Ω resistors?

   1   4 Ω

   2   8 Ω

   3   12 Ω

   A   1, 2, and 3 are correct

   B   Only 1 and 2 are correct

   C   Only 2 and 3 are correct

   D   Only 1 is correct   *(1 mark)*

4 An X-ray tube contains of a stream of electrons which strike a metal target. The current in the tube is 350 mA. Calculate the number of electrons striking the metal target each second.

   electron charge, $e = 1.6 \times 10^{-19}$ C   *(2 marks)*

5 A battery of e.m.f. 9.0 V is connected to a 5.0 Ω resistor. The p.d. across the terminals of the battery is measured to be 8.6 V. Calculate the internal resistance of the battery.   *(3 marks)*

6 A nickel wire of diameter 0.10 mm and length 0.85 m has a current of 0.16 A when there is a p.d. of 1.2 V between its ends. Calculate the conductivity of nickel.   *(4 marks)*

7 A potential divider containing a light-dependent resistor (LDR) and a thermistor is connected as shown in Figure 3.

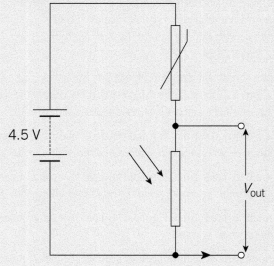

▲ Figure 3

The resistance of the thermistor and LDR under different conditions are given in Table 1.

▼ Table 1

component	thermistor		LDR	
condition	0 °C	25 °C	dark	bright sunlight
resistance / kΩ	33	10	1000	5.4

a The sensors are put a light-proof box at 25 °C. Calculate the p.d. V_{out}. *(3 marks)*

b The circuit is now laid on the laboratory bench, and $V_{out} = 2.4$ V. A cloud suddenly moves across the Sun. Without calculation, explain what will happen to V_{out}. *(3 marks)*

8 In the USA, the mains p.d. is 120 V, and domestic sockets provide a maximum current of 15 A.

a Assuming that a US electric kettle draws a current of 10 A, calculate

(i) the power of the kettle, *(1 mark)*

(ii) the resistance of its heating element. *(1 mark)*

b Heating a litre of cold water to the temperature needed to make tea requires 370 kJ of energy. Calculate the time this takes using a mains connection in the USA. *(2 marks)*

c The kettle is brought from the USA to the UK, where the mains p.d. is 230 V. Assuming that the kettle will still work, calculate the time it takes to boil a litre of cold water. State and justify any assumptions that you make. *(4 marks)*

d In the UK, domestic sockets are limited to 13 A. Suggest why the kettle might work better if left in the USA. *(2 marks)*

9 A temperature sensor circuit is to be constructed using a thermistor in a potential divider. The resistance of the thermistor is 33 kΩ at 0 °C and 680 Ω at 100 °C.

a Draw a circuit diagram for the temperature sensor, including suggested values of battery e.m.f. and resistance of the fixed resistor. You should arrange to have a p.d. reading which increases with increasing temperature and which covers as large a voltage range as possible. *(3 marks)*

b Describe how you would calibrate your temperature sensor circuit over the range 0 °C to 100 °C. *(3 marks)*

c The resistance of the thermistors varies as shown in Table 2.

▼ Table 2

Temperature / °C	0	20	40	60	80	100
Resistance / Ω	33 000	12 000	5300	2500	1300	680

By considering the changes in output of your circuit for 0 °C to 20 °C, and from 80 °C to 100 °C, explain how this resistance variation will affect the sensitivity of your sensor circuit at different temperatures. *(4 marks)*

MODULE 3.2
Mechanical properties of materials

Introduction

Materials have played a crucial role in the development of human society. From the early work with stone, bronze and iron in the ancient world through to twenty-first century materials such as graphene and aerogel, the manufacture of new materials has been a driver of technological change.

In this module you will learn about the mechanical properties of materials and how these properties relate to the microscopic structure of the material. Understanding this relationship helps us predict the properties of materials and design new forms of materials for specific purposes.

Testing Materials introduces the important concepts of stress, strain and the Young Modulus — ideas which are of great importance in engineering. You will learn how to determine the Young Modulus of a material and how to classify materials by their properties. We will focus on the properties of metals, ceramics and polymers, which are all materials with a wide range of uses in the modern world. The chapter introduces many technical terms that help describe the properties of these materials.

You will also consider energy changes when a material is squashed or stretched. Understanding how energy is stored or dissipated in materials when they are deformed has many applications in automotive design, including crumple zones and car suspension systems.

Inside Materials considers the internal structure of materials. You will learn why ceramics, metals and polymers show particular properties. You will review direct evidence for the sizes of the particles and develop a model that explains the different properties of metals, ceramics and polymers. The structure of metals is looked at in detail to explain why alloying changes the property of a metal. An understanding of alloys is important in many areas of design from artificial hip joints to jet aircraft and space vehicles.

Knowledge and understanding checklist

From your Key Stage 4 study you should be able to do the following. Work through each point, using your Key Stage 4 notes and the support available on Kerboodle.

☐ Describe the difference between elastic and plastic deformation.

☐ Calculate work done in stretching a spring.

☐ Describe the relationship between force and extension for a spring, describing the difference between linear and non-linear relationship between force and extension. Calculate the spring constant (force constant) in linear cases.

☐ Recall the typical size (order of magnitude) of atoms and small molecules.

Maths skills checklist

All physicists use maths. In this unit you will need to use many different maths skills, including the following examples. You can find support for these skills on Kerboodle and through MyMaths.

☐ **Change the subject of an equation, including nonlinear equations**, to solve mathematical problems about the energy stored in a spring and the Young Modulus.

☐ **Use an appropriate number of significant figures** when answering problems involving different numbers of significant figures and approximations. For example, when calculating the cross-sectional area of a wire.

☐ **Plot two variables from experimental or other data and use $y = mx + c$**, such as when studying Hooke's law.

☐ **Calculate the gradient from a graph**, for example in experiments to determine the Young Modulus of materials.

☐ **Understand the possible physical significance of the area between a curve and the x-axis, and be able to calculate it or estimate it by graphical methods**, such as when using a graph to calculate the energy stored in a stretched spring.

☐ **Interpret logarithmic plots**, such as when looking at materials selection charts that compare the properties of a range of materials.

MyMaths.co.uk
Bringing Maths Alive

TESTING MATERIALS
4.1 Describing materials
Specification references: 3.2a(iii), 3.2b(i)

Why do adults choose plates made from china but use other materials for baby bowls? Throughout history, materials have been chosen and developed that improve the properties of the finished object. For example, iron axes are superior to stone axes, and cars with lightweight body shells can accelerate more quickly than heavier vehicles. The development of new materials continues – aerogel and graphene were unknown until a few years ago, but are now beginning to find uses in areas such as firefighter suits, composite materials, and photovoltaic cells.

Classes of materials

We are going to consider the mechanical properties of three classes of materials – ceramics, metals, and polymers.

Ceramics are **hard**, meaning that they are difficult to scratch. They are **brittle**, meaning that they will shatter into jagged pieces – think of a broken plate. Ceramics are also **stiff** – they are difficult to stretch or bend. Examples of ceramics include china and other pottery, as well as more modern 'engineering ceramics' such as alumina and silicon carbide.

Metals have a wide range of mechanical properties. Pure metals tend to be soft. Lead can be bent into shape by hand – gold and copper can be easily hammered into shape. Metals that can be shaped easily are called **malleable**, whilst those that can be drawn into wires are called **ductile**. Metal alloys, such as steel, are usually harder than the pure metals used to make them.

Polymers include familiar synthetic materials such as polythene and Perspex, as well as natural materials such as leather and cotton. *Glassy polymers* have properties similar to glass and often replace glass in spectacle lenses. These materials are brittle. *Semi-crystalline polymers* are **tough** – they can undergo considerable deformation and can absorb more energy before breaking (compared to more brittle polymers).

Choosing materials

The choice of materials to use in a product depends to a large extent on the mechanical properties of the material, but other factors such as cost and the look of the finished article are also considered. The properties of each class of material in the hip joint in Figure 2 are used to produce an implant that is effective, long-lasting, and comfortable.

▲ **Figure 1** *Graphene may improve efficiency of photovoltaic cells*

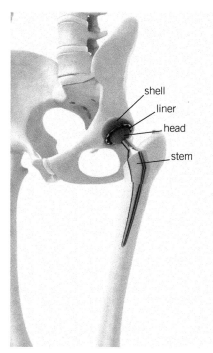

shell
liner
head
stem

▲ **Figure 2** *The hip joint pictured uses a porous metal shell, a polythene liner, a ceramic head, and a metal stem*

Summary questions

1 a Classify each of the following materials as metal, ceramic, or polymer: rubber, acrylic, steel, china.   *(1 mark)*
   b Explain your answer.   *(3 marks)*

2 Use objects around you to name two brittle materials and two tough materials.   *(2 marks)*

3 Identify some of the mechanical properties required for a scalpel used in surgery. Give reasons for your choice.   *(3 marks)*

4 a Suggest desirable mechanical properties for the outer casing of a mobile phone.   *(2 marks)*
   b Compare the properties identified in **a** with the properties required for a mobile phone cover or sleeve.   *(2 marks)*
   c Suggest, using your answers to **a** and **b**, the choice of materials for the casing and the sleeve, giving the classes of materials that these belong to.   *(4 marks)*

4.2 Stretching wires and springs

Specification references: 3.2a(i), 3.2b(i), 3.2b(ii), 3.2c(i), 3.2d(i)

Learning outcomes

Describe, explain, and apply:

→ simple mechanical behaviour: elastic and plastic deformation, and fracture

→ the terms: plastic, elastic, compression

→ force–extension graphs up to fracture

→ the equation for Hooke's law: $F = kx$

→ the equation for energy stored in an elastic material: energy $= \frac{1}{2}kx^2$

→ energy as the area under a force–extension graph for elastic materials

→ an experiment to plot force–extension characteristics for arrangements of springs, rubber bands, and polythene strips.

▲ **Figure 1** *The suspension spring on this vehicle is clearly visible once the wheel has been removed*

For centuries armies relied heavily on storing energy in springy things. The bowmen at Agincourt had yew bows that needed great strength to pull back – crossbows stored even more energy by having a thicker bow that had to be wound back mechanically. A more everyday example of energy stored in a spring is the suspension of a car, designed to absorb the energy of shocks to the wheels from bumps in the ground. In this topic we will look at how materials stretch and how they store energy.

Hooke's law

Hang a weight on the end of a spring and the spring will stretch. In 1678 the English experimenter Robert Hooke found that, for small extensions, the force F is proportional to extension x. This relationship is now known as **Hooke's law**.

The relationship can also be given as

$$F = kx$$

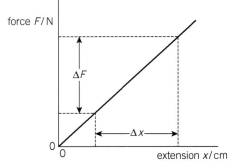

▲ **Figure 2** *A graph showing the relationship $F = kx$*

where k is a constant of proportionality called the spring (or force) constant.

🖩 Worked example: Finding k

Figure 3 shows results from an experiment to measure the extension of a spring as weights are added to it. Use the graph to find a value for the spring constant, k.

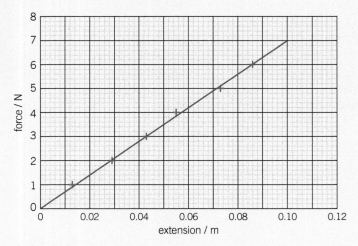

▲ **Figure 3**

Step 1: Identify the equation required.

$$F = k\,x$$

Step 2: Rearrange the equation to find k.

$$k = \frac{F}{x}$$

This is the gradient of a force–extension graph.

Step 3: Find the gradient of the graph.

Taking the points $(0,0)$ and $(0.1,7)$.

$$\text{gradient} = \frac{7.0\,\text{N} - 0.0\,\text{N}}{0.10\,\text{m} - 0.00\,\text{m}} = 70\,\text{N}\,\text{m}^{-1}$$

Every spring or wire has its own value for the spring constant. This value gives a measure of how stiff the specimen is. A large spring constant means the specimen is difficult to stretch. The value of k depends on the material of the wire, its length, and its cross-sectional area. It is important to remember that the spring constant is a value for a specimen rather than for a material, for example, a thin wire will have a much smaller k than that of a thick wire made from the same metal.

 Practical 3.2d(i): Plotting a force–extension graph for a rubber band

The extension of a rubber band can be investigated using the experiment shown in Figure 4.

A piece of stiff wire can be wrapped around the bottom of the weight hanger to provide a pointer. The unstretched length of the band is measured, and then measurements of the length of the band are taken after adding each mass. The extension is found by subtracting the original length from the length when the band is stretched.

The measurements can be repeated as masses are taken off the band and a graph plotted of force (weight) against extension can be plotted.

This experiment can be repeated for a number of similar materials, for example, arrangements of springs and polythene strips.

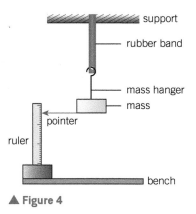
▲ Figure 4

Elastic and plastic deformation

Hooke's law can also be applied to **compression**, that is, squashing the springs rather than stretching them, but it does have its limitations. Extend a spring too far and it will become easier to stretch and will not return to its original length when you let go. The graph in Figure 5 (next page) illustrates how force varies with extension for larger extensions.

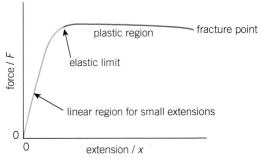

▲ **Figure 5** *The variation of force with extension up to fracture*

Over the first part of the graph, the wire (or spring) stretches or deforms **elastically**. A wire that has deformed elastically will return to its original length when the load is removed. When the extension exceeds the **elastic limit** the wire deforms **plastically**. A wire that has deformed plastically will not return to its original length when the load is removed. The graph is linear for nearly all of the elastic region, curving very slightly near the elastic limit. The plastic region of the graph is non-linear. The wire **fractures** (breaks) at the fracture point.

It is not only wires and springs that follow Hooke's law. For small extensions and compressions all materials will show elastic behaviour, but it is much easier to observe this with a spring than it is with a wire, a glass fibre, or a brick.

Energy stored in a spring

When you stretch a catapult the first little bit is easy – you don't have to pull hard, and you store little energy in the elastic. Let go, and the stone drops feebly from it. The small force from your hand did little work in extending it. Pull harder, store more energy for the extra stretch, and the stone flies off impressively when you let go.

The force F required to stretch a spring by x is kx if the force is proportional to extension. However, the energy stored in an elastically stretched spring is not simply force × displacement, Fx. This is because the force grows steadily larger as the spring is stretched. As the force starts at zero and rises linearly to finish at F the average force is $\frac{F}{2}$.

As a result, the energy stored in an elastically stretched spring $E = \frac{1}{2}Fx$

As $F = kx$, this can also be written as $E = \frac{1}{2}(kx)x = \frac{1}{2}kx^2$

▲ **Figure 6** *The more you stretch a catapult the more energy it stores*

Synoptic link

You will meet the relationship between distance and force in Topic 9.3, Conservation of energy

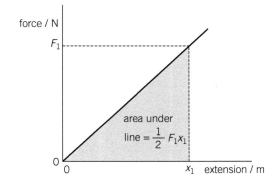

◀ **Figure 7** *Energy stored in a stretched spring* $= \frac{1}{2}kx^2$

Figure 7 shows elastic deformation. When a body deforms elastically the energy stored is equal to the energy transferred stretching the spring. The energy transferred stretching a material can be found from the area under the line.

 Worked example: Finding the energy stored in a spring

A spring has a spring constant k of $70\,\text{N}\,\text{m}^{-1}$. It is extended elastically by $0.055\,\text{m}$. Calculate the energy stored in the spring.

Step 1: Select the appropriate equation.

$$E = \frac{1}{2}kx^2$$

Step 2: Substitute and evaluate.

$$E = \frac{1}{2} \times 70\,\text{N}\,\text{m}^{-1} \times 0.055^2\,\text{m}^2 = 0.11\,\text{J} \ (2\ \text{s.f.})$$

Summary questions

1   Sketch a force–extension graph showing a wire that deforms elastically followed by a region of plastic deformation up to fracture. Label the sections of the graph showing elastic deformation, plastic deformation, and the fracture point.   *(4 marks)*

2   A spring is compressed from a length of 5.2 cm to 4.8 cm when a weight of 22.0 N is placed on it. Calculate the spring constant of the spring. Remember to include the units in your answer.   *(3 marks)*

3   a   Study the graph in Figure 8, which shows force against extension for a spring. State the features of the graph which show that the spring is behaving elastically.   *(2 marks)*
    b   Calculate the spring constant of the spring in Figure 8. Give your answer in $\text{N}\,\text{m}^{-1}$.   *(2 marks)*
    c   Calculate the energy stored in the spring when it has extended by 0.070 m (7.0 cm).   *(2 marks)*

4   For Figure 8, a student suggests that the spring would extend by 42.0 cm when it supported a load of 35.0 N. Deduce how the student must have reached this conclusion, state the assumptions the student has made, and suggest reasons why this may be incorrect.   *(4 marks)*

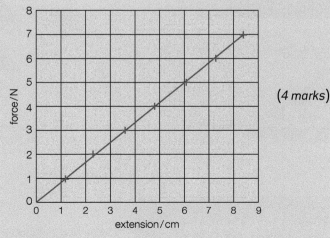

▲ Figure 8

4.3 Stress, strain, and the Young modulus

Specification references: 3.2a(iv), 3.2b(i), 3.2b(ii), 3.2c(ii), 3.2d(ii)

▲ **Figure 1** *Is this web really stronger than this girder?*

Which is stronger, a steel girder or a thread from a spider's web? The answer seems obvious – the girder must be stronger. However, you may have read that spider silk is stronger than steel, so which statement is correct?

Stress

When you compare the strength of a girder to that of a spider thread you are comparing specimens, not materials. It will take a bigger force to break or fracture the girder than the thread. But the girder has a far greater cross-sectional area than the thread. To make a fair comparison we use the concept of **stress**, which is force per unit area.

Fracture stress is the stress at which a material breaks. Stress is found by determining the cross-sectional area of the specimen under investigation, measuring the force on the specimen and dividing the force by the cross-sectional area.

It is also important to know the **yield stress** of materials. This is the stress at which a material begins to deform plastically and become permanently deformed. Imagine you are designing a steel bridge – the breaking stress is an important factor but so is the yield stress as this is the value at which the materials will begin to bend or buckle.

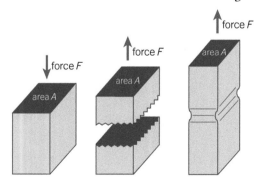

stress is force per unit area

breaking stress is the stress which breaks a material

yield stress is the stress which causes a material to yield

common units of stress:

MN m^{-2} (meganewton per square metre)
or MPa (megapascal)

The units of stress are the same as the units of pressure, pascal. One pascal (1 Pa) is 1 newton per square metre.

Useful rule of thumb – a mass of 1 kg weighs about 10 N on the Earth's surface.

◀ **Figure 2** *Force and stress*

Stress is force per unit area.

$$\text{stress } \sigma \ (\mathrm{N\,m^{-2}}) = \frac{\text{force } F(\mathrm{N})}{\text{cross-sectional area } A(\mathrm{m^2})}$$

 Worked example: Calculating stress

A steel wire of diameter 0.84 mm fractures when under a tension (stretching force) of 152 N. Calculate the fracture stress.

Step 1: Select the appropriate equation for fracture stress.

$$\text{fracture stress } \sigma = \frac{F}{A}$$

Step 2: Calculate the cross-sectional area of the wire in $\mathrm{m^2}$.

$$\text{Cross-sectional area} = \pi \, r^2$$

$$\text{radius of wire} = 0.42 \, \mathrm{mm} = 4.2 \times 10^{-4} \, \mathrm{m}$$

$$\text{cross-sectional area of wire} = \pi \times (4.2 \times 10^{-4})^2$$
$$= 5.541\ldots \times 10^{-7} \, \mathrm{m^2}$$

Step 3: Substitute values back into the equation to evaluate σ.

$$\text{fracture stress } \sigma = \frac{F}{A}$$

$$\frac{152 \, \mathrm{N}}{5.541\ldots \times 10^{-7} \, \mathrm{m^2}} = 2.7 \times 10^8 \, \mathrm{N\,m^{-2}} \ (2 \text{ s.f.})$$

Strain

A long wire will stretch more than a shorter wire of the same cross-sectional area if the same load is hung from both. Therefore, to make a fair comparison, we can calculate the **strain**. Strain is the fractional increase in length. Unlike extension, strain does not depend on the original length of the specimen.

$$\text{Strain } \varepsilon = \frac{\text{extension}}{\text{original length}} = \frac{x}{L}$$

Strain is a ratio of two lengths and is often given as a percentage, for example, a strain of 5%.

 Worked example: Using strain

A rubber cord of original length 3.2 m is extended until it is under a strain of 18%. Calculate the total length of the extended cord.

Step 1: Select and rearrange the appropriate equation.

$$\text{strain} = \frac{\text{extension}}{\text{original length}}$$

$$\therefore \text{ extension} = \text{strain} \times \text{original length}$$

Step 2: Convert percentage strain into a decimal figure.

$$18\% = \frac{18}{100} = 0.18$$

> **Hint**
>
> Remember that the final answer is rounded to the same number of significant figures as the least precise value in the data.

How much does a material stretch? For a given force F a long piece of material will stretch more than a short one. Doubling the length L doubles the extension x

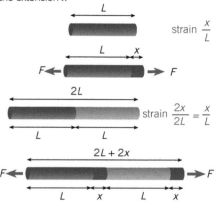

extension x depends on original length

strain = fractional increase in length and does not depend on the original length

▲ **Figure 3** *Stress produces strain*

> **Study tip**
>
> Remember, stress is dependent on *cross-sectional area*, whereas strain is dependent on *length*.

> **Hint**
>
> You must remember to convert percentage strain into the ratio of $\frac{\text{extension}}{\text{original length}}$, $\frac{x}{L}$, before using the value in a calculation.

Step 3: Substitute values and evaluate.

$$\text{extension} = 0.18 \times 3.2\,\text{m} = 0.576\,\text{m}$$

Step 4: Total length = original length + extension

$$= 3.2\,\text{m} + 0.576\,\text{m} = 3.8\,\text{m}\ (2\ \text{s.f.})$$

Stress–strain graphs

The stress–strain graph of a brittle material, such as glass, shows very little plastic deformation. The graph is linear for all its length.

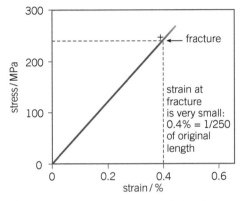

▲ **Figure 4** *A stress–strain graph for glass*

Mild steel is a tough material and undergoes considerable plastic deformation before fracture. This is why tough materials have rounded edges on fracturing rather than the sharp, jagged edges of brittle materials. A rod of mild steel may 'neck' under tension. This means that part of it becomes narrower than the rest.

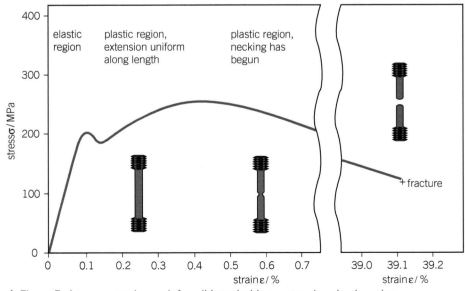

▲ **Figure 5** *A stress–strain graph for mild steel with an extensive plastic region*

Young modulus

The **Young modulus** E gives a measure of the stiffness of a material rather than the stiffness of a particular specimen.

$$E = \frac{stress}{strain} = \frac{\sigma}{\varepsilon} = \frac{F/A}{x/L} = \frac{FL}{xA}$$

The units for Young modulus are $N\,m^{-2}$ or Pa. These are the same as the units of stress. This is because strain is a ratio of two lengths and so has no units – strain is dimensionless. Since F and L are large values, divided by the product of the much smaller x and A, Young modulus is often a very large number.

Many materials stretch in a uniform way. Increase the stretching force in equal steps, and the extension increases in equal steps too, in proportion. That is, the strain is proportional to the stress producing it. This is the same as Hooke's law – the stretching of a spring is proportional to the stretching force you apply.

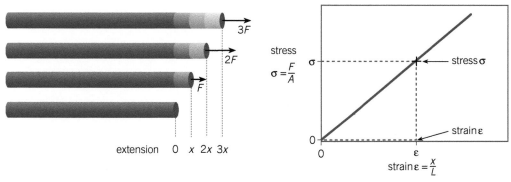

strain $\propto$ stress .... graph is straight line

ratio $\dfrac{stress}{strain}$ is constant

Young modulus $= \dfrac{stress}{strain}$

$$E = \frac{\sigma}{\varepsilon}$$

▲ **Figure 6** *The Young modulus*

 Worked example: The Young modulus

Data about a wire that is under tension and has extended elastically is shown below.

cross-sectional area $= 1.7 \times 10^{-6}\,m^2$

original length $= 2.5\,m$

force on wire $= 220\,N$

extension $= 2.0 \times 10^{-3}\,m$

Calculate the Young modulus of the wire.

Step 1: Select the appropriate equation

$$E = \frac{\sigma}{\varepsilon} = \frac{F/A}{x/L}$$

Step 2: Substitute values and evaluate.

$$E = \frac{220\,N/(1.7 \times 10^{-6}\,m^2)}{(2.0 \times 10^{-3}\,m)/2.5\,m} = 1.6 \times 10^{11}\,N\,m^{-2} \text{ (or Pa) (2 s.f.)}$$

Practical 3.2 d(ii): Determining the Young modulus and fracture stress of a metal

A wire is put under tension and its extension is measured as the force is increased to find its Young modulus. The gradient of a linear stress–strain graph will give you the value of the Young modulus.

Although the method is simple, doing this well is a challenge. The golden rule for any experiment is to carry out a trial experiment first, or make a rough estimate, to find out what to expect.

In the case of steel, a rough estimate or a trial experiment quickly tells you two things:

● for any reasonable length of wire you will have to measure an extension of only a few millimetres

● to stretch the wire with any reasonable force you will have to use a thin wire, a fraction of a millimetre in diameter.

A typical experimental set-up is shown in Figure 7.

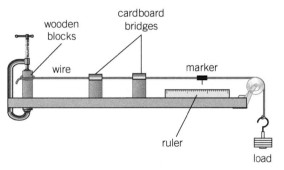

▲ **Figure 7** *Experimental set-up for determining the Young Modulus of a wire*

The diameter and the original length of the wire are measured.

The extension of the wire is measured as more weights are added to the wire, increasing the force. This can be continued until the wire breaks, allowing you to calculate breaking stress and strain.

Stress and strain values can be calculated from the data and a graph of stress (y-axis) against strain (x-axis) drawn. The gradient of the linear region is the best estimate for the Young modulus of the wire. The uncertainty can be estimated by choosing the largest source of uncertainty as a guide.

Summary questions

1 a Explain the difference between fracture stress and yield stress. *(2 marks)*

   b A concrete block of cross-sectional dimensions of 10 cm × 10 cm fractures when a force of 1 MN is applied. Calculate the fracture stress of the concrete in $MN\,m^{-2}$ and in $N\,mm^{-2}$. *(3 marks)*

2 a A spring stretches from an original length of 5.1 cm to a new length of 5.7 cm when a force of 1 N is applied. Calculate the strain of the spring. Give your answer as a decimal and as a percentage. *(3 marks)*

   b What might you expect the length of the spring to be if a 2 N force were applied? Explain your answer, stating any assumptions you make. *(2 marks)*

3 A rubber cord, 5 m long and 50 mm thick, hangs from the top of a zoo cage. A chimp weighing 300 N hangs on the end of the cord. Calculate the extension of the cord. The Young modulus of the rubber is 100 MPa. *(3 marks)*

4 In measuring the Young modulus of copper wire you obtain the following values:

   length of wire = 1.5 ± 0.005 m

   diameter of wire = 0.33 ± 0.02 mm

   extension = 3.5 ± 0.02 mm

   force = 11.5 ± 0.2 N.

   a Show that the largest percentage uncertainty is $\pm 6\%$. *(4 marks)*

   b Calculate the Young modulus of copper. *(2 marks)*

   c Assuming that the uncertainty in the final value is also $\pm 6\%$, calculate the range of possible values of the Young modulus from this result. *(3 marks)*

   d The experiment can be improved if uncertainty in the cross-sectional area can be minimised. This can be done by measuring the mass of the wire and using known density values for the material. Explain how this method can be used to determine the cross-sectional area of the wire and why it may be an improvement over determining the cross-sectional area from the diameter of the wire. *(4 marks)*

   e The experiment is repeated to determine the fracture stress of the copper. Sketch the stress–strain graph you would expect from copper, a ductile metal, up to fracture. *(2 marks)*

Hint

percentage uncertainty = $\dfrac{\text{uncertainty in measurement}}{\text{measured value}} \times 100\%$

4.4 Choosing materials

Specification references: 3.2b(i), 3.2b(iii)

Knowing the fracture stress, yield stress, and Young modulus of different materials is very important to architects, engineers, and designers. It allows them to predict the behaviour of materials under stress, and calculate the requirements of a particular construction project.

Choosing the best material for the job is important in all areas of life. The surgeon's scalpel is made from a particular steel that is strong, hard, and non-corrosive. The clothes you are wearing will be made from natural or synthetic polymers which stretch to fit around your body, showing that the fibres have a degree of elasticity. Bamboo has a low density and is chosen as a material for scaffolding in many countries.

▲ **Figure 1** *Polymers are used extensively in clothing*

▲ **Figure 2** *Bamboo as scaffolding*

A material is in **tension** when a force is acting in a direction to stretch the material. The force in this case is described as a tensile force. On the other hand, a compressive force tends to compress or squash the material. Some materials can be strong in compression but weak in tension. For example, Portland cement concrete has a compressive strength of around 30 MPa and a tensile strength of about 3 MPa.

Selection charts

The choice of a material for a particular use does not only depend on the properties that we have been considering so far. Other physical properties such as **density** (where density $= \dfrac{\text{mass}}{\text{volume}}$) and toughness are also important, as are non-physical properties such as cost and availability. Material selection charts allow quick comparisons to be made between different classes of materials.

As numerical values of stress and strain vary greatly between different classes of materials, scientists use logarithmic scales on charts comparing different materials to display the large range of values.

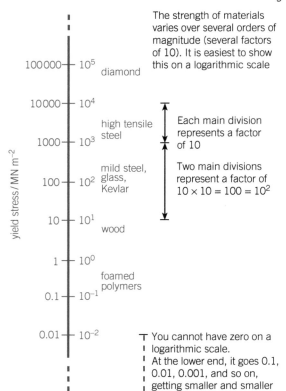

▲ **Figure 3** *Logarithmic scale of stress*

The chart in Figure 4 compares Young modulus with density of different materials. It shows that polymers have a wide range of stiffness values and a narrower range of densities.

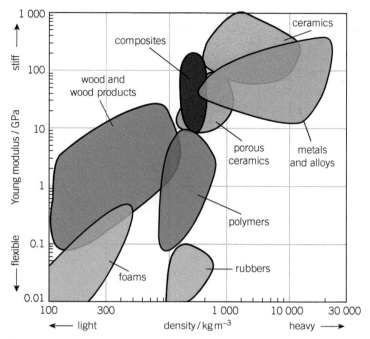

▲ **Figure 4** *Young modulus and density values for different classes of materials*

Polymers also show a wide range of toughness. You will have noticed that some 'plastic' rulers can be bent into all kinds of shapes but that others will shatter – this is an example of the variation in toughness of polymers.

There are also huge variations between strength and toughness for different classes of materials.

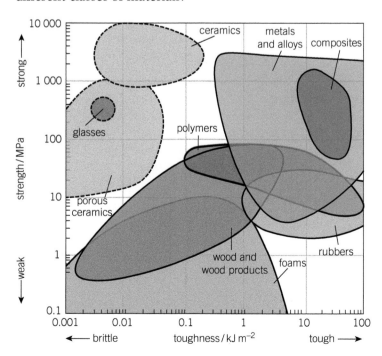

Synoptic link

You will find a section about logarithms in the Maths Appendix.

Summary questions

1 Using the chart of Young modulus and density in Figure 4, show that the stiffest metal has a Young modulus about 30 times that of the least stiff. (*2 marks*)

2 Using the chart of strength and toughness in Figure 5, calculate the ratio of the strength of the strongest metal to that of the weakest. (*2 marks*)

3 A lift has a weight of 5000 N. It carries up to eight passengers.
a Taking the average weight of an adult as 650 N, calculate the tension in the steel cable when the lift is fully loaded. (*2 marks*)
b Calculate the minimum diameter of steel cable needed to support the lift and passengers. Assume the fracture stress of the steel is 1000 MPa. (*2 marks*)
c Explain why a much thicker cable will be used in practice. (*2 marks*)

◀**Figure 5** *Strength and toughness values for different classes of materials*

Physics in perspective

How to build a skyscraper

The Shard is an 87-storey skyscraper topped with a steel and glass spire, transforming the skyline across the Thames from the Tower of London. At a height of 306 m it was the tallest building in the European Union on completion in 2013. Building such a tall tower in a confined space was a challenge for the designers and engineers working on the project. The construction schedule was also demanding – one floor was added to the building each week.

▲ **Figure 2** *The Shard during construction, showing the core, some of the floors of the building, and the glass cladding*

▲ **Figure 1** *The Shard and tower bridge at night*

The building is formed around a core reinforced-concrete tower rising from the foundations. The core accounts for much of the 100 000 tonnes of concrete used in the building. Beneath the core are piles — columns of steel and concrete extending 53 metres down into the ground.

The floors, which you can see in Figure 2, extend from the core tower rather like branches of a tree extending from a trunk. This is called a cantilever arrangement. The upper of a cantilever beam will be in tension and the lower side in compression. This has important design implications.

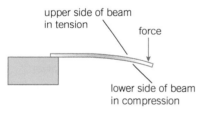

▲ **Figure 3** *A simple cantilever*

upper side of beam in tension

force

lower side of beam in compression

The floors of the Shard are not all constructed using the same materials. The first 40 floors use steel beams extending up to 15 m from the core. These beams provide plenty of room for wiring, air conditioning and other services required by the offices on these levels. Above these floors the building is used as a hotel and above the hotel are residential apartments.

These floors do not need so many services in the floors and ceilings but do need to have effective soundproofing. The pyramid-like shape of the Shard means that the distance from the core to the outer glass wall has reduced to 9 m at this level. Post-tensioned concrete is used for these floors as it has better soundproofing properties than steel beams and can span the 9 m required.

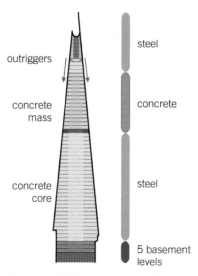

outriggers

steel

concrete mass

concrete

concrete core

steel

5 basement levels

▲ **Figure 4** *The main materials used to construct the shard*

The building is topped with the glass-covered pinnacle that gives the Shard its name. It is certainly at the sharp end of design and construction.

Summary questions

1 In most structures, the stresses and strains within components are designed to remain within the elastic region of the material used. They tend not to stray too close to the elastic limit, or the plastic region which lies beyond.
   a Explain why this is so.
   b Give an example of a structure where plastic deformation of the material is used to advantage.

2 Assume all 100 000 tonnes of concrete used in the Shard went into the core tower.
   a Calculate the smallest cross-sectional area of tower that could support this weight of concrete. The compressive strength of the concrete is 60 MPa.
   b Use your answer from (a) to estimate the maximum height of the tower. The density of concrete is 2400 kg m^{-3}.
   c Compare your value to the height of the Shard and suggest reasons for any difference between the two values.

3 The graphs in Figure 5 show the stress against strain % for the concrete and steel used for reinforcing the core tower and concrete floors of the Shard, up to their yield points.
   a Explain why the graph for steel is symmetrical in tension and compression but the graph for concrete isn't.
   b For concrete, calculate the ratio $\dfrac{\text{strength in compression}}{\text{strength in tension}}$.
   c Calculate the Young modulus for the steel.
   d Calculate the ratio $\dfrac{\text{Young modulus for steel}}{\text{Young modulus for concrete}}$

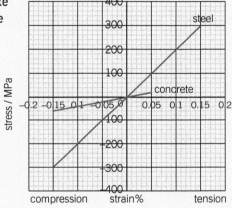

▲ Figure 5

4 For the Shard's core tower construction, liquid concrete was poured around a network of bound steel mesh (this is known as reinforced concrete). Suggest why this composite material is used, and why the surface of the steel used to form the mesh is designed to be rough and knobbly.

5 For the concrete composite floors post-tensioned concrete was used as shown in Figure 6.

steel anchor   steel cable   plastic sleeve   concrete beam   steel anchor

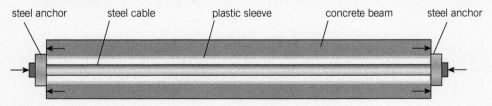

▲ Figure 6

Here steel cables are inserted through plastic sleeves set into the concrete floors. They are stretched with hydraulic levers and then secured by anchor plates, which place the floor beams into a state of compression.
   a Suggest advantages for this method of construction.
   b These concrete composite floors are used for the hotel and living accommodation floors of the Shard, whereas steel floors are used for the commercial offices. Suggest why this might be the case.

Practice questions

1 Which of the following are units of stress?

(1 mark)

A $N kg^{-1}$ **B** $N s$

C $N m$ **D** $N m^{-2}$

2 Which of the properties below is a measure of a material's resistance to stretching or bending? *(1 mark)*

A brittleness   **B** strength

C stiffness   **D** toughness

3 Two wires of the same material are compared. Sample 1 has diameter d. Sample 2 has diameter $2d$. *(1 mark)*

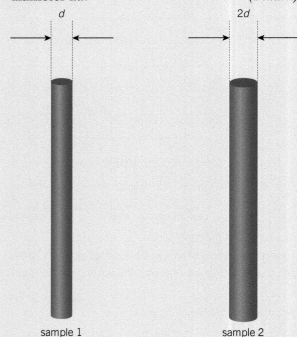

▲ Figure 1

What is the ratio $\dfrac{\text{Young Modulus of sample 1}}{\text{Young Modulus of sample 2}}$?

A $\dfrac{1}{4}$ **B** $\dfrac{1}{2}$

C $\dfrac{1}{1}$ **D** $\dfrac{2}{1}$

4 A weight of 5 N is hung from a steel spring which behaves elastically. The spring extends by 9 cm.

Calculate the energy stored in the spring. Give your answer in J. *(2 marks)*

5 The figure shows the stress–strain graph for a mild steel.

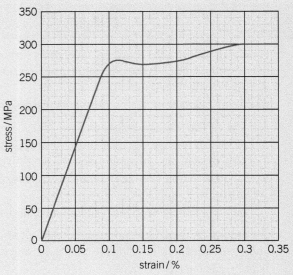

▲ Figure 2

a Calculate the Young Modulus for the linear section of the graph *(2 marks)*

b State how the material is deforming in the non-linear section of the graph.

(1 mark)

6 Two students are discussing the properties of a cast iron frying pan. The first student correctly describes cast iron as a brittle material. The second student argues that cast iron cannot be brittle because it is strong.

a Describe the properties of brittle materials to show why the first student is correct.

(2 marks)

b Give an example of an object that is brittle and weak. *(1 mark)*

c Give an example of a material used in a situation where the property of toughness is important. *(2 marks)*

7 A wire breaks when a force of 700 N is applied. The breaking stress of the wire is 250 MPa. Calculate the diameter of the wire.

(3 marks)

8 Figure 3 shows the force-extension graph for a metal wire.

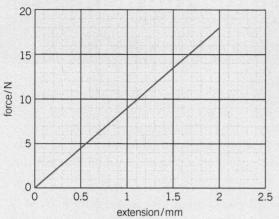

▲ **Figure 3**

a State how the graph shows that the force on the wire is proportional to the extension. (*2 marks*)

b The wire has a Young Modulus of 1.8×10^{11} Pa and an original length of 1.9 m. Use data from the graph to calculate the diameter of the wire. (*4 marks*)

c The wire yields at a force of 18 N and breaks at a force of 18.5 N. At this point the extension of the wire is 2.1mm.

(i) Explain the difference between yield stress and breaking stress. (*2 marks*)

(ii) Use your answer from **(b)** to calculate the yield stress of the specimen. (*2 marks*)

(iii) Explain why the yield stress of a specimen is important in the choice of materials for construction. (*1 mark*)

9 A student is investigating the region of linear strain of a rubber band. The following measurements are taken to calculate the Young Modulus of the rubber.

force = 0.50 N ± 1%

cross-sectional area = 4.0 mm^2 ± 3%

extension = 0.0080 ± 0.0005 m

original length = 0.145 ± 0.001 m

a (i) Explain which measurement contributes most to the overall uncertainty in the calculated value of the Young modulus. (*3 marks*)

(ii) Suggest and explain how the uncertainty in the measurement you have chosen in **(a)**(i) can be reduced. (*2 marks*)

b (i) Using the uncertainties given in the data, show that the maximum calculated value of the Young modulus of the rubber is about 2.5×10^6 Pa. (*4 marks*)

(ii) Without taking into account the uncertainties in the data, the student calculates the Young modulus to be 2.3×10^6 Pa. Use this value and your answer to **(b)**(i) to estimate the percentage uncertainty in the final result. (*2 marks*)

(iii) The student suggests that the uncertainty in the original length can be ignored.

Comment on this suggestion. (*1 mark*)

5 LOOKING INSIDE MATERIALS
5.1 Materials under the microscope
Specification references: 3.2 a(ii)

▲ **Figure 1** *The average height of these giraffes is 10^0 m to the nearest order of magnitude*

How tall are you in metres to the nearest power of ten (or order of magnitude)? Since you are not 0.1 m (10^{-1} m) nor 10 m (10^1 m) tall, you must be closest to 1 m tall (10^0 m). From the diameter of a proton (10^{-15} m) to the size of the observable Universe (10^{27} m) scientists have estimated sizes through measurements, experimentation, and careful thought.

For example, how would you go about measuring the mass of one page of this book using bathroom scales? This is relatively straightforward – measure the mass of the whole book and divide by the number of pages. This is an example of a calculated estimate. It is an estimate because you haven't taken into account the difference in mass for the covers of this book and that bathroom scales have poor resolution.

Estimating the size of atoms and molecules

At the end of the 19th century, Lord Rayleigh performed a simple experiment that enabled him to estimate the thickness of an oil layer floating on water. If the oil spreads out as far as possible the thickness of the layer will be equal to the length of the oil molecule.

d

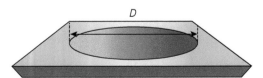

▲ **Figure 2** *Rayleigh's oil drop experiment*

The thickness of the layer, h, is an order of magnitude measure of the largest possible value for the length of an oil molecule.

Rayleigh carried out his experiment by:

1 measuring the diameter d of the oil drop
2 calculating the radius r
3 placing the oil drop on still water to observe it spreading
4 measuring the diameter of the patch of oil D after the oil had spread
5 calculating the radius R of the oil patch.

volume of oil drop $= \frac{4}{3}\pi r^3$

volume of oil patch $= \pi R^2 h$, where h = height of the oil patch

As the volume of drop = volume of patch

$$\frac{4}{3}\pi r^3 = \pi R^2 h$$

Rearranging and cancelling gives

$$h = \frac{4r^3}{3R^2}$$

Uncertainties in estimation

There are a number of sources of uncertainty in this experiment. It is always important to recognise the largest source of uncertainty and to try to minimise it. In this case the largest percentage uncertainty comes from the difficulty of measuring the diameter of an oil drop more precisely than about ± 0.5 mm. For an oil drop of about 1 or 2 mm diameter this is a considerable factor. The value for radius of the drop is cubed in the calculation so any uncertainty in this reading will produce considerable uncertainty in the final result.

 Worked example: The length of an oil molecule

A drop of oil of diameter 1 mm spreads into a disc of approximately 36 cm in diameter. Assuming that the oil spreads into a disc that is uniformly one molecule thick, calculate the length of an oil molecule.

Step 1: Find the radius of the drop and the disc in metres.

$$\text{drop radius} = \frac{1 \times 10^{-3}\,\text{m}}{2} = 5 \times 10^{-4}\,\text{m}$$

$$\text{disc radius} = \frac{0.36\,\text{m}}{2} = 0.18\,\text{m}$$

Step 2: Equate the volume of the drop with the volume of the disc and substitute.

$$\frac{4}{3}\pi r^3 = \pi R^2 h$$

$$\frac{4}{3}\pi \times (5 \times 10^{-4}\,\text{m})^3 = \pi \times 0.18^2\,\text{m}^2 \times h$$

Step 3: Rearrange to make h the subject of the equation and evaluate.

$$h = \frac{4 \times (5 \times 10^{-4}\,\text{m})^3}{3 \times 0.18^2\,\text{m}^2} = 5 \times 10^{-9}\,\text{m} \ (1\ \text{s.f.})$$

Estimating the order of magnitude of the size of a gold atom

A similar method can be used to estimate the size of an atom in a solid by calculating the volume it occupies. For example, if we know the mass of a gold atom and the density of gold we can estimate the diameter of a single atom.

mass of gold atom $= 3.27 \times 10^{-25}\,\text{kg}$

density of gold $= 19\,300\,\text{kg}\,\text{m}^{-3}$

Since density $= \dfrac{\text{mass}}{\text{volume}}$, we can calculate the number of atoms in $1\,m^3$ of gold.

mass of $1\,m^3$ gold $= 19\,300\,kg$

number of atoms in $1\,m^3$ of gold $= \dfrac{\text{total mass of gold}}{\text{mass of one gold atom}}$

$$= \dfrac{19\,300\,kg}{3.279 \times 10^{-25}\,kg}$$

$$= 5.90... \times 10^{28} \text{ atoms}$$

Since $5.90... \times 10^{28}$ gold atoms have a volume of $1\,m^3$,

volume of one atom is $\dfrac{1}{5.90... \times 10^{28}} = 1.69... \times 10^{-29}\,m^3$.

If we make the assumption that gold atoms are cubes (of course they are not, but it simplifies the arithmetic and this is an estimation)

length of side of the cube $= \sqrt[3]{1.69 \quad \times 10^{-29}\,m^3} = 2.6 \times 10^{-10}\,m$ (2 s.f.)

The order of magnitude of a gold atom is therefore $10^{-10}\,m$.

Atoms – from imagination to images

Modern atomic theory developed throughout the 20th century, but it was only at the end of the century that scientists began to make images which showed the atomic structure of matter. Atomic force microscopes (AFMs) and scanning tunnelling microscopes (STMs) can be used to show individual atoms. Scanning electron microscopes (SEMs) show larger scale structures.

The photograph in Figure 3 shows a false-colour STM image of graphite (carbon) atoms (in green) in a regular array, with atoms of gold (orange coloured) piled on the graphite surface. The length of the near side is $1 \times 10^{-9}\,m$. There are roughly ten green blobs representing individual carbon atoms along this side, showing that the length taken up by an individual carbon atom is about $1 \times 10^{-10}\,m$. This agrees with our order of magnitude estimate for the gold atom.

▲ **Figure 3** *STM image of gold atoms on a graphite surface*

 How an AFM works

Rather like an old-fashioned record player, an AFM moves a needle over a sample to detect the contours of the surface. It can detect changes on an atomic scale. A fine point is mounted on the arm and forces between the surface and the tip make the arm bend. A laser beam reflected from the arm detects the bending. One way of using this apparatus is to move the specimen to keep the force on the tip constant. The up and down movement of the specimen as it is scanned under the tip corresponds to the surface profile.

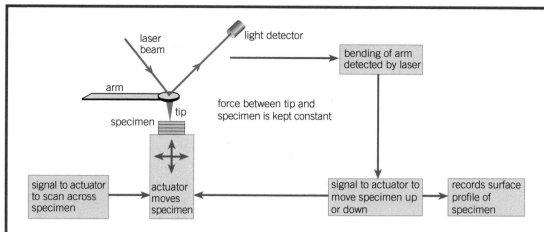

The AFM does not need the specimen's surface to be specially prepared. The specimen doesn't need to conduct electricity or be in a vacuum. The AFM is suitable for biological work.

▲ **Figure 4** *How an AFM works*

Questions
1 Suggest why the specimen moves rather than the tip mechanism.
2 One scientist states that an AFM image of gold on graphite shows a picture of atoms. Discuss whether or not you agree with this statement.

Summary questions

1 You are shown an STM image of graphite (carbon) atoms in a regular array. Explain how you can use the image to make an order of magnitude calculation for the size of a carbon atom, stating any other information you need. *(3 marks)*

2 Suggest reasons why the oil drop experiment gives a *maximum* value for the length of an oil molecule. *(2 marks)*

3 In performing the oil drop experiment, a student measures the diameter of the drop as 0.5 mm and the diameter of the oil patch as 20 cm. Use these values to calculate an order of magnitude estimate for the height of the oil patch. *(3 marks)*

4 The calculation of the diameter of the gold atom using the mass of the atom and the density of gold can only give an order of magnitude answer. Suggest why this is the case. *(2 marks)*

AFMs and STMs tell us about the arrangement of atoms on the surface of a material. However, the atoms within the material may be arranged differently.

Techniques such as X-ray diffraction crystallography can tell us more about the arrangement of atoms beneath the surface of a material. The photograph in Figure 1 shows a famous image of the X-ray diffraction pattern of DNA, produced by Rosalind Franklin in 1953. This, and other images, helped establish the shape of DNA, the crossed bands suggesting its helical structure.

Metal structure and ductility

Metals are **crystalline**. This means that the individual particles are arranged in a regular pattern over distances many times the spacing between the particles. Often the crystalline structure is not obvious from the appearance of the metal – a metal saucepan doesn't look crystalline, but techniques such as X-ray crystallography reveal the hidden structure.

Some pure metals are malleable and ductile. We explain this behaviour with the idea of **dislocations**. These are mismatches in the regular rows of atoms – missing atoms in the otherwise orderly arrangement. It is the movement of dislocations that makes metals ductile. The diagram shows how a dislocation moving through the layers of atoms allows layers to move one atom at a time. This greatly reduces the energy needed to deform the metal. Without mobile dislocations, metals such as gold could not be hammered into shape.

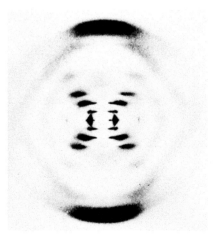

▲ Figure 1 X-ray diffraction image of DNA

▲ Figure 2 SEM image showing crystals of tungsten

▲ Figure 3 'The Mask of Agamemnon' – a beaten gold mask dating from around 1500 BC

Ceramic materials also have dislocations within their structures, but the dislocations are not mobile and so they do not move through the material. Ceramic materials are therefore brittle.

Changing the ductility of metals

Metal **alloys** tend to be less ductile than pure metals. Metal alloys can be formed by the addition of other metallic elements that usually have different sized atoms. These can pin down the dislocations in the metal structure, making slippages between the layers of atoms more difficult.

Atoms in gold are in a regular array: a crystal lattice. To shape the metal, one layer must be made to slide over another.

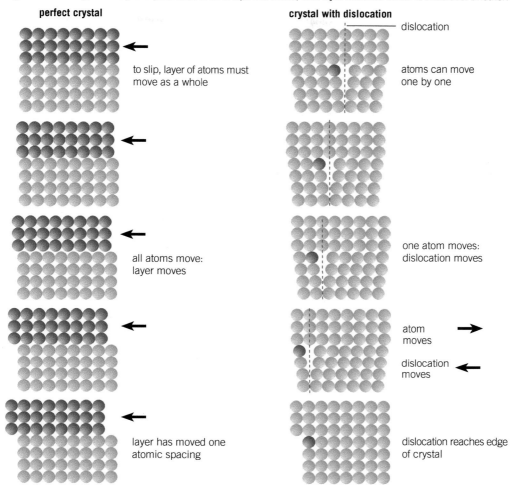

perfect crystal

to slip, layer of atoms must move as a whole

all atoms move: layer moves

layer has moved one atomic spacing

crystal with dislocation

dislocation

atoms can move one by one

one atom moves: dislocation moves

atom moves

dislocation moves

dislocation reaches edge of crystal

in both examples a layer has slipped by one atomic spacing

Wrong model:

Making all the atoms slip together needs considerable energy

This model predicts metals to be 1000 times as strong as they actually are

Better model:

One atom slipping at a time needs much less energy

The dislocation model predicts the strength of metals much better

▲ **Figure 4** *Slipping of particles in metals*

The proportion of alloy atoms in the lattice changes the properties of the alloy — see Figure 5.

Amorphous and crystalline materials

Solids form when liquids cool. The internal structure of the resulting solid may be crystalline, like metals, or **amorphous** (disordered), like glass. Rapid cooling tends to trap particles in an amorphous state, resembling the disordered arrangement in a liquid.

Slow, controlled cooling of a liquid can lead to a single, pure crystal. High-purity silicon is such a crystal, used for making microchips. A single crystal can have a mass of several kilogrammes and contain 10^{26} atoms arranged in a near-perfect array.

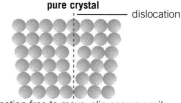

pure crystal

dislocation

dislocation free to move: slip occurs easily move as a whole

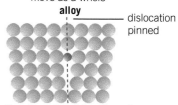

alloy

dislocation pinned

alloy atom pins dislocation: slip is more difficult

▲ **Figure 5** *Particles in metals and metal alloys*

▲ **Figure 6** *Polycrystalline structure of brass (an alloy of copper and zinc) showing the grain boundaries*

Polycrystalline materials

Many solids are neither purely crystalline nor completely amorphous – rather, they are **polycrystalline**. A polycrystalline material consists of a number of grains all orientated differently relative to one another but with an ordered, regular structure within each individual grain.

As a liquid cools, crystals start to form at different points within it. Each crystal grows out into the remaining liquid until it runs into its neighbours. The result is a patchwork of tiny crystals or grains. The interface where these grains meet is known as the **grain boundary**.

Stress concentration and crack propagation

The strength of materials is affected by tiny cracks and flaws in the structure of the material. The stress concentration around such cracks can be hundreds or even thousands of times the applied stress. This can lead to cracks working through a specimen until it fractures. Think about bending a glass rod that has a tiny scratch on the surface.

- The glass becomes strained elastically.
- At the tip of a crack, two neighbouring atoms are pulled apart.
- The next two atoms are pulled apart, and the next two, and so on.
- The crack moves through the material like a zip being undone, propagating through the material.

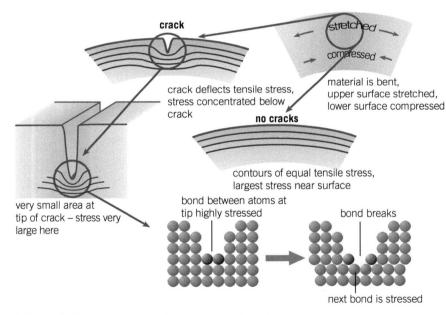

▲ **Figure 7** *Cracks propagate through materials under stress*

Toughness in metals

Many metals are tough. *Toughness* is a measure of the energy needed to extend cracks through a material – it takes more energy to extend a crack in a tough metal than in a brittle substance. This is because they are ductile. When a stress is applied to a crack the metal deforms plastically in the region of the crack which makes the crack broader, reducing the stress around the crack.

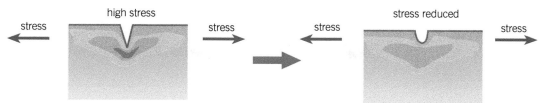

Metals resist cracking because they are ductile. Under stress, cracks are broadened and blunted – they do not propagate.

▲ **Figure 8** *Stopping cracks in metals*

It is important to remember that strength and toughness are not the same thing. A material is **strong** if it has a large breaking stress whereas a tough material is one which will not break by shattering — tough is the opposite of brittle. Glass and steel have similar breaking strengths but steel is much tougher.

Summary questions

1 Describe the differences between an amorphous material and a polycrystalline material. *(2 marks)*

2 A student is heard to say: "Glass shatters because it is weak." Explain why this statement is incorrect and write a few sentences explaining the terms weak, strong and brittle to help the student understand the error. *(4 marks)*

3 Using the concept of mobile dislocations, explain why alloying tends to make metals harder and more brittle. You should include diagrams in your answer. *(4 marks)*

4 Explain why the stress at the tip of a crack in a brittle material can be very large. *(2 marks)*

5 Ceramic materials can have dislocations in their structures, but these do not lead to ductile behaviour. Explain why dislocations produce ductile behaviour in metals but not in ceramics. *(2 marks)*

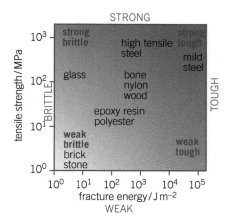

▲ **Figure 9** *Strength and toughness of materials on a logarithmic scale*

5.3 Microscopic structures and macroscopic properties

Specification references: 3.2a(iii)

Microscopic means 'not visible to the unaided eye'. When we talk about the microscopic structures of materials, we mean those structures which are too small to see. At the smallest scale, this means the bonds and patterns of particles making up the material. On the other hand, macroscopic implies something that you can see or a property that is detectable without using sophisticated equipment.

More about ceramics and metals

Ceramics undergo brittle fracture. We have explained macroscopic properties like brittleness and toughness by considering the microscopic structure of the materials. Many of these properties depend upon the bonds between atoms in the materials.

There are three types of bonds between atoms – covalent, ionic, and metallic. However, the bonds in ceramics and ionic compounds have a further property in that they are directional. This means that the atoms are locked in place and cannot slip, making the material hard and brittle.

Ceramics have rigid structures

Covalent structures, for example, silica, diamond, and carborundum

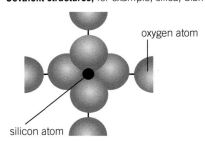
oxygen atom

silicon atom

Atoms share electrons with neighbouring atoms to form covalent bonds. These bonds are directional – they lock atoms in place, like scaffolding.

The bonds are strong – silica is stiff. The atoms cannot slip – silica is hard and brittle.

Metals have non-directional bonds

Metallic structures, for example, gold

negative electron 'glue'

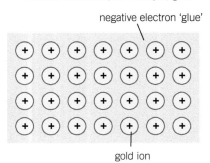

gold ion

Atoms in metals are ionised. The free electrons move between the ions. The negative charge of the electrons 'glues' the ions together, but the ions can easily change places.

The bonds are strong – metals are stiff.

The ions can slip – metals are ductile and tough.

▲ **Figure 2** *Microscopic diagram of ceramics and metals*

Learning outcomes

Describe, explain, and apply:

→ structure of metals, ceramics, and polymers

→ polymer behaviour in terms of chain entanglement/ unravelling.

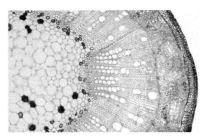

▲ **Figure 1** *The stem of a cotton plant as seen under a microscope*

Synoptic link

Further macroscopic properties of gases, for example, pressure, will be linked back to the microscopic behaviour of the particles of the gas in Topic 14.1, The gas laws in the A level course.

Study tip

Make sure that you can remember an example of a metal, a ceramic, and a polymer. You should also be able to explain their properties using microscopic structures.

Stiffness and elasticity in metals and polymers

Metals behave elastically for small strains, that is, up to strains of around 0.1%. Up to this point the metal extends because the spacing between the positive ions increases. When the tensile force is removed the metal returns to its original length.

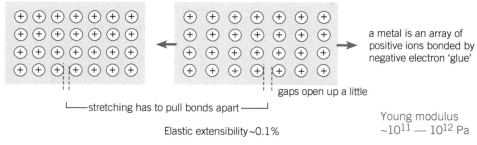

a metal is an array of positive ions bonded by negative electron 'glue'

gaps open up a little

stretching has to pull bonds apart

Elastic extensibility ~0.1%

Young modulus ~10^{11} — 10^{12} Pa

▲ **Figure 3** *Stretching a metal pulls bonds apart*

Polymers such as polythene are long-chained molecules that can extend elastically up to about 1% strains. Polythene is very 'floppy' because it is free to rotate about its bonds. The bonds are strong so although they can rotate they are difficult to break. This gives polythene the macroscopic properties of strength and flexibility.

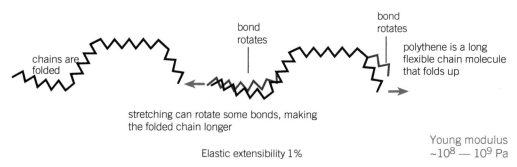

bond rotates

bond rotates

polythene is a long flexible chain molecule that folds up

chains are folded

stretching can rotate some bonds, making the folded chain longer

Elastic extensibility 1%

Young modulus ~10^8 — 10^9 Pa

▲ **Figure 4** *Stretching polythene rotates bonds*

Designing polymers

Not all polymers are flexible. Think of all the different uses for plastics you see around you. Polymers can be stiff if the rotation or unfolding of the chains of molecules is difficult. Adding cross-linkages, where polymer chains are *tied together* at regular intervals along the chains, produces a stiffer material.

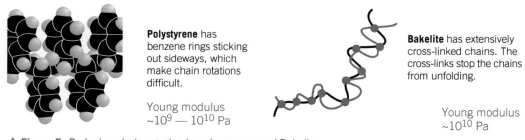

Polystyrene has benzene rings sticking out sideways, which make chain rotations difficult.

Young modulus ~10^9 — 10^{10} Pa

Bakelite has extensively cross-linked chains. The cross-links stop the chains from unfolding.

Young modulus ~10^{10} Pa

▲ **Figure 5** *Reducing chain rotation in polystyrene and Bakelite*

Natural rubber is a runny white liquid of limited uses. In 1839 Charles Goodyear invented the process of *vulcanisation* of rubber in which natural rubber is heated with sulfur. The sulfur atoms form cross-links with the polymer chains. The more sulfur you add, the more cross-links form and the stiffer the rubber. Controlling the microscopic structure of the material allows control of its macroscopic properties. The technique of vulcanisation means that rubber can be chemically adapted for many uses including tyres and shock absorbers.

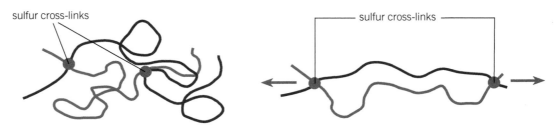

sulfur cross-links

sulfur cross-links

In unstretched rubber, chains meander randomly between sulfur cross-links.

In stretched rubber the chain bonds rotate, and chains follow straighter paths between cross-links. When let go, the chains fold up again and the rubber contracts.

Elastic extensibility > 100%

▲ **Figure 6** *Cross-linkages in rubber*

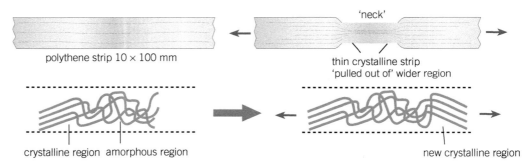

'neck'

polythene strip 10 × 100 mm

thin crystalline strip 'pulled out of' wider region

crystalline region   amorphous region

new crystalline region

Polythene is semicrystalline. Think of polythene as like cooked spaghetti. In amorphous regions the chains fold randomly. In crystalline regions the chains line up.

When stretched plastically, the chains slip past each other. More of the material has lined up chains. More of it is crystalline.

Plastic extensibility > 100%

▲ **Figure 7** *Plasticity in polymers*

Summary questions

1 Polymer chains can become tangled. State the effect(s) this would have on the properties of the material. *(1 mark)*

2 Explain why ions in a metal can slip but atoms in a ceramic do not. *(2 marks)*

3 Suggest why metals only extend elastically up to strains of about 0.1%, but polythene is elastic up to strains of 1%. *(2 marks)*

4 Explain how cross-linking makes polymers stiffer. *(2 marks)*

Module 3.2 Summary

Testing Materials

Describing materials

- describing materials using the terms: hard, brittle, stiff, tough, malleable, ductile, fracture
- the behaviour of metals, ceramics, and polymers
- choosing the right material for the job

Stretching wires and springs

- elastic and plastic deformation and fracture
- force-extension graphs
- Hooke's law, $F = kx$
- energy stored in an elastic material, $E = \frac{1}{2} kx^2$
- energy as the area under a force-extension graph for elastic materials

- practical task: plotting force-extension characteristics for springs, rubber bands, and polythene strips

Stress, strain, and the Young modulus

- using and understanding the terms: tension, stress, strain, fracture stress, yield stress, Young modulus
- stress-strain graphs up to fracture
- stress $\sigma = F/A$
- strain $\varepsilon = x/l$
- Young modulus = σ/ε
- practical task: determining the Young modulus and fracture stress of a metal

Choosing materials

- using materials selection charts and other logarithmic charts to compare mechanical characteristics of materials

Looking inside materials

Materials under the microscope

- evidence of the size of particles and their spacing (Rayleigh's oil drop experiment, scanning tunnelling microscope images)

Modelling material behaviour

- describe and explain the structures of metals and ceramics
- use the model of mobile dislocations allowing slip in metals with brittle materials not having mobile dislocations

- understand the meaning of strong in terms of breaking stress
- interpret images showing the structure of materials

Microscopic structures and macroscopic properties

- describe the microscopic structure of metals, ceramics, and polymers
- explain polymer behaviour in terms of polymer chains entangling and unravelling

Physics in perspective

Graphite to graphene

Graphite and diamond are two forms of pure carbon. Graphite, used as the incorrectly-named pencil lead, has very different properties to those of diamond. This is due to the different arrangements and bonding of the carbon atoms in the two forms — a clear example of microscopic structure producing macroscopic properties.

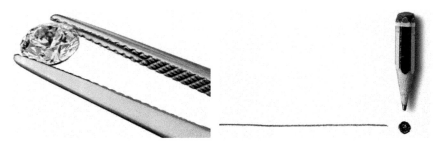

▲ **Figure 1** *The line of graphite from the pencil and the diamond are both carbon*

Graphite and diamond are examples of giant covalent structures. In diamond, as shown in Figure 2, each carbon atom bonds with four others in a rigid, three-dimensional framework.

In graphite, each carbon atom is bonded to three others in a two-dimensional sheet. The distance between each carbon atom in the sheet (the bond length) is 1.42×10^{-10} m or 142 pm. Each layer is attracted to the layers above and below. The distance between the layers is about two and a half times the distance between atoms within a sheet, which explains why graphite is less dense than diamond.

▲ **Figure 2** *The structure of diamond*

The force of attraction between the layers of graphite is much weaker than the attraction between carbon atoms within a layer, which means layers can slide over each other fairly easily. However, the bonds between atoms within a layer of graphite are stronger than those between atoms in diamond. When you write with a pencil you are simply transferring layers of graphite from the pencil to the paper.

Because each carbon atom in graphite makes covalent bonds with three other atoms rather than four, there are delocalised electrons on each layer — electrons can easily move along the sheets of carbon (though not through or between them). This is why graphite has a much higher conductivity than diamond. This suggests that graphite should only conduct electricity in a direction parallel to the layers, but if you take a piece of graphite (from a pencil, say) you will find it conducts in all directions. This is because graphite is not one perfect crystal — it is made from many crystals clumped together in many different orientations, allowing the current to find a route through.

142 pm

335 pm

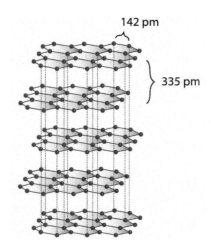

▲ **Figure 3** *The structure of graphite*

Graphene

A perfect, single layer of graphite would only conduct electricity *along* the sheet. Such a layer was produced at Manchester University in 2004, when researchers Andre Geim and Kostya Novoselov used sticky tape to pull thin layers from a sample of graphite. This material, now known as graphene, is one atom thick and shows remarkable properties. It is the first two-dimensional material to be fabricated, an achievement recognised in the award of the 2010 Nobel Prize to Geim and Novoselov.

A sheet of graphene is strong, flexible, and a very good conductor of electricity. It's also light — a single layer of graphene large enough to cover a football field would have a mass of less than 1 g. There is much research happening into ways of putting this remarkable material to use. For example, it is thought that graphene could be used in flexible touchscreens, protective paint, and even solar cells. Many claims have been made about this remarkable material and it will be fascinating to watch new applications develop in the years ahead.

Next time you draw a line with a pencil, remember that a small fraction of the graphite you transfer to the paper will be only a few layers thick, or even a single layer. You may have produced graphene without knowing.

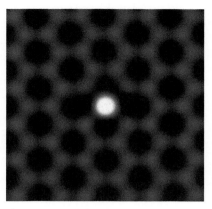

▲ **Figure 5** *SEM image of a silicon impurity atom in a graphene layer. The hexagonal arrangement of carbon atoms can be seen.*

Data: graphene layer
Breaking stress: 130 G Pa.
There is one atom in every $2.6 \times 10^{-20}\,\text{m}^2$ in a graphene layer.
Spacing between layers of graphene = 0.34 nm
Mass of 6.0×10^{23} carbon atoms = 0.012 kg

Summary questions

1 State why graphene is described as a two-dimensional material.

2 It has been said that a single layer of graphene spread over a football pitch would have a mass of less than 1 g. Is this true?
Area of a typical football pitch = 7140 m²

3 a A graphene 'nanoplatelet' has dimensions 7 nm × 50 µm × 50 µm. It consists of layers of graphene of area 50 µm × 50 µm stacked to form a platelet of height 7 nm. Use the data given to calculate the density of the graphene.

   b The density of graphite varies between $2.3 \times 10^3\,\text{kg m}^{-3}$ and $2.7 \times 10^3\,\text{kg m}^{-3}$. Compare these values to your answer from (a) and state what this may suggest about the structure of the two forms of carbon.

4 Many claims have been made about graphene, including that it is the world's strongest material. The breaking stress of structural steel is about $5 \times 10^8\,\text{Pa}$, whilst the breaking stress of graphene is 130 GPa. Explain why the comparison of these two values might be misleading when considering possible uses for graphene.

5 The distance between neighbouring carbon atoms in a graphene layer is about 0.14 nm. Use this value to estimate the area of the SEM image in m².

6 After a talk on graphene, a student was heard to comment that "graphene is basically just graphite — I don't know what all the fuss is about." Describe and explain the similarities and differences between the two materials.

Practice questions

1 Which of the following statements apply to the microscopic structure of metals?

Statement 1   The bonds between metal ions are non-directional

Statement 2   The bonds between metal ions are covalent

Statement 3   The bonds between metal ions can rotate

A   1, 2 and 3 are correct

B   Only 1 and 2 are correct

C   Only 2 and 3 are correct

D   Only 1 is correct   (*1 mark*)

2 The density of silver is $10\,490\,\text{kg}\,\text{m}^{-3}$.

In one mole of silver there are 6.0×10^{23} atoms. The mass of one mole of silver is $0.108\,\text{kg}$.

Use this data to calculate the volume of one atom of silver.   (*3 marks*)

3 Figure 1 shows the microscopic structure of a metal alloy.

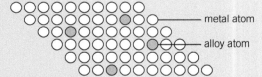

▲ Figure 1

The presence of the alloy atoms reduces the movement of *dislocations* through the metal structure. Explain what is meant by dislocation in this context, and state the effect of the presence of the alloying atoms on the mechanical properties of the material.

     (*2 marks*)

4 Rubber is a long chain molecule. The molecule can be modelled as a series of repeating units of length L joined by bonds that are free to rotate. Rubber can undergo elastic strains greater than 100%.

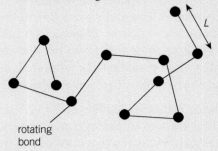

▲ Figure 2

a   A rubber band has an unstretched length of 6 cm. How long will it be when the strain is 100%?   (*1 mark*)

b   Explain why rotating bonds allow the rubber to stretch to such a degree.

     (*2 marks*)

c   A molecule with N rotating bonds linked by units each of length L will have an average total length of $L\sqrt{N}$. The molecule stretches out into an approximately straight line when it is at maximum elastic strain.

Calculate the maximum strain of a rubber molecule of 8 units of length L.   (*3 marks*)

5 The microscopic structure of an ionic crystal such as sodium chloride is an ionic lattice. The bonds between ions are strong and directional. There are no mobile dislocations in the structure.

Explain how this structure makes the material hard and brittle.   (*3 marks*)

6 Polythene bags can be permanently stretched quite a lot before the material tears. Rubber bands stretch easily but relax back to their original length. Explain the behaviour of these materials in terms of the behaviour of molecules.   (*3 marks*)

7 Molecules of DNA have been stretched using 'optical tweezers'. A tensile force of 400×10^{-12} N applied to a single DNA strand produces a strain of 20%. The cross-sectional area of a DNA strand = 2×10^{-17} m².

Calculate an estimate of the Young Modulus of DNA. *(2 marks)*

8 This question relates macroscopic properties of copper to its microscopic structure. The density of copper is 8960 kg m⁻³.

a There are 6.0×10^{23} atoms in 0.063 kg of copper. Show that the volume occupied by one copper atom is about 1.2×10^{-29} m³. *(2 marks)*

b The separation between copper atoms is given by

$$\text{separation} = \sqrt[3]{\dfrac{\text{volume occupied}}{\text{by a copper atom.}}}$$

Calculate the separation between copper atoms. *(1 mark)*

c A copper wire has a cross-sectional area of 0.5 mm². The wire can be pictured as a layer of atoms laid on top of one another. Calculate the number of copper atoms in one layer of wire of cross-sectional area 0.5 mm².

Cross-sectional area occupied by one copper atom is 5.6×10^{-20} m². *(2 marks)*

d A force of about 5×10^{-11} N is required to separate a pair of copper atoms. Use your answers to (c) to estimate the force required to break a copper wire of cross-sectional area 0.5 mm². *(2 marks)*

e Calculate the theoretical breaking stress of copper from your answers. *(2 marks)*

f The accepted value for the breaking stress of copper is about 70 MPa. Compare this with your calculated value in **(e)**, explaining any difference in terms of the microscopic structure of the metal. *(3 marks)*

9 Figure 3 shows the structure of glass.

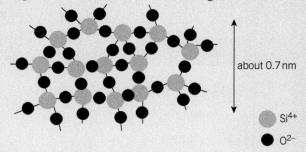

about 0.7 nm

Si⁴⁺

O²⁻

▲ **Figure 3**

The bonding is strong, stiff, and directional within groups of ions, but with random orientations between neighbouring groups. There is little short range order in the structure.

Use features of this micro-structure of glass to suggest **explanations** for the following macroscopic properties of glass.

a Glass fibres are strong but show no plastic deformation before fracture. *(3 marks)*

b A sheet of glass can be broken cleanly and accurately into two pieces, if a scratch is drawn across its surface and the glass is slightly bent. *(3 marks)*

c Solid glass at room temperature is a good electrical insulator, but when heated near its melting temperature it can conduct electricity. *(3 marks)*

May 2009 G481

10 a In a simple cubic arrangement the atoms sit at the corners of imaginary cubes with their curved surfaces just touching. Show that the ratio of filled space to empty space between the atoms is $\dfrac{\pi}{6}$, or about 0.5. This 'packing fraction' is related to the density of the material.

The volume of a sphere of radius r is $\dfrac{4}{3}\pi r^3$. *(3 marks)*

b If the atoms in the simple cubic arrangement in (a) have diameter d, show that the largest impurity atom that can just fit in between the host atoms has a diameter of $0.4d$. *(3 marks)*

MODULE 4.1
Waves and quantum behaviour

Chapters in this Module

6 Wave behaviour

7 Quantum behaviour

Introduction

The behaviour of light has intrigued scientists for centuries — does it travel instantaneously, or does it have a speed that can be measured? Does it behave as a wave or as a particle? As knowledge of the structure of the atom developed at the turn of the twentieth century, similar questions were asked about the nature of the newly-discovered electron.

In this module you will learn about the properties of waves and how light (and all electromagnetic radiation) demonstrates these properties. You will also learn that the behaviour of light cannot be adequately described by picturing it as only a wave or only a particle. You will be introduced to the phasor picture of light that uses the idea of superposition to provide a more comprehensive explanation of the behaviour of light, an explanation that can be extended to the behaviour of electrons.

Wave Behaviour introduces the concepts of wave superposition, path difference

and phase difference. You will learn how to measure the wavelength of light and you will experiment with standing waves on strings and in air. The behaviour of light as it travels from one medium to another is investigated experimentally and explained using wave superposition.

Quantum Behaviour focuses on phenomena that were investigated in the early years of the twentieth century, including the photoelectric effect and electron diffraction. Photons, discrete 'packets' of electromagnetic energy, are introduced and the problems of relating light as photons to light as a wave are considered. At this point, the phasor model is introduced as an alternative way of looking at light. The chapter draws to a close by considering electrons as 'quantum objects' that can show 'wave-like' and 'particle-like' behaviour, which can also be described using phasors.

Knowledge and understanding checklist

From your Key Stage 4 study you should be able to do the following. Work through each point, using your Key Stage 4 notes and the support available on Kerboodle.

- [ ] Describe wave motion in terms of amplitude, wavelength, frequency and period.
- [ ] Define wavelength and frequency, and describe and apply the relationship between these and the wave velocity.
- [ ] Describe the differences between transverse and longitudinal waves.
- [ ] Recall that light is an electromagnetic wave, and recall that electromagnetic waves are transverse and all have the same velocity in space.
- [ ] Recall that wavelength may affect refraction of waves in different substances, and that refraction is related to the difference in velocity of waves in different substances.
- [ ] Recall that atoms and nuclei can generate and absorb electromagnetic radiation.

Maths skills checklist

In this unit, you will need to use the following skills. You can find support for these skills on Kerboodle and through MyMaths.

- [ ] **Use of the small angle approximation $\sin \theta \approx \tan \theta$**, when analysing Young's double-slit experiment.
- [ ] **Understand the relationship between degrees and radians**, and know that phase differences between waves can be expressed in both units. You will also need to find sines of angles on your calculator for angles in degrees and in radians.
- [ ] **Find the slope of a linear graph**, such as in the experiment to determine the Planck constant using LEDs.

MyMaths.co.uk
Bringing Maths Alive

6
WAVE BEHAVIOUR
6.1 Superposition of waves
Specification reference: 4.1a(i), 4.1b(i), 4.1c(i), 4.1d(i) 4.1d(iii), 4.1d(v)

▲ **Figure 1** *An iridescent peacock*

▲ **Figure 2** *Colours on a soap film*

You may have noticed that at the instant that the crests of two water waves pass through each other a bigger crest is formed. This is an example of **superposition**. The principle of superposition states that when two or more waves overlap (superpose), the resultant displacement at a given instant and position is equal to the sum of the individual displacements at that instant and position. This explains a wide range of phenomena.

The colours of a peacock's tail (Figure 1) are iridescent – they shimmer and change as you view it from different angles. This is superposition in action. The colours seen on the surface of a bubble provide another example (Figure 2).

Representing wave motion

The diagrams in Figure 3 and Figure 4 look very similar but there is an important difference between them. Figure 3 is a displacement–displacement diagram, a snapshot of a wave at a single instant in time. The distance between any two points at the same part of the wave cycle is the wavelength, λ. The **amplitude**, A, of the wave is the maximum displacement from the equilibrium position.

Figure 4 shows how the displacement of a point along a wave changes over time. For example, if point P (Figure 3) is at the equilibrium position at time $t = 0$, its displacement will vary with time in the manner shown in Figure 4. The time it takes to return to the same position in the cycle moving in the same direction is the period of the wave, T. The time for one period of the wave is shown.

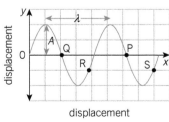

▲ **Figure 3** *A snapshot of a wave at an instant in time*

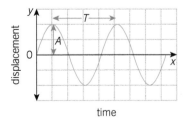

▲ **Figure 4** *The position of particle P on a wave plotted against time*

Time period and frequency

You can use an oscilloscope to measure the time period of a wave or waveform and use this value to calculate the frequency.

Practical 4.1 d(i): Measuring frequency using an oscilloscope

An oscilloscope shows how the potential difference across a component or supply varies with time. The y-value shows the potential difference across the component and the x-value shows the time. The sensitivity of the y-axis (in $V\,cm^{-1}$) and the time base (in $ms\,cm^{-1}$) can be controlled.

When a stable source, such as a signal generator, is connected to the oscilloscope a trace resembling that in Figure 5 is displayed. You can find the time period by noting the setting on the time base and measuring the peak-to-peak or trough-to-trough distance on the display grid.

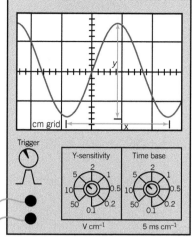

▲ **Figure 5** *Using an oscilloscope*

Synoptic link

You have met the relationship between time period and frequency in Topic 1.4, Polarisation of electromagnetic waves.

 ## Worked example: Finding the frequency of a signal using an oscilloscope

Find the frequency of the signal displayed in Figure 5.

Step 1: Determine the time base set on the oscilloscope.

The illustration shows that the time base is set to $5\,ms\,cm^{-1}$. 1 cm on the x-axis represents a time interval of 5 ms.

Step 2: Measure the distance between two crests, or between two troughs.

Looking on the grid you can see that this distance is 3.4 squares on the cm grid.

Step 3: Use the time base setting to find the time period.

Each cm represents a time interval of 5 ms – therefore, the time period is $5\,ms\,cm^{-1} \times 3.4\,cm = 17\,ms = 17 \times 10^{-3}\,s$.

Step 4: Use the equation frequency $= \dfrac{1}{\text{time period}}$ to find the frequency.

$$\text{Frequency} = \frac{1}{17 \times 10^{-3}\,s} = 59\,Hz$$

Phase and phasors

Phase describes the stage in a wave cycle. *At the top of the wave* is a statement about phase. When two points are at the same stage in the cycle they are **in phase**. You can see from Figure 3 that points

P and Q are in phase, so are points R and S. They are in the same stage of the wave cycle at the instant of time represented by the snapshot.

Phase difference
Two waves doing the same thing at the same moment are in phase. They have no phase difference. Two waves doing exactly opposite things at the same moment are in **antiphase**. These are special cases; if waves are neither exactly in phase nor exactly in antiphase they are said to be **out of phase**.

Phase and phase difference can be measured by a **phase angle**. We can use a rotating arrow, a **phasor**, to show where the wave is in its cycle. The arrow turns through 2π radians (360°) as the wave goes through one cycle. Figure 6 shows this. The vertical displacement of the tip of the clock arrow above or below the midpoint represents the displacement of the wave at that instant in time. If two waves are in phase the difference in phase angle is zero. Waves in antiphase have a phase difference of π radians.

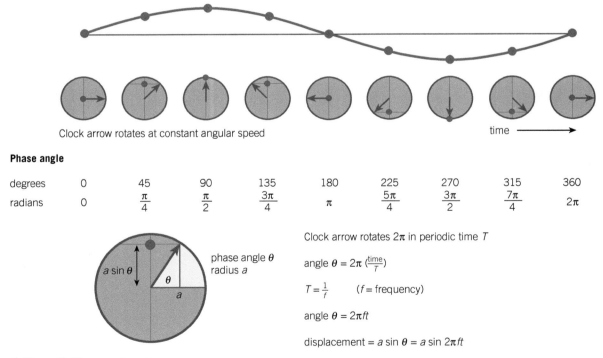

Clock arrow rotates at constant angular speed   time ⟶

Phase angle

degrees	0	45	90	135	180	225	270	315	360
radians	0	$\dfrac{\pi}{4}$	$\dfrac{\pi}{2}$	$\dfrac{3\pi}{4}$	π	$\dfrac{5\pi}{4}$	$\dfrac{3\pi}{2}$	$\dfrac{7\pi}{4}$	2π

phase angle θ
radius a

Clock arrow rotates 2π in periodic time T

angle $\theta = 2\pi \left(\dfrac{\text{time}}{T}\right)$

$T = \dfrac{1}{f}$ (f = frequency)

angle $\theta = 2\pi f t$

displacement = $a \sin \theta = a \sin 2\pi f t$

▲ **Figure 6** *Phase and angle*

We can calculate the displacement when two waves superpose by adding the individual wave displacements together. You can see this in Figure 7. The picture of a phasor arrow is a useful one here. The length of the phasor arrow represents the amplitude of the wave. By adding the phasor arrows together, tip-to-tail, we can find the amplitude of the resultant phasor which is the length of the gap between the tail of the first phasor arrow and the tip of the last. Put another way, the arrow sum of the individual phasor components. Look at the two waves with a phase difference of $\dfrac{\pi}{2}$ radians in

Figure 7. The length of the resultant phasor arrow (in red) tells us the amplitude of the two waves superposed and the angle tells us where the wave is in its cycle.

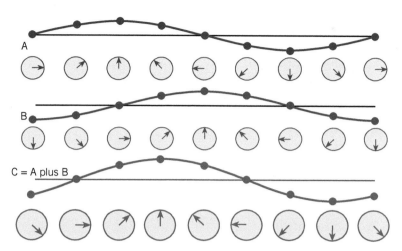

A

B

C = A plus B

Rotating arrows add up

arrows add tip-to-tail

For any phase difference, amplitude of resultant = arrow sum of components

▲ **Figure 7** *Oscillations with $\frac{\pi}{2}$ (90°) phase difference*

Standing waves

A ripple moving across the surface of water is an example of a progressive wave – you can see the crest of the wave moving (or progressing) in time. When two progressive waves of the same frequency are travelling in opposite directions (say, along a string), the waves can appear to stop moving. When this happens, a standing or stationary wave is formed.

Standing waves are an example of superposition. These waves can also be produced when sound waves and electromagnetic waves superpose.

Forming a standing wave on a string

When a string is plucked, waves:

- move along the string in opposite directions
- reflect at the ends of the string
- superpose as they pass through one another.

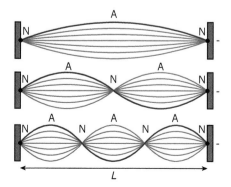

▲ **Figure 8** *Nodes (N) and antinodes (A) on a standing wave on a string for $\lambda = 2L$, $\lambda = L$, and $\lambda = \frac{2}{3}L$*

> **Hint**
>
> Although a string moves (up and down) when a standing wave is produced, the positions of maximum oscillation and minimum oscillation do not change.

There will be maximum amplitude of oscillation at points along the string where the waves meet in phase. These are called **antinodes**. Where the waves travelling along the string meet in antiphase there is zero amplitude. These points are called **nodes**.

Adjacent nodes are half a wavelength apart. The longest standing wave has a wavelength of twice the length of the string. This has the lowest frequency of vibration, called the fundamental frequency. The next possible standing wave has a wavelength equal to the length of the string. You can see this in Figure 8.

Calculating wavelength

To calculate wavelength we use the equation

wave velocity v (m s^{-1}) = frequency f (Hz) × wavelength λ (m)

 Worked example: Using $v = f\lambda$

A ukulele string has a vibrating length of 345 mm. The lowest (fundamental) frequency of its vibration is 440 Hz.

a   Calculate the wavelength of the standing wave.

As the wavelength of the fundamental frequency = twice the length of the string, $\lambda = 2 \times 345$ mm = 690 mm

b   Calculate the velocity of the wave in the string.

Step 1: Select the appropriate formula.

$$v = f\lambda$$

Step 2: Convert values into the correct unit before substituting into the formula.

$$690\,\text{mm} = 0.69\,\text{m}$$

Step 3: Substitute values.

$$v = 440\,\text{Hz} \times 0.69\,\text{m} = 304\,\text{m s}^{-1}\ (3\ \text{s.f.})$$

 Practical 4.1d(iii): Standing waves on a rubber cord

This practical gives you an opportunity to see standing waves. The apparatus is set up as in Figure 9. The rubber cord has been stretched to about twice its original length. Set a signal generator at 10 Hz and gradually increase the frequency. The first standing wave pattern you see has an antinode in the middle of the length of the string and nodes at either end. At double this frequency you will find a second standing wave pattern, which has two humps (see Figure 9). Increase the frequency further and you may see a third standing wave pattern.

The velocity of the progressive waves that form the standing wave can be found by measuring the wavelength of the standing wave, taking the value of the frequency from the signal generator and using the equation $v = f\lambda$.

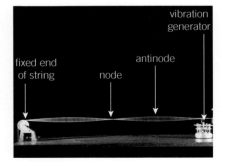

vibration generator

antinode

fixed end of string

node

▲ Figure 9

Standing waves in air

Standing waves can also be formed in air columns – that's how wind instruments such as bagpipes, trumpets, and recorders form their notes.

● A sound wave travels along the tube.

● The wave reflects at the end of the tube.

● Waves travelling up and down the tube superpose with each other.

Sound can be reflected from a closed end of a tube as well as from an open end, so there are two distinct sets of standing waves in tubes. Some instruments have two open ends, a flute, for example. Others effectively have one open and one closed end, for example, a clarinet. Where the waves travelling in opposite directions meet in phase an antinode is formed. A node is formed where the waves meet in antiphase. Figure 10 shows some of the possible standing waves from closed and open pipes. You will see that there is always an antinode at an open end of a tube.

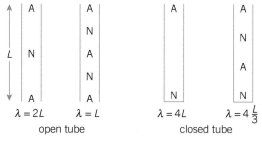

▲ **Figure 10** *Positions of nodes (N) and antinodes (A) in open and closed tubes*

You can also observe standing waves in sound and in microwaves by placing the source of the waves in front of a wall or reflector. Particular separations between the source and reflector will give rise to standing waves as the reflected waves superpose with those from the source.

Practical 4.1d(v): Determining the speed of sound in air using a tube

You can find the speed of sound in air by using a *resonance tube*, a hollow tube with one end immersed in water. A source of sound waves, such as a tuning fork of known frequency or a loudspeaker controlled by a signal generator, is required for this experiment.

By holding a tuning fork at the top of the tube (Figure 11), you can find L_1, the shortest length of tube that increases the amplitude of the sound from the tuning fork. For this length there will be a node at the water surface and an antinode a little above the top of the tube. In other words, the length of the tube plus an *end correction*, k, is equal to one-quarter of the wavelength of the sound wave.

When there is about three times as much tube out of the water the amplitude will increase once more. At this position the length of tube L_2 (plus end correction k) is equal to three-quarters of the wavelength of the note.

By measuring L_1 and L_2 you can work out the wavelength of the sound wave.

$$\frac{\lambda}{4} = L_1 + k \quad \text{and} \quad \frac{3\lambda}{4} = L_2 + k$$

Therefore $L_2 - L_1 = \frac{\lambda}{2}$

Once you have found the wavelength the velocity of the wave can be calculated using the equation $v = f\lambda$.

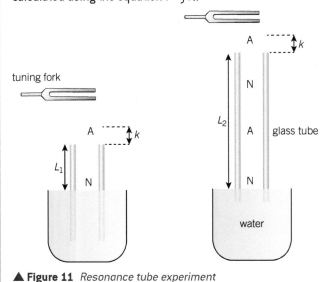

▲ **Figure 11** *Resonance tube experiment*

Summary questions

1 Sketch displacement–displacement graphs of:
   a two equal-amplitude waves in phase; *(2 marks)*
   b two waves of different amplitudes in antiphase; *(2 marks)*
   c two equal-amplitude waves with a phase difference of $\frac{\pi}{2}$ radians. *(2 marks)*
   d Sketch the superposition pattern of the waves you have sketched in **b**. *(1 mark)*

2 A loudspeaker points directly at a wall 3 m away. It emits a sound wave of frequency 680 Hz and a standing wave is formed.
   a Describe how a standing wave is formed. Include an explanation of the terms 'node' and 'antinode'. *(3 marks)*
   b Calculate the distance between each node if the speed of sound is $340\,\mathrm{m\,s^{-1}}$. *(2 marks)*

3 An organ pipe can produce very low frequency sound waves. Calculate the length of pipe, closed at one end, that is needed to produce a sound wave of 30 Hz when the speed of sound is $340\,\mathrm{m\,s^{-1}}$. *(2 marks)*

4 A transmitter emits microwaves of frequency 15 GHz. The speed of microwaves in air is $3.0 \times 10^8\,\mathrm{m\,s^{-1}}$.
   a At what rate does the phasor representing the wave rotate? *(1 mark)*
   b How far does the wave travel in one rotation of its phasor? *(2 marks)*
   c Describe how a student could measure the wavelength of the microwaves using a microwave transmitter, microwave detector, reflector, and a metre rule. *(5 marks)*

6.2 Light, waves, and refraction

Specification reference: 4.1a(iii), 4.1b(i), 4.1 c(ii), 4.1d(ii)

Refraction

Put a couple of drinking straws in a glass of water, look at them from the side, and they appear disjointed and magnified. A similar effect is seen when you look at the straws from above. This is an example of refraction. Waves change speed when they change medium – the material they are travelling through. This change of speed makes the light rays bend and change direction. When you look down on the straws in the water the light you see is refracted and bent at the boundary between the water and the air, giving the impression of a disjointed straw.

Refractive index

The behaviour of light has puzzled many thinkers across the centuries. The first reliable estimate for the speed of light came in the 17th century but many at that time continued to think that light travelled at infinite speed.

A vacuum is a region of space that contains no matter. Scientists now know that light travels at $299\,792\,458\,\text{m}\,\text{s}^{-1}$ in a vacuum. This figure is rounded to $3.00 \times 10^8\,\text{m}\,\text{s}^{-1}$ in this book, but the very precise measurement is extremely important in many areas of physics. In fact, the precise value is used to define the value of the metre. Light travels at very nearly the same speed in air (which is mostly vacuum, when you think about it) but it is very slightly slower in air because it interacts with the electrons in the atoms in the air. The difference is small because air is not very dense so interactions between light and the electrons in the atoms are few and far between. Light travels considerably slower in glass, at around $2 \times 10^8\,\text{m}\,\text{s}^{-1}$, depending on the type of glass. Glass is much denser than air so the number of interactions per metre between light and the electrons in the atoms of glass is much greater.

The ratio of the speed of light in a one medium to the speed of light in another medium is called the **refractive index**.

$$\text{refractive index} = \frac{\text{speed of light in medium 1}}{\text{speed of light in medium 2}}$$

This can be written as refractive index, $n = \dfrac{c_{\text{1st medium}}}{c_{\text{2nd medium}}}$

If the first medium is a vacuum the equation becomes

$$\text{Refractive index of material} = \frac{\text{speed of light in a vacuum}}{\text{speed of light in material}}$$

This is known as the absolute refractive index.

 Worked example: Refractive index of quartz

The speed of light in air is $3.00 \times 10^8\,\text{m}\,\text{s}^{-1}$. The speed of light in quartz is $1.95 \times 10^8\,\text{m}\,\text{s}^{-1}$.

Calculate the refractive index of quartz.

▲ **Figure 1** *Drinking straws in a glass*

Learning outcomes

Describe, explain, and apply:

→ the refraction of light at a plane boundary using the wave model and the changes in speed at the boundary

→ the term refractive index

→ the equations for calculating the refractive index of a material

→ how the refractive index for a transparent block can be determined.

Synoptic link

You will consider the interactions between light and electrons in Chapter 7, Quantum behaviour.

Hint

Refractive index
When you see the refractive index of a material given, this figure is for light travelling from a vacuum or air (1st medium) into the material (2nd medium).

Step 1: Select the appropriate equation.

$$n = \frac{c_{\text{1st medium}}}{c_{\text{2nd medium}}}$$

Step 2: Substitute values.

$$n = \frac{3.00 \times 10^8 \text{ m s}^{-1}}{1.95 \times 10^8 \text{ m s}^{-1}} = 1.54$$

Refraction of light and Snell's law

Refraction can be modelled using both wave-fronts and light rays.

Figure 2 shows a representation of a ray of light travelling from air into glass. The ray bends towards the **normal**, an imaginary line at 90° to the surface of the glass (the air-glass boundary). The diagram also shows the angles of incidence and refraction, measured from the normal.

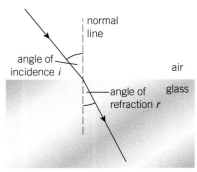

▲ **Figure 2** *Ray representation of refraction at a plane boundary*

The ratio $\frac{\sin i}{\sin r}$ is the refractive index of the material.

This statement is known as Snell's law after its discoverer, the Dutch mathematician Willebrord Snell, sometimes rather wonderfully known by the Latinised version of his name, Willebrordus Snellius.

From the two statements of the refractive index we can write

$$\frac{\sin i}{\sin r} = \frac{c_{\text{1st medium}}}{c_{\text{2nd medium}}}$$

 Worked example: Using Snell's law

The refractive index of pure ice is 1.31. The speed of light in ice is 2.29×10^8 m s^{-1}. The speed of light in air is 3.00×10^8 m s^{-1}. Calculate the angle of refraction in ice for an incident angle of 40°.

Step 1: Select the appropriate equation.

$$\frac{\sin i}{\sin r} = \frac{c_{\text{1st medium}}}{c_{\text{2nd medium}}}$$

Step 2: Rearrange the equation.

$$\sin r = \frac{\sin i \times c_{\text{2nd medium}}}{c_{\text{1st medium}}}$$

Synoptic link

You have met the wave-front and ray models of light in Topic 1.1, Bending light with lenses.

Hint

Dispersion
The refractive index varies with wavelength of the light – that is why prisms and raindrops (forming rainbows) disperse light into the colours of the spectrum.

Step 3: Substitute values and evaluate.

$$\sin r = \frac{\sin 40 \times 2.29 \times 10^8 \, \mathrm{m\,s^{-1}}}{3.00 \times 10^8 \, \mathrm{m\,s^{-1}}} = 0.49$$

Step 4: Find the angle from $\sin^{-1}$.

$$r = \sin^{-1} 0.49 = 29°$$

Practical 4.1d(ii): Determining the refractive index for a transparent block

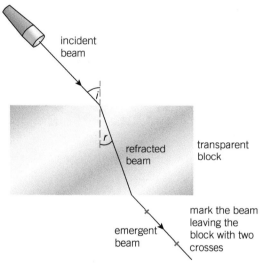

incident beam

refracted beam

transparent block

emergent beam

mark the beam leaving the block with two crosses

▲ **Figure 3** *Determining the refractive index of a transparent block*

You can find the refractive index of a glass block by finding the path of rays as they travel through the block. You need to measure the angle of refraction for a range of angles of incidence, and then calculate $\sin i$ and $\sin r$ for each pair of values. If you plot a graph of $\sin i$ (y-axis) against $\sin r$ (x-axis) you can find the refractive index from the gradient of the line.

The uncertainties in the measurements of i and r will introduce an uncertainty in $\sin i$ and $\sin r$. When you plot the graph there may be an uncertainty in the gradient of the line, and therefore in the value of refractive index that you calculate.

Huygens and wavelets

The 17th century Dutch mathematician Christiaan Huygens provided an explanation of the behaviour of light, including refraction, using the idea that light travelled as a wave.

Huygens imagined light spreading out as tiny *wavelets*. He pictured every point on a wave-front as a source of circular wavelets – just like the circular ripples you get when you let a drop fall into a bowl of water. Huygens suggested that where these wavelets meet in phase they combine to form a new wave-front. Everywhere else the wavelets

▲ **Figure 4** *Christiaan Huygens (1629–1695)*

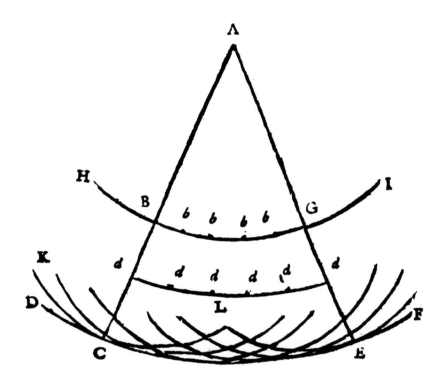

▲ **Figure 5** *Huygens' diagram showing that wave-front DCEF is made of wavelets on wave HBGI starting from all the places like b, b, b*

from one part of the wave will be in antiphase with those from another part of the wave and will cancel out. This simple picture is far-reaching. You have to think of wavelets from the wave as spreading everywhere, but adding up to nothing except where the wave actually goes, another example of superposition.

We can use the model of light as a succession of wave-fronts to explain why light bends when it moves from one material to another at an angle other than along the normal.

We begin by considering what happens when a wave-front enters glass from air parallel to the normal line, as shown in Figure 6. Huygens suggested that waves slow down when they enter the glass so the waves behind *catch up*. The wavelength is shorter in the glass but there is no change of direction.

When light enters a different material at an angle to the normal the wave-fronts become kinked at the boundary because the section of the wave-front travelling through air travels faster than the section of

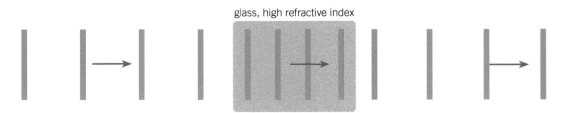

▲ **Figure 6** *Wave-front representation of light travelling from air into glass*

the wave-front in the material. This effect is shown in Figure 7. In time Δt the wave-front moves distance RQ in medium 1 and distance PS in medium 2.

We can use this model and a little geometry to explain why the ratio of speeds is the same as the ratio of the sines of the angles of incidence and refraction.

$$\text{As speed} = \frac{\text{distance}}{\text{time taken}}$$

$$\frac{\text{speed in air}}{\text{speed in glass}} = \frac{\dfrac{RQ}{\Delta t}}{\dfrac{PS}{\Delta t}} = \frac{RQ}{PS}$$

$$RQ = \sin i \times (PQ)$$

$$PS = \sin r \times (PQ)$$

$$\frac{RQ}{PS} = \frac{\sin i \times PQ}{\sin r \times PQ} = \frac{\sin i}{\sin r}$$

Snell's experimental law matches Huygens' analysis based on light wave-fronts slowing down when they travel from air into another transparent material.

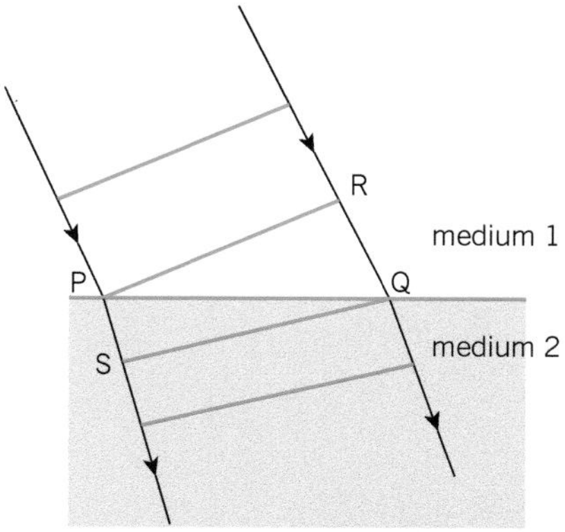

▲ **Figure 7** *Wave-fronts changing direction at a boundary*

Summary questions

1 The velocity of light in air is $3.0 \times 10^8 \, \text{m s}^{-1}$. Diamond has a refractive index of 2.419. Calculate the velocity of light in diamond. Explain your choice of the number of significant figures you give in your answer. *(3 marks)*

2 The refractive index of amber is 1.55. Calculate the angle of refraction in amber for an angle of incidence of 45°. *(3 marks)*

3 Figure 8 shows waves approaching a boundary to a material in which they speed up.
   a Copy and complete the diagram to show the faster-moving waves in the material. *(2 marks)*
   b State what happens to the wavelength and frequency of the waves as they enter the material. *(2 marks)*

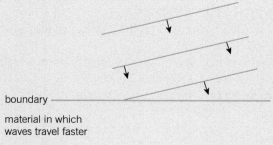

boundary

material in which waves travel faster

▲ **Figure 8**

4 A Pyrex boiling tube containing glycerol is immersed in a beaker of glycerol. Pyrex and glycerol have the same refractive index. Explain why the length of the tube in the glycerol becomes invisible. *(3 marks)*

5 White light passing through a glass prism is dispersed into the colours of the spectrum.
   a State what the refractive index of a material tells you about the behaviour of light as it enters or leaves the material. *(1 mark)*
   b Explain why the dispersion of light through a prism shows that different wavelengths of light travel at different speeds in glass. *(2 marks)*
   c Explain why the dispersion of light shows that the refractive index of a material is wavelength dependent. *(2 marks)*

▲ **Figure 9** *Light dispersed into colours by a prism*

6 Table 1 shows the results from an experiment to determine the refractive index of a glass block. The student taking the results estimated the uncertainty in the readings of the angle of refraction as ± 2°.

▼ **Table 1**

Angle of incidence $i/°$	Angle of refraction $r/°$
0	0
10	5
20	14
30	20
40	26
50	32
60	41

   a Draw a table of $\sin i$ and $\sin r$ for the results, include the uncertainty in $\sin r$. *(4 marks)*
   b Draw a graph of $\sin i$ against $\sin r$. Use this graph to find the refractive index of the material. Show your working clearly and include an estimate of the uncertainty in your value. *(6 marks)*

6.3 Path difference and phase difference

Specification reference: 4.1b(i), 4.1d(iii)

Designers and architects have to take superposition into account when designing concert halls. Some rooms have acoustic *dead spots* where the sound from the stage is of lower volume than expected. This happens when sound waves reflected from the walls and ceiling superpose and the amplitude of the wave is reduced. This effect can also happen in classrooms. The voice of a teacher can be just the right frequency to produce a reduction in amplitude where the sound waves superpose in areas of the room. This has been linked to poor performance of students!

This superposition effect is shown very clearly when two loudspeakers emit a note of the same frequency and amplitude. Walking in front of the loudspeakers you will notice positions where the sound is loud and others where it is quieter. If you position yourself at a region of quiet where you hear little sound and ask for one of the speakers to be turned off, you will find that you hear a louder sound. Switching the speaker back on lowers the volume you hear once more.

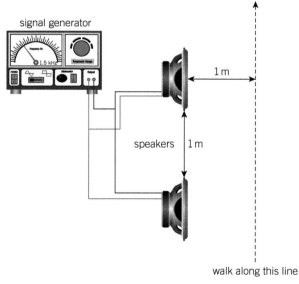

▲ **Figure 1** *Hearing superposition*

Interference of waves

The term **interference** is often used to describe the effect of the superposition of waves. That is, the superposition of waves produces an interference pattern, such as the alternation of loud and quiet positions in the space in front of the speakers in Figure 1. In positions where waves from the two speakers meet in phase a louder sound is heard. Where they meet in antiphase the sound is quieter. If the waves from two speakers are of the same amplitude where they meet in antiphase they will cancel completely, producing silence.

Path difference

You can see from the diagram in Figure 2 that the path length, the distance from speaker 1 to the microphone, is 6λ, whereas the path length between speaker 2 and the microphone, is 8λ. The **path difference**, the difference in path lengths, is $8\lambda - 6\lambda = 2\lambda$. These waves will meet in phase, increasing the amplitude of the superposed wave-form (this is a superposition maximum). If the path difference is, say, $1\frac{1}{2}\lambda$, the waves would meet in antiphase, and would completely cancel out if the amplitude of each wave was the same (this is a superposition minimum). Waves meeting with zero path difference will be in phase. You hear a louder sound when you stand at equal distances from the speakers in the experiment we have been describing.

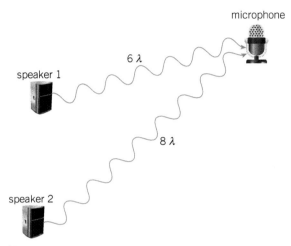

microphone

6λ

speaker 1

8λ

speaker 2

▲ **Figure 2** *An arrangement for waves meeting in phase*

We can state the conditions for waves to meet in phase and in antiphase as

- waves meet in phase when the path difference = $n\lambda$, where n is an integer
- waves meet in antiphase when the path difference = $(n + \frac{1}{2})\lambda$.

 Worked example: Superposition maxima and minima

The two loudspeakers in Figure 2 emit a note of wavelength of $77\,cm$ (corresponding to a frequency of $440\,Hz$). Suggest three values of path difference which will produce superposition maxima and three values of path difference where superposition minima will occur.

Step 1: Identify the correct equations to use.

Superposition maxima occur at a path difference of $n\lambda$, where n is an integer.

Superposition minima occur at a path difference of $(n + \frac{1}{2})\lambda$.

Step 2: Substitute $n = 1$, 2, and 3 into the two equations.

Superposition maxima will occur at $n\lambda$.

$1 \times 77\,\text{cm} = 77\,\text{cm}$, $2 \times 77\,\text{cm} = 154\,\text{cm}$, and $3 \times 77\,\text{cm} = 231\,\text{cm}$

Superposition minima will occur at $(n + \frac{1}{2})\lambda$.

$1.5 \times 77\,\text{cm} = 115.5\,\text{cm}$, $2.5 \times 77\,\text{cm} = 192.5\,\text{cm}$,
$3.5 \times 77\,\text{cm} = 269.5\,\text{cm}$

Coherence

A *stable* superposition pattern is one in which the position of the maxima and minima don't change over time. For example, in the loudspeaker experiment, if you find a minimum position and stand there for a period of time the volume of sound will not change, unless someone moves the speakers or changes the frequency of the note.

Stable superposition patterns can only occur when there is a constant phase difference between waves from the two sources. Waves with a constant phase difference are **coherent**.

The first two sets of waves in Figure 3 are coherent. They are not in phase but they do have a constant phase difference. The second two sets of waves are incoherent.

coherent waves with constant phase difference

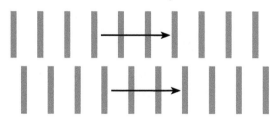

incoherent wave bursts with changing phase difference

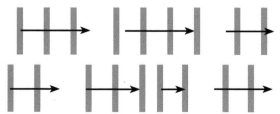

▲ **Figure 3** *Coherence*

Superposition and interference

You will meet the terms *constructive interference* and *destructive interference*. These are both names for the effects of superposition. Constructive interference occurs when waves meet in phase leading to a larger superposition amplitude (a maximum). Destructive interference occurs when waves meet in antiphase leading to a lower superposition amplitude (a minimum).

Active noise-reducing headphones

Noise-reducing (often sold as *noise-cancelling*) headphones allow the wearer to listen to his or her choice of music without needing to turn the volume up to overcome low frequency background noise.

Some of the background noise is blocked simply by the material of the headphones reducing the energy of sound waves passing into the ear. Active noise-reducing headphones use superposition to further reduce the effects of background noise.

A microphone in the ear piece detects the background noise that is not stopped by the material of the headphone ear piece.

Electronic circuitry *flips* the wave-form of the noise so that it is in antiphase (π radians phase difference) with the background noise.

The flipped noise is fed into the headphone speakers. The noise and the flipped noise cancel out.

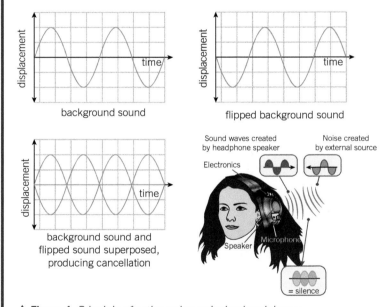

▲ **Figure 4** *Principle of active noise-reducing headphones*

1 Why is it so important that the electronic circuitry is able to flip the background sound extremely quickly?
2 Why do you think these headphones are better at reducing low frequency sounds such as engine noise rather than speech?

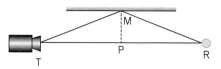

▲ Figure 5

Practical 4.1d(iii): Interference with microwaves

This practical is an example of using superposition to measure the wavelength of waves emitted from a source, in this case a microwave source.

The microwave transmitter T, receiver R, and mirror M are positioned as shown in Figure 5. Waves from the transmitter reach the receiver along the path TR and the path TMR.

As you move the mirror towards and away from the line TR you will detect a series of maxima and minima. Once you've found a position when the receiver shows a maximum value you can measure the distances TR and TMR. You can then find the path difference TMR − TR in terms of λ. As this path difference is for a maximum we can write TMR − TR = $n\lambda$.

When the mirror is moved further away from the line TR the signal at the receiver will decrease to a minimum and then rise again to a maximum as the mirror is moved further. The path difference for this second maximum is found ($n = 2$).

Study tip

Make sure that you know one method of measuring the wavelength of microwaves, visible light, and sound.

Worked example: Interference with microwaves

An experiment is conducted in which a microwave transmitter T, receiver R, and mirror M are positioned as shown in Figure 5. Waves from the transmitter reach the receiver along the path TR and the path TMR. Measurements can be taken to a precision of 0.005 m.

distance TR = 0.900 m   distance MP = 0.270 m

distance TP = 0.450 m

a Calculate the length of the path TMR.

We use Pythagoras' theorem here to calculate the distance between the transmitter and the mirror. We then double this to find TMR.

$$TMR = 2 \times TM = 2 \times \sqrt{TP^2 + MP^2}$$
$$= 2 \times \sqrt{(0.450\,m)^2 + (0.27\,m)^2} = 1.050\,m$$

b Calculate the path difference for waves arriving at the receiver from the two paths. The receiver detects a maximum in this position.

Path difference = TMR − TR = 1.050 − 0.900 = 0.15 m

c The mirror is slowly moved, reducing the distance MP to 0.240 m. The amplitude of the microwaves detected at the receiver falls to a minimum and rises back to a maximum when MP is 0.240 m. Calculate the new length of the path TMR.

We use Pythagoras' theorem once again to find the length of the path TMR.

New path TMR = $2 \times \sqrt{(0.45\,m)^2 + (0.240\,m)^2} = 1.02\,m$

d Calculate the path difference when MP = 0.240 m

Path difference = TMR − TR = 1.020 − 0.900 = 0.12 m

e The amplitude of the microwaves detected at the receiver falls to a minimum and rises back to a maximum when MP changes from 0.270 m to 0.240 m. Use this information to calculate the wavelength of the microwaves.

When distance MP is reduced from 0.270 m to 0.240 m the received signal goes through one cycle of maximum–minimum–maximum. This shows that the path difference TMR − TR has been reduced by one wavelength.

Therefore, the wavelength of microwaves = 0.15 m − 0.12 m
= 0.03 m.

Summary questions

1 Explain the meaning of the terms *path difference* and *phase difference*. *(2 marks)*

2 A laser is a coherent source of light. Explain what *coherent* means in this context. *(1 mark)*

3 Look at Figure 6. When the microphone is in position A an interference maximum is detected.
   a State the path difference necessary for an interference maximum. *(1 mark)*
   b Calculate the path difference between waves from speaker 2 to the microphone and speaker 1 to the microphone. *(1 mark)*
   c The microphone is slowly moved to position B. As the microphone is moved the signal falls to a minimum and then rises to a maximum at B. The path difference between the speakers and the microphone at position B is 0.50 m.
   Calculate the wavelength of the sound from the speakers. Describe how you reached your answer. *(2 marks)*

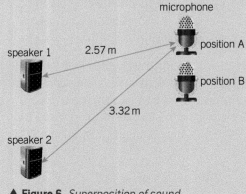

▲ **Figure 6** *Superposition of sound*

6.4 Interference and diffraction of light

Specification reference: 4.1a(ii), 4.1a(iv), 4.1a(v), 4.1b(i), 4.1c(iii), 4.1d(iii), 4.1d(iv)

In the 17th century the poet John Milton described the flames of Hell in these words:

> Yet from those flames
>
> No light; but rather darkness visible

You may think that it is impossible for light to appear dark, but just as sound waves can cancel if they meet in antiphase, light meeting light in antiphase can produce a dark band where the two waves meet. This is just another example of superposition.

Diffraction

When waves pass through a gap of roughly the same width as their wavelength the waves spread out. This is an example of **diffraction**. This can be seen in the aerial photograph of waves entering a harbour, shown in Figure 1.

The amount the waves spread out depends on the width of the gap compared to the wavelength of the waves passing through. For a given wavelength, the smaller the gap the greater the spreading caused by diffraction. Diffraction does not alter the wavelength, speed, or frequency of the waves.

You have experienced sound diffraction without realising it! When someone is talking to you in the middle of an open space the person talking does not need to be facing you for you to hear what is being said. This is because the sound is diffracted from the person's mouth and spreads out. Think how wide your mouth is when speaking – a few centimetres. The wavelength of human speech is of the order of one metre so it is no surprise that the sound diffracts around the speaker as the wavelength is so much bigger than the gap the waves are passing through.

It is more difficult to observe diffraction in light. Light does not seem to bend round corners or spread out through slits. The reason that it is not observed in our everyday lives is that the wavelength is of the order of 10^{-7} m, so a very narrow gap is needed for an appreciable amount of spreading to occur.

Young's double slit experiment

The theory that light travels as waves gained a lot of support from an experiment published by Thomas Young (the same man as the Young modulus is named after) in 1807. He passed light

▲ **Figure 1** Waves diffracting as they enter a bay

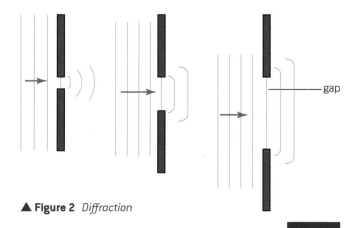

▲ **Figure 2** Diffraction

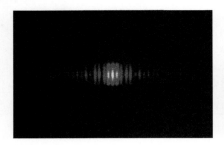

▲ **Figure 4** *Fringe pattern obtained from two slits*

through two pinholes very close together and observed a pattern of dark and bright fringes.

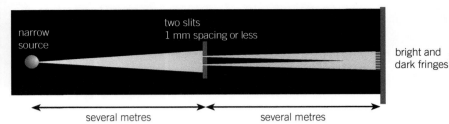

▲ **Figure 3** *Young's double slit experiment*

Modern versions of this experiment use slits rather than pinholes. The observed pattern is due to the superposition of light from the two slits – this is Young's double slit experiment. Where light from both slits meets at the screen in phase, a bright fringe is observed. Light meeting in antiphase produces a dark patch. Figure 5 shows wave-fronts of light from two slits which meet in phase and out of phase at a distant screen.

two simple cases

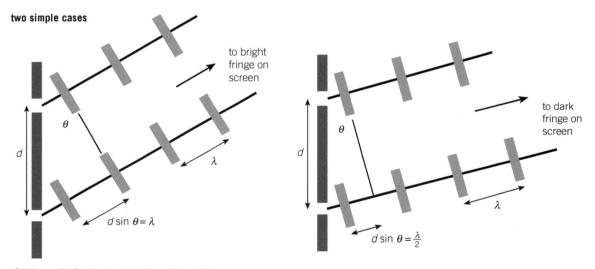

▲ **Figure 5** *Producing bright and dark fringes*

Synoptic link

You have seen from Topic 6.3, Path difference and phase difference, that a bright fringe means that there is a path difference of $n\lambda$ between the light from the two slits.

You may wonder why diagrams of the double slit experiment show parallel rays of light meeting on a distant screen to produce fringe patterns. Of course this is only an approximation as parallel lines never meet. It is entirely reasonable to make this approximation as the distance between the slits and the screen is so large compared to the separation of the slits.

Figure 6 shows a ray representation of the double slit experiment.

The situation in Figure 6 produces a bright fringe. The path difference must therefore be a whole number of wavelengths. In this case we have indicated that the path difference is one wavelength.

As $\sin\theta = \dfrac{\text{opposite}}{\text{hypotenuse}}$

$$\sin\theta = \frac{\lambda}{d}$$

Therefore, for waves with a path difference of one wavelength, $\lambda = d\sin\theta$, where d is the slit separation and θ is the angle made at the slits between one bright fringe and the next.

Figure 7 shows the situation for a path difference of two wavelengths.

For a path difference of $n\lambda$ we can write $n\lambda = d\sin\theta_n$.

Order of maxima

The order of a maximum shows the number of wavelengths path difference between light from two adjacent slits and the screen (or detector). Therefore, the zeroth order ($n = 0$) is the maximum when the path difference between two adjacent slits and the screen is zero. The first order ($n = 1$) is the maximum for a path difference between the slits and the screen of one wavelength. The second order ($n = 2$) is the maximum produced by a path difference of two wavelengths and so on.

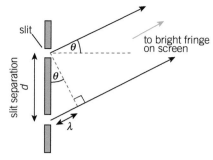

▲ **Figure 6** *Ray representation*

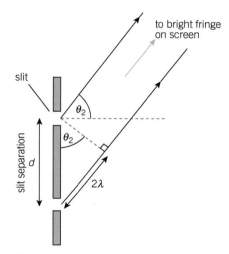

▲ **Figure 7** *Ray representation, two wavelength path difference*

 Worked example: Calculating angles from orders of maxima

Green light of wavelength 500 nm is incident on two narrow slits. The slit separation is 0.5×10^{-3} m.

a Calculate the angle of the first order ($n = 1$) maximum.

Step 1: Select the appropriate equation.

$n\lambda = d\sin\theta_n$ which rearranges to give $\sin\theta_n = \dfrac{n\lambda}{d}$

Step 2: Substitute values and evaluate, remembering that $1\,\text{nm} = 1 \times 10^{-9}\,\text{m}$.

$$\sin\theta_1 = \frac{1 \times 500 \times 10^{-9}\,\text{m}}{0.5 \times 10^{-3}\,\text{m}} = 1 \times 10^{-3}$$

Step 3: Find the angle from the value of the sin.

$$\theta_1 = \sin^{-1} 1 \times 10^{-3} = 0.06°$$

b Calculate the angle of the fourth order ($n = 4$) maximum.

Step 1: Substitute values into $\sin\theta_n = \dfrac{n\lambda}{d}$, where $n = 4$.

$$\sin\theta_4 = \frac{4 \times 500 \times 10^{-9}\,\text{m}}{0.5 \times 10^{-3}\,\text{m}} = 4 \times 10^{-3}$$

$$\theta_4 = \sin^{-1} 4 \times 10^{-3} = 0.23°$$

These angles are very small. This is why the screen must be a good distance away from the slits to allow the bright and dark fringes to be observed and measured.

We can use the equation $n\lambda = d\sin\theta_n$ to determine the wavelength of the light passing through the pair of slits. Figure 8 shows the geometry of rays meeting at a distant screen.

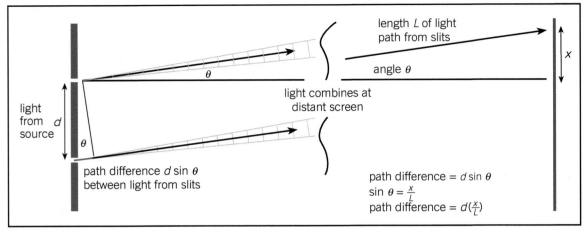

length L of light path from slits

angle θ

light combines at distant screen

light from source d

path difference $d\sin\theta$ between light from slits

path difference $= d\sin\theta$

$\sin\theta = \dfrac{x}{L}$

path difference $= d(\dfrac{x}{L})$

Approximations – angle θ very small; paths effectively parallel; distance L equal to slit–screen distance. Error less than 1 in 1000

▲ **Figure 8** *Geometry*

Synoptic link

Small angle approximations are used in the assumption for Young's double slit experiment. This is covered in the Maths Appendix.

The diagram shows that $\tan\theta = \dfrac{x}{L}$.

For small θ, $\tan\theta \approx \sin\theta$.

By combining $\tan\theta = \dfrac{x}{L}$ and $\sin\theta = \dfrac{\lambda}{d}$ you find $\dfrac{x}{L} = \dfrac{\lambda}{d}$.

Therefore $\lambda = \dfrac{xd}{L}$.

 Worked example: Young's double slit experiment

A double slit experiment is set up using a laser. The screen is 3.00 m away from the slits, which have a separation of 0.4 mm. The spacing between bright fringes on the screen is 5.0 mm. Calculate the wavelength of light used.

Step 1: Select and rearrange the appropriate equation.

$$\lambda = \dfrac{xd}{L}$$

Step 2: Substitute values and evaluate.

$$\lambda = \dfrac{5.0\times10^{-3}\,\text{m}\times4\times10^{-4}\,\text{m}}{3.00} = 6.7\times10^{-7}\,\text{m (2 s.f.)}$$

The diffraction grating

A diffraction grating is a multiple slit version of the two slit system you have been thinking about. The equation $n\lambda = d\sin\theta_n$ also holds for gratings.

Using a grating of many slits increases the brightness of the image on the screen because more light gets through. Gratings also give a sharper fringe pattern. Diffraction gratings can also spread white light out into its component colours because each wavelength of light will produce a maximum at a different angle.

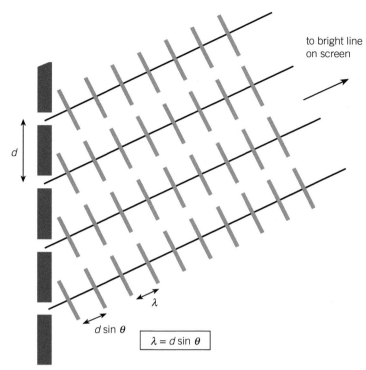

$$\lambda = d \sin \theta$$

▲ **Figure 9** *Waves from many sources all in phase*

Diffraction gratings are often described by their number of lines per mm. To calculate the separation of each line (the equivalent of the slit separation), convert the number of lines per mm into lines per metre and take the reciprocal.

$$\text{line separation (m)} = \frac{1}{\text{number of lines per metre}}$$

 Worked example: Diffraction gratings

A diffraction grating has 300 lines per mm. Calculate the angle to the first order maximum for:

a red light of wavelength 670 nm;

Step 1: Select and rearrange the appropriate equation.

$$\sin \theta_1 = \frac{\lambda}{d}$$

Step 2: Rewrite the equation above in terms of number of lines per metre, not d.

Since line separation $d = \dfrac{1}{\text{number of lines per metre}}$

$\sin \theta = \lambda \times \text{lines per m}$

Step 3: Substitute values and evaluate.

$$\sin \theta = 6.7 \times 10^{-7} \, \text{m} \times 300\,000 \, \text{m}^{-1}$$

$$\sin \theta = 0.201 \text{ therefore } \theta = 11.6°$$

b blue light of wavelength 460 nm.

Work through Steps 1 and 2 as above. Step 3 becomes

$$\sin \theta = 4.6 \times 10^{-7} \, \text{m} \times 300\,000 \, \text{m}^{-1} = 0.138 \text{ therefore } \theta = 7.9°$$

Practical 4.1d(iii) and (iv): Determining the wavelength of light using superposition of light (Young's double slits or a diffraction grating)

There are many ways of determining the wavelength of light using interference effects. The key measurements are:

● the slit separation

● the distance between the slits and the screen

● the fringe separation.

The basic experimental set-up using a laser is shown in Figure 10. The screen should be a few metres away from the double slit.

The greatest source of uncertainty in the double slit experiment is the measurement of the slit separation. If you are using a diffraction grating you will be given the number of lines per mm which will allow you to calculate a confident and precise value of the slit separation.

Performing the experiment with simple apparatus is a worthwhile challenge as you can still get a reasonable estimate for the wavelength of the light you are using and you will be using a method very similar to Thomas Young at the beginning of the 19th century.

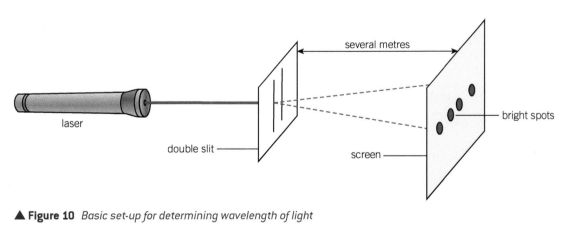

▲ **Figure 10** Basic set-up for determining wavelength of light

Diffraction through a single slit

We can use phasor thinking to explain diffraction through a narrow aperture, as shown in Figure 11.

Single slit

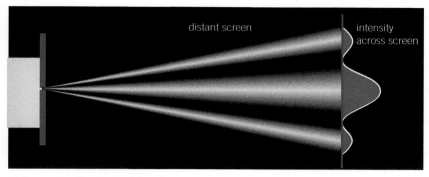

▲ **Figure 11** *Single slit diffraction*

As a wave passes through a slit we imagine that each point in the wave-front is a source of new wavelets – just like Huygens' idea. But we are going to imagine these as rotating phasor arrows.

Where the phasor arrows are all in the same direction we get a large resultant, creating a maximum.

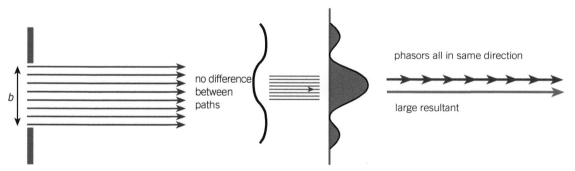

▲ **Figure 12** *Simplified case – distant screen with paths nearly parallel, no path difference*

At an angle θ, the path difference across the whole slit is λ. Each phasor will reach the screen a little out of phase with the phasor from the adjacent point. All the phasors will add together to give a zero resultant. This is a diffraction minimum – it occurs when $\sin\theta = \frac{\lambda}{b}$.

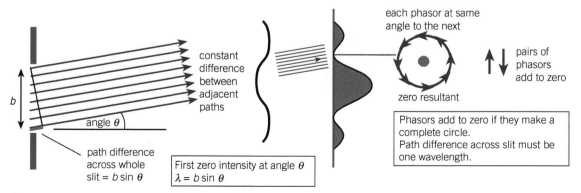

▲ **Figure 13** *Simplified case – distant screen with paths nearly parallel, constant path difference*

The angle (θ in Figure 13) to the first diffraction minimum gives a measure of how much the waves spread out through the gap.

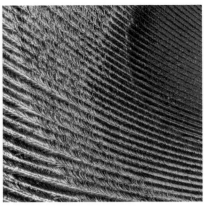

▲ **Figure 14** *Close up of a tail feather of a peacock*

The tail of the peacock

You began this chapter with a picture of a peacock. So, how are the colour changes produced as it moves past? Light strikes the feathers of the bird and is reflected from tiny ridges on the feathers. These ridges act like a diffraction grating. Instead of light passing through a uniform group of slits it is reflected from a uniform group of ridges. There is a path difference between waves from each ridge so at any angle one colour will predominate because light of that wavelength will be meeting at the eye in phase. There are many examples of iridescence in nature and although the details of the effect vary they all depend on the principle of superposition.

Summary questions

1   a   Describe what is meant by diffraction. State whether there is any change in the speed, wavelength, or frequency of waves when they diffract.   *(3 marks)*

    b   A diffraction grating has 80 lines per mm. Light that is perpendicularly incident on the grating produces a first order maximum at an angle of 2.2°. Calculate the wavelength of the light.   *(3 marks)*

2   Light of wavelength 5.4×10^{-7} m passes through a diffraction grating of 300 lines per mm.

    a   Calculate the angle to the second order maximum.   *(3 marks)*

    b   Calculate how many orders of maxima are possible for this set-up.   *(2 marks)*

3   Light of wavelength 600 nm passes through a single slit of width 1×10^{-3} m.

    a   Find the angle θ to the first order minimum where $\sin\theta = \dfrac{\lambda}{b}$, where b is the slit width.   *(2 marks)*

    b   Explain why the result from **a** shows that diffraction would not be observed by the unaided eye in such an arrangement.   *(2 marks)*

4   A class performs a Young's double slit experiment. The gap between the slits is 0.5 ± 0.1 mm. The distance between the slits and the screen is 3.40 ± 0.01 m. The fringe spacing is measured to be 3.5 ± 0.5 mm.

    a   Use these measurements and uncertainties to calculate the biggest and smallest values for the wavelength of light.   *(4 marks)*

    b   Explain which measurement contributes most to the overall uncertainty.   *(2 marks)*

Physics in perspective

Shedding light on light

The debate about the nature of light

Many thinkers have puzzled over the nature of light. In the 17th and 18th centuries the so-called *corpuscularists* favoured a particle-like description, whereas the *undulationists* supported a wave-like description. Christiaan Huygens was an undulationist — his famous illustration of the candle (Figure 1) shows waves emanating from every part of the flame. However, the arguments between the undulationists and the corpuscularists were not always clear-cut.

Isaac Newton (1642–1727) wrote in his book *Optiks*, "Are not the rays of light very small bodies emitted from shining substances?". This suggests that Newton was firmly in the corpuscularist camp, but he went on to suggest that rays have vibrations associated with them of 'several bignesses' or, in modern terms, different wavelengths.

Newton knew of many phenomena that are now explained by waves, and invented a way of measuring them accurately – known as Newton's rings. Newton's rings can be observed when a lens that is convex on one side and flat on the other is placed on a flat glass slide and illuminated from above. A pattern of concentric rings is observed. If the incident light is monochromatic the pattern shows bright and dark rings, with a dark spot in the centre where the surfaces touch. Similar effects are seen when two microscope slides are placed together.

▲ **Figure 1** *Huygens' candle*

Newton did not decisively support one model or the other, but in the years following his death it was generally thought that the great mathematician and experimenter had strongly supported the particle model. This may have made it more difficult for alternative ideas to gain acceptance.

Thomas Young explains interference

Before 1833 you could not be a scientist, for the term had not been invented. Before this date there were only a few individuals who would now be considered 'professional' scientists. So when Thomas Young stood in front of the Royal Society in London in a series of lectures during the early 1800s, the fact that a medical doctor was lecturing about the nature of light was rather less surprising than it seems now. Nevertheless, Thomas Young was an extraordinary individual of many talents – a true polymath.

▲ **Figure 2** *Amongst other interests, Thomas Young (1773–1829) was a linguist, an archaeologist, a poet, and a doctor*

Young used a device of his own invention, a ripple tank, to show how interference effects can be produced in water waves. It is thanks to Young that ripple tanks are used in physics lessons to this day, and that we talk about 'interference' of waves — a phrase he used to describe the superposition effects observed in water.

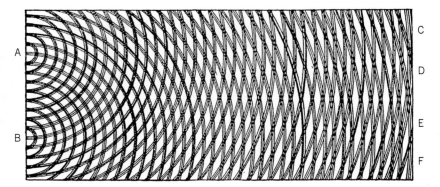

▲ **Figure 3** *Young's illustration of the double slit pattern*

Young compared interference effects in water with those in light, but interference is not easily observed in light. It was not enough for Young to suggest that light could interfere like water waves — he had to give evidence for it. In 1803, he provided the evidence in his now-celebrated double-slit experiment.

At the time, however, some of Young's audiences were less-than-impressed, and it took the work of Auguste Fresnel (1788–1829), a French engineer, to really change scientific opinion.

Interference explains Newton's rings

Figure 4 shows the basis of Young's explanation of Newton's rings. Some of the incident light is reflected from the curved surface of the lens (path 1). Light that passes through the lens is reflected from the surface of the slide (path 2). The superposition of the two rays produces bright fringes when the waves are in phase and dark lines when the waves are in antiphase.

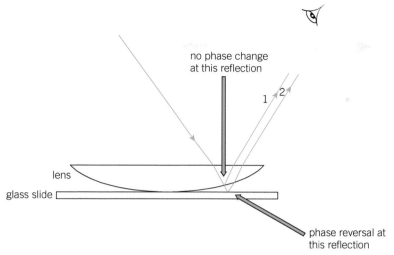

no phase change
at this reflection

1 2

lens

glass slide

phase reversal at
this reflection

▲ **Figure 4** *Light reflecting from two surfaces*

When reflection occurs at a boundary with a material of higher refractive index, there is a phase change of π radians — when light travelling through air reflects at a glass surface, there will be a phase change.

Reflection at a boundary with a material of lower refractive index produces no phase change – when light travelling through glass reflects from the air boundary, there is no phase change.

A dark spot is observed where the two surfaces are very close. There is no path difference, but the phase of light reflecting from the slide is reversed, and so it is in antiphase with the light reflecting from the curved surface of the lens.

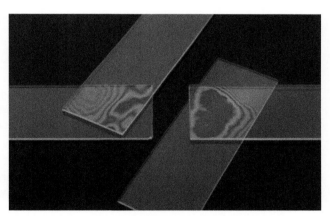

▲ **Figure 5** *Newton's rings effect from two microscope slides*

Using interference in Solar Cells

Understanding interference helps design non-reflective coatings. These are frequently used on lenses and solar cells.

The silicon used in solar cells reflects about 30% of the light incident upon it. This means that 30% of the energy incident on the cell cannot be used to generate electricity. A layer of transparent silicon monoxide acts as non-reflective coating.

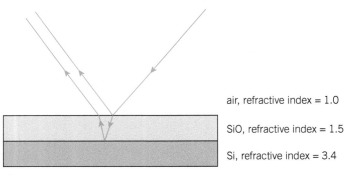

air, refractive index = 1.0

SiO, refractive index = 1.5

Si, refractive index = 3.4

▲ **Figure 6** *Anti-reflection coating*

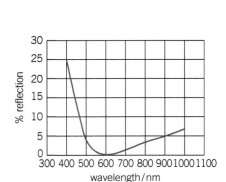

▲ **Figure 7**

There is a phase change of π radians for light reflecting at the air–silicon monoxide boundary *and* at the silicon monoxide–silicon boundary. If the wavelength of light in the silicon monoxide layer is equal to four times the thickness of the layer, the two reflected rays will cancel. The reflective losses can be reduced to about 10% by this method, therefore increasing the efficiency of the solar cell.

Summary questions

1 This question is about the development of the wave model of light.
   a Why is diffraction of visible light not easily observed?
   b Young used a Newton's ring set-up to determine the wavelength of red light as 650 nm. What is the minimum air gap required between the lens and the slide to produce a bright fringe?
   Remember to consider the phase reversal on reflection from the slide.
   c Young and Huygens both considered light to be a longitudinal wave. Explain why the phenomenon of the polarisation of light suggests that light is a **transverse** wave.

2 This question is about anti-reflective coating on a solar cell.
   a In Figure 6, explain why both the light reflecting from the silicon monoxide and the light reflecting from the silicon experience a phase change of π radians.
   b Using ideas about path difference and phase difference, explain why reflected light from the two surfaces cancels out when the thickness of the silicon monoxide layer is equal to one quarter of the wavelength of the light in the material.
   c Silicon monoxide has a refractive index of 1.5. Yellow light of wavelength 580 nm in air is incident on the silicon monoxide layer.
      i Calculate the wavelength of the yellow light in the silicon monoxide.
      ii Calculate the thickness of silicon monoxide required to produce cancellation of the reflected light from the two surfaces.
   d The graph in Figure 7 shows how the percentage reflection varies with wavelength with a coating of thickness of λ/4 for yellow light. Explain why the coating minimises reflection for yellow light but reflection of violet light (λ ~ 400 nm) is barely changed at all.

Practice questions

1 Light enters glass from air. Which of the following statements is/are correct?

Statement 1: The wavelength is shorter in the glass than in air.

Statement 2: The velocity is slower in the glass than in air.

Statement 3: The frequency is lower in the glass than in air.

A 1, 2, and 3 are correct

B Only 1 and 2 are correct

C Only 2 and 3 are correct

D Only 1 is correct *(1 mark)*

2 Figure 1 shows light entering glass from air. The angles of incidence and refraction are given. Use information from the diagram to calculate the velocity of light in the glass.

Velocity of light in air = $3.0 \times 10^8 \, \text{m s}^{-1}$.
(2 marks)

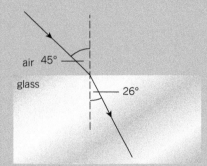

▲ Figure 1

3 A coherent beam of red light is perpendicularly incident on a diffraction grating.

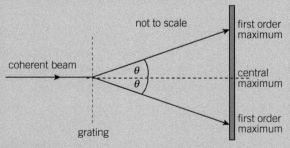

▲ Figure 2

a Explain what is meant by a *coherent* beam of light. *(1 mark)*

b The grating has 80 lines per mm. The angle to the first order maximum is 3°. Calculate the wavelength of the light. *(2 marks)*

4 Make a sketch copy of the graph of the oscillation shown in Figure 3. Add a second waveform of the same amplitude with a phase difference of $\frac{\pi}{2}$ radians. *(2 marks)*

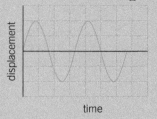

▲ Figure 3

5 Figure 4 represents water waves passing through a gap.

a Copy Figure 4 and show the next three wave-fronts in the water beyond the gap. *(2 marks)*

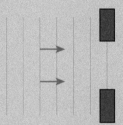

▲ Figure 4

b The velocity of the waves is proportional to $\sqrt{d}$ where *d* is the depth of the water. The depth of water changes from *d* to 2*d*. Calculate how this change of depth affects the wavelength of the water waves, assuming that the frequency remains constant. *(2 marks)*

c State the effect the change in wavelength has on the pattern of the waves passing through the gap. *(1 mark)*

6   A DVD can be used as a reflection grating. Light of wavelength 670 nm is incident on the DVD. It is reflected from a series of tracks as shown in Figure 5. A first order maximum is observed at 31°.

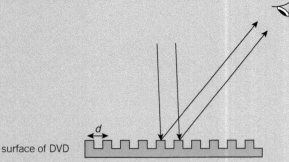

surface of DVD

▲ Figure 5

a   Use the equation $\sin \theta = n\lambda/d$ to calculate the distance d between tracks.   (*2 marks*)

b   Explain why only one maximum is observed for light of this wavelength reflecting from the DVD.   (*2 marks*)

7   This question is about standing waves in a tube closed at one end, such as a clarinet. When the player blows into the instrument a reed vibrates at the closed end. A standing wave is formed in the tube. The tube is 0.6 m long.

a   (i)   Copy Figure 6 and indicate the position of the node(s) and antinode(s) for the longest wavelength standing wave.   (*1 mark*)

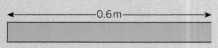

▲ Figure 6

  (ii)   State the wavelength of the longest-wavelength standing wave.   (*1 mark*)

  (iii)   Explain how standing waves form in a tube.   (*3 marks*)

b   (i)   Draw a diagram indicating the nodes and antinodes in the tube when it produces a note of three times the frequency of the lowest possible note.   (*1 mark*)

  (ii)   The speed of sound at room temperature is about 340 m s⁻¹. Calculate the frequency of the sound produced in the tube from the standing wave in **(b)(i)**.   (*2 marks*)

c   The speed of sound increases with temperature. State and explain the effect this has on the frequency of the note produced by the clarinet.   (*3 marks*)

8   This question is about an experiment to determine the wavelength of blue light using a pair of slits. A student makes the following measurements:

Slit separation = $5 \times 10^{-4} \pm 1 \times 10^{-4}$ m
Distance from slits to screen = 2.5 ± 0.01 m

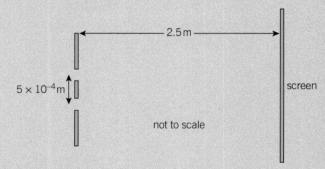

▲ Figure 7

The student measures 10 bright fringes in a distance of 24 ± 1 mm on the screen.

a   (i)   The student states that she only needs to consider uncertainty in the slit separation when calculating the wavelength of light. Explain her reasoning.   (*2 marks*)

  (ii)   Calculate the wavelength of light using the student's results. Include a calculated estimate of uncertainty in your final answer.   (*4 marks*)

The student suggests reducing the uncertainty in the experiment by doubling the slit spacing and doubling the distance between the slits and the screen.

b **(i)** Use your answer to **(a)(ii)** to calculate an estimate of the expected fringe spacing for the new slit separation and slit-screen distance. *(2 marks)*

(ii) Discuss whether these changes have reduced the uncertainty in the value for wavelength and suggest any disadvantages of the new set up. *(3 marks)*

9 This question is about the superposition of sound. Two speakers, A and B, produce a steady, coherent note. As the microphone moves along the line XY it detects a series of amplitude maxima and minima.

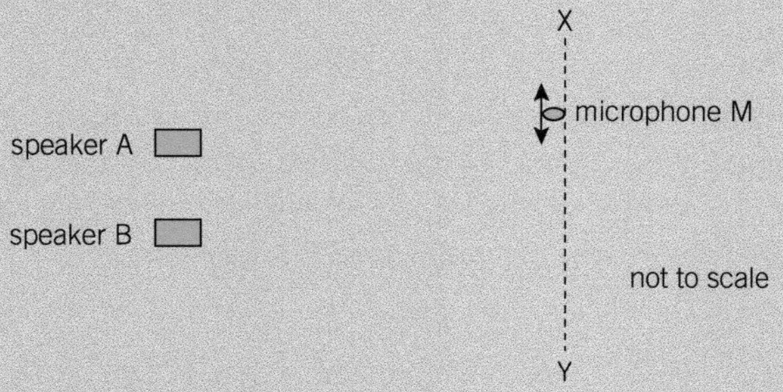

X

↕ microphone M

speaker A ▢

speaker B ▢

not to scale

Y

▲ **Figure 8**

a Use ideas about path difference and phase difference to explain why the microphone detects a series of maxima and minima. *(4 marks)*

b The microphone detects a maximum when path difference BM − AM = 2.4 m. It detects the next maximum when the path difference is 3.2 m. State the wavelength of the sound from the speakers. *(1 mark)*

c The speed of sound v in air varies with temperature as $v = k\sqrt{(T+273)}$ where T is the air temperature in °C. Calculate the percentage difference in the velocity of sound when the air temperature rises from 10°C to 20°C and explain the effect this change will have on the pattern of maxima and minima from the speakers. The frequency of the sound is not changed. *(4 marks)*

10 This question is about standing waves on guitar strings.

Figure 9 shows a guitar whose strings are 0.65 m long.

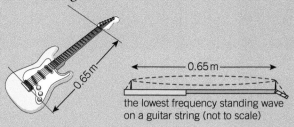

the lowest frequency standing wave on a guitar string (not to scale)

▲ **Figure 9**

a Explain why the wavelength of the standing wave shown in Figure 9 is 1.3 m. *(1 mark)*

b The lowest frequency standing wave on the thickest guitar string is at 82 Hz.

Show that the speed of the wave travelling along the string is about 100 m s⁻¹. *(2 marks)*

c **(i)** The speed v of waves along a string is given by the equation

$$v = \sqrt{\frac{T}{\mu}}$$

Where T is the tension in the string, and μ is the mass of a metre length of the string.

Use this equation to calculate the tension T in the thickest guitar string where $\mu = 8.4 \times 10^{-3}\,\mathrm{kg\,m^{-1}}$. *(2 marks)*

(ii) All strings on the guitar have the same tension and length. Use the equation above to explain why the fundamental frequency of the thinnest string is higher than the fundamental frequency of the thickest string. *(1 mark)*

d Explain clearly how waves travelling along a string can produce standing waves on the string *(3 marks)*

OCR Physics B Paper G492 Jan 2012

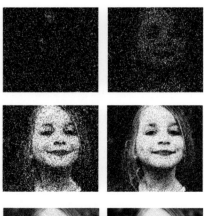

▲ **Figure 1** Six images representing exposures with increasing numbers of photons. With very little light (few photons), the picture is a random pattern of exposed dots, but it becomes more detailed and recognisable as more photons arrive.

You have seen that by modelling light as a wave you can explain refraction, diffraction, and interference, but there are some phenomena that cannot be explained using this model.

What do you think happens when you take a photograph in progressively dimmer light? You may think that the photograph will simply appear fainter, but it does not. In fact, the photograph breaks up into randomly arranged exposed patches on the film or digital sensor, suggesting that light is arriving randomly in small packets. Turn up the brightness and the lumps of light arrive at a greater rate, producing the smooth-looking picture you expect. These packets, or quanta of light, are called **photons**. This effect is shown in Figure 1, which represents the same picture taken with successively more photons.

Photons and chance

The images are a good illustration of how photons hit a detector. Where the picture is bright, there is a good chance of photons arriving at that location. Where it is dim, the chance is low. These are probabilities, not certainties, so it is possible (but not likely) for a photon to appear where the probability is low, or not appear where the probability is high.

A smooth picture is built up through many, many events happening one by one at random with a certain probability. It is a bit like going out in the rain – you cannot tell when or where the next raindrop will fall, but if you stay in the rain you will get wet. Strong sources of electromagnetic radiation are like heavy rain, weak sources are like light rain – the random nature of the arrival of photons (or raindrops) is much more obvious when the rate of arrival is low. If you meet someone whose clothes are spattered with water, you could conclude that they were in a rain shower – meet someone who is soaked from head to foot and you cannot tell if they have been in a very heavy shower of rain (particles) or been hit by a wave!

Quanta

The idea that electromagnetic radiation can be emitted in energy quanta, discrete packets of energy, was suggested by the German physicist Max Planck in 1900. Albert Einstein extended the concept in 1905 when he suggested that electromagnetic radiation is not only emitted in energy **quanta**, but is also absorbed as quanta (photons). The relationship is very simple. Electromagnetic radiation of frequency f is emitted and absorbed in quanta of energy E where

energy E (J) = Planck constant h (J s) × frequency f (Hz)

h, the Planck constant, is equal to 6.6×10^{-34} J s.

As speed of light $c = f\lambda$, we can write $f = \dfrac{c}{\lambda}$ and substitute this into the photon–energy equation to give $E = \dfrac{hc}{\lambda}$.

 Worked example: Photons from a light-emitting diode

An LED emits blue light of wavelength 470 nm at a power of 250 mW. Calculate the number of photons emitted by the LED per second.

Step 1: Identify the correct equation to use.

Since the wavelength is given, and we can find values of c and λ, we can use the equation $E = \dfrac{hc}{\lambda}$.

Step 2: Substitute values into equation.

$$E = \frac{hc}{\lambda} = \frac{6.6 \times 10^{-34}\,\text{J s} \times 3 \times 10^{8}\,\text{m s}^{-1}}{470 \times 10^{-9}\,\text{m}} = 4.2... \times 10^{-19}\,\text{J}$$

This is the energy of a photon of wavelength 470 nm.

Step 3: Identify the energy transferred per second.

Power = energy transfer per second

As power = 250 mW

Energy transfer per second by the LED = 250×10^{-3} J

Step 4: Calculate the number of photons emitted per second.

Number of photons emitted per second = $\dfrac{\text{total energy transferred per second}}{\text{energy of single photon}}$

$$= \frac{250 \times 10^{-3}\,\text{J}}{4.2... \times 10^{-19}\,\text{J}} = 5.9 \times 10^{17} \ (2\ \text{s.f.})$$

A single LED releases 5.9×10^{17} photons each second. This large number helps explain why light appears to be continuous (like a wave motion) rather than a random release of lumps of energy.

Measuring energy in electronvolts

The joule is not a convenient energy unit to use when discussing the energy of photons or sub-atomic particles, as the energy values considered are so small, as you can see in the example above.

The **electronvolt** (eV) is often used as an alternative unit. One electronvolt is the energy transferred when an electron moves through a potential difference of one volt.

As $W = VQ$, and the magnitude of the charge on an electron is 1.6×10^{-19} C

Energy transferred, $W = 1.6 \times 10^{-19}\,\text{C} \times 1\,\text{J C}^{-1} = 1.6 \times 10^{-19}\,\text{J}$

Therefore, $1\,\text{eV} = 1.6 \times 10^{-19}$ J.

> **Synoptic link**
>
> You have met electrical power and the relationship between charge, energy, and potential difference in Topic 3.1, Current, p.d., and electrical power.

 Worked example: Converting from J to eV

A photon has energy 2.4×10^{-19} J. What is its energy in eV?

Step 1: $1\,\text{eV} = 1.6 \times 10^{-19}$ J

Step 2: Energy in eV $= \dfrac{2.4 \times 10^{-19}\,\text{J}}{1.6 \times 10^{-19}\,\text{J eV}^{-1}} = 1.5\,\text{eV}$

The photoelectric effect and Einstein's equation

One of the unresolved problems in physics at the beginning of the 20th century concerned the photoelectric effect – the emission of electrons when light of a sufficiently high frequency strikes a metal surface.

The **intensity** of light is the amount of energy transferred per metre squared per second. In everyday language we think of the intensity as the brightness of a light source. The wave model of light suggested that there would be a delay in the emission of photoelectrons from the metal surface when low intensity light strikes the surface. It was also suggested that a more intense source of light striking the metal surface would result in the emission of photoelectrons with greater kinetic energy.

However, it was found that the kinetic energy of the ejected electrons (photoelectrons) was *not* affected by the intensity of the light striking the metal surface. A more intense light produces greater numbers of photoelectrons but the maximum energy of the electrons depends *only* on the frequency of the light. A low intensity source produces fewer photoelectrons, but there is no measurable delay in the emission. If the frequency is lower than a certain **threshold frequency** f_0, no electrons are released no matter how bright the light source.

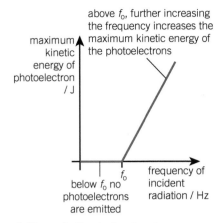

▲ **Figure 2** *Graph showing the relationship between the maximum kinetic energy of the photoelectrons and the frequency of incident light*

Einstein's theory

In 1905 Albert Einstein published ideas that changed the way we think about energy and matter. He produced five papers in that year, an incredible achievement considering that he had a full-time job and was, effectively, an amateur scientist at the time. One of the papers gave an explanation for the photoelectric effect, work that led to Einstein receiving a Nobel Prize in 1921. In his ground-breaking work he pictured light as interacting with matter as particles. This allowed him to explain the results of the photoelectric effect in terms of individual photons interacting with individual electrons on the metal surface. This is completely different from the wave model which pictures the energy of the incident light spread evenly over the metal surface.

It is interesting to note that 1905 also saw the publication of possibly the most famous equation in physics: $E = mc^2$, but it was Einstein's work on the photoelectric effect that led to his award of the Nobel Prize.

▲ **Figure 3** *Einstein in 1905*

Questions

1   How does Einstein's picture of light as photons interacting individually with electrons explain the observation that no photoelectrons are emitted below a minimum frequency of incident light?

2   Einstein's explanation of the photoelectric effect was accepted by most of the scientific community following very careful experiments by the American physicist Robert Millikan in 1916. Why do you think Einstein's explanation was not accepted by all physicists as soon as it became widely known?

It takes energy to remove an electron from the metal surface. The amount of energy required to do so is known as the **work function** of the metal, ϕ. Einstein's equation states that the maximum kinetic energy of the ejected electrons is equal to the energy of an individual photon minus the work function of the metal.

$$E_{k(\text{max})} = hf - \phi$$

When one photon strikes the surface of the metal it transfers energy hf to an *individual* electron. Some of this energy goes to releasing the electron from the surface (ϕ), any remaining energy is transferred as the kinetic energy of the electron.

> **Hint**
>
> The work function is the energy required to remove an electron from the *surface* of the metal. Some photons interact with electrons (a little) below the surface, which require more energy to be released. As we cannot be sure where the electrons emitted came from, we must describe their energies as *maximum* kinetic energy.

 Worked example: The photoelectric effect

The energy required to release an electron from the surface of zinc (the work function of zinc) is 6.9×10^{-19} J.

a Calculate the minimum frequency of light that will release electrons from the surface.

Step 1: Select the appropriate equation.

Minimum photon energy $E = 6.9 \times 10^{-19}$ J $= hf$

Step 2: Rearrange the equation and evaluate.

$$f = \frac{E}{h} = \frac{6.9 \times 10^{-19}\,\text{J}}{6.6 \times 10^{-34}\,\text{J s}} = 1.0 \times 10^{15}\,\text{Hz (2 s.f.)}$$

The wavelength of light of this frequency is 3×10^{-7} m, in the ultraviolet region of the spectrum.

b Ultraviolet light of frequency 1.7×10^{15} Hz is incident on the zinc. Calculate the maximum kinetic energy of the photoelectrons emitted from the surface.

Step 1: Select the appropriate equation.

$$E_{k(\text{max})} = hf - \phi$$

Step 2: Substitute values and evaluate.

$$
\begin{aligned}
E_{k(\text{max})} &= (6.6 \times 10^{-34}\,\text{J s} \times 1.7 \times 10^{15}\,\text{s}^{-1}) - 6.9 \times 10^{-19}\,\text{J} \\
&= 1.122 \times 10^{-18}\,\text{J} - 6.9 \times 10^{-19}\,\text{J} \\
&= 4.3 \times 10^{-19}\,\text{J (2 s.f.)}
\end{aligned}
$$

An electron with this kinetic energy travels at about $1 \times 10^{6}\,\text{m s}^{-1}$.

Using the photoelectric effect to determine the Planck constant

The gradient of the graph in Figure 4 is the Planck constant. The work function can be calculated by finding the threshold frequency (the intercept on the frequency axis) and multiplying by the Planck constant (gradient of the graph).

Worked example: The photoelectric effect

The graph in Figure 4 shows the maximum kinetic energy of photoelectrons emitted from a metal surface when illuminated with light of different frequencies.

a Use the graph to determine the Planck constant.

Step 1: The Planck constant is the gradient of the line. Select suitable pairs of x- and y-values (see graph).

Step 2: Evaluate gradient.

$$\text{Gradient} = \frac{5.4 \times 10^{-19}\,\text{J} - 0.6 \times 10^{-19}\,\text{J}}{13.2 \times 10^{14}\,\text{Hz} - 6.0 \times 10^{14}\,\text{Hz}} = \frac{4.8 \times 10^{-19}\,\text{J}}{7.2 \times 10^{14}\,\text{Hz}} = 6.7 \times 10^{-34}\,\text{J}\,\text{s (2 s.f.)}$$

Note that this value is not quite the expected value of $6.6 \times 10^{-34}\,\text{J}\,\text{s}$. This is because of the inaccuracy in reading values from the graph.

b Use the accepted value of the Planck constant to calculate the work function of the metal.

Step 1: Select the appropriate equation.

Work function = threshold frequency × the Planck constant

Step 2: Find the value for the threshold frequency from the x-intercept of the graph.

Threshold frequency = $5.0 \times 10^{14}\,\text{Hz}$

Step 3: Substitute and evaluate.

Work function = $6.6 \times 10^{-34}\,\text{J}\,\text{s} \times 5.0 \times 10^{14}\,\text{Hz} = 3.3 \times 10^{-19}\,\text{J}$

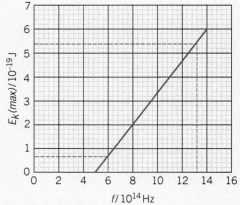

▲ **Figure 4**

Line spectra

Electrons in atoms have different energies and can gain and lose energy by changing **energy levels**. We can picture these energy levels rather like rungs of a ladder. Atoms release photons when an electron falls from a higher energy level (a higher rung) to a lower level (lower rung). The released photon has energy equal to the difference between the energy levels. Figure 5 shows the six energy differences possible for electrons moving between four energy levels.

As $E = hf$, and $c = f\lambda$, higher energy photons will have higher frequency and therefore shorter wavelength.

When a compound or element is heated, electrons in the atoms move into higher energy levels. Photons are released when the electrons fall back to lower levels. An example of this is materials glowing 'red-hot' when they are heated to a sufficiently high temperature.

Atoms of each element have their own set of energy levels. This means they will emit photons of different energies – light of different wavelengths.

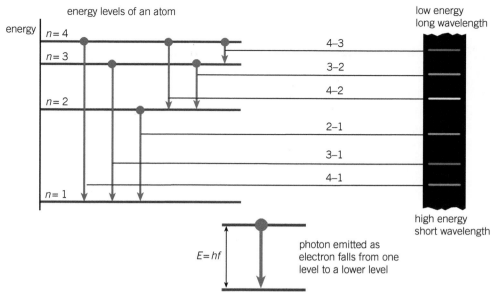

▲ **Figure 5** *Spectral lines and energy levels*

A diffraction grating can be used to spread incident light into its component wavelengths. Line spectra resemble multi-coloured bar codes with each line produced by electrons dropping from a higher energy level to a lower one within the visible spectrum. Atoms from each element have their own pattern of lines – their own bar code.

Figure 6 shows line spectra of several elements. The tungsten filament at the top gives a continuous spectrum which is one reason why many people still prefer 'old-fashioned' filament lamps. The yellow-orange light from sodium gives the familiar colour of many streetlights.

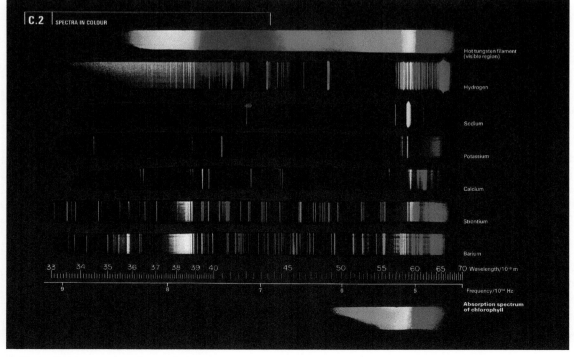

▲ **Figure 6** *Line spectra of several elements*

▲ **Figure 7** *Colourful LEDs*

Light-emitting diodes (LEDs)

LEDs are semiconducting devices that only allow current to travel in one direction. As an electron crosses from one side of the diode to the other it falls to a lower energy level and emits a photon.

The energy change between levels can be calculated by multiplying the p.d. across the diode when it just begins to glow (the 'striking' potential) by the charge on an electron. This value also depends on the semiconducting material. Diodes made from different semiconducting materials will release photons of different energy, that is, light of different wavelengths.

 Worked example: Calculating the wavelength of light emitted from an LED

A diode using zinc selenide as the semiconductor material has a striking potential of 2.5 V. Calculate the wavelength of the light emitted at this potential difference.

Charge on an electron = 1.6×10^{-19} C

Step 1: Select the appropriate equation to obtain the energy transfer.

$$\text{Energy change } W = qV$$

Step 2: Substitute and evaluate.

$$\text{Energy change} = 1.6 \times 10^{-19}\,\text{C} \times 2.5\,\text{V} = 4.0 \times 10^{-19}\,\text{J}$$

Step 3: Select the appropriate equation to find the wavelength of light for a photon energy of 4.0×10^{-19} J.

$$E = hf = \frac{hc}{\lambda} \quad \therefore \quad \lambda = \frac{hc}{E}$$

Step 4: Substitute and evaluate.

$$\lambda = \frac{6.6 \times 10^{-34}\,\text{J s} \times 3.0 \times 10^{8}\,\text{m s}^{-1}}{4.0 \times 10^{-19}\,\text{J}} = 5.0 \times 10^{-7}\,\text{m (2 s.f.)}$$

This is in the blue region of the spectrum.

Practical 4.1d(vi): Determining the Planck constant using LEDs

The striking p.d.s for LEDs emitting different wavelengths of light (photons with different energies) are measured. These measurements need some care – you need to look carefully, ideally in a darkened room, so that you can determine the p.d. at which the LED just begins to glow.

The striking p.d. for each diode is recorded and the energy of the photons emitted is found using the equation

$$\text{Energy of a photon emitted by the LED} = e \times V$$

where e is the charge on an electron and V is the striking p.d. The frequency of the light emitted by each diode is calculated from the wavelengths given by the manufacturers using the equation

$$\text{Frequency} = \frac{\text{wave speed}}{\text{wavelength}} \text{ where wave speed} = 3 \times 10^{8}\,\text{m s}^{-1}$$

A graph of photon energy (y-axis) against frequency (x-axis) is drawn. The Planck constant can be found from the gradient of the graph.

Although you may be given LEDs of specified wavelengths, they actually emit light over a small *range* of wavelengths. This can add to the uncertainty in the experiment and, together with the difficulty of measuring the striking p.d., should be remembered when considering the uncertainty bars on the graph.

Summary questions

1 Single photons of frequencies $f = 3.0\,\text{GHz}$, $f = 6.0 \times 10^{14}\,\text{Hz}$, and $f = 1.2 \times 10^{18}\,\text{Hz}$ are absorbed. Calculate the energy transferred in each case. Give your answers in J and eV.   (3 marks)

2 Choose one of these phenomena:

Line spectra   Photoelectric effect   LEDs

Describe and explain your chosen phenomenon using the idea of the exchange of energy in quanta.   (6 marks)

3 Radiation of wavelength 320 nm strikes a metal surface, releasing photoelectrons with a maximum kinetic energy of $1.4 \times 10^{-19}\,\text{J}$. Use the equation $E_{k(\text{max})} = hf - \phi$ to calculate ϕ (the work function of the metal) and the threshold frequency.   (4 marks)

4 A student investigates the striking potentials of LEDs emitting different colours. Her results are given in Table 1.

▼ Table 1

Wavelength of light / nm	Frequency of photon / 10^{14} Hz	Striking potential / V	Photon energy / 10^{-19} J
470 +/− 30	6.4 +/− 0.4	2.6 +/− 0.2	4.2 +/− 0.4
503 +/− 30		2.5 +/− 0.2	
585 +/− 30		2.1 +/− 0.2	
620 +/− 30		2.0 +/− 0.2	

a Explain why the photon energy = striking potential × charge on an electron.   (1 mark)
b Copy and complete Table 1.   (2 marks)
c Plot a graph of photon energy (J) against photon frequency (Hz). Include uncertainty bars on your graph.   (6 marks)
d Calculate a value for the Planck constant. Include an estimate of the uncertainty of the result and explain how you arrived at this value.   (4 marks)

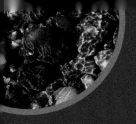

7.2 Quantum behaviour and probability

Specification references: 4.1a(vii), 4.1b(i)

▲ **Figure 1** *Using digital devices*

Think of modern communication technology and you will be thinking of devices that use the understanding that physicists have gained about quantum behaviour. Televisions, phones, and tablet computers all rely on understanding the behaviour of light and electrons at the quantum level. In this topic we will consider a way of understanding this behaviour.

The rule for quantum behaviour

Quantum objects such as photons have their own way of behaving – it isn't wave behaviour and it isn't particle behaviour. The rule for quantum behaviour is very simple – explore all possible paths. This behaviour produces wave-like effects such as interference and particle-like effects such as the way images build up. So, how can we predict or explain phenomena such as the fringes observed in Young's double-slit experiment, an experiment that helped develop the wave model of light?

There are rules for the behaviour of photons.

- A photon is emitted by a source and is detected at a certain place and time.

- Imagine the photon taking every possible path from the source to arrive at that place and time. The longer the path, the earlier the time at which the photon will have to be emitted.

- A path includes the emission, travel, and detection of a photon. For each path a combined phasor arrow can be calculated. The phasor arrow can be thought of as rotating with frequency $f = \dfrac{E}{h}$ (from $E = hf$) for a time equal to the time it takes the photon to travel the length of the path. We call this the *trip time*.

- The angle at which the phasor ends up can be determined for each path.

- Add the phasor arrows for all possible paths, tip-to-tail, to get the resultant phasor.

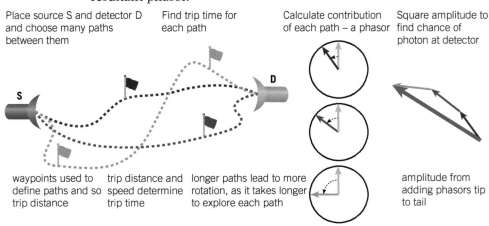

Place source S and detector D and choose many paths between them

Find trip time for each path

Calculate contribution of each path – a phasor

Square amplitude to find chance of photon at detector

waypoints used to define paths and so trip distance

trip distance and speed determine trip time

longer paths lead to more rotation, as it takes longer to explore each path

amplitude from adding phasors tip to tail

▶ **Figure 2** *Carrying out quantum calculations*

Repeating for more paths produces more accurate calculations. With limited time, choose wisely.

Worked example: Frequency of a phasor arrow

Calculate the rate of rotation for the phasor arrow associated with a photon of energy 5.4×10^{-19} J.

Step 1: Select the appropriate equation.

$$f = \frac{E}{h}$$

Step 2: Substitute values and evaluate.

$$f = \frac{5.4 \times 10^{-19} \text{J}}{6.6 \times 10^{-34} \text{J s}} = 8.2 \times 10^{14} \text{Hz}$$

Phasor amplitude, probability, and intensity of light

The length (or amplitude) of the resultant phasor arrow at a particular location gives a measure of the **probability** of a photon arriving at that point – the longer the arrow (larger the amplitude), the higher the probability of arrival. This is an important statement as it shows that where photons end up is not a certainty, but a probability. This explains the random nature of the arrival of photons discussed previously.

Consider these statements about light striking a screen.

a The rate at which energy arrives at a point on a screen is the **intensity** of the light at that point.

b Intensity is proportional to the probability of the arrival of a photon.

c The intensity of light is proportional to the square of the length of the resultant phasor arrow at that point.

Statements **b** and **c** lead to the relationship

probability of arrival of a photon ∝ square of the length of resultant phasor arrow

Hint

We are using the term 'probability' to mean how likely it is that a photon will arrive at a particular place during a given time interval. In this section you will not be required to calculate individual probabilities, but to give a method of comparing the likelihood of a photon arriving at one place compared to another.

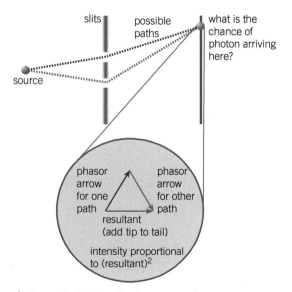

▲ **Figure 3** *Double-slit photon experiment*

Worked example: Phasor arrows and probability

The resultant phasor arrows for the points A and B on the screen of a Young's double-slit experiment are shown in Figure 4. Resultant phasor A is 2 units in length and resultant phasor B is 6 units in length. Calculate the probability of arrival of a photon at point B compared to point A.

Step 1: Select the appropriate relationship.

probability of arrival ∝ square of the length of resultant phasor arrow

Step 2: Substitute values in the relationship.

probability of photon arriving at B $\propto 6^2$
probability of photon arriving at A $\propto 2^2$

Step 3: Use a constant of proportionality to turn the relationships into equations.

probability of photon arriving at B $= k6^2$
probability of photon arriving at A $= k2^2$

Step 4: Divide the equation for B by the equation for A. The constant of proportionality cancels.

$$\frac{\text{probability of photon arriving at B}}{\text{probability of photon arriving at A}} = \frac{6^2}{2^2} = \frac{36}{4} = 9$$

A photon is nine times more likely to arrive at point B than point A.

If a continuous stream of photons passed through the slits, the fringe at point B would be nine times the intensity of that at A.

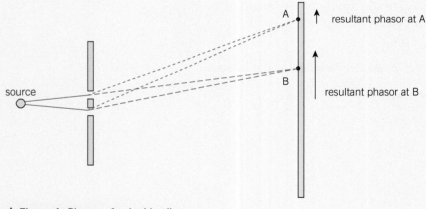

▲ **Figure 4** *Phasors for double slits*

Examples of explaining wave behaviour using phasors

Reflection

Photons obey the law of reflection – they come off the surface of a mirror at the same angle at which they reach it. This can be shown using phasors in Figure 5 (next page). By allowing photons to explore all possible paths from the source S, including ones which are obviously 'wrong', each trip will take a different time and affect the number of rotations of each phasor arrow. Longer trip time = more rotations. Adding the phasor arrows together tip-to-tail at the detector D produces the resultant phasor arrow shown.

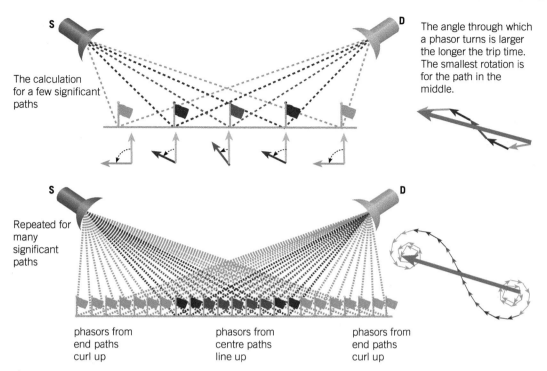

▲ **Figure 5** *Reflection – explorations over a surface*

Look at the coiled pattern produced when the phasor arrows are added together. The phasors from the ends (following the 'obviously wrong' paths) reach the detector at very different **phases** and tend to curl up. These paths do not contribute much to the resultant amplitude. The paths taken near the equal-angled path (the path that follows the law of reflection) have little phase difference and line up. Most of the probability of photons arriving at the detector comes from these paths. The rules of quantum behaviour predict the law of reflection.

Refraction

Photons slow down in glass so a phasor arrow for a photon travelling a given distance in glass will make more rotations than the arrow for a photon travelling the same distance through air.

An example showing how light from a source is focused by a converging lens is shown in Figure 6.

Photons travelling straight through the lens to the detector take the shortest possible route but are slowed down in the glass. Photons taking the longest route (to the tip of the lens in the figure) are not slowed down by the glass. The point at which the lens focuses the light is simply the point at which all the phasor arrows have made the same number of rotations.

> ## Synoptic link
>
> You have met refraction in terms of waves slowing down in glass in Topic 6.2, Light, waves, and refraction.

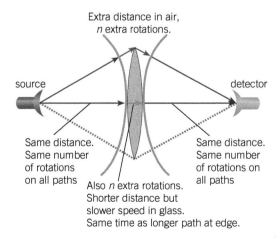

▲ **Figure 6** *Refraction with a lens*

Summary questions

1 Calculate the number of rotations per second for the phasor associated with a photon of energy 3.3×10^{-19} J. *(2 marks)*

2 A photon of frequency 6.0×10^{14} Hz explores a path 6.0 m long.
   a Calculate its trip time. *(2 marks)*
   b Calculate the number of rotations of its phasor arrow during the trip. *(2 marks)*

3 Show how three phasor arrows of equal length can add together to produce a zero resultant. *(1 mark)*

4 A parabolic mirror brings light from a distant source to a sharp focus as shown in Figure 7.

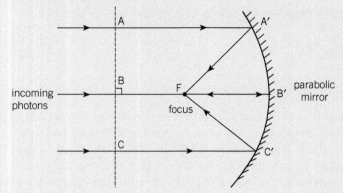

▲ **Figure 7**

   a State the relationship between the trip times for the photons exploring paths AA′F, BB′F, and CC′F. *(1 mark)*
   b State the phase relationship at F between photons from the three paths. Explain your reasoning. (Assume that the photons were in phase along line ABC). *(1 mark)*
   c Use your answers to **a** and **b** to explain why the intensity of reflected light is much lower at positions other than F. *(3 marks)*

5 A converging lens is focusing light, as shown in Figure 8.
   a Express the speed of light in mm s^{-1}. *(1 mark)*
   b How long does it take light to follow the route ACB? *(1 mark)*
   c How long must the light spend between D and E in the glass? *(1 mark)*
   d What is the speed of light in the glass? *(1 mark)*
   e What is the refractive index of the glass? *(2 marks)*
   f The phasors following route ACB make the same number of turns as those following route ADEB. Explain why this is the case. *(2 marks)*

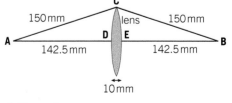

▲ **Figure 8**

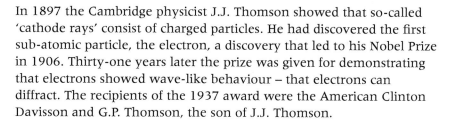

7.3 Electron diffraction

Specification references: 4.1a(viii), 4.1c(v)

In 1897 the Cambridge physicist J.J. Thomson showed that so-called 'cathode rays' consist of charged particles. He had discovered the first sub-atomic particle, the electron, a discovery that led to his Nobel Prize in 1906. Thirty-one years later the prize was given for demonstrating that electrons showed wave-like behaviour – that electrons can diffract. The recipients of the 1937 award were the American Clinton Davisson and G.P. Thomson, the son of J.J. Thomson.

Electron diffraction

Davisson and Thomson independently demonstrated that a thin layer of atoms can act as a natural diffraction grating, the two-dimensional array of atoms producing a diffraction pattern of concentric circles. This shows that the electrons are showing wave-like behaviour. Particle behaviour would result in a central bright region with no concentric circles. Experiments show that electrons diffract through slits in the same manner as light.

de Broglie's equation

In the 1920s the French physicist de Broglie developed an equation which suggested that an electron has a wavelength associated with it. This wavelength λ is related to the Planck constant h, and the momentum of the electron p by the equation

$$\lambda = \frac{h}{p}$$

and momentum is given as mv, where m is the mass of the particle (in this case the electron) and v is its velocity. This equation can therefore be rewritten as

$$\lambda = \frac{h}{mv}$$

> **Learning outcomes**
>
> Describe, explain, and apply:
> → evidence from electron diffraction that electrons show quantum behaviour
> → the de Broglie relationship
> $$\lambda = \frac{h}{p}$$

▲ **Figure 1** *Electron diffraction using a thin sheet of graphite*

> **Hint**
>
> The Planck constant h is fundamental to picturing the wave-like nature of particles, just as it is fundamental to picturing the particle-like nature of light ($E = hf$).

 Worked example: Calculating the de Broglie wavelength of an electron

An electron has a velocity of $1.5 \times 10^6 \, \text{m s}^{-1}$. Calculate its de Broglie wavelength (mass of electron $= 9.1 \times 10^{-31} \, \text{kg}$).

Step 1: Select the appropriate equation – de Broglie wavelength.

$$\lambda = \frac{h}{mv}$$

Step 2: Substitute and evaluate.

$$\lambda = \frac{6.6 \times 10^{-34} \, \text{J s}}{9.1 \times 10^{-31} \, \text{kg} \times 1.5 \times 10^6 \, \text{m s}^{-1}} = 4.8 \times 10^{-10} \, \text{m} \ (2 \ \text{s.f.})$$

Rotating electron arrows

An electron is not exactly like a photon. Electrons have mass and so cannot travel at the speed of light like a photon. Instead, they can travel at different speeds slower than light. However, just like a photon, an electron explores all possible paths between two points and a phasor can be used to track the phase change along every path. Phasors for photons and electrons superpose in the same way.

Rules for the quantum behaviour of electrons

- An electron of mass m and speed v is emitted by a source and is detected at a certain place and time.

- The longer the path, the earlier the time at which the electron will have to be emitted. This depends on the speed of the electron.

- A path includes the whole process of emission, travel, and detection of an electron. For each path a combined phasor arrow can be calculated. The phasor arrow can be thought of as making one turn for each distance $\frac{h}{p}$ along the path.

- Add up the phasor arrows for all possible paths, tip-to-tail, to get their resultant phasor.

- The probability of detection of an electron can be calculated from the square of the resultant phasor.

The image of the head of a cockroach in Figure 2 is taken with an electron microscope. Modern electron microscopes rely on the quantum behaviour of electrons.

▲ **Figure 2** *Electron microscope image of the head of a cockroach.*

Summary questions

1 Calculate the de Broglie wavelength for electrons with momentum $5 \times 10^{-24}\,\mathrm{kg\,m\,s^{-1}}$
   $(h = 6.6 \times 10^{-34}\,\mathrm{J\,s})$.   *(2 marks)*

2 Copy and complete the table ($h = 6.6 \times 10^{-34}\,\mathrm{J\,s}$, mass of electron $= 9.1 \times 10^{-31}\,\mathrm{kg}$).

Momentum, mv	Speed, v	Wavelength, λ
		10 nm
$1.0 \times 10^{-24}\,\mathrm{kg\,m\,s^{-1}}$		
	$2.0 \times 10^{6}\,\mathrm{m\,s^{-1}}$	

(3 marks)

3 3×10^{-19} J is roughly the energy of a photon of visible light (wavelength about 600 nm). The momentum of an electron with this energy is about $7 \times 10^{-24}\,\mathrm{kg\,m\,s^{-1}}$. Compare, using calculations, the wavelength of the electron to the wavelength of light associated with photons of the same energy. Explain, using your comparison, why electrons need to pass through far smaller gaps than photons for interference effects to be observed.   *(3 marks)*

4 The images in Figure 2 suggest that electrons were showing both wave-like and particle-like behaviour in the experiment.
   a The image is composed of many discrete spots. State why this suggests that electrons are interacting with the screen like particles.   *(1 mark)*
   b The image is a series of bright and dark fringes. State why this suggests that the electrons are showing wave-like behaviour.   *(1 mark)*
   c Explain the pattern on the screen using the phasor model of electron behaviour.   *(4 marks)*

Wave behaviour

Superposition of waves

- how standing waves are formed
- the terms: phase, phasor, amplitude, superposition
- identify the wavelength of standing waves from diagrams
- practical task — using an oscilloscope to find the frequencies of waves
- practical task — using a rubber cord to demonstrate superposition
- practical task — determining the speed of sound by the formation of stationary waves in a resonance tube

Light, waves, and refraction

- describe the refraction of light using the wave model and the changes in speed at the boundary between the two media
- the term refractive index, n

- Snell's Law: $n = \dfrac{\sin i}{\sin r}$
  $$= \dfrac{\text{speed of light in medium 1}}{\text{speed of light in medium 2}}$$
- absolute refractive index
  $$= \dfrac{\text{speed of light in a vacuum}}{\text{speed of light in medium}}$$
- Practical task — determining the refractive index of a transparent block

Path difference and phase difference

- the terms: interference, coherence, path difference
- practical task — determining the wavelength of microwaves

Interference and diffraction of light

- diffraction of waves through an aperture, a double slit and a diffraction grating
- using the equation $n\lambda = d \sin \theta$
- practical task — determining the wavelength of light using superposition of light

Quantum behaviour

Quantum behaviour

- evidence that photons exchange energy in quanta
- the equation $E = hf$ applied to line spectra, LEDs, and the photoelectric effect
- measuring energy in electronvolts
- the terms: work function, threshold frequency
- practical task — using different coloured LEDs to determine the Planck constant

Quantum behaviour and probability

- quantum behaviour in terms of the probability of arrival obtained by adding phasor arrows — probability of arrival of a photon found by combining amplitude and phase for all paths
- probability of arrival of a photon $\propto$ square of the length of the resultant phasor arrow
- intensity at a point as the rate at which energy arrives at the point

Electron diffraction

- electron diffraction as evidence that electrons show quantum behaviour
- the de Broglie relationship $\lambda = \dfrac{h}{mv}$

What does it all mean?

A peculiar thing about quantum behaviour is that although everything agrees with experimental results, there is more than one picture of what lies behind the calculations, and no agreement between scientists about which is the 'right' picture. In this chapter we have chosen the story that seems to be the simplest — a story first devised by physicist Richard Feynman.

You must keep in mind that the picture we have painted is not a definitive description of *how things are*. The story of photons following all possible paths is an easy way of remembering how to do the calculations, but it's important to not take this interpretation too literally. For example, don't imagine photons trying paths one at a time, one after the other — if anything, they must be imagined trying all paths at once.

Feynman's "try all paths" idea was not altogether new — Huygens long ago sowed the seeds. Huygens thought of his wavelets as spreading out everywhere they could, and building up the new position of a wave by all arriving at that point in phase. The reason the wavelets are not observed everywhere is that in most places they add up to nothing at all.

Shocking or not?

People originally found quantum behaviour peculiar. Niels Bohr, one of the founders of quantum theory wrote "Anyone who is not shocked by quantum theory has not fully understood it."

Quantum behaviour can be modelled by combining phasors (amplitude and phase in one spinning arrow) to calculate probabilities. When there are countless numbers of photons in a beam, the probability just determines the brightness, which varies smoothly from place to place. And the brightness varies just as if waves were superposing to give the result, because of the quantum adding-up of phasors. Wave-like behaviour is what you observe when the photons are numerous.

This argument may feel uncomfortable, as it is explaining something almost relatively simple (wave behaviour) in terms of something more challenging (phasors). However, if we didn't explain anything in this way, then the simplest concepts would never be explained.

Waves or particles?

The modern story told so far avoids the question of whether electrons and photons are truly waves or particles. Many quantum physics books attempt to scratch the surface of this tricky question, but in doing so only make it itch even more — the best solution at this point

▲ **Figure 1** *Neils Bohr with Einstein. The two men could never agree about the probabilistic nature of the quantum world.*

is to answer that they are neither waves nor particles. They are simply quantum objects.

So far we have used the electron as the only example of a quantum object with mass, but quantum behaviour doesn't stop with electrons. In 1999, a team working at the University of Vienna showed that 'buckyballs', spherical molecules composed of sixty carbon atoms, can diffract through a suitable grating in the same manner as electrons.

The mass of a buckyball is more than a million times that of an electron, but in carefully controlled conditions buckyballs can behave as quantum objects. In 2011, quantum behaviour was demonstrated in molecules composed of four hundred and thirty atoms with lengths of up to 6 nanometres. Experiments such as these are of more than academic interest — they may well help towards the development of quantum computers.

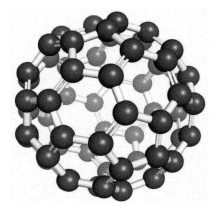

▲ **Figure 2** *A representation of a C-60 or carbon fullerene molecule, affectionately known as a buckyball*

Summary questions

1 It is suggested that Huygen's idea of wavelets superposing has similarities to the Feynman picture of phasors exploring all possible paths. Explain what these similarities are, comparing the addition of wavelets to the adding of phasor arrows. You can include diagrams in your answer.

2 This question is about the quantum behaviour of buckyballs.
   a Calculate the de Broglie wavelength of a buckyball of mass 1.2×10^{-24} kg moving with a velocity of 220 m s^{-1}.
   b Buckyballs of velocity 220 m s^{-1} pass through a grating with slit separation of 1×10^{-7} m. Show that the separation between interference maxima at a detector 1.3 m away from the grating is about 3×10^{-5} m.
   The diameter of a buckyball is about 1 nm.
   c Suggest why the interference maxima and minima will blur together when the separation between maxima at the detector is less than this diameter.
   d Calculate the velocity of the buckyballs at which the maxima and minima blur together.

Practice questions

1 An electron travelling at $1 \times 10^6 \, \text{m s}^{-1}$ has a de Broglie wavelength λ_1.

A proton has a mass of roughly 1800 times that of an electron.

What is the de Broglie wavelength of a proton travelling at $1 \times 10^6 \, \text{m s}^{-1}$?   *(1 mark)*

A $\dfrac{\lambda_1}{1800}$ **B** $1800 \, \lambda_1$

C $\dfrac{1800}{\lambda_1}$ **D** $\lambda_1 \sqrt{1800}$

2 A laser emits light of wavelength 400 nm. The energy is radiated at a rate of 18 mW ($18 \, \text{mJ s}^{-1}$). Calculate the number of photons emitted per second by the laser.   *(4 marks)*

3 An electron has kinetic energy = 1.6 keV.

Use the equation kinetic energy = $\dfrac{\text{momentum}^2}{2m}$ to calculate the momentum of the electron and use your value to calculate its de Broglie wavelength.   *(4 marks)*

4 The energy required to release an electron from the surface of magnesium is 3.7 eV. Radiation of wavelength 170 nm is incident on the surface of the metal. Calculate the maximum kinetic energy of the photoelectrons released.   *(3 marks)*

5 Calculate the velocity of an electron with a de Broglie wavelength of $6.6 \times 10^{-10} \, \text{m}$.   *(3 marks)*

6 This question is about photons and phasors. Figure 1 shows two paths of photons from slits S_1 and S_2 to a point on the screen P_1.

The path difference $S_2 P_1 - S_1 P_1$ is $\dfrac{\lambda}{3}$ where λ is the wavelength of the light.

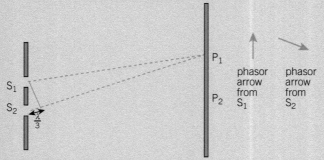

▲ **Figure 1**

a The angles of the phasor arrows arriving at P_1 from slits are shown on the right hand side of the diagram. State why the phase difference is $\dfrac{2\pi}{3}$ radians.   *(1 mark)*

b Draw a scale diagram to find the length of the resultant phasor arrow.   *(2 marks)*

c Phasors meet at P2 with no phase difference. Calculate the ratio

$\dfrac{\text{length of the resultant arrow at } P_2}{\text{length of resultant arrow at } P_1}$.   *(2 marks)*

d Calculate the ratio

$\dfrac{\text{probability of photon arriving at } P_1}{\text{probability of photon arriving at } P_2}$.   *(2 marks)*

e A third slit is added, as shown in Figure 2.

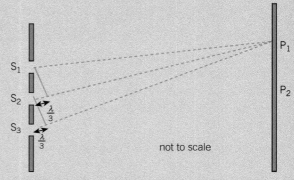

▲ **Figure 2**

Explain the effect this has on the probability of a photon arriving at P1. You may include a diagram of the phasor arrows at P1 in your answer.   *(3 marks)*

7 This question is about the photoelectric effect.

When light above the threshold frequency f_0 is incident on a metal surface, photoelectrons are emitted. The maximum kinetic energy of the photoelectrons is given by Einstein's equation:

$$E_{k(max)} = hf - \phi$$

where ϕ is the minimum energy required to remove an electron from the surface of the metal and f is the frequency of the light incident on the surface.

Red light of frequency 4.5×10^{14} Hz is incident on a metal surface. The maximum energy of the ejected photoelectrons is 0.2 eV.

a State how the number and energy of the emitted photoelectrons will change when the intensity of the light incident on the surface is doubled. Explain your answer in terms of photons interacting with electrons in the surface of the metal.

(4 marks)

b Violet light of frequency 7.5×10^{14} Hz ejects electrons with maximum energy of 1.4 eV. Use the data for red and violet light to calculate a value for the Planck constant h. *(4 marks)*

c Calculate the minimum frequency for the release of photoelectrons from this surface. *(2 marks)*

8 In a simple wave model to explain the diffraction of waves at a gap, the gap of width b is divided into three equal parts as shown in Figure 3a.

The centre of each part is treated as a source of waves.

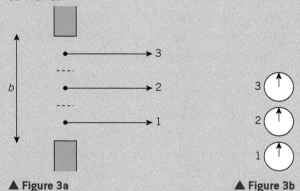

▲ Figure 3a   ▲ Figure 3b

a The phasors for the waves from each of the three parts of the gap reaching a distant screen in the straight-on direction are shown in Figure 3b.

  (i) The paths taken by the waves in Figure 3a are all equal in length. Explain how the phasors in Figure 3b confirm this. *(1 mark)*

  (ii) Each phasor has an amplitude A. Write down the amplitude of the resultant phasor at the distant screen. *(1 mark)*

b At an angle θ to the straight-on direction, the path difference between neighbouring paths is Δx, as show in Figure 4a. For one particular value of θ, the resultant intensity is **zero**.

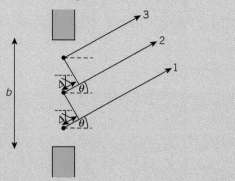

▲ Figure 4a   ▲ Figure 4b

  (i) Explain why the phasor for path 1 has rotated 120° more than the phasor for path 2, when $\Delta x = \frac{1}{3}\lambda$, where λ is the wavelength of the waves. *(2 marks)*

  (ii) Draw arrows on a copy of Figure 4b to represent the phasors for waves 2 and 3. Explain, using a diagram, why the three phasors have a zero resultant. Label your phasors in the diagram 1, 2, and 3. *(1 mark)*

  (iii) Use Figure 4a and the fact that $\Delta x = \frac{1}{3}\lambda$ to show that $\lambda = b\sin\theta$ where b is the total width of the gap. Show your working clearly. *(1 mark)*

c Use the equation $\lambda = b\sin\theta$ to calculate the angle θ at which a minimum signal occurs when microwaves of wavelength 2.4 cm are incident on a gap of width 6.0 cm. *(2 marks)*

OCR Physics B Paper G492 Jan 2011

MODULE 4.2
Space, time, and motion

Chapters in this Module
8 Motion

9 Momentum, force, and energy

Introduction

Newtonian dynamics, which is the study of the relationships between force and motion, forms the basis of all of Physics. In this module you will learn how physicists analyse and measure motion. You will see how the concepts of momentum, force, and energy provide a framework to explain phenomena, from the tiny interactions of molecules in a gas to the vast movements of stars and planets.

Motion develops the mathematical methods with which motion is described and analysed. Starting from graphs and definitions met at GCSE, the relationships between displacement, velocity, and acceleration are developed. Skills covered include the interpretation of gradients and areas under s-t and *v-t* graphs. Other analytical approaches to motion involve the use of vectors and the algebra of the

kinematic equations of motion. Iterative modelling is introduced as a method of analysing change, preparing for its later use in both AS and A level work.

Momentum, force and energy is a logical consequence of chapter 8. Chapter 8 is about the *description* of motion — chapter 9 is about its *explanation*. The principle of conservation of momentum is used to analyse collisions and explosions. These concepts lead to Newton's laws of motion and force, providing the underlying concepts that explain changes in motion and the path of projectiles. This chapter will also cover work and the principle of conservation of energy, including calculations in kinetic energy and gravitational potential energy. Power is introduced as the rate at which work is done, or the rate at which energy is transferred.

Knowledge and understanding checklist

From your Key Stage 4 study you should be able to do the following. Work through each point, using your Key Stage 4 notes and the support available on Kerboodle.

- [ ] Relate changes and differences in motion to appropriate distance-time and velocity-time graphs.
- [ ] Apply formulae relating distance, time, and speed for uniform motion, and for motion with uniform acceleration.
- [ ] Recall examples of ways in which objects interact (e.g., by gravity, by contact) and describe how such examples involve interactions between pairs of objects that produce a force on each object.
- [ ] Apply Newton's first law to explain the motion of objects and apply Newton's second law in calculations relating force, mass, and acceleration.
- [ ] Use vector diagrams to illustrate resolution of forces, a net (resultant) force, and equilibrium situations.
- [ ] Use the relationship between work done, force, and distance moved along the line of action of the force and describe the energy transfer involved.
- [ ] Explain the definition of power as the rate at which energy is transferred.
- [ ] Calculate energy efficiency for any energy transfer, and describe ways to increase efficiency.

Maths skills checklist

In this unit, you will need to use the following maths skills. You can find support for these skills on Kerboodle and through MyMaths.

- [ ] **Change the subject of an equation, including nonlinear equations**, such as when using kinematic equations, or calculations involving Newton's second law and energy transfers.
- [ ] **Use an appropriate number of significant figures** in any calculation using data expressed to a certain number of significant figures.
- [ ] **Plot two variables from experimental or other data and use $y = mx + c$** when analysing the motion of falling objects.
- [ ] **Determine a rate of change from the gradient from a graph (including tangents)** in order to calculate velocities and accelerations.
- [ ] **Understand the possible physical significance of the area between a curve and the x-axis, and be able to calculate it or estimate it by graphical methods**, such as with curved displacement-time or force-time graphs.

MyMaths.co.uk
Bringing Maths Alive

MOTION

8.1 Graphs of motion

Specification reference: 4.2a(vii), 4.2b(i), 4.2b(ii), 4.2d(i)

Synoptic link

You will learn more about directional movement in Topic 8.2, Vectors.

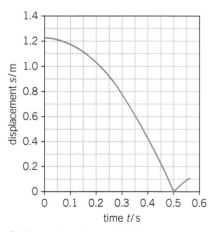

▲ **Figure 1** *A dropped ball*

Displacement and velocity

Imagine you go out for a run. How can you possibly have run a displacement of 0 m?

The terms *displacement* and *velocity* may seem to be just formal scientific words with exactly the same meaning as *distance travelled* and *speed*, but as you shall see later they give extra information about the direction of movement. In this case, the distance you travelled is the amount of ground you covered. The **displacement** is how far you are from your starting point, and in which direction. This would of course be zero if you end up back home.

In the same way, **speed** is just how fast you are going, whereas **velocity** also specifies the direction.

Displacement–time graphs

The graph in Figure 1 shows the displacement of a ball dropped onto the floor. Displacement is measured from the floor, with the direction up being positive.

The graph shows the timeline of the fall. The ball was released at time 0 s, when its displacement, just over 1.2 m, is the height from which the ball was dropped. In the first 0.1 s the displacement changes quite slowly as the line is nearly horizontal. By the time the ball reaches the floor (displacement = 0 m), the steepness of the line shows that displacement is changing rapidly. The ball then bounces, but the rebound line is less steep, showing that it is moving more slowly.

To calculate the average velocity, you use the equation

$$\text{average velocity } v \text{ (m s}^{-1}) = \frac{\text{change in displacement } s \text{ (m)}}{\text{total time taken } t \text{ (s)}}$$

A constant velocity is shown on a displacement–time graph by a straight line and the velocity is given by the gradient of the line. In Figure 1 the velocity is changing all the time, so you need to draw a tangent to the curve at the time when you need to find the velocity.

Delta notation

Changes in the value of a parameter are common in physics, particularly in this chapter, so we will use the standard shorthand of the Greek letter delta (Δ).

Δs (read it as delta-es) means the change in s. If s changes from 2.10 m to 1.64 m, then $\Delta s = -0.46$ m.

Worked example: Calculating velocity from a displacement–time graph

Calculate the velocity of the ball at time $t = 0.20\,\text{s}$ in Figure 1.

Step 1: Draw a tangent to the curve at $t = 0.20\,\text{s}$.

Step 2: Draw the gradient triangle (the red lines in Figure 2).

Using such a large triangle reduces the percentage uncertainty in the readings of the change in s, Δs, and the change in t, Δt.

Step 3: Read Δs and Δt from the triangle.

Δs = final s – initial s = 0.22 m – 1.40 m = –1.18 m

Δt = final t – initial t = 0.60 s – 0.02 s = 0.58 s

Step 4: Calculate the velocity.

$$v = \text{gradient of tangent} = \frac{\Delta s}{\Delta t} = \frac{-1.18\,\text{m}}{0.58\,\text{s}} = -2.034...\,\text{m s}^{-1}$$

Step 5: Round the answer to an appropriate number of significant figures, and do not forget the units.

$$v = -2.03\,\text{m s}^{-1}\ (2\ \text{s.f.})$$

Two significant figures is acceptable in the answer – if you round to 1 s.f. you are losing information, and using 3 s.f. is not justified because you cannot read the graph to that level of precision.

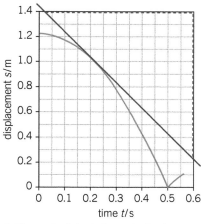

▲ **Figure 2** *Calculating velocity from a displacement–time graph*

Velocity–time graphs

Figure 3 shows a journey with varying velocity.

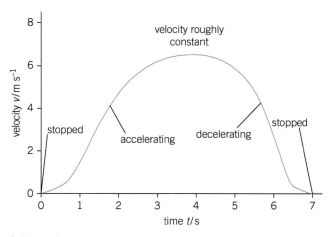

▲ **Figure 3** *Journey with varying velocity*

This graph shows the timeline for the journey of an object travelling in a straight line, and we can also use it to calculate the displacement. If the direction changed you would never be able to work out what the displacement was, as the object could have ended up back where it started.

You can see that the mean velocity is more than $2\,\mathrm{m\,s^{-1}}$ and less than $6\,\mathrm{m\,s^{-1}}$, probably between $3\,\mathrm{m\,s^{-1}}$ and $4\,\mathrm{m\,s^{-1}}$, so you can estimate the displacement by assuming a mean velocity of $3.5\,\mathrm{m\,s^{-1}}$ for the $7\,\mathrm{s}$:

$$v = \frac{s}{t} \Rightarrow s = \text{mean velocity} \times \text{time} = 3.5\,\mathrm{m\,s^{-1}} \times 7\,\mathrm{s} = 24.5\,\mathrm{m}$$

But how can you use the graph to find the displacement more accurately? Figure 4 shows one tiny part of this journey when the velocity was $4.0\,\mathrm{m\,s^{-1}}$.

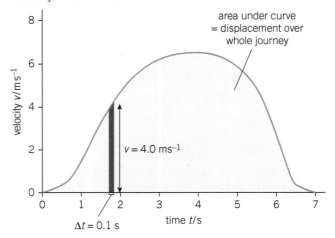

▲ **Figure 4** *Using a velocity–time graph to find displacement*

The area of the strip in Figure 4 = height of strip × width of strip

$$= v \times \Delta t$$
$$= \text{distance } s \text{ travelled in the short time } \Delta t$$
$$= 4.0\,\mathrm{m\,s^{-1}} \times 0.1\,\mathrm{s} = 0.4\,\mathrm{m}$$

If a thin strip like this is drawn every $0.1\,\mathrm{s}$ over the whole graph from the beginning to the end of the motion, then adding all the areas together (the total area under the curve) gives the total distance travelled.

You can (correctly) observe that the thin strips do not exactly fit under the graph, as they make the curve into a saw-like edge. The solution – make the strips thinner, for example, $0.01\,\mathrm{s}$, $1\,\mathrm{ms}$, or $1\,\mathrm{ns}$. This increases the accuracy of the estimate but it just takes longer to calculate for the increased number of strips.

In exactly the same way as the area under this velocity–time graph gives the displacement (providing the direction has stayed the same), so the area under a speed–time graph gives the distance travelled. Here a change in direction makes no difference.

🖩 **Worked example: Finding distance travelled from a speed–time graph**

Find the distance travelled in the $9\,\mathrm{s}$ journey described by Figure 5.

Step 1: Divide the polygon into shapes whose area can be easily calculated. There are usually several ways to do this. Figure 6 shows one possibility.

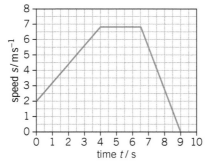

▲ **Figure 5** *Using a speed–time graph to find distance travelled*

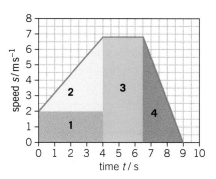

▲ **Figure 6** *Speed–time graph divided into regular shapes*

 Step 2: Sum the areas of all the shapes.

Total area = area 1 + area 2 + area 3 + area 4

$$= (2\,\text{m s}^{-1} \times 4\,\text{s}) + \left(\tfrac{1}{2} \times [6.8 - 2.0]\,\text{m s}^{-1} \times 4\,\text{s}\right)$$
$$+ (6.8\,\text{m s}^{-1} \times 2.5\,\text{s}) + \left(\tfrac{1}{2} \times 6.8\,\text{m s}^{-1} \times 2.5\,\text{s}\right)$$
$$= 8\,\text{m} + 9.6\,\text{m} + 17\,\text{m} + 8.5\,\text{m} = 43.1\,\text{m}$$

Calculating acceleration

In Figure 3, increasing velocity and decreasing velocity were described as **acceleration** and deceleration. These need definition to allow calculation.

$$\text{Just as velocity (m s}^{-1}) = \frac{\text{change in displacement (m)}}{\text{time taken (s)}}$$

$$\text{so acceleration } a\text{ (m s}^{-2}) = \frac{\text{change in velocity } \Delta v \text{ (m s}^{-1})}{\text{time taken } \Delta t \text{ (s)}}$$

$$a = \frac{\Delta v}{\Delta t}$$

In both cases, these are given by the gradient of a graph. $v = \frac{\Delta s}{\Delta t}$ is the gradient of the s–t graph, and $a = \frac{\Delta v}{\Delta t}$ is the gradient of the v–t graph.

Hint

Take care to include the units on the axes. This will help avoid *power of 10* errors by missing km, ms, and so on.

Hint

Remember that area under a v-t graph is in m, not m², because the units multiply like this: m s⁻¹ × s = m.

Worked example: Using a velocity–time graph to calculate acceleration

Calculate the acceleration at time $t = 1.75\,\text{s}$ for Figure 3.

Step 1: Since the *accelerating* portion of the graph is not straight, a tangent must be drawn to find the acceleration.

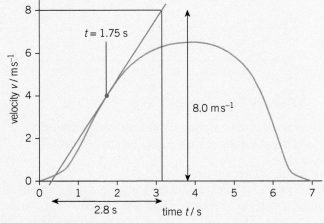

▲ **Figure 7** *Acceleration*

Step 2: Draw a large gradient triangle and read the values of Δv and Δt.

$$\Delta v = 8.0 - 0 = 8.0\,\text{m s}^{-1} \text{ and } \Delta t = 3.1 - 0.3 = 2.8\,\text{s}$$

Hint

Negative acceleration is an alternative expression for deceleration.

Practical 4.2d(i): Investigating the motion of a trolley using light gates

Figure 8 shows a dynamics trolley on a ramp tilted at an angle θ.

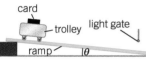

▲ **Figure 8** *Trolley on a ramp*

When released, the trolley will accelerate down the slope. Measurements of displacement s and time t from release are straight-forward, but measuring velocity v is harder. The method shown in Figure 8 is to use a light gate containing an infrared beam that is interrupted by the card fastened to the moving trolley. The light gate is connected to a computer/data logger which will measure the time the card takes to pass through the beam (Δt). This then allows for the calculation of v using $v = \dfrac{\Delta s}{\Delta t}$, where Δs is the width of the card.

The motion of objects can also be investigated using ticker timers, data loggers, or video analysis.

Step 3: Calculate the acceleration by substituting appropriate values in the equation $a = \dfrac{\Delta v}{\Delta t}$, rounding the answer to an appropriate number of significant figures.

$$a = \frac{\Delta v}{\Delta t} = \frac{8.0\,\text{m s}^{-1}}{2.8\,\text{s}} = 2.857...\,\text{m s}^{-2} = 2.9\,\text{m s}^{-2}\ (2\ \text{s.f.})$$

Summary questions

1 A car moving along a straight road accelerates from rest at a uniform rate for 5 seconds, travels at a constant velocity for 5 seconds, and decelerates uniformly to rest over the next 10 seconds. Sketch the velocity–time and displacement–time graphs for this journey. Sketch the graphs one above the other, with the same scale on the time axis. *(5 marks)*

2 Figure 9 is a speed–time graph for a car journey along a country road, where the direction changes frequently.

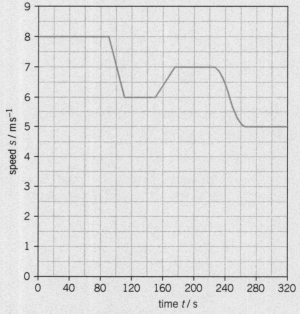

▲ **Figure 9** *Journey along a country road*

Calculate the distance travelled in the 5 minutes shown in the graph, and explain why the car's displacement in this time has a value less than this. *(5 marks)*

3 The data below is for an object accelerating in a straight line at a non-uniform rate. Plot a velocity–time graph and use it to calculate the acceleration at time $t = 5\,\text{s}$ and the distance s travelled in 10 seconds. *(5 marks)*

$t\,/\,\text{s}$	0	2	4	6	8	10
$v\,/\,\text{m s}^{-1}$	0	3	8	15	24	35

8.2 Vectors

Specification reference: 4.2a(i), 4.2b(i), 4.2c(i), 4.2c(ii)

What are vectors?

Vectors appear everywhere in the world around us, from ocean currents and air flow, to forces and fields.

Vectors are quantities that have both magnitude and direction. These can be represented by an arrow of the appropriate length pointing in the correct direction.

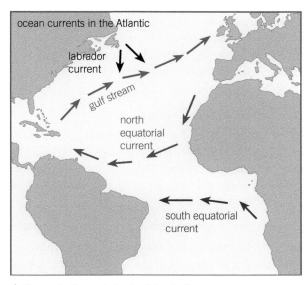

▲ **Figure 1** *Currents in the Atlantic Ocean*

Not all quantities have direction. You have met speed before and it does not tell you in which direction you are moving. Quantities like speed that do not have a direction are called **scalar** quantities.

Distance and displacement

Distance is a scalar quantity as it has no direction, only magnitude (a numerical value together with its unit). If you wish to state not only how far away something is but in which direction, then you need a **vector** quantity which has the same magnitude as distance, but also has direction. In this case the vector quantity is displacement.

Adding vectors

Each vector is represented by an arrow pointing in the appropriate direction. Choose an appropriate scale, and draw each arrow with a length proportional to its magnitude.

Draw the two vector arrows tip-to-tail. The resultant vector is given by the line from the tail of the first vector to the tip of the last.

Synoptic link

You have met scalar quantities such as speed and distance in Topic 8.1, Graphs of motion.

Worked example: Adding displacements using graphical representations of vectors

A woman walks 25 m due east and then 15 m due north. What is her displacement from the starting position?

Step 1: Using graph paper, choose a suitable scale (see Figure 2).

In this example, one large square (1 cm) = 5 m. Put a compass rose in the corner to avoid mistakes (see Figure 2).

Step 2: Plot the displacements, one after the other.

It doesn't matter which you start with, the answer will be the same. The example shows the 25 m due east being plotted first. Then, from the tip of the first vector, draw the second vector with the correct magnitude in the correct direction. Remember to put an arrowhead to show which way it is going, and label the magnitude of the displacement.

Step 3: Draw in the resultant vector.

Join the tail of the first vector to the tip of the second vector. This gives the resultant vector, which is the displacement of the woman from her starting position. Label the angle between the resultant displacement and one of the reference directions – here you can choose either N or E.

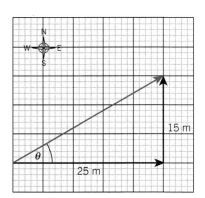

▲ **Figure 2** *Adding displacements*

Step 4: Measure the resultant displacement and the angle.

Using a ruler and protractor, the magnitude of the displacement on the graph paper is 6.0 cm, corresponding to $\frac{6\,cm}{1\,cm} \times 5\,m = 30\,m$, and the angle is 30°.

The resultant displacement is therefore 30 m in a direction E 30° N (or a direction N 60° E, or a bearing of 060°).

Notice that the symbol θ (the Greek letter theta) has been used. This is the usual symbol chosen for an angle.

Geometry versus trigonometry

You may have spotted that the displacement in this example is easier to calculate using Pythagoras' theorem and a bit of trigonometry. This has the advantage that you do not need to draw the vector addition triangle with great accuracy. A sketch will do, if it is clearly labelled.

 Worked example: Addition of vectors using trigonometry

By using trigonometry in the example above, Step 4 becomes, using Pythagoras' theorem:

(resultant magnitude)^2 = $(25\,m)^2$ + $(15\,m)^2$ = $850\,m^2$

resultant magnitude = $\sqrt{850\,m^2}$ = 29.1...m = 29 m (2.s.f.)

You can use sin, cos, or tan to calculate θ, as the triangle is right-angled and you know all three sides. Tan is usually easiest.

$$\tan\theta = \frac{\text{opposite}}{\text{adjacent}} = \frac{15\,\text{cm}}{25\,\text{cm}} = 0.60 \Rightarrow \theta \text{ is the inverse tangent of } 0.60$$

arctan (0.60) or $\tan^{-1}(0.60) = 30.9°... = 31°$ (2 s.f.)

The resultant displacement is therefore more accurately 29 m in a direction E 31° N (or a direction N 59° E, or a bearing of 059°).

Study tip

Although trigonometry is more accurate than a scale drawing, you should learn to use both methods.

Synoptic link

See the Maths Appendix for further uses of trigonometry in physics.

Speed and velocity

In everyday speech, speed and velocity have the same meaning. In physics, they have an important difference. Speed is a scalar quantity, with no direction indicated. It is what is registered by a car speedometer. Velocity is a vector, where the direction is important, as shown in the weather map of Figure 3.

Figure 3 refers, quite correctly, to wind velocities, not speeds. This is because the arrows show the direction, and their size shows the magnitude. Wind speed maps that show only the magnitude are not common.

▲ **Figure 3** *Wind velocities around the UK*

 Worked example: Velocity and displacement

Calculate the mean velocity of the woman in the previous worked example. She took 10 s to walk the 25 m east and a further 7.0 s to walk the 15 m north.

Step 1: Select the correct equation to use in this context.

$$\text{Mean velocity} = \frac{\text{displacement}}{\text{time taken}}$$

Do not be tempted to find the mean velocity of the walk east, then the mean velocity of the walk north and then add them with a vector triangle. This would be wrong!

Step 2: Since the overall displacement has been found through geometry or trigonometry, we can substitute this value into the equation.

$$\text{Mean velocity} = \frac{29.1...\,\text{m}}{10\,\text{s} + 7.0\,\text{s}} = \frac{29.1...\,\text{m}}{17\,\text{s}} = 1.7\,\text{m s}^{-1} \text{ (2 s.f.)}$$ in the direction of the displacement, which is E 31° N.

Relative velocity and subtracting vectors

When measuring velocity you need to be clear what it is measured relative to. If you are flying an aircraft, you need to know your velocity relative to the ground in order to set the correct course, but if there is another aircraft flying nearby, you need to know if you will collide!

To find the velocity of another aircraft relative to yours, you subtract the velocity of your aircraft (relative to the ground) from that of the other one (relative to the ground). Because X − Y = X + (−Y), you subtract a vector by adding one of the same magnitude in the opposite direction.

Worked example: The relative velocity of two aircraft

Figure 4 shows two aircraft on possible collision courses, drawn to scale. Calculate if these two aircraft will collide.

You need to find the velocity of the other aircraft relative to your aircraft to decide this.

Step 1: Add a velocity opposite to that of your aircraft to both aircraft, as shown in Figure 5.

Step 2: Calculate relative velocities by adding the two velocity vectors for each aircraft.

Relative velocity of your plane
= 225 − 225 = 0 m s⁻¹ (since you are not moving relative to yourself).

For the relative velocity of the other aircraft, we cannot calculate this using trigonometric means as the angles are not given. This means that we cannot be sure if this is a right-angled triangle or not.

Measuring the red vector arrow in Figure 6 relative to the scale given gives us an approximation of 170 m s⁻¹.

Figure 6 shows the resultant velocity of the other aircraft relative to your aircraft. The magnitude of the velocity (the speed), by measurement on the vector addition triangle, is 170 m s⁻¹. This is not as important here as the direction, as the magnitude of the velocity tells you only how long it will be before any possible collision.

The direction of the relative velocity vector suggests the other aircraft will pass in front of you. However, it may be best to take evasive action!

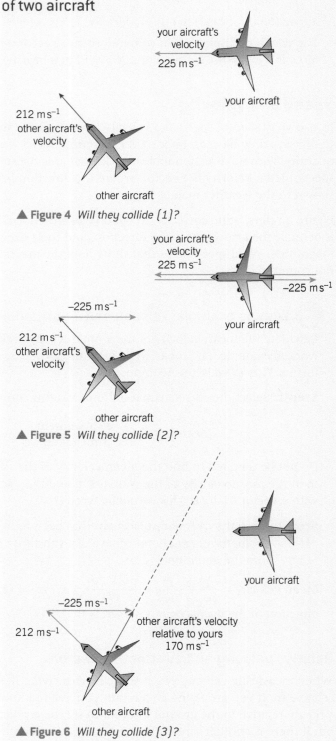

▲ **Figure 4** *Will they collide (1)?*

▲ **Figure 5** *Will they collide (2)?*

▲ **Figure 6** *Will they collide (3)?*

The father of relativity

If you ask most people who first had the idea of relativity, most will say Einstein. Certainly Einstein's theory of Special Relativity in 1905 was a ground-breaking development, but the principle of relativity had first been proposed 273 years earlier by Galileo.

Galileo argued that the natural motion of any object was not slowing down to a stop, as the ancient Greek thinker Aristotle stated, but moving in a straight line at a steady speed, that is, with a constant velocity. He also noted that if you were on a ship travelling at a constant velocity and not rocking from side to side you would be unable to tell that you were moving – a dropped object would seem to fall as if it were on stationary land rather than moving sideways at the same time.

What Galileo realised was that there was no fixed 'zero' of position from which measurements should be made, and that if two places of observation were moving at constant velocity relative to each other, the laws of physics would be the same. Making calculations of what is happening in one such place observed from the other one are called Galilean transformations.

1   The surface of the Earth is not moving at a constant velocity, as it is spinning. Why does it seem that dropped objects fall vertically downwards?

▲ **Figure 7** *Galileo Galilei*

Acceleration

Acceleration was previously defined as the rate at which velocity is changing, that is, $a = \frac{\Delta v}{\Delta t}$. In everyday terms, the acceleration of a car is often described as, for example, 0 to 60 mph in 11.3 s.

We do not normally measure velocity in mph, but in m s^{-1}, and 60 mph = $26.8\,\text{m s}^{-1}$, so the acceleration of this car,

$$a = \frac{26.8\,\text{m s}^{-1} - 0\,\text{m s}^{-1}}{11.3\,\text{s}} = 2.37\,(\text{m s}^{-1} \div \text{s}) = 2.37\,(\text{m s}^{-1} \times \text{s}^{-1}) = 2.37\,\text{m s}^{-2}$$

Acceleration is the rate of change of velocity and, like velocity, it is a vector. Because acceleration is a vector, it can also be represented using a vector arrow, just like with displacement and velocity.

> **Synoptic link**
>
> You have met acceleration in Topic 8.1, Graphs of motion.

Acceleration and *g*

One well-known acceleration is that due to gravity, which is given the special symbol *g*. It has the same value, with small fluctuations, anywhere on the Earth's surface, that is, $g = 9.81\,\text{m s}^{-2}$ downwards.

High accelerations are often quoted in terms of *g*: a dragster car can accelerate from 0 to $44\,\text{m s}^{-1}$ in 0.86 s, giving an acceleration of $51\,\text{m s}^{-2}$.

As $\frac{51\,\text{m s}^{-2}}{9.8\,\text{m s}^{-2}} = 5.2$, this is equal to an acceleration of 5.2*g*.

> **Synoptic link**
>
> You will see how to measure *g* in Topic 8.4, Speeding up and slowing down.

Worked example: A thrown ball

A ball is thrown horizontally at $5.0\,\mathrm{m\,s^{-1}}$. Ignoring any air resistance, calculate its velocity $0.25\,\mathrm{s}$ later. The acceleration due to gravity, $g = 9.8\,\mathrm{m\,s^{-2}}$.

Step 1: Use the definition of acceleration to calculate the change in velocity.

$$a = \frac{\Delta v}{\Delta t} \Rightarrow \Delta v = a\Delta t$$

Step 2: Substitute values in the equation.

$\Delta v = a\Delta t = g\Delta t = 9.8\,\mathrm{m\,s^{-2}} \times 0.25\,\mathrm{s} = 2.45\,\mathrm{m\,s^{-1}}$ vertically downwards

Step 3: Add v_{vertical} to the vector diagram.

Add this change in velocity to the initial velocity of $5.0\,\mathrm{m\,s^{-1}}$ horizontally, as shown in Figure 8. On your vector diagram, label the resultant vector arrow and the angle it makes with the horizontal (or vertical).

Step 4: Use measurement on Figure 8, or geometry and trigonometry, to find the magnitude and direction of the final velocity.

Using Pythagoras' theorem,

$$(\text{resultant magnitude})^2 = (5.0\,\mathrm{m\,s^{-1}})^2 + (2.45\,\mathrm{m\,s^{-1}})^2$$
$$= 31.0...\,\mathrm{m^2\,s^{-2}}$$

Resultant velocity $= \sqrt{31.0\ \ \mathrm{m^2\,s^{-2}}} = 5.6\,\mathrm{m\,s^{-1}}$ (2 s.f.) in the direction θ

where $\tan\theta = \dfrac{2.45\,\mathrm{m\,s^{-1}}}{5.0\,\mathrm{m\,s^{-1}}} = 0.49 \Rightarrow \theta = 26°$ (2 s.f.) below the horizontal.

Note that the direction can be been seen from the labelled diagram.

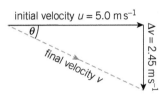

initial velocity $u = 5.0\ \mathrm{m\,s^{-1}}$
θ
final velocity v
$\Delta v = 2.45\ \mathrm{m\,s^{-1}}$

▲ **Figure 8** *A thrown ball*

Hint

Remember that acceleration a is a vector. a has the same direction as the velocity change Δv.

Resolving a vector into components

Just as two vectors can be added to give a resultant vector, so a single vector can be resolved, or split, into two components at right angles to each other. The directions of these components are often horizontal and vertical.

In Figure 9, adding the components s_{H} and s_{V} tip-to-tail will give the displacement s. Note that either ADC or ABC could be the 'tip-to-tail' triangle for this addition. However, as s_{H} and s_{V} both transform the object originally at point A, the resolution into components is usually represented in the way shown in Figure 9.

$$\cos\theta = \frac{\text{AB}}{\text{AC}} = \frac{\text{magnitude of } s_{\text{H}}}{\text{magnitude of } s} \Rightarrow s_{\text{H}} = s\cos\theta$$

$$\sin\theta = \frac{\text{BC}}{\text{AC}} = \frac{\text{magnitude of } s_{\text{V}}}{\text{magnitude of } s} \Rightarrow s_{\text{V}} = s\sin\theta$$

Horizontal and vertical motion can be calculated separately as the acceleration due to gravity acts vertically.

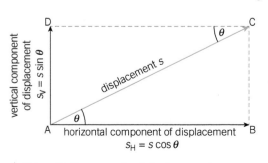

vertical component of displacement $s_{\text{V}} = s\sin\theta$

D   C
θ
displacement s
θ
A   horizontal component of displacement   B
$s_{\text{H}} = s\cos\theta$

▲ **Figure 9** *Components of displacement*

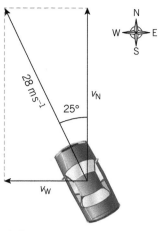

 Worked example: Calculating components of velocity

You are driving at a speed of $28\,\mathrm{m\,s^{-1}}$ in a direction N 25° W. Resolve this velocity into north–south and east–west components.

Step 1: Sketch the situation described in the question, adding rectangles to show the resolution of velocity (Figure 10).

Make sure you keep the directions the same as in the original sketch. You will often find it easiest to superimpose this resolution diagram on top of the sketch. Label the two components clearly.

Step 2: Use trigonometry to find the magnitudes of the two components. The opposite faces of the rectangle are equal, so you need only consider the triangle in which the angle is labelled.

$$\cos(25°) = \frac{\text{adjacent}}{\text{hypotenuse}} = \frac{v_N}{28\,\mathrm{m\,s^{-1}}} \Rightarrow v_N = \cos(25°) \times 28\,\mathrm{m\,s^{-1}}$$
$$= 0.906... \times 28\,\mathrm{m\,s^{-1}} = 25\,\mathrm{m\,s^{-1}}\ (2\ \text{s.f.})$$
$$\sin(25°) = \frac{\text{opposite}}{\text{hypotenuse}} = \frac{v_W}{28\,\mathrm{m\,s^{-1}}} \Rightarrow v_W = \sin(25°) \times 28\,\mathrm{m\,s^{-1}}$$
$$= 0.422... \times 28\,\mathrm{m\,s^{-1}} = 12\,\mathrm{m\,s^{-1}}\ (2\ \text{s.f.})$$

▲ **Figure 10** *Components of velocity*

Summary questions

1 On a treasure map, you must take 20 paces north from the big tree, then 6 paces east, to dig for the treasure. Draw an addition vector diagram for these displacements and use it to find the displacement of the treasure from the tree. (You can ignore the vertical displacement.) *(3 marks)*

2 A rower can row at $2.0\,\mathrm{m\,s^{-1}}$ in still water. She needs to row straight across a river where the current is flowing at $0.8\,\mathrm{m\,s^{-1}}$.

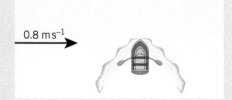

▲ **Figure 11** *Crossing the river*

Draw a vector addition diagram to show her resultant velocity relative to the river banks, and use it to calculate the direction in which she must row. *(4 marks)*

3 An aircraft is flying relative to the air at $200\,\mathrm{m\,s^{-1}}$. The aircraft is heading north-west and there is a $50\,\mathrm{m\,s^{-1}}$ wind relative to the ground heading west. Use components of velocity to find the velocity of the aircraft relative to the ground. *(4 marks)*

8.3 Modelling motion

Specification reference: 4.2b(iii), 4.2c(xi), 4.2d(iii)

Learning outcomes

Describe, explain, and apply:

→ graphical models in terms of vector additions to represent displacement and velocity

→ computational models to represent changes of displacement and velocity in small time steps

→ terminal velocity using an experiment with paper cases in air.

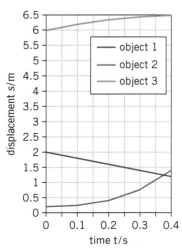

▲ **Figure 1** *Hurricane approaching Florida – a complex system*

▲ **Figure 2** *Displacement–time graph of data in Table 1*

You have so far met descriptions of the behaviour of simple systems, and will meet ideas which explain these behaviours later on. However, real situations can be very complex, with many interacting variables to be considered. This is true of hurricanes.

One approach to analysing such systems is to use the large memory and fast processors of modern supercomputers to perform many simple calculations of tiny parts of the system. These can then be added together to describe the motion of the complete complex system.

These approaches are all based on dividing motion into tiny intervals or increments of time and defining what happens between one such increment and the next.

Iterative computational models

Iterative means step-by-step. An iteration is a single step, usually a tiny time increment. Iterative models track what is happening as time progresses. Computers are normally used for these models but you can best see how they work by working through a few steps manually.

Working through an iterative model

The table shows the displacements of three different objects 1, 2, and 3 after four time increments of 0.1 s. You can assume that all motions are along the same straight line, and all displacements are relative to the same fixed point.

▼ **Table 1** *Displacement data*

Time t / s	Displacement s_1 / m	Displacement s_2 / m	Displacement s_3 / m
0.0	2.0	0.20	6.00
0.1	1.8	0.24	6.20
0.2	1.6	0.40	6.35
0.3	1.4	0.76	6.45
0.4	1.2	1.40	6.50

You can identify the types of motion by looking at the changes in displacement Δs for successive time increments Δt. For object 1, Δs is negative, so it is approaching the fixed point. It is also constant, so the velocity is constant. For object 2, Δs is positive and increasing, so it is accelerating away from the fixed point. For object 3, Δs is positive and decreasing, so it is decelerating away from the fixed point. The motion of these three objects is shown in Figure 2.

Constructing an iterative computational model

In constructing a model, you write a set of instructions defining how v and s will change at regular time intervals Δt.

Velocity v and displacement s after each time interval Δt are calculated using the rules:

$$\text{new } s = \text{previous } s + \Delta s$$
$$\text{new } v = \text{previous } v + \Delta v$$

Figure 3 shows how the extra Δs and Δv in the time increment Δt between some time t and one time interval later, $t + \Delta t$, are added to the previous values. A blue copy of the object with its velocity at time t is duplicated below the object at time $t + \Delta t$ to show how the new velocity is obtained.

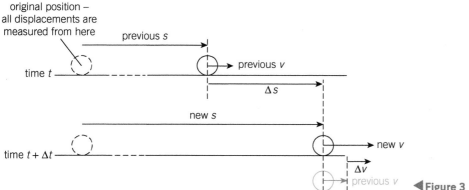

▲ Figure 3 An iterative model

In order to apply this model you need to calculate the change in $s (\Delta s)$, and the change in $v (\Delta v)$.

These come from the definitions of velocity and acceleration.

$$v = \frac{\Delta s}{\Delta t} \Rightarrow \Delta s = v\Delta t \quad \text{and} \quad a = \frac{\Delta v}{\Delta t} \Rightarrow \Delta v = a\Delta t$$

Modelling free fall

To show how an iterative model can work, we will work though three iterations (steps) using a time interval Δt of 0.1 s.

 Worked example: A crude model for free fall

Use an iterative model with time increment 0.1 s to investigate the changes in velocity and displacement for an object falling freely from rest.

In free fall, the acceleration a is constant at $g = 9.8\,\text{m s}^{-2}$. This allows us to use the equation $\Delta v = a\Delta t$ to calculate the change in velocity Δv in each time interval. Each Δv will be the same. As we work through the model you will see how the *updated* values of v affect the values of Δs and therefore s.

▼ Table 2

t/s	$a/\text{m s}^{-2}$	$v/\text{m s}^{-1}$	s/m	
0.0	9.8	0	0	← **Step 1:** Put in the initial conditions of the model
0.1	9.8	$0 + 9.8 \times 0.1 = 0.98$	$0 + 0 \times 0.1 = 0$	← **Step 2:** Use the previous values of a and v to calculate the new v and s
0.2	9.8	$0.98 + 9.8 \times 0.1 = 1.96$	$0 + 0.98 \times 0.1 = 0.098$	← **Step 3:** Repeat Step 2
0.3	9.8	$1.96 + 9.8 \times 0.1 = 2.94$	$0.098 + 1.96 \times 0.1 = 0.294$	← **Step 4:** Repeat Step 2

Note that the velocity goes up in equal steps of $0.98\,\mathrm{m\,s^{-1}}$ every $0.1\,\mathrm{s}$ but that the increase in displacement gets larger and larger in each time interval.

In the model, the changes in v and s occur instantly at the end of each time interval, and then do not change until the end of the next time interval, as shown in the graphs of this model (Figure 4).

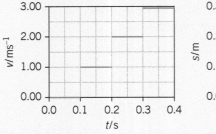 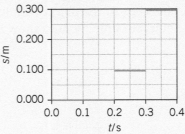

▲ **Figure 4** *Graphs for the free fall model*

In this worked example, you will have noticed how monotonous Step 2, Step 3, and Step 4 become. What would your response be if you were asked to do it for 30 intervals of $0.01\,\mathrm{s}$ instead of 3 intervals of $0.1\,\mathrm{s}$? Or for 300 intervals of $0.001\,\mathrm{s}$? This is one reason why computers are often used.

Pros and cons of the iterative model

Advantages to iterative models

If friction acts on a falling object, the acceleration is not constant but decreases. The faster an object is moving, the greater the upwards frictional force acting on it. Using algebra and calculus to tackle this problem is difficult, but an iterative model can be used just as easily as for the constant acceleration case above.

Problems with iterative models

The graphs in Figure 4 do not look realistic. In reality, when an object is in free fall, the velocity and displacement change all the time, not in sudden jumps, and all the calculated values occur earlier than the graphs suggest as the model uses the previous values of a and v to make the calculation.

One way to improve the model is to use a time interval Δt smaller than the $0.1\,\mathrm{s}$ used above. Figure 5 shows the effect of reducing Δt five-fold, to $0.02\,\mathrm{s}$.

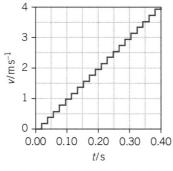 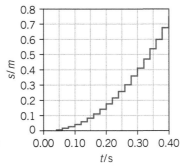

▲ **Figure 5** *An improved model*

Graphical models

An incremental model can be based on vector addition rather than the spreadsheet approach above, and this can be useful when the direction of movement is changing. Although this can be analysed with a computational model using components, a graphical approach is visually clearer.

 Worked example: Tracking the path of an object falling under gravity

For clarity, this will be applied to an object initially moving horizontally. If it were not, all the vectors would be in the same vertical line.

What path would be predicted for an object initially moving horizontally at $5.0\,\text{m}\,\text{s}^{-1}$, analysed with a time increment of $0.1\,\text{s}$? In this question, use $g = 10\,\text{m}\,\text{s}^{-2}$.

Step 1: Find the velocity change Δv caused by gravity each $0.1\,\text{s}$.

As $g = 10\,\text{m}\,\text{s}^{-2}$, $\Delta v = a\Delta t = 10\,\text{m}\,\text{s}^{-2} \times 0.1\,\text{s} = 1.0\,\text{m}\,\text{s}^{-1}$ vertically downwards

Step 2: Choosing a suitable scale, draw a horizontal arrow near the top to represent $5.0\,\text{m}\,\text{s}^{-1}$ horizontally. Divide the horizontal axis of the graph paper into equal intervals each the same width as the initial velocity vector.

At the tip of the initial velocity arrow, draw a $\Delta v = 1.0\,\text{m}\,\text{s}^{-1}$ arrow vertically downwards. Draw the resultant of these two velocities. In this example, the resultant velocities are shown in red whereas the two velocities being added will be in a different colour for each stage.

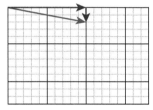

▲ **Figure 6** *Graphical model after Step 2*

Step 3: Continue the resultant arrow on for the same horizontal distance as the length of the initial velocity vector. This is the same length and direction as the previous resultant. Add another velocity change Δv at the tip, and add as before. This is the vector version of new v = previous $v + \Delta v$.

Step 4: Repeat Step 3 again and again until you run out of graph paper.

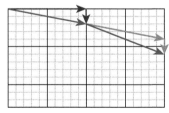

▲ **Figure 7** *Graphical model after Step 3*

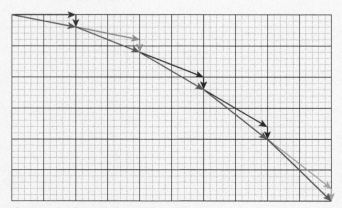

▲ **Figure 8** *Graphical model after Step 6*

Synoptic link

Projectiles will be studied in Topic 9.4, Projectiles.

As this worked example shows, the object thrown horizontally follows a curved path.

Modelling more complex situations

In complex situations there are often simple ideas that can help with the modelling. When an object falls through air, it is common knowledge that the air acts to reduce the acceleration of gravity, and the faster the object falls, the greater the effect of air resistance. Engineering experiments suggest that a good model for this situation is that the downwards acceleration at any time is given by

$$a = 9.8\,\text{m s}^{-2} - Kv^2$$

where v is the velocity of the object through the air and K is a constant which depends on the shape, size, and density of the object. For example, a thin needle of steel will have little air resistance and so a small value of K, but a balloon full of air will have considerable air resistance and a large value of K.

In this worked example you will see how the use of this engineering equation combines with the previous ideas to allow useful predictions of behaviour.

 Worked example: Modelling an object falling through air

Use an iterative model with time increments of 0.1 s to investigate the changes in velocity and displacement over two seconds for an object falling from rest with a value of $K = 0.2\,\text{m}^{-1}$ in the engineering equation $a = 9.8\,\text{m s}^{-2} - Kv^2$. Display your results using a sketch graph.

Step 1: Set up the initial conditions in the first row of Table 3. The object has not started to fall, and there is no air resistance yet, so $a = g$.

Remember that each calculation uses the values of a and v from the previous step.

Step 2: Substitute values into $a = 9.8\,\text{m s}^{-2} - Kv^2$. Use the initial value of v to calculate a at 0.1 s using
$$a_{\text{new}} = 9.8\,\text{m s}^{-2} - (0.2\,\text{m}^{-1}) \times (v_{\text{previous}})^2.$$

Use the initial value of a to calculate v at 0.1 s using
$$v_{\text{new}} = v_{\text{previous}} + a_{\text{previous}} \times \Delta t.$$

Use the initial value of v to calculate s at 0.1 s using
$$s_{\text{new}} = s_{\text{previous}} + v_{\text{previous}} \times \Delta t.$$

Step 3: Repeat Step 2, using the values of v and a at 0.1 s to calculate the new values at 0.2 s.

Step 4: Repeat again, using the values of v and a at 0.2 s to calculate the new values at 0.3 s.

This completes the first four rows of the table.

▼ **Table 3**

t / s	a / m s⁻²	v / m s⁻¹	s / m
0.0	9.8	0	0
0.1	$9.8 - 0.2 \times 0^2 = 9.8$	$0 + 9.8 \times 0.1 = 0.98$	$0 + 0 \times 0.1 = 0$
0.2	$9.8 - 0.2 \times (0.98)^2$ $= 9.61$	$0.98 + 9.8 \times 0.1$ $= 1.96$	$0 + 0.98 \times 0.1$ $= 0.098$
0.3	$9.8 - 0.2 \times (1.96)^2$ $= 9.03$	$1.96 + 9.61 \times 0.1$ $= 2.92$	$0.098 + 1.96 \times 0.1$ $= 0.294$

Using a computer to repeat this several more times gives Figure 9, which shows how the values of acceleration, velocity, and displacement change over the first 2 seconds using this model.

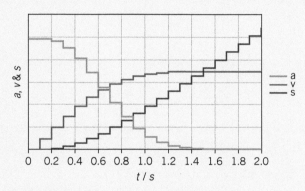

▲ **Figure 9** *Graphs for free fall with air resistance*

Even with the large value of $\Delta t = 0.1$ s, the essential features of free fall with air resistance are clear. The acceleration decreases to zero, the velocity becomes constant, and the displacement graph changes from a curve to a straight line. The constant velocity reached when falling against resistance is referred to as **terminal velocity**.

 Practical 4.2d(iii): Investigating falling cupcake cases

The paper cases used for baking cupcakes fall in quite a regular way without spinning round or capsizing, but they are greatly affected by air resistance and so reach terminal velocity within about a metre of fall.

With simple measurements of displacement and time, and an appropriate choice of experimental conditions, such as the height of the drop and over which part of its fall the cupcake case is being timed, it is possible to investigate how the terminal velocity reached depends on factors such as the mass of the falling object.

Summary questions

1   A car accelerates from rest along a straight, horizontal road at a constant 3.4 m s⁻². Construct an iterative model with $\Delta t = 0.1$ s to find the distance travelled after 0.8 s.
(4 marks)

2   In the Worked example *Modelling free fall with air resistance*, the object ends up moving at a constant velocity of 7.00 m s⁻¹. If the model is repeated with a much smaller time increment, it has the same constant velocity at the end. Explain this.   *(2 marks)*

3   The rower in Figure 10 starts to row across the river at 2.0 m s⁻¹ relative to the water, without attempting to correct her course.

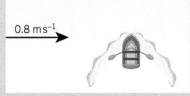

▲ **Figure 10** *Crossing the river*
(4 marks)

Use a graphical model to add the displacements at a time interval of $\Delta t = 1$ s to plot her path over the first four seconds.

8.4 Speeding up and slowing down

Specification reference: 4.2c(iii), 4.2d(ii)

You have seen that vectors and iterative computation can model the motion of objects. This topic deals with the use of algebra, which is often the most flexible approach to problem solving. Algebra is also particularly appropriate when dealing with the physical principles of force and energy, and relating them to motion. To do this, you need to use the kinematic equations that describe motion.

Kinematic equations

There are four equations relating the following variables:

- displacement, s (m)
- initial velocity, u (m s^{-1})
- final velocity, v (m s^{-1})
- uniform acceleration, a (m s^{-2})
- time, t (s).

Each equation is used only for uniformly accelerated motion in a straight line.

$$v = u + at \qquad s = \frac{u + v}{2}\, t \qquad s = ut + \frac{1}{2}\, at^2 \qquad v^2 = u^2 + 2as$$

The kinematic equations are often called the *suvat* equations, because they relate to those five variables.

Where do the equations come from?

The first two equations are simple re-arrangements of definitions you have already met.

$$a = \frac{\Delta v}{t} = \frac{v - u}{t} \Rightarrow at = v - u \Rightarrow v = u + at$$

Average velocity = $\frac{s}{t}$ and, if the acceleration or deceleration is uniform,

average velocity = $\frac{(u + v)}{2}$.

$$\frac{u + v}{2} = \frac{s}{t} \Rightarrow s = \left(\frac{u + v}{2}\right)t$$

The third equation, often the most useful in practice, can be shown by referring to a graph of accelerated motion.

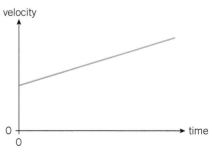

▲ **Figure 1** *Uniformly accelerated motion*

▣ Worked example: Relating an equation to a graph

The graph in Figure 1 shows uniformly accelerated motion. Label u, v, and t, and explain how a and s may be found. Then derive an equation for s in terms of u, a, and t.

Step 1: Draw horizontal and vertical construction lines from each end of the line of the graph.

Step 2: Label u, v, and t on the graph and use this to label the sides.

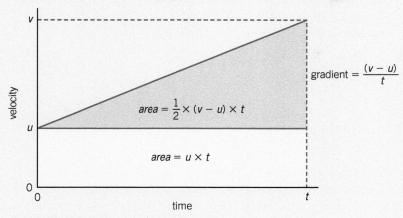

▲ **Figure 2** *Rectangle and triangle marked out*

Step 3: The displacement, s, is the area under the line.

$$\Rightarrow s = \text{area of rectangle} + \text{area of triangle}$$

$$= \text{base} \times \text{height of rectangle} + \frac{1}{2} \times \text{base} \times \text{height of triangle}$$

$$= u \times t + \frac{1}{2} \times t \times (v - u)$$

This is almost there, but it has v and not a.

But the first *suvat* equation is $v = u + at \Rightarrow (v - u) = u + at - u = at$

Step 4: Substitute $v - u = at$ into $s = u \times t + \frac{1}{2} \times t \times (v - u)$

$$s = ut + \frac{1}{2} t \times (at) = ut + \frac{1}{2} tat = ut + \frac{1}{2} at^2$$

Note that ut = area of rectangle = the distance the object would have travelled if it had not accelerated, and $\frac{1}{2} at^2$ = area of triangle = the extra distance travelled by virtue of the gain in speed.

The fourth *suvat* equation can be obtained algebraically by substituting $s = \left(\frac{u + v}{2}\right)t$ into $v = u + at$ in order to eliminate t, but you will not need to do this. In Topic 9.3, Conservation of energy, you will see that it can be explained in energy terms more easily.

 Worked example: How deep is the well?

A stone is dropped down a deep well. A splash is heard 1.5 s after the stone is released.

How deep is this well?

Step 1: List the variables.

displacement s = ?; initial velocity u = 0; final velocity v = ?; acceleration $a = g = 9.8 \, \text{m s}^{-2}$ and time t = 1.5 s.

Study tip

Be methodical in using equations

Before any calculation, write down the values you have for s, u, v, a, and t. Once you have listed these you should be able to see which equation(s) to use. Remember to take account of direction for all vector quantities and that g is a downward acceleration.

Step 2: Eliminate the unknown you do not wish to find.

We need s, so we can eliminate v. The equation we need is the one with s, u, a, and t.

$$s = ut + \frac{1}{2}at^2$$
$$s = 0 \times 1.5\,\text{s} + \frac{1}{2} \times 9.8\,\text{m s}^{-2} \times (1.5\,\text{s})^2$$
$$= 0 + 4.9\,\text{m s}^{-2} \times 2.25\,\text{s}^2 = 11\,\text{m}$$

The well is 11 m deep.

In this example, there was an obvious equation to use. In truth, however, it was not the only way to do it. Even though we had eliminated v as unnecessary, there was no reason why you could not first use $v = u + at$ to calculate v, and then use $s = \left(\frac{u+v}{2}\right)t$ to find s. It just takes a little longer.

Practical 4.2d(ii): Measuring the acceleration due to gravity, g

This can be done using the equation $s = ut + \frac{1}{2}at^2$.

For an object falling from rest, $u = 0$ and $a = g$, so the equation to be used is $s = \frac{1}{2}gt^2$.

Figure 3 shows one arrangement that can be used.

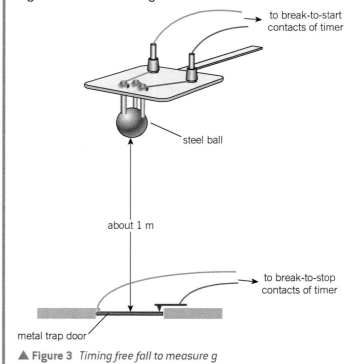

▲ **Figure 3** *Timing free fall to measure g*

The steel ball is usually held in place at the top using an electromagnet. Switching off the electromagnet circuit releases the ball. This breaks the contacts at the top of the measured drop, and starts an accurate timer. Striking the trap door at the bottom of the measured drop breaks those contacts to the timer circuit, stopping the timing.

There are alternative ways of measuring g – light gates can be set up to find the velocity at two different points along the fall of a dense object, attached to a timing card, or the fall of a compact object such as a golf ball can be video recorded and the recording analysed frame-by-frame.

Car stopping distance and speed

In stopping a car, there are two factors to consider. One is the human reaction time, or *thinking time* before you press your foot hard on the brake pedal. The other is the time taken for the car to come to a halt. For each of these times, there is a distance the car will travel – the *thinking distance* is the distance the car travels before your press the pedal, and the *braking distance* is the distance the car travels during deceleration. The stopping distance is the sum of these two distances, and this must be less than the distance to the hazard you have just seen if an accident is not to happen.

Two extracts on car stopping distances from the Highway Code are shown in Figure 4.

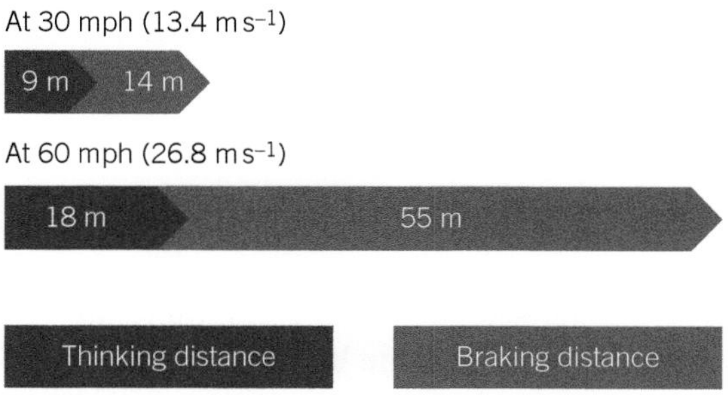

At 30 mph (13.4 m s^{-1})

| 9 m | 14 m |

At 60 mph (26.8 m s^{-1})

| 18 m | 55 m |

| Thinking distance | Braking distance |

▲ **Figure 4** *Thinking distance and braking distance*

As the speed of the car increases, the thinking distance increases in proportion – this is because the speed has doubled, from 30 mph to 60 mph in this case, so it travels twice as far in the same thinking time.

However, the braking distance goes up about four times because there are two factors at play:

- the average speed has doubled, so it travels twice as far each second
- Δv has doubled from 30 mph to 60 mph, so the time taken for the same deceleration to reduce the speed to zero becomes twice as long.

The thinking and stopping distances in the Highway Code are based on assumed values of human reaction time and a car's deceleration. These factors can be affected by a number of factors such as tiredness or icy road conditions.

Worked example: Car stopping distance

The Highway Code suggests safe braking distances for cars going at different speeds. Calculate the braking deceleration that the Highway Code assumes. At 30 mph, which is 13.4 m s⁻¹, the Highway Code indicates the breaking distance to be at least 14 m.

Step 1: List the variables.

distance $s = 14$ m; initial velocity $u = 13.4$ m s⁻¹;
final velocity $v = 0$ m s⁻¹; acceleration $a = ?$; and $t = ?$

Step 2: Eliminate the unknown you do not need to find.

This is t.

Step 3: Identify the equation required.

The equation we need is the one with s, u, v, and a.

$$v^2 = u^2 + 2as$$

Step 4: Substitute values into the equation.

$$(0 \, \text{m s}^{-1})^2 = (13.4 \, \text{m s}^{-1})^2 + 2 \times a \times 14 \, \text{m}$$

$$0 = 179.56 \, \text{m}^2\text{s}^{-2} + 28 \, \text{m} \times a$$

$$-179.56 \, \text{m}^2\text{s}^{-2} = 28 \, \text{m} \times a$$

$$a = \frac{-179.56 \, \text{m}^2\text{s}^{-2}}{28 \, \text{m}} = -6.4 \, \text{m s}^{-2} \ (\text{2 s.f.})$$

The minus sign shows that it is decelerating, that is, $\Delta v = v - u$ is in the opposite direction to s and u.

Applying the kinematic equations to two-dimensional motion

As displacement, velocity, and acceleration are all vectors, there are often cases where they do not all lie on the same line. To use the equations in these cases, it is necessary to resolve one or more of these values into appropriate directions.

When resolving, you often have a choice as to which perpendicular directions you choose to resolve along, as shown in this worked example.

 Worked example: Trolley on a ramp

A trolley is released from rest to roll a distance of 2.5 m along a ramp inclined at 15° to the horizontal. Calculate the speed it is going when it reaches the 2.5 m mark. You can ignore friction.

Step 1: Choose two directions at right angles to resolve components along. You could resolve vertically and horizontally, but here it is easier to choose one of the directions to be along the ramp, as that is the way the trolley is moving.

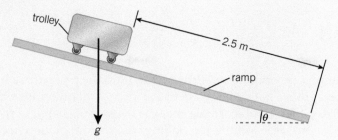

▲ **Figure 5** *Trolley on an inclined ramp 1*

The components of the acceleration due to gravity, g, are $g_{perpendicular}$, which has no effect, and $g_{parallel}$, which accelerates the trolley along the ramp.

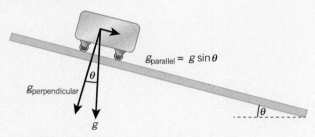

▲ **Figure 6** *Trolley on an inclined ramp 2*

Step 2: List the variables.

Along the ramp, $s = 2.5$ m; $u = 0$; $v = ?$; $a = g_{parallel}$; $t = ?$

Step 3: Find $g_{parallel}$.

$$g_{parallel} = g\sin\theta = 9.8 \text{ m s}^{-2}\sin(15°) = 2.53... \text{ m s}^{-2}$$

The rest of this worked example follows the example on for car stopping distance.

Step 4: Eliminate the unwanted variable.

This is t.

Step 5: Identify the correct equation.

This is the one with s, u, v, and a.

$$v^2 = u^2 + 2as$$

Step 6: Substitute values into the equation.

$$v^2 = u^2 + 2as = 0^2 + 2 \times 2.53... \text{ m s}^{-2} \times 2.5 \text{ m} = 12.6... \text{ m}^2\text{s}^{-2}$$

$$v = \sqrt{(12.6... \text{ m}^2\text{s}^{-2})} = 3.6 \text{ m s}^{-1} \text{ (2 s.f.)}$$

Note here that $v = -3.6 \text{ m s}^{-1}$ is also a solution to $v = \sqrt{(12.7 \text{ m}^2\text{s}^{-2})}$. In this case the question asked for the speed, so the direction is not required.

Summary questions

1 Use the braking distance data in Figure 4 to show that the deceleration to a final velocity $v = 0$ is the same at 60 mph as it was at 30 mph.   (*3 marks*)

2 A car is travelling in a straight line. It has an initial velocity of 12 m s^{-1}. It accelerates at 3.5 m s^{-2} for 4 s, then decelerates at 1.5 m s^{-2} for 3.5 s. By considering the acceleration and deceleration stages separately, calculate the distance travelled since the start of the acceleration.   (*4 marks*)

3 A ball is thrown upwards at a velocity of 11 m s^{-1} at an angle of 50° to the horizontal. By resolving the velocity into appropriate components, find the time taken to reach its highest point.   (*3 marks*)

4 A ball is thrown vertically upwards at 12 m s^{-1} to dislodge a kite stuck in a tree 5 m above the thrower. Calculate the time before the ball reaches the kite. Explain why this question has two solutions.   (*5 marks*)

Galileo takes on gravity

In a famous experiment published in 1638, Galileo described how he timed bronze balls rolling down a straight, smooth groove in an inclined plank. He reasoned that the pattern of changing speed would be the same as for a free-falling ball, the only difference being that its slower rate of change would make it easier to measure. He had already deduced that uniform acceleration would produce displacements proportional to the square of the time taken, and he wished to verify this.

▲ **Figure 1** *A 19th century painting showing Galileo (the tallest figure, just left of the centre) demonstrating his revolutionary concepts to Don Giovanni de Medici and members of his court.*

Galileo wrote, in *Dialogues Concerning Two New Sciences*, translated by Crew & de Salvio

"We rolled the ball along the channel, noting, in a manner presently to be described, the time required to make the descent. We repeated this experiment more than once in order to measure the time with an accuracy such that the deviation between two observations never exceeded one-tenth of a pulse beat. Having performed this operation and having assured ourselves of its reliability, we now rolled the ball only one-quarter the length of the channel; and having measured the time of its descent, we found it precisely one-half of the former. Next we tried other distances, comparing the time for the whole length with that for the half, or with that for two-thirds, or three-fourths, or indeed for any fraction; in such experiments, repeated a full hundred times, we always found that the spaces traversed were to each other as the squares of the times, and this was true for all inclinations of the place, i.e. of the channel, along which we rolled the ball.

For the measurement of time, we employed a large vessel of water placed in an elevated position; to the bottom of this vessel was soldered a pipe of small diameter giving a thin jet of water, which we collected in a small glass during the time of each descent, whether for the whole length of the channel or for a part of its length; the water thus collected was weighed, after each descent, on a very accurate

*balance; the differences and ratios of these weights gave us the differences and
ratios of the times, and this with such accuracy that, although the operation was
repeated many, many times, there was no appreciable discrepancy in the results"*

Summary questions

1   Galileo described in detail how he lined the groove in the plank with parchment and polished it, and used a
    smooth, very round bronze ball. Explain why he needed to take these steps.

2   Galileo had earlier reasoned that the displacement is proportional to the square of the time for constant
    acceleration. Give the equation that describes this kind of motion.

3   Figure 2 illustrates Galileo's apparatus. The end of the 12-cubit plank was lifted by different heights between 1 and
    2 cubits, and various fractions of the measured length are marked.

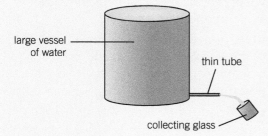

Galileo's 'stop clock'

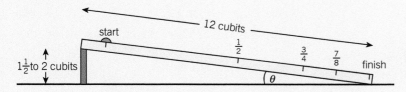

The inclined plank

▲ Figure 2

Study tip

Cubits were an old unit used for
measuring distance. One cubit is
equal to roughly half a metre.

a   Calculate the maximum and minimum values of the angle θ between the plank and the horizontal that he
    would have used.

b   Galileo would have done this experiment with assistants. Describe a procedure that they might have followed.

c   Accurate timing was not possible in the 17th Century, but weighing could be done to great precision. Galileo
    writes 'We repeated this experiment more than once in order to measure the time with an accuracy such that
    the deviation between two observations never exceeded one-tenth of a pulse beat'. Suggest how he might have
    calibrated his 'stop clock' in order to convert his weights into 'pulse beats'.

4   Using a value for θ of 8°,
    a   calculate the component of the acceleration of gravity down the slope $g = 9.8 \, \text{m s}^{-2}$
    b   use this value of acceleration to calculate the time taken to travel a distance of 12 cubits along the
        plank 1 cubit = distance from the elbow to the tip of the middle finger
    c   given that Galileo wrote that 'the deviation between two observations never exceeded one-tenth of a pulse
        beat', write down the expected time from **(b)** together with its uncertainty.

5   Galileo is often described as the person who did most to establish scientific method. Use examples from the
    extract of his book given above to explain how his approach to this experiment supports this claim.

Practice questions

1   A car travelling in a straight line accelerates at a constant rate from rest to a velocity of $20\,\mathrm{m\,s^{-1}}$ over a time of 4.0 s, continues at this velocity for 2.0 s, and then decelerates at $2.5\,\mathrm{m\,s^{-2}}$ for 3.0 s.

Which of following shows a velocity–time graph for this motion?   (*1 mark*)

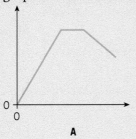
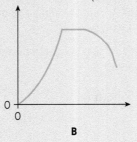

A   **B**

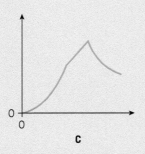
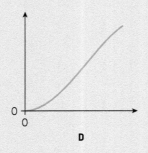

C   **D**

▲ Figure 1

2   Which of the graphs, **A**, **B**, **C** or **D**, given in question 1, shows a displacement–time graph for the motion described?   (*1 mark*)

3   A stone is thrown with a velocity of $8.0\,\mathrm{m\,s^{-1}}$ down a well 20 m deep. What is its speed to 2 s.f. when it hits the bottom? ($g = 9.8\,\mathrm{m\,s^{-2}}$)

   a $20\,\mathrm{m\,s^{-1}}$

   b $21\,\mathrm{m\,s^{-1}}$

   c $390\,\mathrm{m\,s^{-1}}$

   d $460\,\mathrm{m\,s^{-1}}$ (*1 mark*)

4   A woman dives from a high diving board into a swimming pool. She starts the dive by leaping upwards. Sketch a velocity–time graph for the dive from the moment she leaves the board to the moment that she enters the water. Label the graph fully.

                               (*3 marks*)

5   Table 1 gives data values for an object falling freely from rest under gravity.

▼ Table 1

t/s	s/m
0	0
0.5	1.2
1.0	5.0
1.5	11
2.0	20
2.5	31

Plot a displacement–time graph from these data. By finding the velocity at two appropriate places, determine the acceleration due to gravity.   (*6 marks*)

6   A hiker travels 8.4 km due north, and then 5.2 km due west. Represent the vector addition on graph paper and calculate the displacement from the starting point.   (*4 marks*)

7   A stone is thrown horizontally at $15\,\mathrm{m\,s^{-1}}$ from the top of a vertical cliff, 50 m above the sea. Calculate the distance from the bottom of the cliff to the place where the stone hits the water. $g = 9.8\,\mathrm{m\,s^{-2}}$.   (*3 marks*)

8   A block of ice slides down a smooth plane which makes an angle of 18° with the horizontal. The block starts from rest.

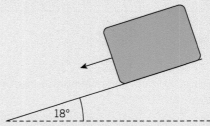

▲ Figure 3

   a   Calculate the component of g parallel to the plane.   (*1 mark*)

   b   Calculate the time taken for the block to slide 48 cm from rest along the plane, assuming there are no frictional forces.   (*1 mark*)

9 Figure 1 is the velocity-time graph for a short train journey between two stations joined by a straight track.

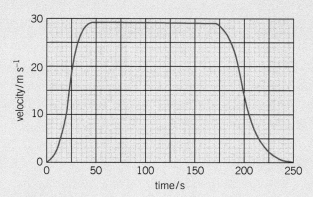

▲ Figure 2

a Describe, in words, the journey shown in Figure 2. *(3 marks)*

b The graph is not symmetrical. Suggest and explain a reason why the graph for the last 20 s of travel is not a mirror-image for the first 20 s of travel. *(2 marks)*

c Calculate the acceleration at time $t = 215$ s. Show your working clearly. *(3 marks)*

d Use the graph to find the distance between the two stations. *(3 marks)*

10 An aircraft is scheduled to fly from London to Belfast, a distance of 510 km in a direction N 40° W (a bearing of 320°). The aircraft has a cruising speed of 240 m s⁻¹ in still air.

a Calculate the northwards and westward component of the aircraft's velocity when travelling on a still day. *(2 marks)*

b On the day of the flight, there is a wind of velocity 15 m s⁻¹ towards the east.

Using a clearly-labelled scale drawing, find the direction in which the aircraft must fly to reach Belfast without any change of course, and the magnitude of the velocity of the aircraft relative to the ground. *(5 marks)*

c Find the extra time the aircraft will take to fly to Belfast in this wind, compared with the time it would take on a still day. *(2 marks)*

11 Two students are measuring the acceleration due to gravity, g, by timing the fall of a ball bearing.

a They time the ball falling five times for a height of 0.65 m and obtain the results in Table 2.

▼ Table 2

time/s	0.361	0.372	0.354	0.378	0.367

Use the data to find the mean time taken for the ball to fall this height. Give the uncertainty. *(2 marks)*

b The students repeat the experiment for a range of heights, obtaining the results shown in Table 3.

▼ Table 3

s/m	t/s	Δt/s
0.65		
0.70	0.38	0.03
0.75	0.40	0.03
0.80	0.42	0.04
0.85	0.43	0.04
0.90	0.44	0.05

Copy Table 3, including your results from part **(a)**, and add an extra column headed t^2/s^2. Fill in the values for this column. *(2 marks)*

c Explain why a graph of s (on the y-axis) against t^2 (on the x-axis) would be expected to give a straight line through the origin, and explain how the gradient is related to g, the acceleration due to gravity. *(3 marks)*

d Use the data in the table from part (b) to plot a graph of s against t^2, including uncertainty bars for t, and use this to obtain a value for g and its uncertainty. You can assume all the uncertainty bars are of the same size. *(6 marks)*

e Identify any systematic error in the experiment, and suggest and explain one way in which the two students could improve their experiment. *(3 marks)*

9 MOMENTUM, FORCE, AND ENERGY
9.1 Conservation of momentum
Specification references: 4.2a[ix], 4.2b[i], 4.2c[iv], 4.2c[vi]

▲ **Figure 1** *A collision*

▲ **Figure 2** *A cannon being fired*

What does a game of snooker have in common with firing a cannon? They both illustrate one of the most useful and important principles of physics, which can be applied in contexts ranging from nuclear physics to sending space probes around the Solar System.

Collisions and explosions

The collision of two snooker balls and the firing of a cannon both illustrate the principle of conservation of momentum. Each of these interactions is best analysed with 'before' and 'after' diagrams.

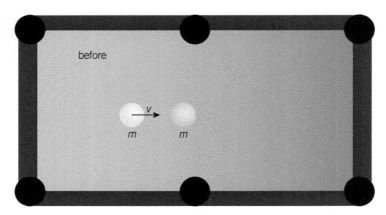

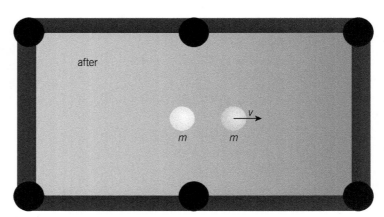

▲ **Figure 3** *Collision in snooker*

The two snooker balls have identical mass, m. In a 'stun' shot the white ball, initially with velocity v, stops on impact and the yellow ball moves off with the same velocity v. If the balls had different masses, this would not have happened – a very light white ball hitting a massive yellow one would rebound, whilst the yellow one would move only very slowly as a consequence.

In the cannon, the cannon and ball are both stationary before firing. When the gunpowder explodes, the lighter ball moves off at a great speed v_{ball}, whilst the heavier cannon recoils with a smaller velocity v_{cannon}.

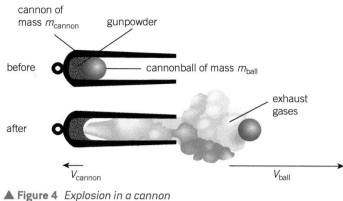

▲ **Figure 4** *Explosion in a cannon*

Conservation of momentum

In the examples above, both the mass of the interacting objects and their velocities need to be considered when describing the interaction. The linking factor in these two events is **momentum**, p, which is defined as

$$\text{momentum } p \text{ (kg m s}^{-1}) = \text{mass } m \text{ (kg)} \times \text{velocity } v \text{ (m s}^{-1})$$

 Worked example: Momentum of a bullet

Calculate the momentum of a 3.0 g bullet travelling at $180 \, \text{m s}^{-1}$.

Step 1: Identify the equation.

$$p = mv$$

Step 2: Substitute values into the equation, making sure to use SI base units.

$$p = 3.0 \times 10^{-3} \, \text{kg} \times 180 \, \text{m s}^{-1} = 0.54 \, \text{kg m s}^{-1}$$

Note the units of momentum are $\text{kg} \times \text{m s}^{-1} = \text{kg m s}^{-1}$.

> **Hint**
>
> Like velocity, momentum is a vector.

For the two snooker balls, the total momentum before the collision = $mv + 0 = mv$, where v is taken to be positive in the direction left→right. The total momentum after the collision = $0 + mv = mv$, the same as the total momentum before the collision.

For the cannon firing, the masses are different, but the velocities after firing are also different, and in opposite directions – if the cannon is 200 times more massive than the cannonball ($m_{cannon} = 200 \, m_{ball}$), then its recoil velocity will be 200 times smaller ($v_{ball} = -200 \, v_{cannon}$). The total momentum before the explosion = 0, as nothing is moving. After the explosion, ignoring the momentum of the exhaust gases,

$$\begin{aligned}\text{Total momentum} &= m_{cannon} \times v_{cannon} + m_{ball} \times v_{ball} \\ &= 200 \, m_{ball} \times v_{cannon} + m_{ball} \times (-200 \, v_{cannon}) \\ &= 200 \, m_{ball} \times v_{cannon} - 200 \, m_{ball} \times v_{cannon} = 0\end{aligned}$$

These examples illustrate an important law in physics – *the principle of conservation of momentum*. This principle states that, for *any* interaction, total momentum before = total momentum after.

The two examples above were simplified to make the issues clear. In reality, there are other interactions taking place in both cases. The principle holds true even in those cases – if all of the momentum changes in any interaction are included, then momentum is conserved.

Worked example: Recoil of a cannon

One type of 17th century cannon, a demi-culverin, had a mass of 1500 kg and could fire a 3.6 kg cannonball at a speed of 120 m s^{-1}. Calculate the recoil velocity of the cannon, and explain why it would be larger than this in practice.

Step 1: Calculate the momentum p_{ball} of the cannonball after the explosion.

$$p_{ball} = m_{ball} \times v_{ball} = 3.6\,\text{kg} \times 120\,\text{m s}^{-1} = 432\,\text{kg m s}^{-1}$$

Step 2: Apply the principle of conservation of momentum to the total momentum before and after firing.

Total momentum before = 0, so total momentum after also must = 0

Step 3: Calculate the momentum of the cannon after explosion, p_{cannon}.

From Step 2, $p_{cannon} + p_{ball} = 0$

$p_{cannon} + 432\,\text{kg m s}^{-1} = 0$ so $p_{cannon} = -432\,\text{kg m s}^{-1}$

Step 4: Use the definition of momentum to find the recoil velocity v_{cannon}.

$$p_{cannon} = m_{cannon} \times v_{cannon}$$

$$-432\,\text{kg m s}^{-1} = 1500\,\text{kg} \times v_{cannon}$$

$$v_{cannon} = \frac{-432\,\text{kg m s}^{-1}}{1500\,\text{kg}} = -0.29\,\text{m s}^{-1}\ (2\ \text{s.f.})$$

In practice, the gases produced in the explosion also have momentum. As the total forward momentum of cannonball + exhaust gases must be larger than 432 kg m s^{-1}, the momentum of the cannon must be of the same size, so it will recoil faster than 0.29 m s^{-1}.

Other types of collisions

You have now seen the principle of conservation of momentum applied to two situations. This principle can be applied for every type of collision, even when two bodies move off in the same direction afterwards, when two bodies collide head-on, or when two bodies make a glancing collision and move off at different angles to their path of approach.

 Worked example: A glancing collision

A white snooker ball of mass 0.16 kg moving at 6.0 m s^{-1} hits a yellow ball of the same mass. After the collision, the two balls move off as shown in Figure 5. Calculate the final speed v of the yellow ball.

→

This can be solved by resolving the velocities in two directions, x and y.

Take x-direction to be the initial path of the white ball (where positive is left to right).

Take y-direction to be at right-angles to the x-direction (where positive is up).

By applying the principle of conservation of momentum to the y-direction, the momentum component for the white ball will be negative.

Step 1: Find the component of momentum for each ball after the collision in the y-direction.

White: $p_y = mv \cos(30°)$
$\qquad = 0.16\,\text{kg} \times -3.0\,\text{m s}^{-1} \times 0.866$
$\qquad = -0.41568\,\text{kg m s}^{-1}$

Yellow: $p_y = mv \cos(60°)$
$\qquad = 0.16\,\text{kg} \times v \times 0.5 = 0.08\,\text{kg} \times v$

Step 2: Apply the principle of conservation of momentum in the y-direction.

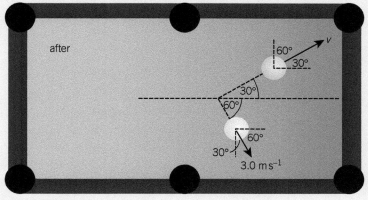

▲ **Figure 5** *Glancing collision*

$$p_y \text{ before} = 0 = p_y \text{ after}$$

$$0 = (-0.41568\,\text{kg m s}^{-1}) + (0.08\,\text{kg} \times v)$$

$$0.08\,\text{kg} \times v = 0.41568\,\text{kg m s}^{-1} \Rightarrow v = \frac{0.41568\,\text{kg m s}^{-1}}{0.08\,\text{kg}} = 5.2\,\text{m s}^{-1}\ (2\ \text{s.f.})$$

This question can also be solved if you apply the principle of conservation of momentum in the x-direction. You should reach the same answer from the following calculation.

$$0.16\,\text{kg} \times 6.0\,\text{m s}^{-1} = (0.16\,\text{kg} \times 3.0\,\text{m s}^{-1} \times \cos(60°)) + (0.16\,\text{kg} \times v \times \cos(30°))$$

Summary questions

1 Calculate the momentum of   **a** a 1000 kg car travelling at 60 mph $(27\,\text{m s}^{-1})$;   *(1 mark)*

   b a person walking at a typical walking pace. (You will need to estimate values.)   *(3 marks)*

2 In a nuclear reaction, an atom of mass 4.0×10^{-25} kg changes by emitting a particle of mass 6.6×10^{-27} kg at a speed of $1.6 \times 10^7\,\text{m s}^{-1}$. Calculate the speed at which the new atom recoils. State any assumptions that you make.   *(3 marks)*

3 In a head-on collision between a car of mass 900 kg and a van of mass 1300 kg, the two vehicles, locked together, skid in the direction the car was travelling. The van was travelling at $11\,\text{m s}^{-1}$ before impact. Calculate whether the car was exceeding the speed limit $(13.4\,\text{m s}^{-1})$ before the two vehicles crashed.   *(3 marks)*

9.2 Newton's laws of motion and momentum

Specification references: 4.2a(viii), 4.2a(ix), 4.2b(i), 4.2c(v)

▲ **Figure 1** *Statue of Sir Isaac Newton in Trinity College, Cambridge. Of Newton, Wordsworth wrote, "Newton with his prism and silent face, the marble index of a mind forever voyaging through strange seas of thought, alone."*

Study tip

Newton's first law can be summarised as:

A body will have a constant velocity, or stay at rest, unless an external force acts upon it.

There is often confusion between mass and force, which is made worse by using units such as kg, stones, and pounds, which are measures of **mass**, or the amount of matter in a body, and calling these units of weight. **Weight** is a **force**, and all forces are pushes and pulls – the weight of a mass is the pull of gravity acting on it.

In 1687 Isaac Newton published his *Principles of Natural Philosophy*, arguably the most important scientific book ever written, in which he used mathematics to explain the relationship between force and motion and the movements of planets and moons. He started this work with three assumptions. These he called his *three laws of motion*.

Newton's first law — inertia

Since the time of Aristotle, most people believed that things will not move unless they are pushed or pulled, and that moving bodies eventually 'lost impetus' and slowed down. Galileo showed that this was not true, and Newton's first law summarises this idea of **inertia** – that the momentum of a body will not change unless a force acts on it.

Newton's second law — force

A weaker force and a stronger force will both change the momentum of an object, but the stronger force will make it happen faster. This means that force is proportional to the rate at which momentum changes. Since $p = mv$, we can say

$$F \propto \frac{\Delta p}{\Delta t} \propto \frac{\Delta(mv)}{\Delta t}$$

This is Newton's second law. As momentum change Δp is a vector quantity, force F must also be a vector quantity in the same direction.

The newton

Although the equation $F \propto \dfrac{\Delta(mv)}{\Delta t}$ implies the need for a constant of proportionality, providing that we use SI base units (kg, $\mathrm{m\,s^{-1}}$, and s) then $F = \dfrac{\Delta(mv)}{\Delta t}$ where F is in newtons (N).

Force and acceleration

Providing that the mass of a body stays constant then we can write Newton's second law in a different form. As $p = mv$, if m stays constant then $\Delta p = m\Delta v$. The equation now becomes

$$F = \frac{\Delta p}{\Delta t} = \frac{m\Delta v}{\Delta t} = m\frac{\Delta v}{\Delta t} = ma$$

$F = ma$ is often the most convenient form of Newton's second law.

Worked example: Force accelerating a car

A luxury car of mass 2400 kg accelerates from 0 to 60 mph ($26.8\,\mathrm{m\,s^{-1}}$) in 4.7 s. Calculate the mean resultant force on the car during that time.

There are two ways of doing this.

Method 1: Using momentum change

Step 1: Calculate the change in momentum.

$$\text{Initial momentum} = 0 \text{ (it's not moving)}$$

$$\text{Final momentum } p = mv = 2400\,\mathrm{kg} \times 26.8\,\mathrm{m\,s^{-1}}$$
$$= 64\,300\,\mathrm{kg\,m\,s^{-1}}$$

$$\text{Momentum change } \Delta p = 64\,300\,\mathrm{kg\,m\,s^{-1}} \; 0$$
$$= 64\,300\,\mathrm{kg\,m\,s^{-1}}$$

Step 2: Divide by the time it took for that momentum change.

$$F = \frac{\Delta p}{\Delta t} = \frac{64\,300\,\mathrm{kg\,m\,s^{-1}}}{4.7\,\mathrm{s}} = 14\,000\,\mathrm{N}\ (2\ \mathrm{s.f.})$$

Method 2: Using acceleration

Step 1: Use the *suvat* equations to find the acceleration, *a*.

$$s = ?,\ u = 0,\ v = 26.8\,\mathrm{m\,s^{-1}},\ a = ?,\ \text{and } t = 4.7\,\mathrm{s}$$

Step 2: Discard the variable which is not needed (*s*) and use the equation with the other four variables.

$$v = u + at \Rightarrow a = \frac{v - u}{t} = \frac{26.8\,\mathrm{m\,s^{-1}} - 0}{4.7\,\mathrm{s}} = 5.70...\,\mathrm{m\,s^{-2}}$$

Step 3: Substitute values into $F = ma$.

$$F = ma = 2400\,\mathrm{kg} \times 5.70...\,\mathrm{m\,s^{-2}} = 14\,000\,\mathrm{N}\ (2\ \mathrm{s.f.})$$

> **Study tip**
>
> Newton's second law can be summarised as:
>
> *The momentum change per second produced by an external force is in the same direction as the force, and proportional to it.* $F = \dfrac{\Delta p}{\Delta t} = ma$

> **Hint**
>
> Remember that acceleration,
>
> $a = \dfrac{\text{change in velocity } \Delta v}{\text{change in time } \Delta t}$

> **Synoptic link**
>
> You have met acceleration in Topic 8.1, Graphs of motion.

> **Synoptic link**
>
> You first met *suvat* equations in Topic 8.4, Speeding up and slowing down.

Impulse

The statement of Newton's second law $F = \dfrac{\Delta p}{\Delta t}$ can be rearranged to give $\Delta p = F\Delta t$. $F\Delta t$ is called the **impulse** of the force. If the force is larger, or the time it acts for is larger, then the impulse is greater. This means there is a greater momentum change produced.

Crumple zones in cars are designed to increase the time Δt taken in a collision. Since $\Delta p = F\Delta t$, for the same change in momentum the force F on the car is reduced as the time Δt is increased.

Seat belts and air bags have exactly the same role in protecting the driver and passengers – seat belts gradually stretch and air bags gradually deflate, increasing the time taken for a person to come to a halt if the car stops very suddenly.

The relationship between impulse and momentum change can be applied in cases where F is not constant, as in kicking a football (Figure 2).

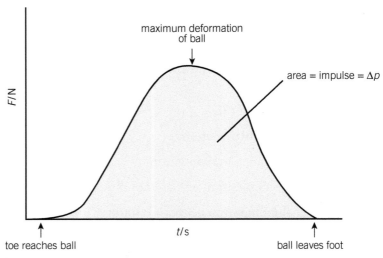

▲ **Figure 2** *Kicking a football*

Synoptic link

Using the total area under a *v–t* graph to find distance travelled was covered in Topic 8.1, Graphs of motion.

When the toe first meets the ball, the ball just starts to deform, so the force is not great. The force rises to a maximum as the ball accelerates away and drops to zero as the ball leaves the foot. For a constant force, the impulse $F\Delta t$ would be the area under a line Δt long and F high. In this case, the impulse is the total area under the graph, just as the distance travelled is the total area under a $v–t$ graph.

Newton's third law

This is exactly equivalent to the principle of conservation of momentum. Figure 3 shows an interaction between two bodies **A** and **B** where **A** gains momentum and **B** loses momentum. The nature of the force between **A** and **B** is not important – it could be a gravitational or electromagnetic force, or simply from contact.

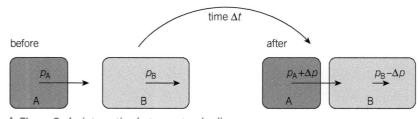

▲ **Figure 3** *An interaction between two bodies*

Momentum is always conserved, so the momentum lost by B is that gained by A. A's momentum change is $+\Delta p$ and B's momentum change is $-\Delta p$. The momentum changes occur over the same time Δt, so from $F_A = \dfrac{\Delta p}{\Delta t}$ and $F_B = \dfrac{-\Delta p}{\Delta t}$, $F_A = -F_B$.

This is true for *all* interactions. Forces always come in pairs, equal in size and opposite in direction. If **A** pulls **B** in one direction, then **B** pulls **A** with an identical force in the opposite direction.

Study tip : Draw a diagram

To avoid confusion, always draw clear diagrams showing all the forces present. Remember that the two forces in a Newton's third law pair act on different objects.

Mass and weight

A free-falling object of mass m near the Earth's surface accelerates downwards at $g = 9.8\,\text{m s}^{-2}$. From Newton's second law, the force needed to produce this acceleration is $F = ma = mg$. This force is called the *weight* of the object. A supported object, such as a book on a shelf, is not accelerating and so by Newton's first law has zero resultant force – the shelf must be exerting an equal upwards force.

The weight mg is caused by the gravitational pull of the Earth. By Newton's third law,

downwards pull of the Earth   =   upwards pull of the book
on the book   on the Earth

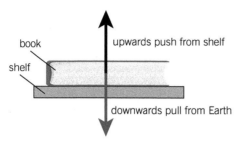

▲ **Figure 4** *Forces on a book on a shelf*

 Worked example: Does the Earth move?

A 65 kg diver dives from a 10 m diving board into a pool. He takes 1.4 s to cover the 10 m. How far does the Earth move upwards? Mass of Earth = 6.0×10^{24} kg.

Step 1: Find the force on the diver.

The force is the diver's weight $= mg = 65 \text{ kg} \times 9.8 \text{ m s}^{-2}$
$$= 637 \text{ N downwards}$$

Step 2: Identify Newton's third law pair.

Downward force on diver = upward force on the Earth
$$= 637 \text{ N upwards}$$

Step 3: Find the upwards acceleration of the Earth using Newton's second law.

$$F = ma \Rightarrow a = \frac{F}{m} = \frac{637 \text{ N}}{6.0 \times 10^{24} \text{kg}} = 1.06... \times 10^{-22} \text{ m s}^{-2} \text{ upwards}$$

Step 4: Use the *suvat* equations to find the displacement s.

$s = ?$, $u = 0$, $v = ?$, $a = 1.1 \times 10^{-22}$ m s^{-2}, $t = 1.4$ s. Eliminate v which is not needed.

$s = ut + \frac{1}{2}at^2 = 0 + \frac{1}{2}at^2 = 0.5 \times 1.1 \times 10^{-22} \text{ m s}^{-2} \times (1.4 \text{ s})^2$
$$= 1.0 \times 10^{-22} \text{ m (2 s.f.)}$$

Yes, the Earth does move, but only by a distance of 1.0×10^{-22} m, which is approximately 10^{12} times smaller than the size of a hydrogen atom.

Summary questions

1 A car of mass 1100 kg travelling at 15 m s^{-1} collides with a wall and stops in a time of 0.1 s. Calculate the deceleration of the car and use this value to find the average force acting on it during the collision. *(3 marks)*

2 When a gun fires a bullet, it recoils. Explain this using:

    a   the principle of conservation of momentum; *(1 mark)*

    b   Newton's third law. *(2 marks)*

3 When a car stops suddenly, you feel yourself being thrown forward. Explain in terms of Newton's laws what is actually happening. *(2 marks)*

4 During the collision in question 1, a passenger of mass 75 kg is decelerated uniformly to rest by a seatbelt. The belt stretches so much that the passenger travels a total distance of 2.1 m between the front of the car hitting the wall and the passenger coming to rest. Calculate the impulse of the force acting on the passenger and use this to calculate the mean force acting on him. *(4 marks)*

5 A rocket engine ejects 500 g of exhaust gases each second for 3 s. The velocity of the gases, relative to the rocket, is 250 m s^{-1}. The mass of the rocket is 2.5 kg.

    a   Calculate the force acting on the rocket. *(2 marks)*

    b   Explain why the acceleration will not remain constant for the 3 s of the rocket's flight. *(2 marks)*

9.3 Conservation of energy

Specification references: 4.2a(v), 4.2b(i), 4.2c(vii), 4.2c(viii), 4.2c(ix)

In a lecture, the Nobel-winning physicist Richard Feynman said, 'There is a fact, or if you wish a law, governing all natural phenomena that are known to date. There is no exception to this law – it is exact so far as is known. The law is called the *conservation of energy*. Energy is a mathematical principle, a numerical quantity which does not change when something happens. It is not a description of a mechanism, or anything concrete; it is just a strange fact that we can calculate some number and when we finish watching nature go through her tricks and calculate the number again, it is the same'.

If you look in a dictionary it will tell you that energy 'is a measure of the ability to do work'. But what is meant by *work*?

Force and work

In the previous topic, you saw that a force F acting on an object over a time Δt had an effect called *impulse*, where impulse = $F \times \Delta t$, which was the change in the momentum of the object. Another important measure of what a force does is obtained when you multiply the force F by the displacement Δs of the object it acts on. This effect is given the name **work** ΔE and is measured in joules (J), when F is in N and s in m, giving the equation

Work ΔE (J) = force F (N) × displacement Δs (m)

 Worked example: Work done in accelerating a car

In the previous topic you saw that the resultant force accelerating a car from 0 to 60 mph ($26.8 \, \mathrm{m\,s^{-1}}$) in 4.7 s was 13 700 N. Calculate the work done during this acceleration.

Step 1: Write down the *suvat* variables for this situation.

$$s = ?, \ u = 0, \ v = 26.8 \, \mathrm{m\,s^{-1}}, \ a = ?, \ t = 4.7 \, \mathrm{s}$$

a can be eliminated as it is not required here.

Step 2: Select the correct *suvat* equation and substitute in values.

$$s = \frac{u+v}{2}\,t = \frac{(0 + 26.8 \, \mathrm{m\,s^{-1}})}{2} \times 4.7 \, \mathrm{s} = 13.4 \, \mathrm{m\,s^{-1}} \times 4.7 \, \mathrm{s} = 62.98 \, \mathrm{m}$$

Step 3: Calculate the work done.

$$E = F \times s = 13\,700 \, \mathrm{N} \times 62.98 \, \mathrm{m} = 862\,826 \, \mathrm{J}$$
$$= 860 \, \mathrm{kJ} \ (2 \ \mathrm{s.f.})$$

Conservation of energy

Energy is the capacity to do work. The fuel in the car, burning in air, does 860 kJ of work on the car to accelerate it, so the fuel has less capacity to do work than it had. If the car decelerates to a halt, the brakes exert a force on the car and do work on the brake linings,

Learning outcomes

Describe, explain, and apply:

→ the principle of conservation of energy

→ the terms: work, energy

→ the equation work done $\Delta E = F\Delta s$

→ the equation $E_k = \frac{1}{2}mv^2$

→ the equation for gravitational energy, $\Delta E_{grav} = mg\Delta h$.

▲ **Figure 1** *Richard Feynman*

heating them. The moving car has the capacity to do work (heating the brake linings) which it did not have when it was still. The energy of any moving object is called its **kinetic energy**.

In this example, we can say that the fuel is acting as an energy store, and that doing 860 kJ of work in speeding up the car reduces the fuel and air energy store by 860 kJ. The moving car is now an energy store. Would it now have 860 kJ? In reality, the kinetic energy store of the moving car will be less than 860 kJ as there are numerous other forces present, all doing work.

Kinetic energy and work done

To obtain a useful equation for the kinetic energy store of a moving object, consider a uniformly accelerating car.

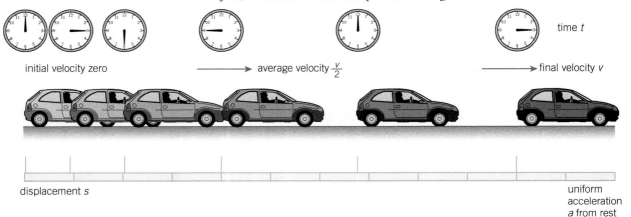

initial velocity zero

average velocity $\frac{v}{2}$

time t

final velocity v

displacement s

uniform acceleration a from rest

▲ **Figure 2** *Calculating kinetic energy increase*

The accelerating force is given by Newton's second law, $F = ma$.

The kinetic energy gained by the car is the work done in accelerating it, as it had no kinetic energy beforehand, as it was stationary.

$$\text{Kinetic energy } E_k = Fs = mas$$

We can get $a\,s$ from the *suvat* equation $v^2 = u^2 + 2as$.

$$\text{as } u = 0,\ v^2 = 0 + 2as = 2as \Rightarrow \tfrac{1}{2}v^2 = a\,s$$

$$\text{Kinetic energy, } E_k = m(a\,s) = m(\tfrac{1}{2}v^2) = \tfrac{1}{2}mv^2$$

 Worked example: Calculating the kinetic energy of a bee

A bumble bee has a mass of 0.50 g, and flies at 80 cm s^{-1}. Calculate its kinetic energy.

Step 1: Convert the data to SI base units.

$$m = 0.5 \times 10^{-3}\,\text{kg},\ v = 80 \times 10^{-2}\,\text{m s}^{-1}$$

Step 2: Substitute values into the kinetic energy equation.

$$E_k = \tfrac{1}{2}mv^2 = 0.5 \times (0.5 \times 10^{-3}\,\text{kg}) \times (80 \times 10^{-2}\,\text{m s}^{-1})^2 = 0.000\,16\,\text{J}$$
$$= 0.16\,\text{mJ}$$

Gravitational potential energy and work done

Another way in which a force can do work is to lift something. Gravity gives masses weight, which near the Earth's surface will make them accelerate downwards at $a = g = 9.8\,\mathrm{m\,s^{-2}}$ in the absence of any other forces. As $F = ma$, the force with which an object of mass m is pulled downwards (its weight) $= mg$.

The force that is needed to support an object in the Earth's gravitational field is its weight mg and that is the force you need to apply to lift it at a steady speed. When it is lifted through a displacement h – the symbol 'h' is used rather than 's' to remind us it is a height, measured vertically – then

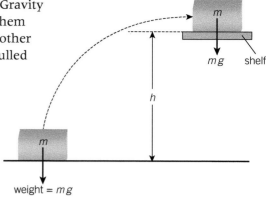

$$\text{Work done} = \text{force} \times \text{displacement} = (mg) \times h = mgh$$

The increase in gravitational potential energy, $\Delta E_{\mathrm{grav}} = mgh$

Gravitational potential energy acts as a store of energy – the work done in lifting it can be transferred from this store if it is allowed to fall.

▲ **Figure 3** *Calculating gravitational potential energy increase*

 Worked example: Energy exchange as an apple falls

A 0.10 kg apple falls 3.5 m from a tree. Calculate the speed of the apple when it reaches the ground.

Step 1: State the law of conservation of energy in this situation.

Kinetic energy gained = change in gravitational potential energy

Step 2: Using the appropriate equations, calculate each term in this equation.

$$E_{\mathrm{k}} = \tfrac{1}{2}mv^2 = 0.5 \times 0.10\,\mathrm{kg} \times v^2 = (0.050\,\mathrm{kg}) \times v^2$$

$$\Delta E_{\mathrm{grav}} = mgh = 0.10\,\mathrm{kg} \times 9.8\,\mathrm{m\,s^{-2}} \times 3.5\,\mathrm{m} = 3.43\,\mathrm{J}$$

Step 3: Substitute the terms into the conservation of energy equation and solve.

$$(0.050\,\mathrm{kg}) \times v^2 = 3.43\,\mathrm{J} \Rightarrow v^2 = \frac{3.43\,\mathrm{J}}{0.05\,\mathrm{kg}} = 68.6\,\mathrm{m^2\,s^{-2}}$$

$$v = \sqrt{(68.6\,\mathrm{m^2\,s^{-2}})} = 8.3\,\mathrm{m\,s^{-1}}\ (2\ \text{s.f.})$$

The question above could equally be solved using *suvat* equations (in particular $v^2 = u^2 + 2as$) as multiplying this equation throughout by $\tfrac{1}{2}m$, putting $a = g$, and putting $s = h$ results in

$$\tfrac{1}{2}mv^2 = \tfrac{1}{2}mu^2 + mgh$$

Final kinetic energy = initial kinetic energy + change in gravitational potential energy

Summary questions

1 Using a reasonable estimate of the velocity, calculate the kinetic energy of a 1100 kg car speeding down a motorway. *(2 marks)*

2 A catapult fires a 0.05 kg stone vertically into the air. It is accelerating the stone over a distance of 20 cm with an average force of 1.6 N. Use the principle of conservation of energy to find the distance the stone rises into the air. *(4 marks)*

3 Show that kinetic energy is related to momentum by the equation $E_k = \dfrac{p^2}{2m}$. *(3 marks)*

4 A trolley of mass 0.8 kg is accelerated from rest along a horizontal surface as shown in Figure 4.

▲ **Figure 4**

The accelerating weight has a mass of 600 g. Ignoring friction, calculate the velocity of the trolley when the weight has fallen 55 cm. *(5 marks)*

9.4 Projectiles

Specification references: 4.2a[ii], 4.2a[iii]

An object projected outwards is called a **projectile**. An early example of a projectile, which was much analysed, is a cannonball. People have long known that it is in the nature of things to fall downwards if they can, so that a projectile would return to the Earth somewhere, but before Galileo and Newton, the exact way in which cannonballs moved was not understood. Figure 1 illustrates the common misconception at the time – that the gunpowder gave the cannonball 'impetus', which carried it upwards. Eventually, the impetus was all used up, and the cannonball fell. Gunnery was not a precise science in the 16th century.

Horizontal and vertical components

The weight mg of a body can be resolved into perpendicular components at an angle θ as shown in Figure 2.

As the angle θ becomes larger, the component $mg\cos\theta$ becomes smaller and smaller. In the limit, when $\theta = 90°$

$$mg\cos\theta = mg\cos(90°) = mg \times 0 = 0$$

Weight is a vertical force – it has no horizontal component and cannot change the horizontal velocity component of a moving object. Air resistance will affect this component, but in this section we are assuming that air resistance is negligible. For any projectile, the initial velocity u can be resolved into vertical and horizontal components as shown in Figure 3.

The vertical component $u\sin\theta$ is accelerated downwards by the weight of the object – this vertical acceleration $= g$ downwards.

The horizontal component $u\cos\theta$ is not affected by gravity, and continues at this value as long as the projectile is in motion. This means that forces perpendicular to one another act *independently*.

A serve in tennis is hit from a height of approximately 3 m. If simply dropped from that height, you would estimate that the ball would reach the ground in around 0.8 s. If the server hits the ball horizontally at that height, the ball still takes just the same amount of time to reach the ground. Its horizontal (component of) velocity simply carries it towards you as it falls.

In reality, the server usually hits the ball at a downward angle, giving it quite a large downward component of velocity. So the ball reaches the ground near you even sooner, just as it would if it had been thrown downwards.

Figure 4 shows the independent vertical and horizontal motion of a tennis ball struck horizontally.

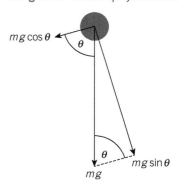

▲ **Figure 1** *Medieval projectile analysis*

▲ **Figure 2** *Components of weight*

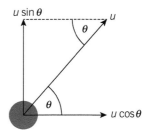

▲ **Figure 3** *Resolving the initial velocity of a projectile*

207

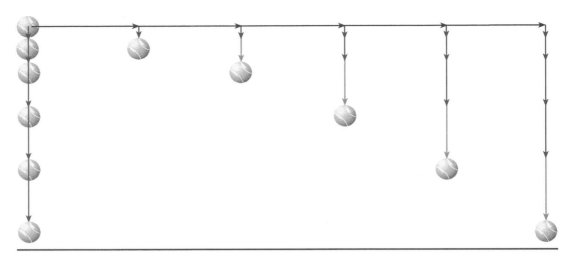

Vertical component of velocity

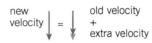

Horizontal component of velocity

stays constant

▲ **Figure 4** *Falling and going sideways*

▲ **Figure 5** *Fountains send water along parabolic paths*

Detailed mathematical analysis using the equation $s = ut + \frac{1}{2}at^2$ applied to the horizontal and vertical components of displacement gives the result that the shape of the trajectory of a projectile is a parabola. Or, rather, it has this shape if the forces that are the result of the ball moving through and spinning in the air, for example, air resistance, are unimportant. This means that a parabola is a good approximation of the path of a moving cricket ball, a fair approximation for a moving football or tennis ball, and a very bad approximation for a moving badminton shuttlecock, since air resistance plays a huge part in the movement of a shuttle.

The path of a tennis ball during a lob can be analysed using iterative modelling. In a short interval of time, the downward acceleration will have added a vertical downward component to the velocity. The resultant velocity is the vector sum of the two. So the ball's motion is tilted down a little, and it slows down a little. In the next moment, the same thing happens, and again, and again, so that the ball's path always curves downwards. Sooner or later, the downward acceleration will have removed all of the upward component of velocity, and the ball will begin travelling downwards.

In Figure 6 (next page), in each time interval Δt the blue arrow is the displacement vector assuming that the ball had the same velocity as in the previous Δt, and the red arrow is the extra vertical displacement $\frac{1}{2}g(\Delta t)^2$ produced by gravity in that time. The vector sum gives the resultant displacement shown in purple.

downward displacement from acceleration of free fall in each instant

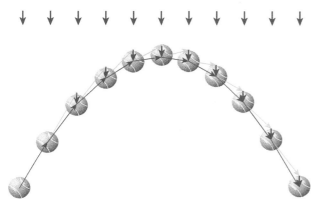

▲ **Figure 6** *Lobbing a ball in the air*

 ## Worked example: Range of a projectile

A ball is thrown with an initial velocity of $18\,\mathrm{m\,s^{-1}}$ at an angle of $60°$ to the horizontal. Calculate how far it travels before hitting the ground. You can assume it was thrown from a height of $1.5\,\mathrm{m}$ above the ground.

Step 1: Decide on the components required to answer this question.

Use vertical motion to find the time taken to reach the ground, then horizontal motion to find how far it has travelled.

Step 2: Identify the equations required to resolve the components of u.

$$\text{Vertical component} = u\sin\theta$$

$$\text{Horizontal component} = u\cos\theta$$

Step 3: Substitute values and evaluate components.

Vertical component $= u\sin\theta = (18\,\mathrm{m\,s^{-1}}) \times \sin(60°) = 15.5...\,\mathrm{m\,s^{-1}}$

Horizontal component $= u\cos\theta = (18\,\mathrm{m\,s^{-1}}) \times \cos(60°) = 9.0\,\mathrm{m\,s^{-1}}$

Step 4: Use values from Step 3 to calculate vertical motion using $suvat$.

Taking downwards as positive, $s_v = 1.5\,\mathrm{m}$, $u_v = -15.5...\,\mathrm{m\,s^{-1}}$, $a = g = 9.8\,\mathrm{m\,s^{-2}}$, $t = ?$

$$s_v = u_v\,t + \tfrac{1}{2}a t^2$$

$$1.5\,\mathrm{m} = (-15.5...\,\mathrm{m\,s^{-1}})t + 0.5 \times (9.8\,\mathrm{m\,s^{-2}})t^2$$
$$= (4.9\,\mathrm{m\,s^{-2}})\,t^2 - (15.5...\,\mathrm{m\,s^{-1}})t$$

$(4.9\,\mathrm{m\,s^{-2}})t^2 - (15.5...\,\mathrm{m\,s^{-1}})t - 1.5\,\mathrm{m} = 0$ is a quadratic equation $ax^2 + bx + c = 0$ where $a = 4.9\,\mathrm{m\,s^{-2}}$, $b = -15.5...\,\mathrm{m\,s^{-1}}$, and $c = -1.5\,\mathrm{m}$.

The solution to $ax^2 + bx + c = 0$ is $x = \dfrac{-b \pm \sqrt{b^2 - 4ac}}{2a}$

In this case

$$t = \frac{-(-15.5...\,\mathrm{m\,s^{-1}}) \pm \sqrt{(-15.5...\,\mathrm{m\,s^{-1}})^2 - 4(4.9\,\mathrm{m\,s^{-2}})(-1.5\,\mathrm{m})}}{2(4.9\,\mathrm{m\,s^{-2}})}$$

$$= \frac{-15.5...\,\mathrm{m\,s^{-1}} \pm \sqrt{273\,\mathrm{m^2\,s^{-2}}}}{9.8\,\mathrm{m\,s^{-2}}}$$

$$t = \frac{15.5...\,\mathrm{m\,s^{-1}} + 16.5...\,\mathrm{m\,s^{-1}}}{9.8\,\mathrm{m\,s^{-2}}} = -\frac{0.9...\,\mathrm{m\,s^{-1}}}{9.8\,\mathrm{m\,s^{-2}}} \text{ or } \frac{32.09...\,\mathrm{m\,s^{-1}}}{9.8\,\mathrm{m\,s^{-2}}}$$

$$= -0.093...\,\mathrm{s} \text{ or } 3.27...\,\mathrm{s}$$

The negative solution means that, if the ball had been thrown from ground level, it would have followed this path had it been thrown (slightly faster) 0.093... s earlier.

Step 5: Using t from Step 4, find the range for horizontal motion.

The speed is constant, so $s_h = u_h t = 9.0\,\mathrm{m\,s^{-1}} \times 3.27...\,\mathrm{s} = 29\,\mathrm{m}$ (2 s.f.)

Summary questions

1 Explain how the imagined path of the cannonball shown in Figure 1 illustrates the old 'impetus' theory of motion. Explain how the path of a real projectile differs from this. *(3 marks)*

2 A projectile is launched from ground level at 25 m s^{-1} at an angle of 45° to the horizontal. Calculate the range of the projectile and show that a 1° change in angle in either direction will reduce the range. *(6 marks)*

3 A basketball player launches a ball from a height of 2.0 m towards a hoop as shown in Figure 7. Show, using calculations, that the ball will go through the hoop. *(4 marks)*

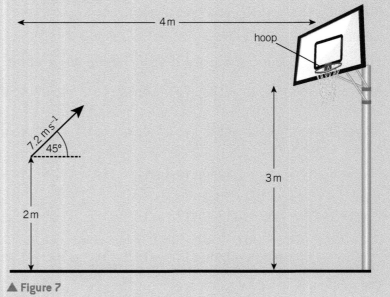

▲ Figure 7

If the weights in Figure 1 were being held up by a stand instead of by a woman, no work would be done. This is because work is done only when a force moves its point of application.

$$\text{No movement} = \text{no work}$$

Work is done in lifting the weights, but not in just holding them up. The weightlifter in Figure 1 would probably disagree – she would say she is working hard to hold the weights in this position.

This is because biological systems are different. If you stand for a long time holding a heavy object – this could actually be your own weight – you do expend energy due to muscle fibres continually relaxing and contracting. Forces are moving their points of application – inside your muscles – and so work is being done. You can tell that your muscles are working, because their low efficiency means that you get hot, and you also get fatigued, which indicates your body's need for more energy.

Work done by a force at an angle

When a constant force F moves an object by a displacement s, then the work done $= Fs$ only if the force is in the direction of displacement. You first met work done in the direction of displacement in Topic 9.3, Conservation of energy. Figure 2 shows a constant force F acting on an object which can move only in the direction shown in orange, which is at an angle θ to the direction of the force. The force moves the object from P_1 to P_2.

Learning outcomes

Describe, explain, and apply:

→ calculations of work done, including cases where the force is not parallel to the displacement

→ power as the rate of transfer of energy

→ the term: power

→ the equations for power, where $P = \dfrac{\Delta E}{t}$ and $P = Fv$.

▲ **Figure 1** *Doing work*

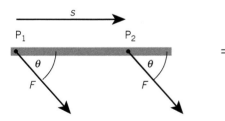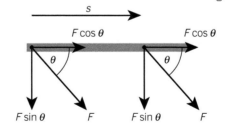

▲ **Figure 2** *Force and work*

When F is resolved into components perpendicular to s and parallel to s, as in Figure 2, you can see that the object is not moving in the direction of the perpendicular component $F\sin\theta$ at all, so this component does no work. Work is being done only by the parallel component $F\cos\theta$ so

$$\text{Work done} = (F\cos\theta) \times s = Fs\cos\theta$$

Although both force and displacement are vector quantities, work, like energy, is a scalar quantity and has no direction. The way in which two vector quantities F and s separated by an angle θ multiply together to

Synoptic link

You first met vector and scalar quantities in Topic 8.2, Vectors.

give a scalar quantity is known as a *scalar product*, and is common in physics.

A force often produces a displacement in a different direction. All that is required is that the displacement is constrained to be in a particular direction. One example is sailing, where the shape of the boat, and the presence of a keel, define the direction in which it must travel.

Worked example: Work done in sailing

Figure 3 shows a boat sailing forwards under the action of wind acting at 56° to the direction of travel. The wind pushes on the sail with a force of 500 N. Calculate the work done in sailing the boat forwards a distance of 25 m.

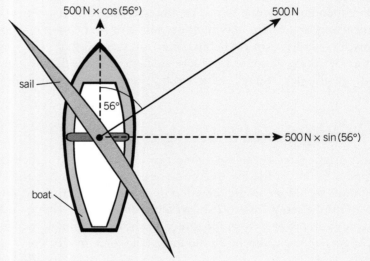

▲ **Figure 3** *Work done in sailing*

Step 1: Resolve the force into components to find the component in the direction of movement.

$$\text{Forwards component} = F\cos\theta = 500\,\text{N} \times \cos(56°)$$
$$= 500\,\text{N} \times 0.559... = 279.5...\,\text{N}$$

Step 2: Multiply this component by the displacement in this direction.

$$\text{Work done} = \text{forwards component of force} \times \text{displacement}$$
$$= 279.6\,\text{N} \times 25\,\text{m} = 7000\,\text{J (2 s.f.)}$$

Note that this two-stage process is exactly the same as using the equation work done = $Fs\cos\theta$.

Although the component of the force perpendicular to the direction of the motion does no work, this does not mean it has no effect. In sailing, it makes the boat tilt over, so the crew need to lean out to prevent it overturning.

Work done and the principle of conservation of energy

The principle of conservation of energy is such a powerful tool in physics that whenever energy seems to go missing, it raises the question – to where must the energy have been transferred? This energy transfer can often be explained as work done against resistive forces, and usually results in heating.

A force is frequently moving its point of application, doing work in circumstances where it is not clear where the energy is transferred. In Figure 4 the skydiver is falling at his terminal velocity (about $60\,\mathrm{m\,s}^{-1}$ or $130\,\mathrm{m.p.h.}$) and is clearly losing gravitational potential energy. As he's not accelerating, his kinetic energy remains constant.

As the skydiver falls, he can feel the resistive force exerted on him by the air through which he is falling, and by Newton's third law he must be exerting an equal force on the air.

▲ **Figure 4** *Skydiver at terminal velocity*

Synoptic link

You met the relationship between gravitational potential energy and kinetic energy in Topic 9.3, Conservation of energy, and Newton's third law in Topic 9.2, Newton's laws of motion and momentum.

 Worked example: Work done on the air

Calculate the amount of energy transferred to the air each second by a $70\,\mathrm{kg}$ skydiver at his terminal velocity of $60\,\mathrm{m\,s}^{-1}$. Assume that about two-thirds of the original gravitational potential energy is transferred to the air, with the rest heating the skydiver's suit.

Step 1: Calculate using $s\,u\,v\,a\,t$ equations how far the skydiver falls in this time.

He is not accelerating, so $v = u$ and $a = 0$, so $s = \dfrac{u + v}{2}t$ and $s = ut + \frac{1}{2}at^2$ both become $s = ut$.

$$\text{height } h = s = ?, u = 60\,\mathrm{m\,s}^{-1}, t = 1\,\mathrm{s}$$

$$h = s = ut = 60\,\mathrm{m\,s}^{-1} \times 1\,\mathrm{s} = 60\,\mathrm{m}$$

Step 2: Calculate the change in gravitational potential energy in this time.

$$\Delta E_{\mathrm{grav}} = mgh = 70\,\mathrm{kg} \times 9.8\,\mathrm{m\,s}^{-2} \times 60\,\mathrm{m} = 41\,160\,\mathrm{J}$$

Step 3: Use the information in the question to find how much energy is transferred to the air.

$$\text{Work done on the air} \approx \tfrac{2}{3} \times 41160\,\mathrm{J} = 30\,000\,\mathrm{J}\ (1\,\mathrm{s.f.})$$

Study tip

For calculations involving rough estimates, answers should be rounded to one significant figure.

Power

Power is the rate at which work is done, that is, the rate at which energy is transferred. It was the focus of the Worked example above – the energy transferred to the air in one second is the power *dissipated* by the gravitational potential energy change of the skydiver heating the air. The word 'dissipated' is frequently used to refer to energy transferred to the surroundings, usually heating them.

Assuming that the displacement resulting from a force is in the direction of the force itself, we can simply relate the power dissipated to the force.

$$\text{Power} = \frac{\text{energy transfer}}{\text{time}} = \frac{\text{work done}}{\text{time}} = \frac{Fs}{t} = F\frac{s}{t} = Fv$$

Worked example: Drag force on the Eurostar train

A Eurostar train is driven by two engines, each providing an output power of 5.6 MW. Calculate the drag force on the Eurostar when it is travelling at a constant speed of 186 m.p.h. ($83\,\mathrm{m\,s^{-1}}$).

Step 1: Calculate the force provided by the engines at this velocity.

$$P = Fv \Rightarrow F = \frac{p}{v} = \frac{2 \times 5.6 \times 10^6\,\mathrm{W}}{83\,\mathrm{m\,s^{-1}}} = 134\,939.7...\,\mathrm{N}$$

Step 2: Apply Newton's first law.

The train is not accelerating, so the resultant force on it must be 0.

Forward force provided by engines = drag force in the opposite direction

so drag force = $134\,939.7...\,\mathrm{N} = 130\,\mathrm{kN}$ (2 s.f.)

 Dinorwig Power Station

For many of the UK power stations, particularly the coal-burning and nuclear power stations, changing the output power takes many hours. Sudden changes in demand for electricity – this can be when everyone switches on an electric kettle to make a cup of tea at the end of a popular television programme – result in demands which cannot be met rapidly. To help with situations like this, the pumped-storage power station in Dinorwig in North Wales was constructed.

The turbines which generate electricity are inside Elidir mountain, as are 16 km of underground tunnels leading up to the storage reservoir near the top of the mountain. During times of low electricity demand, the electricity generators are used as motors to pump water from the lower lake through a height of 500 m to the storage

reservoir, which has a capacity of $7.0 \times 10^6\,\mathrm{m^3}$. Water has a density of $1000\,\mathrm{kg\,m^{-3}}$, so the total mass of water in the storage reservoir = $7.0 \times 10^9\,\mathrm{kg}$ when it is full.

The water in the storage reservoir may be rapidly run back down to the lower lake through the turbine tunnels to cope with surges in electricity demand, and can deliver a power of 1.7 GW within 16 s of being turned on.

1 Calculate the total gravitational potential energy difference when all the water runs from the storage reservoir to the lower lake, and use this value to calculate the time for which Dinorwig can deliver a power of 1.7 GW. The generators operate on average with an efficiency of 75%.

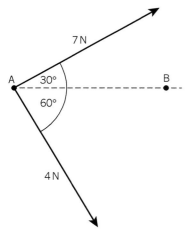

▲ **Figure 5** *An object under the action of two forces*

Summary questions

1 A car drives at a steady speed of $27\,\mathrm{m\,s^{-1}}$ along a motorway against drag forces of 3000 N. Calculate the power dissipated against the drag forces. *(2 marks)*

2 Figure 5 shows two forces acting on an object which moves in the direction shown by the dotted line.

Calculate the work done in moving the object from A to B, a distance of 60 cm. *(3 marks)*

3 A car of mass 900 kg accelerates from 0 to $26.8\,\mathrm{m\,s^{-1}}$ in 9.2 s. The thrust from the engine is constant at 3500 N during this period. By comparing the kinetic energy gained by the car with the total work done by the engine, calculate the work done against resistive forces. *(5 marks)*

Module 4.2 Summary

Motion

Graphs of motion

- measuring: speed, velocity, acceleration, and displacement
- graphs of accelerated motion, including gradient and area under graph
- investigating the motion of a trolley using light gates or other techniques

Vectors

- the terms: vector, scalar, resultant, component
- the representation of displacement, velocity, and acceleration as vectors
- the addition of two vectors by drawing a 'tip to tail' triangle
- finding relative velocity
- the resolution of a vector into two components at right angles to each other

Modelling motion

- computational models to represent changes of displacement and velocity
- graphical models in terms of vector additions to represent motion
- using time intervals in iterative models with relation to accuracy of prediction
- investigating falling cupcake cases

Speeding up and slowing down

- kinematic (*suvat*) equations for straight-line motion with uniform acceleration:
  $v = u + at, s = ut + \frac{1}{2}at^2, v^2 = u^2 + 2as$
- measuring g

Momentum, force and energy

Newton's laws of motion and momentum

- momentum $p = mv$
- conservation of momentum
- the terms force, mass, and impulse
- Newton's laws of motion
- Newton's third law as a consequence of the principle of conservation of momentum
- the equation: $F = \dfrac{\Delta p}{\Delta t} = ma$ where the mass is constant

Conservation of energy

- the principle of conservation of energy
- the terms: work, energy
- the equation: $E_k = \frac{1}{2}mv^2, \Delta E_{grav} = mg\Delta h,$ work done $\Delta E = F\Delta s$

Projectiles

- the trajectory of a body moving under constant acceleration
- the independent effects of perpendicular components of a force

Work and power

- the terms: work and power
- the equation: work = (component of force in the direction of the displacement) × displacement
- power is the rate of transfer of energy
- the equations: $P = \dfrac{\Delta E}{\Delta t}$ and $P = Fv$

Physics in perspective

Racing car design

▲ **Figure 1** *Ralph de Palma winning the 1914 Vanderbilt Cup in his 115 brake horsepower Mercedes Grand Prix race car*

▲ **Figure 2** *Using a wind tunnel to investigate air flow around a car*

When motor racing started, the aim of car designers was to use ever more powerful engines. Figure 1 shows a 1914 Mercedes, which had a mass of about a tonne. It was fast, with a top speed of 115.0 mph and an engine power of 115 brake horsepower. It won many races at that time, but it was hardly what you would call streamlined.

Drag and supercharging

As speeds increased in the 1920s and 1930s, designers realised that they needed to tackle drag force. This force is generally described by the equation

$$F_D = \frac{1}{2} \rho C_D A v^2$$

where ρ is the density of fluid through which the object is moving (in this case, air), and v and A are the velocity and cross-sectional area of the moving object. C_D, the drag coefficient, is a dimensionless number, usually less than 1. Increasing v greatly increases the drag, but increasing v is what car designers wanted — thankfully, both C_D and A can be reduced. A is reduced by lowering the profile of the car, and C_D is reduced by finding ways to streamline the vehicles so that the air flows smoothly over them.

Automobile designers also introduced supercharging, which involved pumping extra air into the engine so that it could burn fuel faster and generate greater power, up to 180 kW. A modern development of this, which uses waste heat and so takes less energy from the engine, is called turbocharging.

Acceleration and stability

Acceleration is change in velocity. In motor racing, it comes up in 3 main situations.

- Speeding up — the easiest way to improve this is increasing the engine power. This increases the forces involved in driving the car, and so the driver must be protected against the force with which the seat pushes him or her forwards during acceleration.

- Slowing down — the braking system must be able to safely bring the car, along with the driver, to a stop in a reasonable time.

- Cornering may not involve a change of speed, but it does involve a change in direction and so also a change in velocity. The force needed to push the car into a new direction has to be provided by friction with the ground. So the wheel size and tyres used must maximise the contact between the vehicle and the ground.

As cars became more and more streamlined they became less stable. Just as aircraft reaching a certain speed take off, so too do a car's wheels leave the ground if it is travelling fast enough. Furthermore, race tracks became smaller in size, resulting in tighter bends and more laps to be driven, so more corners had to be negotiated at speed. In the 1960s and

1970s, car design began integrating technology developed in the aircraft industry, particularly from the use of wind tunnels to investigate air flow around vehicles. As a result, cars started to sprout wings, fins and skirts.

The combined effect of the changes introduced was to increase the drag coefficient C_D considerably, but in such a way that the resultant drag force was downwards — this is usually referred to as 'downforce'. The overall behaviour of the car became rather like that of an aircraft wing, but upside down. This increased the 'grip' of the tyres on the ground, an effect further increased by the use of wide wheels with smooth, soft tyres called 'slicks'.

▲ **Figure 3** *A modern Formula 1 car, with wide, smooth 'slicks'*

Safety and Formula 1

Formula 1 refers to the highest class of single-seat car racing, and the 'formula' is the set of regulations which must apply to all cars. By the 1970s, speeds had increased dramatically but safety had not, and deaths were common. As a result, drivers campaigned for improved safety, which gave rise to modern Formula 1 Regulations.

Besides the use of protective clothing and regular inspections of vehicles and track, a number of safety measures were integrated into the cars. These included crumple zones, automatic cut-off of fuel lines and a 'survival cell' in which the driver's cockpit was built. Public safety was also considered, with the introduction of double guard rails and wide verges to the track.

Other regulations restrict the performance of vehicles — these include increasing the minimum mass of vehicles, restricting the engine capacity (and so keeping the power down), and banning some aerodynamic developments which increased the downforce. These restrictions make the race fairer and also less risky for the drivers. As car designers continually come up with technological improvements, the Formula 1 regulations have to be refined each year.

Summary questions

1 Ralph de Palma won the 1914 Vanderbilt Cup in a car of mass 1080 kg with a top speed of 51.5 m s^{-1} and an engine output power of 85.8 kW. The race consisted of 35 laps on a circuit of length 13.6 km.
   a Calculate the shortest time that this race could have taken and explain why the race would have taken longer than this.
   b After the race, de Palma demonstrated the car's maximum speed by driving in a straight line on horizontal ground. Draw a labelled diagram showing all the forces acting on the car at this steady speed. You can assume that the rear wheels are providing the driving force.
   c Assuming that 80% of the car's output power is delivered to work done against resistive forces, calculate the value of the resultant resistive force when driving at a constant 51.5 m s^{-1}.

2 The drag force acting on a car is given by $F_D = \frac{1}{2}\rho C_D A v^2$. Assuming that a family car has a drag coefficient of 0.32 and a maximum cross-sectional area of 2.1 m^2, calculate the drag force while driving down the motorway at 70 mph (31.3 m s^{-1}). The density of air, $\rho = 1.2$ kg m^{-3}.

Practice questions

1 Which of the following combinations of SI base units is equivalent to the joule, J?

   A $kg\,m\,s^{-1}$

   B $kg\,m\,s^{-2}$

   C $kg\,m^2\,s^{-2}$

   D $kg\,m^2\,s^{-3}$ (*1 mark*)

2 An object of mass 2.5 kg travelling at $6.4\,m\,s^{-1}$ collides with and sticks to a stationary object of mass 5.5 kg. The combined masses move forwards.

After the collision, which of the following statements are correct?

   1   The kinetic energy of the combined masses is 51.2 J.

   2   The momentum of the combined masses is 16 N s.

   3   The velocity of the combined masses is $2\,m\,s^{-1}$.

   A   1, 2, and 3 are correct

   B   Only 1 and 2 are correct

   C   Only 2 and 3 are correct

   D   Only 1 is correct   (*1 mark*)

3 A rifle of mass 3.2 kg fires a 3.0 g bullet horizontally with a velocity of $370\,m\,s^{-1}$. Ignoring the effect of any exhaust gases produced in the firing, calculate the speed at which the rifle recoils.   (*3 marks*)

4 Two identical 4.0 kg masses, heading directly towards each other, each with a speed of $0.8\,m\,s^{-1}$, collide and take a time of 0.15 s to come to rest.

   a   Calculate the force acting on each mass during the collision.   (*2 marks*)

   b   Calculate the energy dissipated in the collision.   (*2 marks*)

5 A roller-coaster car of mass 900 kg is stationary at the top of the track, and then falls a height h to the lowest point, as in Figure 1.

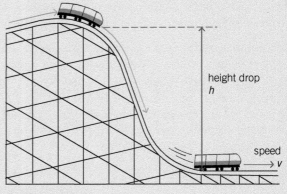

▲ **Figure 1**

   a   Calculate the maximum velocity of the roller-coaster car at the bottom of the track if $h = 60\,m$.
        Acceleration due to gravity, $g = 9.8\,m\,s^{-2}$.   (*3 marks*)

   b   Explain why the velocity calculated in **(a)** is likely to be less than this value.   (*1 mark*)

6 A car of mass 1100 kg accelerates from rest to $27\,m\,s^{-1}$ in 5.6 s.

   a   Calculate the mean resultant force acting on the car.   (*2 marks*)

   b   Calculate the mean power output of the car.   (*2 marks*)

7 In a nuclear decay reaction, a nucleus of mass 3.6×10^{-25} kg emits a particle of mass 6.6×10^{-27} kg with kinetic energy 8.0×10^{-13} J.

   a   Calculate the kinetic energy of the nucleus after the reaction.   (*4 marks*)

   b   The same nucleus can also decay by emitting a particle of mass 9.1×10^{-31} kg with kinetic energy 1.4×10^{-15} J. Explain how the kinetic energy of the nucleus after the reaction would compare with the value obtained in **(a)**.   (*4 marks*) c   In a nuclear reaction, the kinetic energy of any particle with kinetic energy E_k given by the equation $E_k > (10^{15}\,\mathrm{m^2\,s^{-2}}) \times$ (mass of particle in kg) cannot be obtained from the standard $E_k = \frac{1}{2}mv^2$: a relativistic equation must be used. Check whether either of the emitted particles in **(a)** and **(b)** is travelling fast enough for this condition to apply.   (*2 marks*)

8   In an experiment, a cylindrical wooden rod was dropped vertically from different heights into a container of water, and the depth to which it sunk before floating back upwards was measured.

a   For a drop height of 12 cm, the following results were obtained.

▼ Table 1

depth / cm	5.2	5.8	3.4	5.7	5.4	5.5

Use the data to obtain a mean value for depth at this drop height together with its uncertainty. Justify any decision you make in selecting data.   (*3 marks*)

b   Add your result from **(a)** to a copy of Table 2, and use this data to plot a graph on a copy of Figure 2. Add a line of best fit to the graph.   (*2 marks*)

▼ Table 2

height /cm	2	4	6	8	10	12
depth /cm	0.4	1.5	2.6	3.6	4.6	

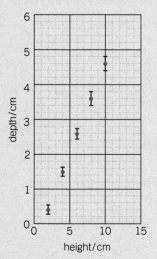

▲ Figure 2

c   It is suggested that the depth reached by the rod is directly proportional to the height from which the road was dropped. Explain why this is not supported by the data.   (*2 marks*)

d   Suggest and explain reasons why the data do not support the expected relationship.   (*4 marks*)

9   A tennis ball of mass 59 g, moving at $50\,\mathrm{m\,s^{-1}}$, is struck by a racket and travels straight back at $37\,\mathrm{m\,s^{-1}}$. The racket is in contact with the ball for a time of 3.5 ms.

a   Use the momentum change of the ball to calculate the mean force exerted by the racket in returning the ball.   (*4 marks*)

b   The ball leaves the racket horizontally at a height of 1.2 m above the ground. The playing area of the tennis court is 23 m long, and the player receiving the ball is standing 3.0 m from the back of this area. By calculating the distance the ball travels before striking the ground, find whether the ball will be 'in', reaching the ground inside the playing area. Assume that air resistance is negligible. ($g = 9.8\,\mathrm{m\,s^{-2}}$)   (*4 marks*)

c   Explain in terms of the components of velocity of the ball why hitting the ball slightly upwards at $37\,\mathrm{m\,s^{-1}}$ may have resulted in the ball being 'out' for a certain range of angles. No calculation is required.   (*3 marks*)

MODULE 5.1
Creating models

Chapters in this module

Introduction

In 1687 Isaac Newton published his *Principia*, a work that reflected his scientific thinking over two decades and formed the foundation of physical science for two centuries. His mathematical description of gravity described a cosmic mechanism in which the movement of every body could, in principle, be analysed and predicted – a clockwork Universe.

One crucial advance made by Newton and others was the development of the differential equation – mathematical descriptions or models of how physical quantities change with time or position. Modelling is a deep and powerful part of physics, which allows us to simplify and better understand the world around us.

Modelling decay begins by considering how radioactive decay varies with time, using a single decay probability to model larger numbers of nuclei. You will be introduced to iterative calculations that model the differential equation of radioactive decay, and you will see how radioactive decay of individual nuclei leads to the predictable behaviour of large numbers of nuclei. You will also investigate the discharging and charging of capacitors, and model their behaviour using the same iterative methods.

Modelling oscillations introduces simple harmonic oscillators – the 'simple' means it has been simplified to make the mathematical description more straightforward. You will be able to experiment with oscillators and develop iterative models of their behaviour, and you will also study resonance.

The gravitational field introduces some of the most important ideas ever formulated in physics. You will see how Newton's majestic mathematical work explained earlier observations, and how it provides a template still used to this day. You will consider the concepts of field strength and potential, and use graphical and numerical methods to predict the form of the gravitational field around objects.

Our place in the Universe gives a brief introduction to a wider model of the relationship between space and time – Einstein's special theory of relativity. You will also consider the evidence for the Big Bang theory, which may prompt you to investigate some of the currently unanswered questions in astronomy – there is still a lot to learn about our place in the Universe.

Knowledge and understanding checklist

From your Key Stage 4 or first-year A Level study you should be able to do the following. Work through each point, using your Key Stage 4/first-year A Level notes and the support available on Kerboodle.

- ☐ Describe electrical current as the rate of flow of charge.
- ☐ Define the half-life of a sample of a radioisotope.
- ☐ Recall Hooke's law and the energy stored in a stretched wire.
- ☐ Understand the terms: amplitude, wavelength, frequency, time period.
- ☐ Calculate velocity from the gradient of a displacement–time graph and acceleration from the gradient of a velocity–time graph.
- ☐ Calculate work done as force × distance.
- ☐ Define gravitational field strength and gravitational potential energy.
- ☐ Define kinetic energy.

Maths skills checklist

Maths is a vitally important aspect of Physics. In this unit, you will need to use the following maths skills. You can find support for these skills on Kerboodle and through MyMaths.

- ☐ **Using calculators in finding exponential and logarithmic functions** to solve for unknowns in decay problems. You will need to be able to keep track of repetitive, iterative calculations and handle the concept of probability.
- ☐ **Converting radians into degrees** and vice versa, such as when dealing with small angle approximations.
- ☐ **Solving algebraic equations**, including non-linear equations such as the relativistic factor and those with logarithmic and exponential terms, when considering decay.
- ☐ **Applying concepts that underlie calculus** by solving equations involving rates of change in modelling decay and simple harmonic motion.
- ☐ **Taking tangents from curves** when finding field strengths from potential gradients. You will need to be able to sketch sin and cos graphs and understand the phase relationship between the two.
- ☐ **Interpreting the area under a graph curve** where it has physical significance – for example, the energy stored on a capacitor as the area under a Q–V graph.
- ☐ **Interpreting logarithmic plots**, for example, to find the time constant for capacitor discharge or the decay constant of a radionuclide.

MyMaths.co.uk
Bringing Maths Alive

10 MODELLING DECAY
10.1 Radioactive decay and half-life

Specification reference: 5.1.1a(iv), 5.1.1b(ii), 5.1.1b(iv), 5.1.1c(ii), 5.1.1d(iv)

Radioactivity

At the end of the 19th century, physicists began to investigate previously unknown 'radiations' given off by certain compounds. The French physicist Henri Becquerel brought this phenomenon to the attention of the scientific world in 1896, and an intense period of experiment and discovery followed. In 1898 Pierre and Marie Curie isolated the elements polonium and radium. These elements were shown to produce many times more radiation than the uranium compounds previously tested. The Curies and Becquerel were awarded the Nobel Prize in 1903 for the development of the physics of **radioactivity**, a word the Curies invented to describe the new phenomenon.

▲ Figure 1 Marie and Pierre Curie in a cartoon published in 1904 celebrating the discovery of radium

The cartoon in Figure 1 reflects the public interest in the revolutions occurring in science. It is interesting that Pierre is shown holding the phial of newly discovered radium whilst Marie is simply watching over his shoulder — possibly a reflection of the attitude of the times. In fact Marie Curie was a great scientist in her own right, and she did much of the original work without Pierre, and went on to gain a second Nobel Prize in 1911. She also pioneered the use of X-rays in medicine, developing mobile X-ray units and working just behind the front line in the First World War, an early example of the benefits of the new science. Sadly, the illness that led to her death in 1934 was probably caused by years of exposure to radiation, showing the dangers inherent in the discoveries.

In the early years of the 20th century an understanding developed that the nuclei of some atoms are unstable and break down (**radioactive decay**) to emit particles of three distinct types, alpha (α), beta (β), and gamma (γ). Later work showed that alpha particles are helium nuclei, beta particles are fast-moving electrons, and gamma rays are high-energy photons. However, the original names have stuck.

Activity and half-life

The **activity** A of a source is the number of nuclei decaying per second. The unit of activity is the becquerel, Bq. $1\,\text{Bq} = 1\,\text{decay}\,\text{s}^{-1}$.

The **half-life** $T_{1/2}$ of a radioactive source is the time required for the number of nuclei in a sample to fall to half the original value.

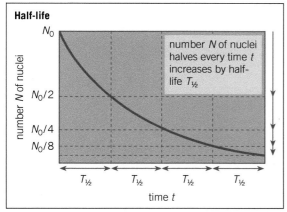

Half-life

number N of nuclei

number N of nuclei halves every time t increases by half-life $T_{\frac{1}{2}}$

time t

▲ **Figure 2** *Graph showing first four half-lives of a sample*

Activity

number N of nuclei

$\text{slope} = \text{activity} = \dfrac{dN}{dt}$

halves every half-life

time t

▲ **Figure 3** *Graph showing activity over four half-lives*

In any time t the number N is reduced by a constant factor.
In one half-life $T\frac{1}{2}$ the number N is reduced by a factor 2.
In L half-lives the number N is reduced by a factor 2^L
(e.g. in 3 half-lives N is reduced by the factor $2^3 = 8$).

Half-life can also be defined as the time it takes for the activity of a sample to fall to half the original value. This is a more useful definition, because it is easier to measure the activity of a sample rather than count the number of nuclei it contains.

 ### Worked example: Polonium half-life

The half-life of a polonium isotope is 138 days. A sample of the isotope has an activity of 1408 Bq. Calculate the activity of the sample after 552 days.

Step 1: Calculate the number of half-lives.

$$\frac{552}{138} = 4\,\text{half-lives}$$

Step 2: Calculate the activity after 4 half-lives.

$$\text{Final activity} = \frac{\text{original activity}}{2^4} = \frac{1408\,\text{Bq}}{16} = 88\,\text{Bq}$$

Hint

The activity of a sample is the number of nuclei that decay in one second. The count rate of a sample is the number of particles (α, β, or γ) emitted from the sample that the measuring system detects per second. The count rate will be smaller than the activity because not all the particles emitted will be detected. However, count rate is proportional to activity – if the activity falls to half its original value, so will the count rate.

Worked example: Half-life from graphs

Figure 4 shows how the count rate of a radioactive source varies with time. Use the graph to find the half-life of the source.

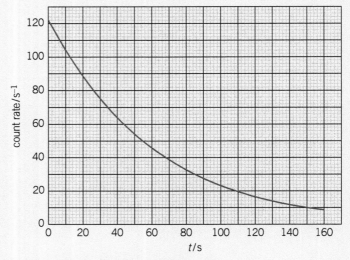

▲ **Figure 4** *Variation in count rate with time*

Step 1: Draw a horizontal line from a given count rate (say $80\,\text{counts}\,\text{s}^{-1}$) until the line meets the curve, and find the x-value at this point by drawing a vertical line to the axis. Repeat this for an activity value of half the original (in this case, $40\,\text{counts}\,\text{s}^{-1}$).

At $80\,\text{counts}\,\text{s}^{-1}$, $t = 26\,\text{s}$. At $40\,\text{counts}\,\text{s}^{-1}$, $t = 67\,\text{s}$.

Step 2: The time interval is the half-life.

$$T_{1/2} = 41\,\text{s}$$

It is good practice to use the graph to determine a number of half-life values and then find the mean. In this case, for example, the mean time taken for the activity to fall from $40\,\text{counts}\,\text{s}^{-1}$ to $20\,\text{counts}\,\text{s}^{-1}$ is $42\,\text{s}$.

 Practical 5.1.1d(vi): Determining the half-life of protactinium

Protactinium-234 is a radioactive isotope that emits beta particles, which can be detected with suitable apparatus.

Before measuring the activity of the source, the **background count** is measured. There are many sources of background radiation – rocks, soil, and the air. Failing to take background radiation into account would lead to a systematic error in the results.

When the protactinium source is in position, the detection apparatus records the counts every 10 s for a period of about 3 minutes.

The count rate is found by dividing the counts in each period of 10 s by ten. The corrected count is found by subtracting the background count rate from the measured count rate.

A graph of corrected count rate (y-axis) against time (x-axis) can be drawn and the half-life determined as in the worked example. We will consider a second method of analysing the results in Topic 10.2, Another way of looking at radioactive decay.

Radioactive decay and randomness

In radioactive decay a particle (for example, an alpha particle) has a given **probability** of being emitted from a nucleus in a given time. This event is absolutely **random**. Say you could observe a group of nuclei, knowing that each nucleus had a probability of decay of 0.5 in 1 s. If the number of nuclei was large enough you could confidently predict that half of them would decay in 1 s – but you could not predict which nuclei would decay. Of course, this is a thought experiment, but you can do something similar by throwing 100 coins in the air and counting how many land showing heads. Probability suggests that, on average, 50 will show heads, but you can't predict which particular coins will show heads.

- The fall of an individual coin is random in the same way as the decay of an individual nucleus is random.
- The outcome of many coins falling is predictable in the same way as the decay of a large number of nuclei is predictable.

The decay constant and activity

A nucleus has a fixed probability of decaying in a given time.

The probability that a nucleus will decay during a time interval of 1 s is given by the **decay constant**, λ (unit: s^{-1}). From our previous definition of activity you can see that

activity A = probability of a nucleus decaying in one second × number of nuclei present = λN

 Worked example: Activity

A radioactive source contains 4×10^{21} radioactive nuclei. The decay constant for the nuclei in the source is $2.3 \times 10^{-18}\,s^{-1}$. What is the activity of the source?

Step 1: Activity = $\lambda N = 2.3 \times 10^{-18}\,s^{-1} \times 4 \times 10^{21} = 9.2 \times 10^{3}\,s^{-1}$ or $9.2 \times 10^{3}\,Bq$

Modelling decay

The number ΔN of nuclei that decay in a small time interval Δt is given by

$$\Delta N = -\lambda N \Delta t$$

so

$$\frac{\Delta N}{\Delta t} = -\lambda N$$

The negative sign indicates that after each decay fewer original nuclei remain.

You can see that the activity (the number of decays per second) is proportional to the number of nuclei remaining.

$$\frac{\Delta N}{\Delta t} \propto -N$$

This is an example of an **exponential** change, where the rate of change of a quantity is proportional to the value of the quantity. In this case, the negative sign shows that we are considering an exponential *decay*.

> **Hint**
>
> 'Exponential' does not mean 'quick'. For example, an exponential increase means that the rate of increase grows over time, but this does not necessarily mean that it is increasing rapidly.

Making an iterative model

You can use the equation

$$\frac{\Delta N}{\Delta t} = -\lambda N$$

to make an iterative model of radioactive decay.

Consider a radioactive source of 1000 nuclei with a decay constant of $0.6\,s^{-1}$.

- The number of nuclei decaying in the first second $= \lambda N$
  $= 1000 \times 0.6 = 600$
- The number of undecayed nuclei remaining after one second
  $= 1000 - 600 = 400$
- The number decaying in the next second $= 400 \times 0.6 = 240$
- This leaves 160 undecayed nuclei.

You can iterate (repeat) this calculation, as shown in Table 1. The value for the number of nuclei remaining after 1 s is picked out in red so you can see that this becomes the value for the number present at the beginning of the next time interval. The first few calculations are shown.

▼ **Table 1** *Iterative calculation of number of nuclei decaying*

Time elapsed /s	Number N of nuclei	Number ΔN of nuclei decaying in period Δt of 1 s, $\Delta N = \lambda N \Delta t$	Number of original remaining $= N - \Delta N$
0	1000	$0.6 \times 1000 \times 1 = 600$	$1000 - 600 = 400$
1	400	$0.6 \times 400 \times 1 = 240$	$400 - 240 = 160$
2	160	$0.6 \times 160 \times 1 = 96$	$160 - 96 = 64$
3	64	$0.6 \times 64 \times 1 = 38$	$64 - 38 = 26$
4	26	$0.6 \times 26 \times 1 = 16$	$26 - 16 = 10$

The results of this calculation are represented in Figure 5(a).

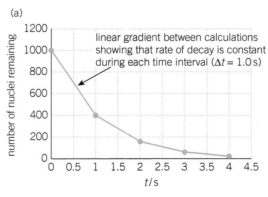

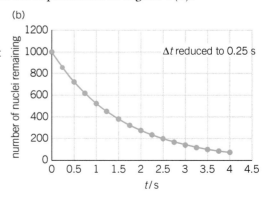

▶ **Figure 5** *Graphs showing results of iterative calculations (a) made in Table 1, with $\Delta t = 1.0$ s, and (b) with $\Delta t = 0.25$ s*

Notice that the line between each calculated value is straight because the model assumes that the activity is constant during each time interval. In reality this is not the case. The rate of decay falls continuously as the number of nuclei remaining falls. Our model can be improved by reducing the time interval between calculations. The second graph (Figure 5(b)) shows the effect of recalculating every 0.25 s.

As Δt is reduced the line will become a smoother curve and will more accurately model the continuously changing rate of decay.

Summary questions

1 A radioactive isotope has a half-life of about 3.7 days. Calculate how many days it will take for the activity of the sample to fall to one-eighth of its original value. *(2 marks)*

2 A sample contains 7×10^{21} nuclei of a radioactive isotope. The decay constant of the isotope is $2.7 \times 10^{-6}\,s^{-1}$. Calculate the activity of the sample. *(2 marks)*

3 The activity of a sample of a few milligrams of a radioactive waste material, containing about 10^{15} nuclei, is 10^6 Bq. Calculate the decay constant. *(2 marks)*

4 The half-life of the radioactive waste material from a nuclear reactor is 6 months.
   a Show that the activity will decrease by a factor of about 1000 in 5 years. *(2 marks)*
   b Calculate the factor by which the activity will decrease in 10 years. *(2 marks)*

5 The data in Table 2 gives the average rate of decay in disintegrations per minute from a radioactive sample, measured over a period of 1 minute at successive times.

▼ Table 2

Time /hour	0	1	2	3	4	5	6	7	8	9	10
Rate	8190	6050	4465	3300	2430	1800	1330	980	720	535	395

Rate = rate of disintegration in counts per minute

   a State approximately how many decays occur in the first hour. *(1 mark)*
   b Plot a graph of the rate of disintegration against time. *(4 marks)*
   c Use the graph to estimate the half-life. *(2 marks)*
   d Find the radioactive decay constant. Give the units. *(2 marks)*
   e Calculate the number of nuclei that decay per second for each million nuclei in the sample. *(2 marks)*
   f Find the approximate number of undecayed nuclei present at the start. *(2 marks)*

6 Here are two statements about radioactive decay:
   • Radioactive decay is random.
   • The number of nuclei in a sample that will decay in 1 s can be accurately predicted.
   a Using your understanding of the decay constant, explain how both statements are true for a sample containing many nuclei with a short half-life. *(3 marks)*
   b Explain why the second statement may not hold true for a sample containing few nuclei with a half-life considerably more than one second. *(2 marks)*

10.2 Another way of looking at radioactive decay

Specification reference: 5.1.1a(iv), 5.1.1b(v), 5.1.1c(i)

Learning outcomes

Describe, explain, and apply:

→ the number of nuclei decaying being proportional to the number remaining, $\frac{dN}{dt} = -\lambda N$

→ the equation $N = N_0 e^{-\lambda t}$

→ exponential curves plotted with linear or logarithmic scales

→ calculating activity and half-life of a radioactive source from data

→ the equation $T_{1/2} = \ln\frac{2}{\lambda}$.

Going further with modelling

In Topic 10.1, Radioactive decay and half-life, you saw that the gradual, smoothed-out decrease of radioactivity from a source can be described very simply by

$$\frac{\Delta N}{\Delta t} = -\lambda N \text{ (where } \lambda N = \text{activity)}$$

You can use the equation to produce a model of radioactive decay. The model assumes that the rate of decay is constant during each time interval, so the accuracy of the model depends on the duration of the time interval Δt. The smaller the time interval, the better the model.

The term $\frac{dN}{dt}$, the derivative or differential of N with respect to time, is the limit of the ratio $\frac{\Delta N}{\Delta t}$ as the time interval Δt is made shorter and shorter. (In reality, as the time interval is reduced more and more, it covers only single, isolated decays, not a smooth rate of decay, so this is another simplified model.)

An equation like this, involving the rate of change of a quantity, is called a **differential equation**. Such equations are immensely important in physics.

 ### Extension: Solving the equation $\frac{dN}{dt} = -\lambda N$

You do not have to solve differential equations in this course, but if you are studying mathematics you may find it interesting to see how the solution that you use is obtained.

We solve the differential equation by integration.

Separating the variables and choosing limits of integration gives

$$\int_{N_0}^{N} \frac{dN}{N} = -\int_{0}^{t} \lambda dt$$

The limits of integration show that N_0 nuclei are present at time $t = 0$. At time t, N nuclei remain. Integrating this expression between the limits gives

$$\ln\left(\frac{N}{N_0}\right) = -\lambda t$$
$$\therefore N = N_0 e^{-\lambda t}$$

To think about: Explain why we can state that the activity A of a sample $= A_0 e^{-\lambda t}$ simply from knowing that $A \propto N$, where N is the number of nuclei in the sample.

The solution to the equation $\frac{dN}{dt} = -\lambda N$ is

$$N = N_0 e^{-\lambda t}$$

This solution relates N, the number of nuclei remaining, to N_0, the original number of nuclei, and t, the time elapsed. The symbol e represents an irrational number, 2.718... (an unending number like π, and just as important).

As activity A is proportional to the number of nuclei present, we can also write

$$A = A_0 e^{-\lambda t}$$

Hint

You will need the number e a great deal in this second year of your physics course. You can see it on your calculator if you enter e^1.

Worked example: Activity

A radioactive sample has an activity of 2.50 kBq. The decay constant of the nuclei in the sample is $5.70 \times 10^{-6}\,\text{s}^{-1}$. Calculate the activity of the source after 24 hours.

There is quite a lot to do in this question. All the calculations can be performed in one step, but you need to make sure that you are confident in using your calculator. In this example we are going to perform the calculation in two stages to show the method clearly.

Step 1: The appropriate equation is $A = A_0\,\text{e}^{-\lambda t}$. First evaluate $-\lambda t$. The decay constant is given in the unit s^{-1}, so the time interval needs to be converted into seconds.

$-\lambda t = -5.70 \times 10^{-6}\,\text{s}^{-1} \times 24 \times 60 \times 60\,\text{s} = -0.492\,48$. We do not round this figure because it is an intermediate value.

Step 2: $A = 2.50 \times 10^3\,\text{Bq} \times \text{e}^{-0.49248} = 2.50 \times 10^3\,\text{Bq} \times 0.611\,11 = 1.53 \times 10^3\,\text{Bq}$ (3 s.f.)

Another look at $A = A_0\,\text{e}^{-\lambda t}$

If we take natural logarithms of both sides of the equation, we obtain

$\ln A = \ln(A_0\,\text{e}^{-\lambda t})$

$\therefore \ln A = \ln A_0 + \ln \text{e}^{-\lambda t}$

As the natural logarithm of e^x is x, we can write

$\ln A = \ln A_0 - \lambda t$

which gives

$\ln \dfrac{A}{A_0} = -\lambda t$

The activity of a sample will fall to half the original activity after one half-life $T_{1/2}$, that is, when $t = T_{1/2}$, $A = \dfrac{A_0}{2}$. So, for this special case

$\ln \dfrac{\frac{A_0}{2}}{A_0} = -\lambda T_{1/2}$

$\therefore \ln 0.5 = -\lambda T_{1/2}$

$\therefore T_{1/2} = \dfrac{\ln 2}{\lambda}$

This is a very useful result, because it links the decay constant and the half-life.

 Worked example: Calculating the half-life from the decay constant

An isotope of polonium has a half-life of 3.05 minutes. Calculate its decay constant in s^{-1}.

Step 1: $T_{1/2} = \dfrac{\ln 2}{\lambda} \quad \therefore \quad \lambda = \dfrac{\ln 2}{T_{1/2}}$

Step 2: To calculate the decay constant in seconds, remember to convert time from minutes into seconds.

$\lambda = \dfrac{\ln 2}{T_{1/2}} = \dfrac{0.693...}{3.05 \times 60} = 0.00379\,\text{s}^{-1}$ (3 s.f.)

A graph of the natural log of the activity of a source ($\ln A$) against time (t) has the equation $\ln A = \ln A_0 - \lambda t$. The gradient of the graph is $-\lambda$.

You may wonder how we can determine half-lives of thousands of years when we can't wait for the activity to halve. Instead, the change in activity can be measured over a short period and a graph of $\ln A$ against time drawn. The decay constant found from the gradient.

Worked example: Determining half-life from a graph of $\ln A$ against t

The graph shows how the natural logarithm of count rate of a radioactive source varies with time. Use the graph to find the decay constant and the half-life of the source.

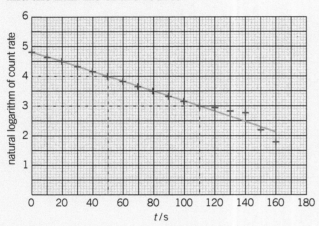

▲ **Figure 1** *Natural logarithm of count rate plotted against time for the worked example Half-life from graphs, in Topic 10.1, Radioactive decay and half-life, Figure 4*

Step 1: The decay constant is the negative of the gradient.

$$\text{Gradient} = \frac{3.0 - 4.0}{110\,s - 50\,s} = -0.0166...\,s^{-1}$$

so decay constant $\lambda = (+)\ 0.017\,s^{-1}$ (2 s.f.)

Step 2: You can use the unrounded decay constant $\lambda = \dfrac{1}{60\,s}$.

Half-life $T_{1/2} = \dfrac{\ln 2}{\lambda} = 0.693... \div 0.0166...\,s$
$= 42\,s$ (2 s.f.)

The half-life can also be determined by finding the time interval over which the logarithm of the count rate falls by $\ln 2$.

Summary questions

1 The half-life of nitrogen–16 is 7.4 s. Calculate the decay constant of the isotope. *(1 mark)*

2 The decay constant of protactinium is $9.63 \times 10^{-3}\,s^{-1}$. A sample of the isotope has an activity of 400 Bq. Calculate the expected activity of the sample after 10 minutes have passed. *(2 marks)*

3 The half-life of carbon–14 is about 1.8×10^{11} seconds. State the percentage of a sample that will remain after:
   a 2 half-lives   *(1 mark)*   b 10^{10} seconds. *(3 marks)*

4 Technetium–99 is a medical tracer used to diagnose heart and muscle function. It is prepared in an excited state, which decays with a half-life of 6 hours, each atom emitting a gamma ray of energy 140 keV (1 eV = 1.6×10^{-19} J). Suppose that a decay rate of $1000\,s^{-1}$ is required for diagnosis.
   a Calculate the probability of decay, per second, of an excited technetium–99 nucleus. *(2 marks)*
   b Find the number of technetium–99 atoms that must enter the patient's heart in order to give an initial decay rate of $1000\,s^{-1}$. *(2 marks)*
   c In the end, all these nuclei will decay. What is the total energy in joules released in the patient? *(2 marks)*
   d The mass of a technetium atom is 99 atomic mass units (a.m.u.), where 1 a.m.u. = 1.66×10^{-27} kg. What mass of technetium–99 needs to be used in the tracer? *(2 marks)*
   e What activity will remain after 2 days? *(2 marks)*

10.3 Capacitors in circuits

Specification reference: 5.1.1a(i), 5.1.1a(ii), 5.1.1b(vi), 5.1.1c(v)

Capacitor structure and function

You have seen that the exponential decay of radioactive sources can be modelled using a simple iterative process. Many other processes in physics show exponential decay. For example, electrical components called **capacitors** lose charge through a mathematically analogous process. In Topic 10.4, Modelling capacitors, you will consider how to model this process, but first you need to learn about the basic characteristics of these components.

Capacitors store charge. The basic design of a capacitor is very simple – just a pair of electrical conductors separated by a thin layer of insulator. The conductors are often referred to as 'plates' because early capacitors used large metal plates. Nowadays, capacitors are usually made of sheets of metal foil with an insulating layer between them, and come in all shapes and sizes (Figure 1). When a potential difference (p.d.) is applied across the plates, for example if a cell is connected across the two plates,

▲ **Figure 1** *Capacitors and resistors in a circuit*

negative charge flows from the negative terminal of the cell to one plate, making it negative. An equal negative charge flows from the other plate to the cell through the positive terminal, making that plate positive. This process of charge separation is called charging the capacitor (Figure 2).

Charge stored on a capacitor

Figure 3 shows how the charge Q on a capacitor varies with the p.d. across its plates. The graph is a straight line through the origin, showing that the charge on the plates is proportional to the p.d.

Capacitance C is the charge separated per volt: $C = \dfrac{Q}{V}$. So C is the gradient of the graph in Figure 3. The unit of capacitance is the farad, F, where $\mathrm{F} = \mathrm{C\,V^{-1}}$. One farad is a very large capacitance, so it is common to use microfarads (μF) or picofarads (pF).

> **Worked example: Charge on a capacitor**
>
> A 0.22 μF capacitor can be charged with a p.d. of up to 50 V. Calculate the charge on the capacitor at this p.d.
>
> **Step 1**: Rearrange $C = \dfrac{Q}{V}$ to give $Q = CV$.
>
> **Step 2**: Substitute the values, remembering that the capacitance has been given in microfarads.
>
> $$Q = 0.22 \times 10^{-6}\,\mathrm{F} \times 50\,\mathrm{V} = 1.1 \times 10^{-5}\,\mathrm{C}$$

Learning outcomes

Describe, explain, and apply:

→ capacitance as the ratio $C = \dfrac{Q}{V}$

→ energy on a capacitor
$E = \dfrac{1}{2}QV = \dfrac{1}{2}CV^2$

→ energy of a capacitor as the area below a Q–V graph.

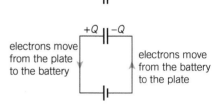

▲ **Figure 2** *Capacitor symbol (top) and charging a capacitor*

Study tip

Remember that charge does not flow through a capacitor. The plates are separated by an insulator.

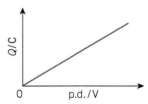

The gradient of the graph of charge against p.d. gives the capacitance

▲ **Figure 3** *Relationship between p.d. and charge Q for a capacitor*

Hint

Take care not to confuse the symbol for capacitance C with the symbol for coulomb, the unit of charge, C.

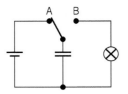

▲ **Figure 4** *Circuit for charging and discharging a capacitor*

Synoptic link

You have already met the link between potential difference, charge, and energy in Topic 3.1, Current, p.d., and electrical power.

Energy stored on a capacitor

When the switch in Figure 4 is in position A, the capacitor will charge. When the switch is moved to position B, negative charge flows from the negative plate of the capacitor to the positive plate, **discharging** the capacitor. The lamp in the circuit glows during the discharge, showing that energy has been transferred from the capacitor to the lamp. Capacitors have many uses as energy storage devices, such as in camera flash circuits and computers.

When a charge Q moves through a p.d. V, the energy transferred $= QV$.

When a capacitor is charged to a potential difference V the energy stored is $\frac{1}{2}QV$. You can see why this is (and why the energy released is not QV) by considering the energy released during discharge. As illustrated in the graphs in Figure 5, the area under the line can be considered as a series of strips of area $V\delta Q$, where δQ represents a small charge flowing from the capacitor (unlike Figure 3, this graph has charge on the x-axis to show the strips of equal width δQ more easily). The energy δE delivered by the same charge δQ falls as V falls. The energy stored by the capacitor is the area of all the strips, that is, the area of the triangle $\frac{1}{2}QV$.

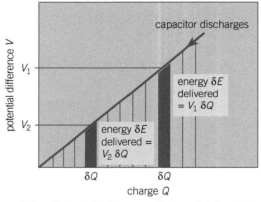

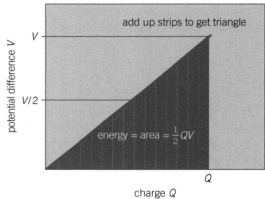

Energy delivered at p.d. V when a small charge δQ flows $\delta E = V\delta Q$.
Energy δE delivered by same charge δQ falls as V falls.

Energy delivered = charge × average p.d.
Energy delivered $= \frac{1}{2}QV$.

▲ **Figure 5** *Energy stored by a capacitor* $= \frac{1}{2}QV$

Energy $= \frac{1}{2}QV$

Substituting in $Q = CV$,

Energy $= \frac{1}{2}CV^2 = \frac{1}{2}Q^2/C$

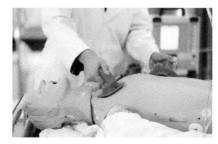

▲ **Figure 6** *A defibrillator in use on a dummy*

 Worked example: p.d. of a defibrillator

A defibrillator is used to restore regular heartbeat in patients whose heart is beating randomly instead of pumping blood around the body, perhaps following a heart attack. It delivers a controlled electric shock from a charged capacitor through paddles on the chest (Figure 6).

A 40.0 µF capacitor delivers 300 J to the patient. Calculate the p.d. across the capacitor before discharge.

Step 1: The question states the energy and the capacitance, so to find the p.d. use $E = \frac{1}{2}CV^2$.

$$300\,\text{J} = \frac{1}{2} \times 40 \times 10^{-6}\,\text{F} \times V^2$$

$$V^2 = \frac{300\,\text{J}}{\left(\frac{1}{2}\right) \times 40 \times 10^{-6}\,\text{F}} = 15 \times 10^6\,\text{J}\,\text{F}^{-1}$$

$$V = \sqrt{15 \times 10^6\,\text{J}\,\text{F}^{-1}} = 3870\,\text{V} \ (3\ \text{s.f.})$$

Summary questions

1 Use the equations $Q = CV$ and $E = \frac{1}{2}QV$ to show that

   a $E = \frac{1}{2}CV^2$ *(1 mark)*   **b** $E = \frac{1}{2}Q^2/C$.   *(1 mark)*

2 Copy and complete Table 1 showing energy stored, p.d., and charge for a number of different capacitors. The first line has been completed for you.   *(5 marks)*

   ▼ Table 1

Capacitance /F	p.d. /V	Charge /C	Energy /J
2.2×10^{-3}	20	0.044	0.44
		0.015	0.11
2.2×10^{-4}			0.058
2.2×10^{-6}		1.2×10^{-4}	

3 A 470 µF capacitor is charged to a potential difference of 6.00 V.

   a Calculate the energy stored on the capacitor.   *(1 mark)*

   b The capacitor discharges until it reaches a potential difference of 3.00 V. Calculate the energy that has been released.   *(2 marks)*

4 The graph in Figure 7 shows how the charge on a capacitor varies with the p.d. across its plates.

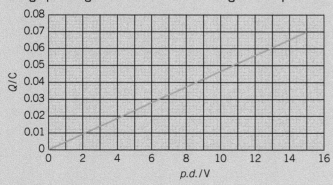

   ▲ Figure 7

   a Use the graph to calculate the capacitance of the capacitor.   *(2 marks)*

   b Use the graph to calculate the energy stored by the capacitor at a p.d. of 10.0 V.   *(2 marks)*

   c A student tells a friend that the capacitor is 'fully charged' at 15.0 V. Her friend disagrees and says that a capacitor can never be 'fully charged'. Who is right? Explain your answer.   *(2 marks)*

   d Capacitors have a maximum working potential difference marked on them. Suggest and explain a possible consequence of exceeding the maximum potential difference.   *(2 marks)*

10.4 Modelling capacitors

Specification reference: 5.1.1a(iii), 5.1.1b(i), 5.1.1c(iii), 5.1.1c(iv), 5.1.1c(v), 5.1.1d(iii)

Capacitor discharge

In Topic 10.3, Capacitors in circuits, you saw that the charge on a capacitor is proportional to the p.d. across its plates. In this section you will see how the charge on a capacitor varies with time.

 Practical 5.1.1d(iii): Investigating discharge of a capacitor

The circuit in Figure 1 can be used to observe how the p.d. across a capacitor changes with time as it discharges through a resistor.

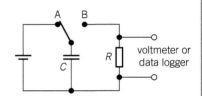

▲ Figure 1 Circuit for observing the change in potential difference across a capacitor as it discharges

The capacitor is charged by connecting the switch to A. The capacitor will begin discharging when the switch is moved to B. Begin timing as soon as the discharge begins. If a 2200 μF capacitor and a 10 000 Ω resistor are used, you should take readings at least every 5 s. Smaller values for either quantity require more frequent readings – that's when a data logger is useful. You can analyse the data and produce graphs showing the variation of:

● charge on the capacitor with time,

● p.d. across the capacitor with time,

● current in the circuit with time.

The graph in Figure 2(a) shows the variation of p.d. with time as a capacitor discharges. It resembles the graph of count rate against time for a radioactive sample in Topic 10.1, Radioactive decay and half-life, Figure 4. They are both examples of exponential relationships.

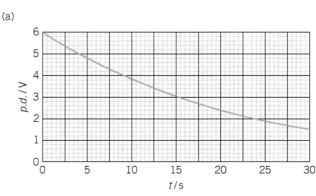

(a)

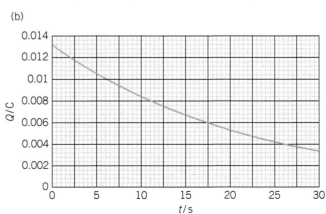

(b)

▲ Figure 2 Measurements as a capacitor discharges

The p.d. across a capacitor is proportional to the charge on the capacitor $\left(V = \dfrac{Q}{C}\right)$ so the variation of charge with time will also show exponential decay (Figure 2(b)).

The capacitor is discharging through a fixed resistance R. The current through the resistor is proportional to the p.d. across the resistor $\left(I = \dfrac{V}{R}\right)$ so the current in the circuit will also show exponential decay.

Modelling capacitor discharge

The techniques you used to model radioactive decay can be used to construct a mathematical model of capacitor discharge. Consider a capacitor in the simple circuit in Figure 3. For a small time interval Δt, assuming that the current remains steady during this time interval, the charge ΔQ leaving the capacitor in Δt is

$$\Delta Q = -I\Delta t = -\frac{V}{R}\Delta t = -\frac{Q}{RC}\Delta t$$

Rearranging, $\dfrac{\Delta Q}{\Delta t} = -\dfrac{Q}{RC}$

As the resistance and the capacitance are constant, the rate of change of charge is proportional to the amount of charge remaining – an exponential relationship.

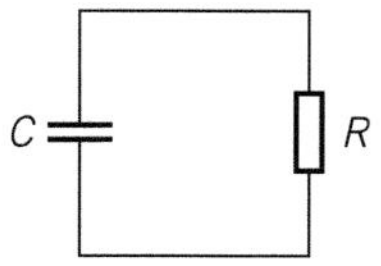

This equation can be solved iteratively using the process depicted in Figure 4. The charge on the capacitor at the end of the time interval is the starting point for the next iteration of the calculation. Spreadsheets help make these calculations quick and easy.

▲ **Figure 3** *A simple circuit for capacitor discharge*

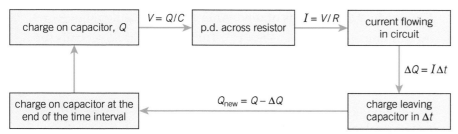

▲ **Figure 4** *Flowchart for iterative calculation of the changing charge on a discharging capacitor*

Table 1 gives an example of a few iterations based on the following initial values:

capacitance = $2200\,\mu F$ resistance = $10\,000\,\Omega$
initial charge = $0.0132\,C$ initial p.d. = $6.0\,V$ time interval $\Delta t = 10\,s$

▼ **Table 1** *Four iterative calculations of the charge on a discharging capacitor*

Time elapsed / s	Charge on the capacitor / mC	p.d. across resistor / V	Current / mA	Charge lost from capacitor / mC	Charge on capacitor at the end of the time interval / C
0	13.2	6.0	0.6	6.0	7.2
10	7.2	3.3	0.33	3.3	3.9
20	3.9	1.8	0.18	1.8	2.1
30	2.1	1.0	0.1	1.0	1.1
40	1.1				

The values have been rounded to two significant figures.

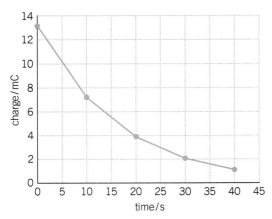

▲ **Figure 5** *Four iterations of the model of the changing charge on a discharging capacitor*

You can see in Figure 5 that the gradient decreases with each iteration and that the change in gradient also decreases with each iteration. These outcomes suggest that the model approximates to exponential decay, but it is far from the smooth curve obtained in a carefully conducted experiment. This is because the model holds the current constant during each time interval, whereas in reality the current falls continuously throughout each interval. We can improve the model by reducing the time interval Δt, the length of time the current is held constant. As Δt tends to zero,

$$\frac{dQ}{dt} = -\frac{Q}{RC}$$

This differential equation is identical in form to the decay equation you met in Topic 10.2, Another way of looking at radioactive decay. Similarly, the solution to this equation is

$$Q = Q_0 e^{\frac{-t}{RC}}$$

where Q_0 is the charge on the capacitor at the beginning of the discharge.

As the p.d. across a capacitor is proportional to the charge on its plates,

$$V = V_0 e^{\frac{-t}{RC}}$$

As current in the circuit is proportional to the p.d. across the resistor,

$$I = I_0 e^{\frac{-t}{RC}}$$

Time constant $\tau = RC$

The product RC has units of time. After time $t = RC$, the ratio $\frac{t}{RC} = 1$. At this time, $Q = Q_0 e^{-1}$, so $\frac{Q}{Q_0} = e^{-1}$. Therefore, after RC seconds the charge on the capacitor has been reduced to $\frac{1}{e}$ of its original value. The product RC is known as the **time constant** τ of the discharge circuit.

$\frac{1}{e}$ is about 0.37. After τ seconds the charge on the capacitor will be reduced to about 37% of its original value. After $2 \times \tau$ seconds the charge remaining will be $\frac{1}{e^2}$ or roughly 37% of 37%, about 14%. After $3 \times \tau$ about 5% of the charge remains.

It is interesting to see how the time constant is related to the time it takes for the capacitor to lose half its charge.

If $Q = \frac{Q_0}{2}$,

$$\frac{Q_0}{2} = Q_0 e^{\frac{-t}{RC}}$$

$$\frac{1}{2} = e^{\frac{-t}{RC}}$$

Taking natural logs of both sides,

$$\ln\frac{1}{2} = \frac{-t}{RC} \therefore 0.693 = \frac{t}{RC}$$

$\therefore$ time for charge to halve = $0.693\,RC$

🖩 Worked example: Discharging a capacitor

A 4700 μF capacitor is at a p.d. of 4.5 V. It discharges through a 2000 Ω resistor.

a Calculate the time constant of the circuit.

Step 1: time constant $\tau = RC$
= 4700 × 10⁻⁶ F × 2000 Ω = 9.4 s

b Calculate the p.d. across the capacitor after 15 s.

Step 2: $V = V_0 e^{\frac{-t}{RC}}$

$$V = 4.5\ \text{V} \times e^{\frac{-15}{9.4}}$$

$$= 4.5\ \text{V} \times 0.2027...$$

$$= 0.91\ \text{V (2 s.f.)}$$

Note that the intermediate value was not rounded. If we had rounded $e^{\frac{-15}{9.4}}$ to 0.2 we would have reached the final value of 0.90 V (to 2 s.f.), which is incorrect. You may prefer to do everything in one stage, but it is still important to show all stages of your working.

Worked example: Quick test for exponential decay

To investigate how the temperature difference between a hot object and cooler surroundings affects the rate of cooling, a student takes measurements of the temperature above room temperature of a small beaker of hot water at intervals of 1 minute. The results are given in Table 2. Describe a test to decide whether the temperature above room temperature is decreasing exponentially. Carry out the test and draw a conclusion from your results.

Step 1: Describe the test you are going to use.

An exponential change will show the constant ratio property. If the temperature is decreasing exponentially then the ratio $\frac{60}{54}$ will be the same as $\frac{54}{49}, \frac{49}{45}$, and $\frac{45}{41}$. Calculating and comparing these ratios will indicate whether the change is exponential.

Step 2: Perform the test.

To 2 s.f., the ratios are $\frac{60}{54} = 1.1, \frac{54}{49} = 1.1, \frac{49}{45} = 1.1, \frac{45}{41} = 1.1$

Step 3: Draw a conclusion.

The results are consistent with exponential cooling.

Hint

Exponential changes have the constant ratio property. If a change is exponential, the quantity will change by the same ratio in equal intervals of time. The time constant for a capacitor circuit and the half-life of a radioactive compound are examples of the constant ratio property – in equal intervals of time the number of nuclei changes by equal ratios, in this case 0.5.

▼ **Table 2** *Temperature changes for a beaker of water*

Time / minutes	Temperature above room temperature /°C
0	60
1	54
2	49
3	45
4	41

Charging a capacitor

 Practical 5.1.1d(iii): Investigating charging a capacitor

The circuit in Figure 6 can be used to observe how the p.d. across a capacitor changes with time as it is charged.

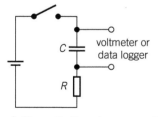

▲ **Figure 6** *Charging a capacitor*

Make sure that the capacitor is fully discharged by briefly connecting a lead across the terminals of the capacitor before beginning to charge it.

The capacitor will begin to charge when the switch is closed. Record the p.d. across the capacitor at equal time intervals. As with the discharge experiment, a data logger allows readings to be taken at short time intervals.

Figure 7 shows the charging curve of a capacitor in a circuit like that in Figure 6. The capacitor is being charged from a 6.0 V supply. The sum of the p.d.s across the capacitor and the resistor (which are in series) is 6.0 V (Figure 8). As the p.d. across the capacitor rises, the p.d. across the resistor falls.

As the p.d. across the resistor falls so does the current in the circuit, and so the rate of charging the capacitor also falls.

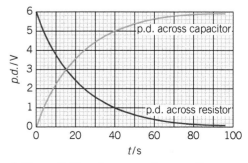

▲ **Figure 7** *Charging curve of a capacitor*

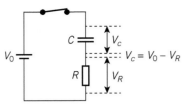

▲ **Figure 8** *Capacitor circuit*

The current in the circuit is given by the equation

$$I = I_0 e^{\frac{-t}{RC}}$$

As the current in the resistor is proportional to the p.d. across the resistor, V_R:

$$V_R = V_0 e^{\frac{-t}{RC}}$$

As $V_C = V_0 - V_R$,

$$V_C = V_0 - V_0 e^{\frac{-t}{RC}} = V_0(1 - e^{\frac{-t}{RC}})$$

Finally, as the p.d. across a capacitor is proportional to its charge,

$$Q = Q_0(1 - e^{\frac{-t}{RC}}), \text{ where } Q_0 = CV_0$$

> ### Study tip
>
> The rate of increase of charge on the capacitor as it is charged is *not* an example of exponential growth. It is incorrect to say 'the charge increases exponentially', as this would mean that the charge is increasing at an increasing rate.

 Worked example: Charging a capacitor

An uncharged $220\,\mu F$ capacitor is charged using a supply p.d. of $1.50\,V$. The time constant of the circuit is $0.2\,s$. Calculate the charge on the capacitor after $0.5\,s$.

$$Q = Q_0(1 - e^{\frac{-t}{RC}}) = CV_0(1 - e^{\frac{-t}{RC}})$$

$$= 220 \times 10^{-6}\,F \times 1.5\,V \times (1 - e^{\frac{-t}{RC}}) = 3.0 \times 10^{-4}\,C \text{ (2 s.f.)}$$

Summary questions

1 A $470\,\mu F$ capacitor discharges through a $5000\,\Omega$ resistor.
   a Calculate the time constant τ of the capacitor and resistor. (*1 mark*)
   b As an approximation, it is sometimes said that a capacitor is fully discharged after 5τ. Calculate the percentage of the original charge that will remain on the capacitor after 5τ. (*2 marks*)

2 Capacitance $4700\,\mu F$, resistance $5000\,\Omega$, initial p.d on capacitor = $4.5\,V$.

   Using a calculator or spreadsheet, make an iterative model of the first 30 s of the discharge of this capacitor using a time interval Δt of 5 s. Plot a charge–time graph of your model. Explain how you could improve the model. (*10 marks*)

3 Show that the time it takes to reduce the charge on a capacitor to a tenth of its original value is about 2.3 *RC*. (*3 marks*)

4 A $470\,\mu F$ capacitor discharges through a resistor. The charge on the capacitor falls to 40% of its original value after $0.30\,s$. Calculate the value of the resistor. (*3 marks*)

5 An uncharged capacitor is charged by a $3.00\,V$ battery in series with a $10.0\,k\Omega$ resistor. After $10.0\,s$ the p.d. across the capacitor has risen to $2.64\,V$. Calculate the capacitance of the capacitor. (*3 marks*)

Physics in perspective

How old is the Earth?

The age of the Earth is often said to be about four and a half billion years, but how do we know?

Early suggestions

Until recent times, most estimates of the age of the Earth were very precise and very different from modern values. For example, in the 17th century Archbishop Usher stated that Creation occurred on Sunday 23rd October, 4004 B.C.

Isaac Newton calculated a rather greater age by applying his own law of cooling, which states that the rate of fall of temperature is proportional to the temperature difference between a hot body (the Earth) and its surroundings. This analysis suggested that it must have taken 50 000 years for the Earth to cool to its present surface temperature, but Newton himself rejected this value for a shorter age that fitted in better with his interpretation of certain passages from the Bible.

▲ Figure 1   How do we date the Earth?

Clocks in rocks – radiometric dating

Modern estimates of the age of rocks and of the Earth itself use radiometric dating, which relies on our understanding of radioactivity and how the rate of decay of a radioisotope falls over time. These estimates are consistent with values obtained from studies of continental drift and evolution.

When a radioisotope decays it produces a new isotope, the *daughter product*. The original isotope is known as the parent. The amount of the daughter isotope (D) produced by the decay of the parent is given by the equation

$$D = N_0 - N$$

Where N_0 is the original number of parent nuclei and N is the number of parent nuclei present at the time of the measurement. As $N = N_0 e^{-\lambda t}$,

$$D = \frac{N}{e^{-\lambda t}} - N = N e^{\lambda t} - N = N(e^{\lambda t} - 1)$$

This relationship makes certain assumptions. One assumption is that all the daughter product present in the rock is due to the decay of the parent product. Another assumption is that neither daughter nor parent products have escaped from the rock in the time since its formation.

Closure temperature

Some minerals expel the daughter product of a radioactive decay when they're heated. This suggests that, for these minerals, the radioactive clock is set back to zero when a given temperature is exceeded, as all of the daughter product leaves the rock. When the mineral then cools beneath what is called the *closure temperature*, it becomes crystalline and any further daughter product produced remains in the rock. Therefore, reasonable estimates of the time since solidification and crystallisation

▲ Figure 2   Volcanic landscape in Iceland, an island country with some of the youngest rocks on Earth

can be made as any daughter product must have been formed after the closure temperature was reached. This is how we can confidently describe some rocks as 'young' – it means that there has been a relatively short span of time since the mineral cooled.

Potassium–argon dating

Potassium–40 is a radioisotope that decays in two ways. 89% of the isotope decays into calcium–40, but this is not useful as a clock because there will be calcium–40 present in the rock that is not a daughter product of potassium decay. 11% of the potassium–40 decays into argon–40, which is a much more useful indicator of age.

Argon is a noble gas, so it is very unreactive. This means that it will easily escape from the mineral in its hot, liquid form, and we can assume that there is no argon present when the clock starts at closure temperature. The dating equation for this decay includes a factor of 0.11 because the argon in the rock represents only 11% of the decayed potassium–40.

The equation is

$$^{40}\text{Ar} = 0.11 \times {}^{40}\text{K}(e^{\lambda t} - 1)$$

Although argon is unreactive there is still a possibility that the minerals gain or lose argon over time. More sophisticated analysis gives more reliable results but the technique of measuring ratios of parent and daughter isotopes is the basis of radiometric dating.

What about the Earth?

The oldest rocks on Earth have been dated to be around 4.4 billion years old. The Earth cannot be younger than the rocks on its surface, so we know that it must be at least 4.4 billion years old. But what is the oldest it could be?

The heavy elements we find on Earth, some of which we know are also present in the Sun, were formed in one or more supernova explosions. Uranium–238 and uranium–235 are amongst the isotopes formed in such explosions. We can begin to get a feel for the oldest possible age of the Solar System by finding the time elapsed since the supernova exploded. We do this by looking at the relative abundances of these two isotopes of uranium.

The present-day ratio $^{235}\text{U} / {}^{238}\text{U}$ is about 0.00715.

Assuming that ^{235}U and ^{238}U were produced in equal quantities we can write

$$0.00715 = \frac{e^{-\lambda_{235}t}}{e^{-\lambda_{238}t}} = e^{(\lambda_{238} - \lambda_{235})t}$$

Substituting for the values of the decay constants, $\lambda_{235} = 98.5 \times 10^{-11}$ yr^{-1} and $\lambda_{238} = 15.5 \times 10^{-11}$ yr^{-1}, gives a time since the supernova explosion of about 6 billion years. So the Earth cannot be older than about 6 billion years, nor can it be younger than about 4.4 billion years.

▲ **Figure 3** *Part of the Canadian Shield landscape, where rocks have been dated to be billions of years old*

Summary questions

1 Newton's Law of Cooling can be written as

$$\Delta\theta_t = \Delta\theta_0\, e^{-kt}$$

where $\Delta\theta_0$ is the temperature difference between the hot body and the surroundings when the time $= 0$, and $\Delta\theta_t$ is the temperature difference between the hot body and the surroundings at time t.

The law applies to bodies in a steady draught and the value of the cooling constant k depends on a number of factors, including the specific thermal capacity of the object, its surface area, mass, and shape.

a An iron nail is heated to red hot, at which point the temperature of its surface is $700\,°C$ above its surroundings. It is cooled in a draught for $70\,s$, after which its temperature is $100\,°C$ above its surroundings. Calculate the value of the cooling constant k.

b Describe assumptions and estimates that Newton may have used to reach a value of the age of the Earth from the law of cooling and the measured temperature of its surface.

2 $^{238}_{92}U$ decays to the stable end product $^{206}_{82}Pb$. The half-life of the process is 4.5×10^9 years.

A rock sample has a ratio $\dfrac{^{206}_{82}Pb}{^{238}_{92}U} = 0.6$. Use the relationship $D = N(e^{\lambda t} - 1)$

to calculate an estimate for the age of the rock. State an assumption that you are making in your estimate.

3 a Explain an advantage of using potassium–argon decay in radiometric dating.

b The ratio of argon to potassium in a sample of rock is 0.33. The half-life of potassium–40 is 1.3×10^9 years. Calculate the age of the rock.

4 Use information given in these Physics in perspective pages to show that the Solar System cannot be older than about 6 billion years.

Practice questions

1 The graph in Figure 1 shows the discharge of a capacitor.

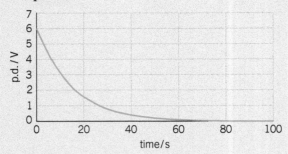

▲ Figure 1

Which of the following statements is/are true?

Statement **1**:   The time constant for the circuit is between 10 and 20 s.

Statement **2**:   The gradient represents the current in the circuit

Statement **3**:   The area under the graph represents the initial energy stored on the capacitor

A   1, 2 and 3 are correct

B   Only 1 and 2 are correct

C   Only 2 and 3 are correct

D   Only 1 is correct   (*1 mark*)

2 The mass of an americium–241 atom is 4.0×10^{-25} kg. The decay constant of americium–241 is 5.0×10^{-11} s^{-1}. Calculate the activity of a sample containing 8.0×10^{-11} kg of americium–241.   (*2 marks*)

3 The decay constant of carbon–14 is about 3.8×10^{-12} s^{-1}.

a Explain what 'decay constant' means in this context.   (*2 marks*)

b Calculate the half-life of the radioisotope.   (*2 marks*)

c Calculate the proportion of a sample of carbon–14 that will remain after 14 000 years.   (*2 marks*)

4 The graph in Figure 2 shows the natural log of activity A against time t for a short-lived radioisotope.

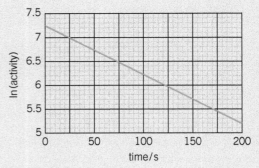

▲ Figure 2

a Use the equation $\ln A = \ln A_0 - \lambda t$ to explain how the decay constant can be found from the graph.   (*2 marks*)

b Calculate the value of the decay constant.   (*2 marks*)

c Calculate the half-life of the radioisotope.   (*2 marks*)

5 Strontium–90 has a decay constant of 7.6×10^{-10} s^{-1}. Calculate the time in years for the activity of a sample to fall to 10% of its original level. 1 year $= 3.2 \times 10^7$ s.   (*3 marks*)

6 Table 1 shows the first few steps of an iterative calculation to model the decay of a radioisotope. The decay constant of the radioisotope is 0.01 s^{-1}.

▼ Table 1

Time elapsed / s	Number of nuclei N	Number decaying $\Delta N = \lambda N \Delta t$	Number remaining $= N - \Delta N$
10	10 000	$0.01 \times 10\,000 \times 10$ $= 1000$	$10\,000 - 1000$ $= 9000$
20	9000	900	8100
30	8100		
40			

a Copy and complete Table 1.   (*2 marks*)

b Use the equation $N = N_0 e^{-\lambda t}$ to calculate the number of nuclei remaining at 40.0 s. Compare your value with that in the last cell of the table and give a reason for the difference in values obtained from the two models of decay. Explain how the iterative model can be improved to produce a closer match to the value obtained from the equation.   (*5 marks*)

7 A nuclear fusion device (a tokamak) is required to deliver 1 MJ of energy to a gas discharge, using capacitors discharged through the gas. Calculate the capacitance required if the largest workable p.d. for the capacitors is 10 kV. *(2 marks)*

8 An uncharged 220 µF capacitor is charged using the circuit in Figure 3.

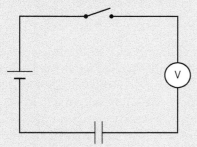

▲ **Figure 3**

When the switch is first closed, the voltmeter reads 3.0 V. After 100 s the voltmeter reads 1.9 V.

a Calculate the resistance of the voltmeter. *(3 marks)*

b Calculate the time constant of the circuit. *(1 mark)*

c Calculate how long it takes for the reading on the voltmeter to fall to 0.5 V. *(2 marks)*

9 A student sets up the circuit represented in Figure 4 using a high-resistance voltmeter.

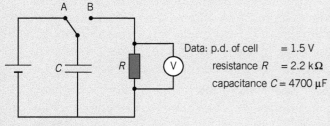

Data: p.d. of cell = 1.5 V
resistance R = 2.2 kΩ
capacitance C = 4700 µF

▲ **Figure 4**

a Calculate the time constant for the circuit. *(2 marks)*

b The student charges the capacitor and begins measuring the p.d. across the resistor as soon as the switch is moved to position B. Her results are given in

Table 2. The voltmeter used gave readings to a precision of 0.01 V.

▼ Table 2

Time / s	p.d. across resistor / V
0.0	1.50
5.0	0.92
10.0	0.58
15.0	0.35

By using the constant ratio property or another method, test whether the p.d. across the resistor falls exponentially. Comment on your answer. *(3 marks)*

10 Power for memory backup in computers can be provided by a charged capacitor. Calculate the capacitance required to support an initial current of 1 µA at a p.d of 5 V. The p.d. must not drop by more than 10% in a quarter of an hour. *(3 marks)*

11 A 10 000 µF capacitor is charged to 6 V. It is discharged through a 6000 Ω resistor. Then it is recharged, and discharged through a 6 V lamp rated at 1 W.

a What is the initial charge on the capacitor? *(2 marks)*

b Estimate the initial current when it starts discharging through the 6000 Ω resistor. *(2 marks)*

c Use the answer to **(b)** to estimate the fraction of the charge that has been discharged in the first 10 seconds. *(2 marks)*

d Calculate the product RC. Include units in your answer and explain the significance of RC. *(4 marks)*

e Calculate the energy stored on the capacitor when charged at 6 V. *(2 marks)*

f Calculate the resistance of the 6 V, 1 W lamp. *(1 mark)*

g Estimate by calculation how long the capacitor will take to discharge through the lamp. Explain your reasoning and the limitations of your estimate. *(4 marks)*

MODELLING OSCILLATIONS
11.1 Introducing simple harmonic oscillators

Specification reference: 5.1.1a(vi), 5.1.1b(iii), 5.1.1c(x)

▲ **Figure 1** *Chandelier in Pisa cathedral*

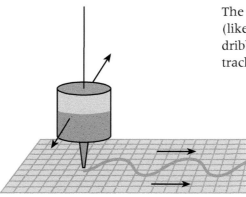

sand falling from a swinging pendulum leaves a trace of its motion on a moving track

▲ **Figure 2** *Visualising harmonic motion*

A famous chandelier

If you visit Pisa to view the famous Leaning Tower, you may notice that many tourists are also fascinated by a long chandelier hanging in the nearby cathedral (Figure 1). The sixteenth-century mathematician and astronomer Galileo is said to have made a fundamental discovery about the behaviour of pendulums by observing the motion of such a chandelier in Pisa cathedral. He found that the time of swing did not depend on how big the swing was. This constant timekeeping property is at the heart of pendulum clocks. Although these days we rarely use pendulum clocks, many modern timekeeping devices still rely on oscillations that can be described using ideas that stem from the day that Galileo considered the swinging chandelier in Pisa.

Simple harmonic oscillators

The time for one complete swing of an oscillator is called the **time period**. The displacement of the oscillator varies about an **equilibrium position**. For a pendulum, the equilibrium position is at the bottom of the swing, the position in which there is no net force on the oscillator. The **amplitude** of an oscillation is the greatest displacement from the equilibrium position.

The crucial property of a pendulum is that the time period of a complete swing does not depend on the amplitude of the oscillations (to a good approximation, for small amplitudes). As Galileo observed, the swings of a pendulum are (approximately) isochronous (from the Greek for "same time"). This property is shared by a number of oscillating systems called **simple harmonic oscillators**.

The displacement of a simple harmonic oscillator varies sinusoidally (like a sine wave) over time. This can be seen when sand or water dribbles from a swinging pendulum and leaves a trace on a moving track (Figure 2).

Simple harmonic motion and circular motion

The motion of a particle in a circle produces a mathematical description of simple harmonic motion. In Figure 3, point P is travelling around a circle of radius A. P* is the projection onto the line RS. As P moves around in a circle, P* will oscillate between points R and S.

At the instant represented in Figure 3, displacement $x = A\cos\phi$.

A particle travelling around a full circle (2π radians) in time T will have an angular frequency $\omega = \dfrac{2\pi}{T}$ rad s^{-1}.

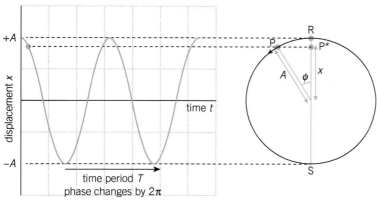

▲ **Figure 3** *Circular motion projected onto a line*

T is the time period of the oscillation. Because $T = \dfrac{1}{\text{frequency } f}$ we can write

angular frequency $\omega = 2\pi f \,\text{rad s}^{-1}$.

From Figure 3 you can see that if point P is at R when $t = 0$, then after t seconds the position of P* will be

$$x = A\cos 2\pi f t = A\cos \omega t$$

If the starting point of the oscillation is taken to be when $x = 0$, the equation describing the displacement at a given time becomes

$x = A\sin 2\pi f t$ or $x = A\sin \omega t$

 Worked example: A mass on a spring

A mass hanging from a spring oscillates up and down with a time period of 2.5 s.

a   Calculate the frequency of the oscillation.

Step 1: frequency $= \dfrac{1}{2.5\,\text{s}} = 0.4\,\text{Hz}$

b   The amplitude of the oscillation is 0.07 m. Calculate the position of the mass 3.9 seconds after its maximum displacement.

Step 1: $x = A\cos 2\pi f t = 0.07\,\text{m} \times \cos(2\pi \times 0.4\,\text{s}^{-1} \times 3.9\,\text{s})$

Step 2: $x = -0.065\,\text{m}$ (2 s.f.)

The negative sign shows that the mass is below the equilibrium point at this instant.

The acceleration of a simple harmonic oscillator

An oscillator is always pulled back to the equilibrium position by a restoring force that accelerates the oscillator. The oscillator will show simple harmonic motion if:

- the acceleration is proportional to the displacement from the equilibrium position
- the acceleration is always directed towards the equilibrium position.

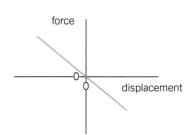

▲ **Figure 4** *The force on a simple harmonic oscillator is proportional to −displacement, so a ∝ −x*

Synoptic link

You will see how this equation can be derived in Topic 12.1, Circular motion.

In symbols we can write
$a \propto -x$ or $a = -kx$, where k is a constant.

The general relationship for all simple harmonic oscillators is
$a = -\omega^2 x = -4\pi^2 f^2 x$ (since $\omega = 2\pi f \, \mathrm{rad \, s^{-1}}$).

 Worked example: Acceleration

A simple harmonic oscillator has an amplitude of 0.020 m. The frequency of the oscillation is 1.5 Hz.

a   State the acceleration of the oscillator as it passes through the equilibrium point.

Step 1: The acceleration is zero at the equilibrium point, as $x = 0$ and acceleration is proportional to $-x$.

b   Calculate the maximum acceleration of the oscillator.

Step 1: $a = -4\pi^2 f^2 x = -4 \times \pi^2 \times (1.5 \, \mathrm{Hz})^2 \times 0.020 \, \mathrm{m}$
$= -1.8 \, \mathrm{m \, s^{-2}}$ (2 s.f.)

Alternatively, you could use the maximum negative displacement to give a maximum acceleration of $+1.8 \, \mathrm{m \, s^{-2}}$.

Summary questions

1   Calculate the time period of an oscillator of frequency 440 Hz.   *(1 mark)*

2   A simple harmonic oscillator has a frequency of 10 Hz and an amplitude of 50 mm. Timing begins as it passes through the position of zero displacement.
   a   Calculate the displacement at 0.070 s.   *(2 marks)*
   b   Calculate the maximum acceleration.   *(2 marks)*

3   A simple harmonic oscillator has a time period of 0.2 s and a maximum acceleration of 4.9 m s⁻². Calculate the amplitude of the oscillation.   *(2 marks)*

11.2 Modelling simple harmonic oscillation

Specification reference: 5.1.1a(v), 5.1.1b(vii), 5.1.1c(ix)

Displacement, velocity, and acceleration of simple harmonic oscillators

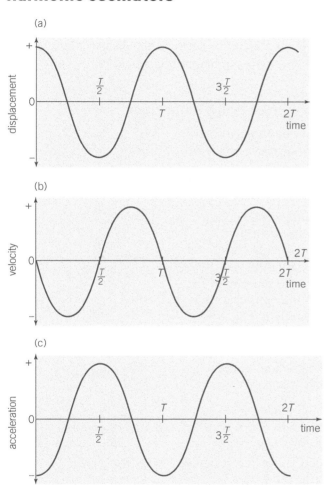

(a)

(b)

(c)

▲ **Figure 1** *Graphs of (a) displacement, (b) velocity, and (c) acceleration against time for a simple harmonic oscillator*

Figure 1(a) shows a displacement–time graph that represents the motion of a simple harmonic oscillator.

Velocity is the rate of change of displacement. The gradient of the displacement–time graph at any instant is the velocity at that instant. We can use gradients taken from a displacement–time graph to plot a velocity–time graph.

Similarly, acceleration is the rate of change of velocity. The gradient of the velocity–time graph at any instant is the acceleration at that instant.

As you saw in Topic 11.1, the equation for the displacement x of the oscillator shown in the graph is $x = A\cos 2\pi ft$, where A is the amplitude of the oscillation.

The velocity of the oscillator at any time t is given by

$$v = \frac{dx}{dt} = -2\pi fA \sin 2\pi ft$$

The acceleration of the oscillator at any time t is given by

$$a = \frac{dv}{dt} = -4\pi^2 f^2 A \cos 2\pi ft$$

Since $A \cos 2\pi ft = $ displacement x, $a = -4\pi^2 f^2 x$.

The graphs show that the phase difference between the displacement curve and the velocity curve is one quarter of a cycle or $\frac{\pi}{2}$ radian. The phase difference between the velocity curve and the acceleration curve is also $\frac{\pi}{2}$ radian. The phase difference between the displacement and acceleration curves is therefore π radian, or half a cycle.

 Worked example: velocity–time graph

Figure 2 shows a displacement–time graph for a simple harmonic oscillator.

Draw a graph representing the variation of velocity with time for the oscillator.

▲ **Figure 2** *Graph of displacement against time for a simple harmonic oscillator*

You may know enough mathematics to answer this question without much thought. You can also reach the answer by recalling that the velocity is zero at maximum displacement, that the velocity is maximum at zero displacement, and that the graph is sinusoidal. But the crucial relationship to use is that the velocity is the gradient of the displacement–time graph.

Note that the amplitude of the velocity–time graph is not important in this case, but it must be constant.

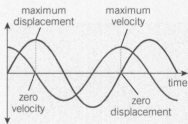

▲ **Figure 3** *Graphs of displacement and velocity against time for a simple harmonic oscillator*

Modelling simple harmonic oscillators

Think of a mass held between two rigid walls by a pair of springs. For now, forget about the need to support the mass and forget about friction. These can be put back in later.

Imagine holding the mass to one side. The spring with the larger extension exerts the larger force. Now let go. The mass is accelerated towards the centre, speed builds up, and the mass travels past the centre. Now the situation is reversed so that the mass slows down until it stops, experiencing a net force in the opposite direction. You can see this in Figure 4.

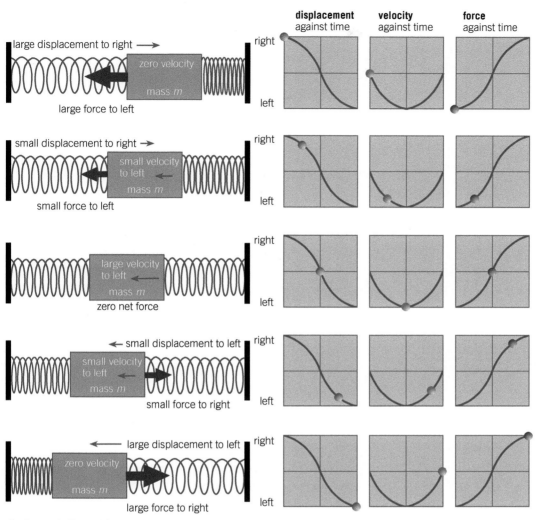

displacement against time	velocity against time	force against time

▲ **Figure 4** *Everything about harmonic motion follows from the fact that the restoring force (and so the acceleration) is proportional to the negative of the displacement*

The sequence of actions then repeats in reverse, taking the mass back to where it started, ready to start a new cycle of the motion.

With mass m and force constant k the acceleration is given by:

$$a = \frac{F}{m} \text{ and } F = -kx, \text{ so } a = -\frac{kx}{m}$$

Note that the acceleration of the mass is proportional to the negative of the displacement, $a \propto -x$, the defining characteristic of all simple harmonic oscillators.

The equation for the oscillation of a mass on a spring can be written as

$$\frac{\Delta^2 x}{\Delta t^2} = -\frac{kx}{m}$$

This equation can be solved iteratively to find the displacement and velocity of the oscillator at a given time. One method is:

At time $t = 0\,\text{s}$, the mass is at maximum displacement (Figure 4). At this point its velocity is zero. The flow chart in Figure 5 shows the series of calculations that lead to the value of the displacement a time Δt later.

Synoptic link

You met force constants for springs (also called spring constants) in Topic 4.2, Stretching wires and springs.

Hint

Velocity is the rate of change of displacement: $v = \dfrac{\Delta x}{\Delta t}$
Acceleration is the rate of change of velocity: $a = \dfrac{\Delta v}{\Delta t}$.
You can see that acceleration is the rate of change of the rate of change of displacement: $a = \dfrac{\Delta^2 x}{\Delta t^2}$.

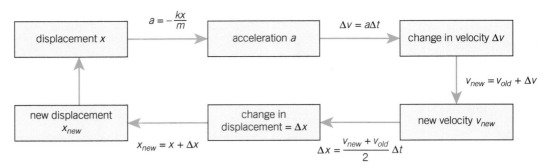

▲ **Figure 5** *Flow chart showing calculations for the simple harmonic oscillation in Figure 4*

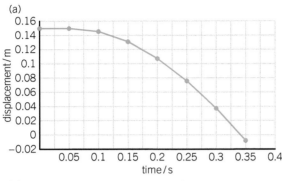

(a)

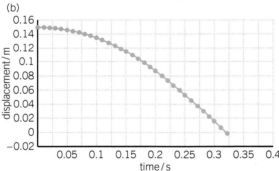

(b)

▲ **Figure 6** *Oscillation modelled with a time interval between calculations of (a) 0.05 s and (b) 0.01 s*

Synoptic link

Although this model does not perfectly match reality, simple mathematical models play an important part in physics, as you saw with exponential decay in Chapter 10.

In this model, v_{old} is the velocity at the beginning of the time interval (which is the same as the velocity at the end of the previous time interval). The new displacement is used as the starting point for the next **iteration** of the calculation.

Here are some data about a mass oscillating between springs:

- mass = 0.8 kg
- force constant $k = 20\,\mathrm{N\,m^{-1}}$
- amplitude of oscillation = 0.15 m
- time interval between calculations $\Delta t = 0.05\,\mathrm{s}$.

Applying the method shown in the flow chart to these data gives the graph in Figure 6(a). The line resembles a sinusoidal curve. This suggests that the model is a reasonable match to reality, although you can see that we have simplified it by holding the displacement constant during the first 0.05 s. We can improve this aspect of the model by recalculating with smaller time intervals (Figure 6(b)).

However, when the model is run for a longer time it produces an ever-increasing amplitude, as you can see in Figure 7. This is *not* what happens in simple harmonic oscillation and shows a limitation of this simple mathematical model.

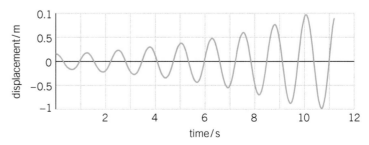

▲ **Figure 7** *Increase of displacement with time – the model is not exact*

Summary questions

1 The graph in Figure 8 shows how the velocity of a harmonic oscillator varies with time. Copy the graph and add lines to show how the displacement and the acceleration vary with time. Label the lines clearly. *(2 marks)*

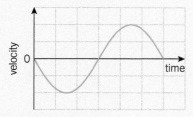

▲ Figure 8

2 Part of a building structure is oscillating with approximately simple harmonic motion with a period of 10 s and amplitude 1 m.
   a Calculate the maximum speed of the oscillation. *(2 marks)*
   b Calculate the maximum acceleration of the oscillation. *(2 marks)*

3 The graph in Figure 6a shows the results from a simple model of a harmonic oscillator.
   a Use data from the graph to calculate the frequency of the oscillation. *(2 marks)*

   The model is run with double the amplitude. The resulting graph is shown in Figure 9.

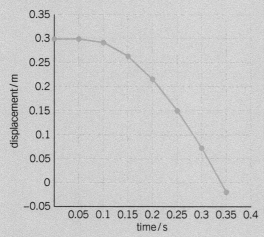

▲ Figure 9

   b State how the maximum acceleration of the second model compares with that of the first. Explain your answer using the mathematics of the model. *(2 marks)*
   c State how the maximum velocity of the second model compares with that of the first. Explain your answer. *(2 marks)*
   d Describe the limitations of this mathematical model, referring to the graph to support your ideas. *(2 marks)*

11.3 Using the models of simple harmonic motion

Specification reference: 5.1.1c(vi), 5.1.1c(vii), 5.1.1d(i)

Time period and frequency

You have seen that the acceleration of all simple harmonic oscillations is given by the equation

$$a = -4\pi^2 f^2 x$$

We can apply this general equation to specific oscillating systems to find the **frequency** and time period of the system.

Looking again at the oscillations of a mass on a spring

You have seen that the acceleration of a mass m oscillating on a spring of force constant k is given by

$$a = -\frac{k}{m} x$$

Substituting $a = -4\pi^2 f^2 x$,

$$4\pi^2 f^2 x = \frac{k}{m} x$$

Rearranging,

$$f = \frac{1}{2\pi} \sqrt{\frac{k}{m}}$$

$$f = \frac{1}{T}, \text{ so}$$

$$T = 2\pi \sqrt{\frac{m}{k}}$$

This equation shows that the time period of a mass–spring oscillator will increase with increasing mass. This relationship can be experimentally tested.

 Practical 5.1.1d(i): Measuring the period of a mass on a spring

Linking three or four springs together in series produces a longer, less stiff spring — reducing the stiffness results in a longer time period, which is more easily measured. Use expendable springs joined and hung from a stand. Attach a mass, raise it a little above the equilibrium position, and release to start the oscillation (Figure 1). The time for, say, 10 oscillations can be measured to find the time period of the system.

This process can be repeated for different masses.

The equation predicts that T is proportional to $\sqrt{m}$, or, in other words, that T^2 is proportional to m. This can be tested by plotting a graph of T^2 (y-axis) against m. A straight line through the origin shows proportionality. Alternatively, a graph of $\log T$ against $\log m$ will give a straight line of gradient 0.5, showing that $T \propto m$.

A similar experiment can be performed to investigate the relationship between the time period of a simple pendulum and the length of the pendulum string.

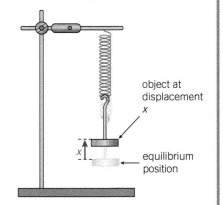

▲ **Figure 1** *Investigating the oscillations of a mass on a spring*

Worked example: Force constants

A mass of 0.30 kg oscillates on a spring. The frequency of the oscillation is 0.70 Hz. Calculate the force constant of the spring (also called the spring constant).

Step 1: Rearrange the equation for the time period of a mass on a spring.

$$T^2 = 4\pi^2\frac{m}{k} \quad \therefore k = 4\pi^2\frac{m}{T^2}$$

Step 2: Substitute and evaluate.

$$k = 4\pi^2 \times 0.30\,\text{kg} \times (0.70\,\text{Hz})^2 = 5.8\,\text{kg s}^{-2}\ (2\ \text{s.f.})$$

You might be surprised to see the units of the force constant k written as kg s^{-2} rather than N m^{-1}. The two units are equivalent as $1\,\text{N} = 1\,\text{kg m s}^{-2}$.

Looking again at the simple pendulum

The oscillation of a simple pendulum fascinated Galileo. We can now consider this system in more detail (Figure 2). When the mass at the end of the pendulum string (the bob) is displaced by x, a **restoring force** F acts in the direction of the equilibrium position. This restoring force is the horizontal component of the string tension T. For small displacements

$$F = -T\frac{x}{L}$$

The tension is approximately equal to mg (it will actually be a shade larger)

$$T = mg$$

Therefore

$$F = -mg\frac{x}{L}$$

As $F = ma$,

$$a = -g\frac{x}{L}$$

For small displacements x, the simple pendulum follows the relationship $a \propto -x$, so the pendulum will oscillate with simple harmonic motion.

$$a = -4\pi^2 f^2 x = -g\frac{x}{L}$$

$$\therefore 4\pi^2 f^2 = \frac{g}{L}$$

$$\therefore f = \frac{1}{2\pi}\sqrt{\frac{g}{L}}$$

$$T = \frac{1}{f} = 2\pi\sqrt{\frac{L}{g}}$$

Note that the time period of the pendulum is independent of the mass of the bob but does depend on g, the gravitational field strength.

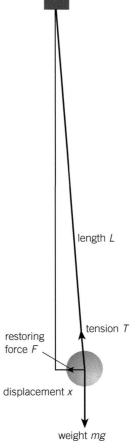

length L

tension T

restoring force F

displacement x

weight mg

▲ **Figure 2** *The simple pendulum*

Worked example: Chandelier

A chandelier hangs from a rope 15 m long. Calculate the time period of the simple harmonic oscillation of the chandelier. ($g = 9.8\,\mathrm{N\,kg^{-1}}$)

Select the correct equation and substitute the values into it.

$$T = 2\pi\sqrt{\frac{L}{g}} = 2\pi\sqrt{15\frac{m}{9.8}\mathrm{N\,kg^{-1}}} = 7.8\,\mathrm{s}\ (2\ \mathrm{s.f.})$$

Summary questions

1 A mass m oscillates between two springs. The time period of the oscillation is T.

The mass is replaced with a mass $2m$. Explain how this will affect the time period of the oscillation. *(2 marks)*

2 A student performed an experiment in which she measured the time for ten oscillations of a simple pendulum of varying length. She divided the measured time by ten to find the time for one oscillation in each case.

a Explain why the student timed ten oscillations. *(2 marks)*

Here are her results:

Length / m	0.60	0.80	1.00	1.20	1.40	1.60	1.80
Time period / s	1.5	1.8	2.0	2.2	2.4	2.5	2.7

b Plot a graph of length (x-axis) against (time period)^2 (y-axis). Explain how the graph shows whether the behaviour of the pendulum follows the equation for the time period of a pendulum:

$$T = \frac{1}{f} = 2\pi\sqrt{\frac{L}{g}}$$

(6 marks)

c Plot a graph of log length (x-axis) against log time period (y-axis). Explain how the graph shows whether the behaviour of the pendulum follows the equation for the time period of the pendulum. *(6 marks)*

3 A tube is floating in a liquid (Figure 3). In equilibrium a length L of the tube is below the surface. The tube is pushed down a distance x and released. It is modelled to oscillate with simple harmonic motion. The acceleration of the system is given by $a = -g\frac{x}{L}$.

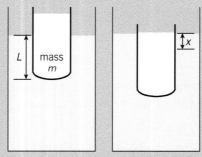

▲ **Figure 3** *A tube floating in a liquid*

a Use the general equation for the acceleration of a simple harmonic oscillator to find an expression for the time period of the tube oscillating in the water. *(2 marks)*

b Calculate the length of tube L required to produce an oscillation of period 1.0 s. *(2 marks)*

c Suggest why this oscillator would not function as a useful timekeeper. *(2 marks)*

11.4 Resonance

Specification reference: 5.1.1a(vii), 5.1.1a(viii), 5.1.1b(iii), 5.1.1b(viii), 5.1.1c(viii), 5.1.1c(xi), 5.1.1d(ii)

Shaky bridges

A tragic and apparently unlikely accident happened in Angers, France, in 1850. A group of soldiers was crossing a suspension bridge in a high wind when the bridge began to vibrate dramatically. The oscillation led the soldiers to march in time with the regular swings of the bridge. The vibration increased until the bridge collapsed, sending more than 200 men to their deaths.

Free and forced oscillations

Free oscillations have constant amplitude. A freely oscillating pendulum will swing at its particular, **natural frequency** with the same amplitude for all time.

Marching soldiers on a bridge provide an example of a periodic driving force. Each time they take a step they exert a force on the bridge. A periodic driving force will make the system (in this case, the bridge) oscillate at the frequency of the driving force. These are **forced oscillations**.

If the driving frequency matches the natural frequency of the system, large-amplitude oscillations are produced. This is **resonance**. This is what happened in Angers. The frequency of the marching feet matched a natural frequency of oscillation of the bridge, and the amplitude of the oscillation increased. Similarly, in 1940 the Tacoma Narrows bridge near Seattle collapsed when the wind induced resonant oscillations in the structure (Figure 1), and the Millennium Bridge across the Thames in London has been nicknamed the Wobbly Bridge because, when first built, it swayed by up to 7 cm when pedestrians began to synchronise their steps with the bridge's natural frequency.

Energy of a harmonic oscillator

The energy stored in a mechanical oscillator such as a mass on a spring continually shifts between the kinetic energy of the mass and the elastic potential energy stored in the spring. When the displacement of the oscillator is a maximum, the velocity of the oscillator is zero. At this point all the energy in the system is elastic potential energy, given by $E_p = \frac{1}{2}kx^2$.

At this maximum displacement, $x = A$, the amplitude of the oscillation. Therefore, the total energy of the oscillator $= \frac{1}{2}kA^2$.

When the oscillator passes through the equilibrium position, $x = 0$. At this point all the energy is the kinetic energy E_k of the oscillator.

At any time the total energy $E_{total} = \frac{1}{2}kA^2 = E_k + E_p$

Learning outcomes

Describe, explain, and apply:

→ kinetic and potential energy changes in simple harmonic motion

→ free and forced vibrations, damping, and resonance

→ the variation of the amplitude of a resonator with driving frequency

→ $F = kx, E = \frac{1}{2}kx^2,$
$E_{total} = \frac{1}{2}mv^2 + \frac{1}{2}kx^2$

→ qualitative observations of forced and damped oscillations for a range of systems.

▲ **Figure 1** *The Tacoma Narrows suspension bridge oscillates wildly and then collapses when the wind induces resonance, 1940*

Synoptic link

You met the equations for energy stored in a spring in Topic 4.2, Stretching wires and springs.

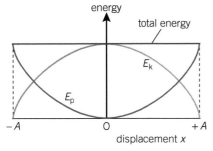

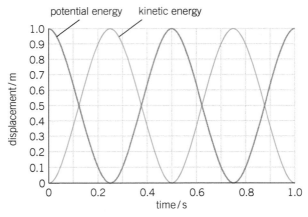

potential energy   kinetic energy

▲ **Figure 2** *Variation of kinetic energy and potential energy with displacement in an oscillator*

Figure 2 shows how the energy of an oscillator varies between kinetic and potential as the displacement changes, and Figure 3 shows how the energy of an oscillator changes between kinetic and potential over time. The oscillation in the graph has a time period of 1 s, so you are looking at one complete oscillation. Notice that the kinetic energy and potential energy both peak **twice** during each oscillation.

▲ **Figure 3** *Variation of kinetic energy and potential energy in an oscillator over time*

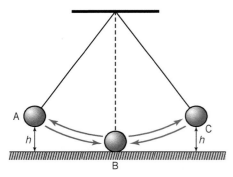

▲ **Figure 4** *Variation of kinetic energy and potential energy in an oscillator over time*

🖩 Worked example: A pendulum

For the simple pendulum in Figure 4, the maximum height h above the equilibrium position is 0.020 m and the mass of the pendulum bob is 0.050 kg.

a Describe the changes between gravitational potential energy (mgh) and kinetic energy as the pendulum bob moves between point A and point B.

Step 1: Include the details the question requires and no more (there is no need to describe movement to point C).

At point A the bob is instantaneously stationary at its maximum height, so it has zero kinetic and maximum gravitational potential energy. As the bob approaches point B it accelerates, gaining kinetic energy, whilst h is decreasing so the gravitational potential energy also decreases. At point B the kinetic energy is a maximum and, as $h = 0$, the gravitational potential energy relative to the equilibrium point is zero.

b Calculate the total energy of the system.

Step 1: total energy of the system = maximum potential energy

Step 2: $E_{total} = 0.05\,\text{kg} \times 9.8\,\text{N kg}^{-1} \times 0.02\,\text{m} = 9.8 \times 10^{-3}\,\text{J}$

c Calculate the velocity of the bob as it passes through the equilibrium position.

Step 1: At this point, kinetic energy = total energy of system
$$= 9.8 \times 10^{-3}\,\text{J}$$

Step 2: $\frac{1}{2}mv^2 = 9.8 \times 10^{-3}\,\text{J}$

Therefore $v = \sqrt{\dfrac{2 \times 9.8 \times 10^{-3}\,\text{J}}{0.05}}\,\text{kg} = 0.63\,\text{m s}^{-1}$ (2 s.f.)

Damped oscillations

A free oscillator will oscillate indefinitely. As the amplitude of the oscillation remains constant, it follows that the total energy $\frac{1}{2}kA^2$ will also remain constant.

The oscillators we observe, such as pendulums, do not oscillate indefinitely. Energy gradually leaks away from them because of **damping** – the action of forces such as friction and air resistance.

Extension: More about damping

Figure 5 shows an example of **light damping**. The maximum displacement of the oscillator is reduced in each oscillation, but the time period of the oscillation is roughly constant – that's why pendulum clocks keep good time even as the amplitude of the swing decreases.

An oscillation is **critically damped** if the oscillator stops at the equilibrium position without completing a cycle, and **heavily damped** if the oscillator returns to the equilibrium position much more slowly than a lightly damped or critically damped oscillator (Figure 6).

The amount of damping is related to the quality factor Q of the oscillation. The quality factor is defined as $Q = 2\pi \times \dfrac{\text{energy stored per cycle}}{\text{energy lost per cycle}}$

As a rule of thumb, Q is roughly equal to the number of free oscillations before the oscillator transfers all its energy to the surroundings. A simple pendulum has a quality factor of about 1000, which is the same as a guitar string. A car suspension, on the other hand, has a quality factor of 1.

> **To think about:**
>
> 1 Why is a very low Q factor an advantage in a car suspension? Would you design a car suspension to be lightly, heavily, or critically damped? Explain your reasoning.
> 2 How long would you expect a guitar string oscillating at 110 Hz to vibrate after it has been set in motion? What about a string oscillating at 330 Hz?

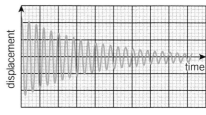

▲ Figure 5 *Variation of displacement with time for light damping*

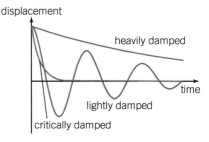

▲ Figure 6 *Variation of displacement with time for different degrees of damping*

Resonance

Resonance occurs when the periodic frequency driving an oscillation matches the natural frequency of the oscillator. The amplitude of oscillations of the oscillator will rise until the energy losses per cycle are equal to the energy supplied by the driver each cycle. If the system is only lightly damped the increase in the amplitude can be dramatic. The graphs in Figure 7 show the effect of resonance. Note that when the driving frequency exceeds the natural frequency, the amplitude of oscillation has a lower value than without a driver.

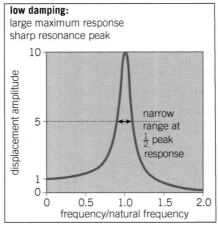

low damping:
large maximum response
sharp resonance peak

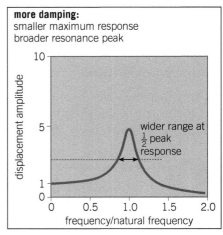

more damping:
smaller maximum response
broader resonance peak

▲ Figure 7 *Damping and resonance*

Resonance is extremely important throughout physics and engineering. Atoms resonate to light of the right frequency. Buildings, particularly towers (and bridges!) resonate to vibrations caused by the wind. Tuning in the radio uses resonance. Resonance is everywhere.

 ## Practical 5.1.1d(ii): Observing resonance

It is easy to observe resonance in the laboratory. One simple method is to suspend a heavy book between lab stands (Figure 8). Try making the book oscillate by blowing at it through a drinking straw. This is very difficult unless you blow periodically at the natural frequency of the suspended book.

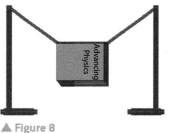

▲ Figure 8

The effect of resonance can be measured using the oscillations of a mass suspended between two springs (Figure 9). The oscillation is driven by a vibrator connected to a signal generator. The amplitude of the oscillations is measured and a graph of amplitude against frequency is plotted.

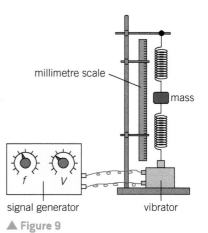

▲ Figure 9

Summary questions

1 A spring stores 0.025 J when extended by 4 cm. Calculate the force constant k of the spring. *(2 marks)*

2 A 500 N person sitting on the wing of a car deflects the suspension over that wheel by 50 mm.
  a Show that the force constant for the spring is about 10^4 N m^{-1}. *(1 marks)*
  b A bump in the road suddenly deflects the wheel by 100 mm. Calculate the energy stored. *(2 marks)*
  c The mass of the wheel is 25 kg. Calculate the natural frequency of oscillation. *(2 marks)*

3 Figure 10 shows the absorption of infrared radiation by a thin slice of a sodium chloride crystal, plotted against the wavelength of the radiation. This is a resonance effect — the electric field of the radiation makes the charged sodium and chloride ions oscillate near their natural frequency.
  a State the frequency at which maximum absorption occurs. *(2 marks)*
  b State the natural frequency of oscillation of ions in sodium chloride. *(1 mark)*
  c Taking the mass of an oscillating ion to be about 5×10^{-26} kg, estimate the force constant for these oscillations. *(2 marks)*

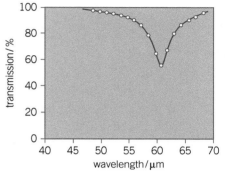

▲ **Figure 10** *Absorption of infrared by a thin slice of a sodium chloride crystal*

Physics in perspective

Resonance all around us

An oscillator resonates when the frequency of the oscillating force matches the natural frequency of the oscillator. Examples of resonance and of the effects of forced oscillation are all around us — here we consider just a few.

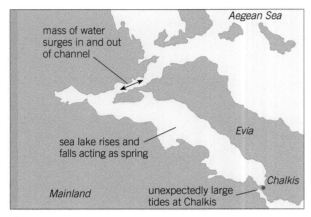

▲ **Figure 1** *Chalkis lies on a narrow channel that connects a large sea lake to the rest of the Mediterranean Sea.*

Mysterious tides

The Mediterranean Sea shows almost no tidal variation, whilst the town of Chalkis (or Chalcis) in Greece sees large tides — an observation that puzzled the philosopher Aristotle two and a half thousand years ago. The explanation of this phenomenon, known as 'trela nera' or 'crazy waters', is still a subject of debate.

Chalkis is at the bottom of a 'bottle' – a large sea lake with a narrow channel opening at the far end. Comparing the system to an oscillating mass on a spring, the 'spring' is the rising and falling of the water in the lake under gravity, and the 'mass' is the water surging in and out of the channel. The natural time period of the system (the time for the mass of water to travel up and down the channel) is very close to the 12.4 hours with which the tidal tug of the moon varies. The lake resonates and the water at Chalkis rises and falls much more than would be expected.

Wine glasses and opera singers

Wine glasses can oscillate — put a little liquid in the bottom of a wine glass and rub a wet finger along the top edge of the glass, and it will sound a note of a frequency determined by the glass and the amount of liquid it contains. This is the natural frequency of the system. This is the principle of the 'armonica' – an instrument invented by the scientist and politician Benjamin Franklin, in which glass dishes of different sizes sound different notes.

▲ **Figure 2** *Franklin's Armonica — a set of nestling glass dishes which rotate and are played by rubbing fingers against the wet glass*

▲ **Figure 3** *A wine glass is shattered using a powerful loudspeaker*

If a note sounded near a delicate wine glass matches the glass's natural frequency, the amplitude of the oscillations can grow until the glass shatters — it shakes itself to bits. Trained singers have been filmed performing this example of resonance. It is not so much the volume of their voice that shatters the glass, but rather the pitch.

Microwave ovens

Microwave ovens are **not** an example of the use of resonance, but they are a good example of the use of forced oscillations. Microwave radiation passing through the food forces the water molecules to oscillate at the frequency of the radiation. The water molecule can oscillate in a number of different ways, two of which are shown in Figure 5. Each mode of vibration has a different resonant frequency.

▲ **Figure 4** *Microwave ovens use forced oscillations, but not resonance*

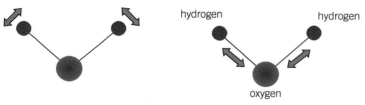

hydrogen   hydrogen

oxygen

▲ **Figure 5** *Two of the vibrational modes of water, H_2O (not to scale)*

The resonant frequencies of water are around 20 GHz — far higher than the 2.5 GHz radiation used in microwave ovens. If your microwave were tuned to the resonant frequency, the absorption of energy by water would be so strong that only few millimetres of the food would be heated, leaving the interior cold.

Microwave ovens seem to take a long time to defrost food compared to, say, heating it through. This is because the resonant frequency for water molecules in ice is even further from the microwave frequency than that of liquid water.

Summary questions

1 Give two examples of resonance not covered in the text. For each example, state the oscillator and the driving force. For example, in the case of the wine glass, the driving force is the sound waves and the oscillator is the glass.

2 The time period of the water surges at Chalkis is about 12.4 hours. Calculate the resonant frequency of the oscillation in Hz.

3 a It is said that wine glasses are easier to break by resonant vibrations if the glass contains tiny flaws or imperfections. Suggest and explain why this might be the case (Hint: this involves thinking about the microscopic structure of glass.)
  b Explain why a glass will shatter when the resonant vibrations grow above a given amplitude.

4 a 'Resonant oscillations are a special case of forced oscillations.' Explain this statement by describing the meaning of the terms in italics.
  b Explain why microwave ovens don't use radiation that matches the natural vibrational frequencies of water molecules.

Practice questions

1   Here are three statements about simple harmonic oscillators. Which of the statements is/are true?

   1   The acceleration of oscillator ∝ −displacement.

   2   The time period is independent of the amplitude of oscillation.

   3   The potential energy of the oscillator ∝ displacement.

   A   1, 2, and 3

   B   Only 1 and 2

   C   Only 2 and 3

   D   Only 1   (1 mark)

2   A 0.25 kg mass is held horizontally between two springs such that the effective spring constant is $10 \, \text{N m}^{-1}$.

   a   Calculate the force and acceleration at a displacement of

      (i)   +100 mm

      (ii) +50 mm.   (4 marks)

   b   Calculate the time period of the oscillation.   (2 marks)

3   The piston in a car engine oscillates with simple harmonic motion. The amplitude of the motion is 6.0 cm. The frequency of the motion is 90 Hz.

   a   Calculate the time period of the motion.   (1 mark)

   b   Calculate the maximum acceleration of the piston.   (2 marks)

4   The oscillations of a tuning fork are described by the equation $x = 0.001\,40 \sin(1650 t)$ where t is time in seconds.

   a   State the amplitude of oscillation.   (1 mark)

   b   Calculate the frequency of the fork.   (2 marks)

5   A horizontal plate has sand scattered on it. The plate is oscillating vertically. If the acceleration of the plate is greater than the acceleration due to gravity the sand will lose contact with the plate at certain points of the oscillation. The plate is oscillating at 15 Hz. Calculate the maximum amplitude of oscillation before the sand loses contact with the plate. State and explain at which point of the oscillation the sand will first lose contact with the plate.   (4 marks)

6   Figure 1 shows a mass suspended from a spring attached to a vibration generator that can drive the mass to oscillate at a chosen frequency. The amplitude of vibration of the generator is 2 mm.

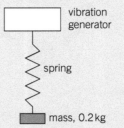

vibration generator

spring

mass, 0.2 kg

▲ Figure 1

The natural frequency of the system is 2.4 Hz.

   a   Calculate the force constant of the system.   (2 marks)

   b   Sketch a graph to show how the amplitude of oscillation of the mass varies with frequency of the vibration generator between 0.1 Hz and 7.0 Hz. At resonance the amplitude of oscillation is 8.0 mm.   (4 marks)

7   The graph in Figure 2 shows that the total energy of an undamped oscillator is constant with displacement. Copy the graph and add lines showing how the kinetic energy and potential energy of the oscillator vary with displacement. The amplitude of the oscillations is A.

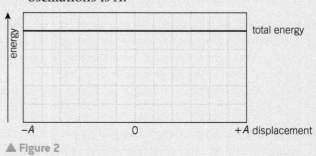

total energy

energy

−A   0   +A displacement

▲ Figure 2

(2 marks)

8 A scene in a movie shows a car dropped on a flat surface bouncing on its suspension springs. The time of oscillation is about 0.5 s. The mass of the car is about 500 kg. Take $g = 10 \, \text{N} \, \text{kg}^{-1}$ for your estimates.

 a Estimate the spring constant of the suspension. (*2 marks*)

 b By how much does the car compress the suspension when it is at rest? (*2 marks*)

 c If the drop initially compresses the springs by 100 mm beyond the equilibrium point, estimate the energy initially going into the oscillations. (*2 marks*)

 d Estimate by calculation the maximum vertical velocity of the car as it oscillates. (*2 marks*)

9 Many bicycles incorporate a spring on the front fork to improve the ride over rough ground.

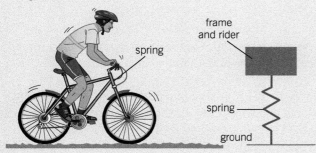

▲ Figure 3

 When the bicycle goes over a bump the rider and frame supported by the spring oscillate vertically. For these oscillations, the frame and rider can be modelled as a mass m supported by a spring of force constant k, as shown in Figure 3.

 a The spring compresses by 1.3×10^{-2} m when the rider mounts the bicycle, increasing the force on the spring by 3.6×10^2 N.

 Show that the force constant of the spring is about $3 \times 10^4 \, \text{N} \, \text{m}^{-1}$. (*1 mark*)

 b Calculate the frequency of free oscillations of the mass supported on the spring.

 $m = 62$ kg (*2 marks*)

c Explain why the spring on the bicycle will need to be damped if it is to be safely ridden across bumpy ground. (*3 marks*)

d The graph of Figure 4 shows the variation of displacement with time for the mass when the bicycle is ridden along a bumpy road.

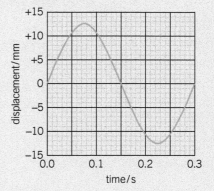

▲ Figure 4

Draw a graph on a copy of Figure 5 to show the variation of kinetic energy for the same time period. (*2 marks*)

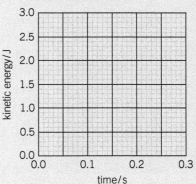

▲ Figure 5

OCR Physics B Paper G494 June 2012

▲ **Figure 1** *Isaac Newton*

The clockwork Universe

Until the mid-16th century nearly everyone assumed that the Earth was the still centre of the turning skies, which carried the Sun, Moon, planets, and stars around the Earth once a day.

You will know that this is not correct. One of the first things you learnt at school is that the Earth goes around the Sun. You might dismiss the old idea without much thought. However, although mistaken, the idea is common sense. For example, you do not feel as though you are on the part of the planet's surface that is rotating at about 1000 km per hour. Nor do you feel that you are sitting on a planet that is orbiting the Sun at 100 000 km per hour! Common sense is not always a reliable guide. We don't notice our motion through space because our acceleration is small and everything around us is travelling too.

The 16th and 17th centuries saw models develop that put the Sun at the centre of the Solar System – even if this did seem to disagree with everyday experience. Isaac Newton (Figure 1) then produced a mathematical framework for the sun-centred Solar System and compiled a coherent system of ideas that explained all observed motions, both on Earth and in the sky — his model described a 'clockwork Universe'. It has been said that Newton's contribution to scientific thought is the greatest by any single individual, ever. We still study his ideas today, and this chapter in particular relies almost completely on work Newton undertook in the 17th century.

Centripetal acceleration and centripetal force

A major challenge for Newton was to explain how forces produce a circular path. His conclusion: nothing pushes an object along a circular path. What it takes is a single, external force acting perpendicular to the velocity at every instant.

An object moving along a circular path has an acceleration towards the centre (Figure 2),

$$a = \frac{v^2}{r}$$

Velocity is a vector. A change in velocity can be a change in direction rather than speed. So, for example, as an object moves at uniform speed along a circular path its velocity changes continually. In other words, the body is accelerating whilst its speed is unchanged. Look carefully at Figure 2. The change in velocity in each small step of time Δt is always towards the centre of the circular motion and at right angles to the velocity at that moment. For small changes, the angle $\Delta \theta$ in radians through which the velocity turns is equal to $\dfrac{\Delta v}{v}$, from the vector diagram adding the change in velocity Δv to the velocity v. Now notice that this angle $\Delta\theta$ can be calculated in a different way. It is just the angle through which the radius turns in time Δt. In this time, a particle travels a distance $v\Delta t$ along an arc of the circular motion, so that the angle $\Delta\theta$ is just equal to the arc divided by the radius r, that is, $\Delta\theta = \dfrac{v\Delta t}{r}$. Equating the two ways of calculating $\Delta\theta$,

$$\frac{v\Delta t}{r} = \frac{\Delta v}{v}$$

It follows that the acceleration $\dfrac{\Delta v}{\Delta t}$, directed towards the centre of the motion, is given by

$$a = \frac{v^2}{r}$$

This is known as the centripetal acceleration, from the Latin for 'centre-seeking'. The force producing it is called a centripetal force. The centripetal force of gravity keeps the planets in their orbits.

Using the familiar statement of Newton's second law we can substitute for a to obtain an equation for centripetal force,

$$F = ma = \frac{mv^2}{r}$$

We can also find the centripetal acceleration of an object by considering its angular velocity $\omega = 2\pi f$.

We can find the speed of an object moving in a circle from
$v = \dfrac{\text{circumference}}{\text{time}}$ for one revolution,

$$v = \frac{2\pi r}{t}$$

As $t = \dfrac{1}{f}$, where f is frequency,

$$v = 2\pi rf$$

Because $2\pi f = \omega$, the angular velocity in rad s^{-1},

$$v = \omega r$$

Substituting for v in $a = \dfrac{v^2}{r}$ gives

$$a = \omega^2 r$$

From Newton's second law, $F = ma = mr\omega^2$

The force always acts at right angles to the motion of the body.

Imagine a ball on a string being swung in a horizontal circle. The force towards the centre, the centripetal force, is the tension in the string. If the string is cut, the ball will fly off in a straight line in the direction its velocity vector was pointing when the string was cut. The same principle dries clothes in a spin dryer (Figure 3).

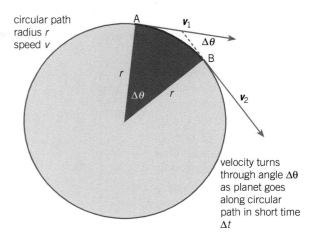

circular path radius r speed v

velocity turns through angle $\Delta\theta$ as planet goes along circular path in short time Δt

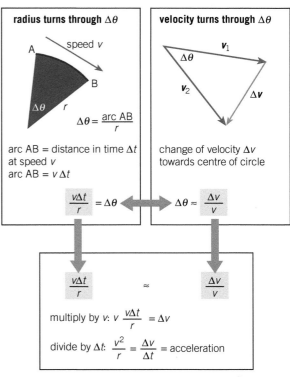

radius turns through $\Delta\theta$

speed v

$\Delta\theta = \dfrac{\text{arc AB}}{r}$

arc AB = distance in time Δt at speed v
arc AB = $v\Delta t$

$\dfrac{v\Delta t}{r} = \Delta\theta$

velocity turns through $\Delta\theta$

change of velocity Δv towards centre of circle

$\Delta\theta \approx \dfrac{\Delta v}{v}$

$\dfrac{v\Delta t}{r} \quad \approx \quad \dfrac{\Delta v}{v}$

multiply by v: $v\dfrac{v\Delta t}{r} = \Delta v$

divide by Δt: $\dfrac{v^2}{r} = \dfrac{\Delta v}{\Delta t} = $ acceleration

▲ Figure 2 *Acceleration in circular motion* $a = \dfrac{v^2}{r}$

Synoptic link

You can review angular velocity in Chapter 20, Maths for physics.

Synoptic link

The equation $a = \omega^2 r$ is equivalent to $a = -\omega^2 x$, which you encountered in Topic 11.1, Introducing simple harmonic oscillators.

Hint

In Chapter 11, Models 2: Oscillations and simple harmonic motion, we used circular motion to produce a mathematical model of simple harmonic motion. The product $2\pi f$ is referred to as *angular frequency* when considering oscillations because nothing is actually moving in a circle. When we consider objects moving in circular motion (like the mass on the end of a string) we use the term *angular velocity*.

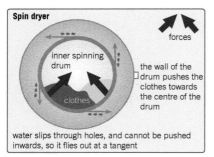

Spin dryer

inner spinning drum

clothes

forces

the wall of the drum pushes the clothes towards the centre of the drum

water slips through holes, and cannot be pushed inwards, so it flies out at a tangent

▲ **Figure 3** *Clothes lose water in a spin dryer because no force is accelerating the water in a circle*

▲ **Figure 4** *On rides such as this you experience a centripetal force from the push of the chair against you*

Worked example: Hammer throwing

A hammer thrower swings a mass of 7.3 kg in a circle. Each revolution takes 1.5 s. The radius of the circle traced by the mass is 2.1 m.

a Calculate the speed of the mass.

Step 1: Use the equation for speed (not angular velocity).

$$\text{speed} = \frac{\text{distance}}{\text{time}}$$

Step 2: $\text{speed} = \dfrac{2\pi r}{t} = \dfrac{2\pi \times 2.1\,\text{m}}{1.5\,\text{s}} = 8.8\,\text{m s}^{-1}$ (2 s.f.)

b Calculate the angular velocity of the mass.

Step 1: $\text{angular velocity} = 2\pi f = \dfrac{2\pi}{t} = \dfrac{2\pi\,\text{rad}}{1.5\,\text{s}} = 4.2\,\text{rad s}^{-1}$ (2 s.f.)

c Calculate the centripetal force on the mass.

Step 1: $F = \dfrac{mv^2}{r} = \dfrac{7.3\,\text{kg} \times 8.8^2\,\text{m}^2\,\text{s}^{-2}}{2.1\,\text{m}} = 270\,\text{N}$ (2 s.f.)

The centripetal force on the mass – the tension in the hammer wire and the arms of the thrower – is over three times its weight.

Why don't we notice the spin of the Earth?

We don't notice our centripetal acceleration because it is small, and we don't notice our speed because everything around us (including the air) is travelling at the same speed with us. A point on the equator is at a distance of 6400 km from the centre of the Earth. It revolves around the centre once a day at a speed of about 470 m s^{-1}. The centripetal acceleration at the equator is $a = \dfrac{v^2}{r} = \dfrac{470^2\,\text{m}^2\,\text{s}^{-2}}{6.4 \times 10^6\,\text{m}} = 0.035\,\text{m s}^{-2}$. This is far smaller than the acceleration of gravity (9.8 m s^{-2}), and it is not surprising that we don't notice it in everyday life.

Feeling the force

Many fairground rides subject you to centripetal forces. You feel as if you will be flung off the ride, but the force from the seat acting towards the centre accelerates you in a circle (Figure 4). Trainee astronauts and fighter pilots feel far more dramatic centripetal forces when they are put in a capsule at the end of an arm more than 10 m long (Figure 5). The arm and the capsule swing round at high speed and can create a centripetal acceleration of about ten times the acceleration due to gravity. The force the astronaut experiences is therefore about ten times body weight.

◀ **Figure 5** *Equipment used to train cosmonauts in Russia – such machines can produce very large centripetal accelerations*

Summary questions

1 State what provides the force to make each of these travel in circular paths:
   a a child sitting on a roundabout; (*1 mark*)
   b a toy train on a circular track; (*1 mark*)
   c a discus just before the athlete releases it for the throw. (*1 mark*)

2 A bicycle is moving at a constant speed of $6 \, m \, s^{-1}$ in a circle of radius 9 m. Calculate the acceleration of the bicycle. (*3 marks*)

3 Astronauts and fighter pilots train to withstand large accelerations in a capsule rotated on the end of a boom 15 m long like the one in Figure 5. The trainee must reach an acceleration of $100 \, m \, s^{-2}$.
   a Calculate the speed of the capsule. (*2 marks*)
   b Calculate the angular velocity of the capsule. (*2 marks*)

4 A child of mass 20 kg is playing on a swing. Her centre of mass is 4 m below the pivot supporting the swing when she moves through the lowest point of her swing at $5 \, m \, s^{-1}$. Show that a centripetal force of 125 N is required and explain why the seat exerts a total force of about 325 N. (*4 marks*)

5 Rutherford's 1911 model of the hydrogen atom pictures the electron orbiting the proton with an orbital radius of $5.4 \times 10^{-11} \, m$ and a speed of $2.2 \times 10^6 \, m \, s^{-1}$.
   a Calculate the force needed to keep the electron (mass $9.1 \times 10^{-31} \, kg$) in orbit. (*2 marks*)
   b State what could provide this force. (*1 mark*)

6 The radius of the Earth's orbit around the Sun is roughly $1.5 \times 10^{11} \, m$. One year is about $3.2 \times 10^7 \, s$. Calculate the centripetal acceleration of the orbiting Earth. Compare it with the centripetal acceleration of a point on the rotating Earth calculated above and comment on the value. (*5 marks*)

12.2 Newton's law of gravitation

Specification reference: 5.1.2c(iv)

Learning outcomes

Describe, explain, and apply:

→ Newton's law of gravitation

→ the radial component of the gravitational field,

$$F_{grav} = -\frac{GmM}{r^2}$$

$$g = \frac{F_{grav}}{m} = -\frac{GM}{r^2}.$$

Newton and the apple

1687 saw the publication of possibly the greatest text in the history of science, Newton's *Philosophiae naturalis Principia Mathematica*, or *Mathematical Principles of Natural Philosophy*. The theory of gravity discussed in the work developed from ideas that he had considered two decades previously.

Newton had thought about the force keeping the Moon in orbit being gravity, weakened by the extra distance from Earth. If gravity decreases as the square of the distance, it would be reduced to a quarter when the distance doubles, to a ninth when the distance trebles, and so on. He told the tale that observing the fall of an apple from a tree led him to a way of testing this 'inverse-square' law for gravity.

- He knew how rapidly a falling apple accelerates to the ground.
- He could calculate the acceleration of the Moon towards Earth using $a = \frac{v^2}{r}$.
- Knowing that the radius of the Moon's orbit is about 60 times the radius of the Earth, he could check whether the apple's acceleration was 60^2 times the acceleration of the Moon.

Work through Figure 1 and you will see that Newton's hunch was correct. Falling apples and the orbiting Moon are both explained by the same force – gravity.

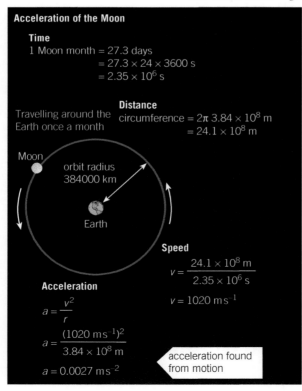

Acceleration of the Moon

Time
1 Moon month = 27.3 days
= 27.3 × 24 × 3600 s
= 2.35 × 10⁶ s

Distance
Travelling around the Earth once a month
circumference = 2π 3.84 × 10⁸ m
= 24.1 × 10⁸ m

Moon
orbit radius
384000 km
Earth

Speed
$$v = \frac{24.1 \times 10^8 \text{ m}}{2.35 \times 10^6 \text{ s}}$$
$$v = 1020 \text{ ms}^{-1}$$

Acceleration
$$a = \frac{v^2}{r}$$
$$a = \frac{(1020 \text{ ms}^{-1})^2}{3.84 \times 10^8 \text{ m}}$$
$$a = 0.0027 \text{ ms}^{-2}$$

acceleration found from motion

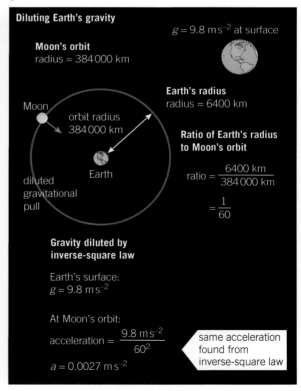

Diluting Earth's gravity

$g = 9.8 \text{ ms}^{-2}$ at surface

Moon's orbit
radius = 384000 km

Earth's radius
radius = 6400 km

Moon
orbit radius
384000 km
Earth
diluted gravitational pull

Ratio of Earth's radius to Moon's orbit
$$\text{ratio} = \frac{6400 \text{ km}}{384000 \text{ km}}$$
$$= \frac{1}{60}$$

Gravity diluted by inverse-square law

Earth's surface:
$g = 9.8 \text{ ms}^{-2}$

At Moon's orbit:
$$\text{acceleration} = \frac{9.8 \text{ ms}^{-2}}{60^2}$$
$$a = 0.0027 \text{ ms}^{-2}$$

same acceleration found from inverse-square law

▲ **Figure 1** *The acceleration of the Moon is diminished Earth gravity – the Moon is falling just as an apple does*

The law of gravitation

The core principle of Newton's theory of gravity is: *All particles in the Universe attract all other particles.*

Don't be fooled by the apparent simplicity of this revolutionary statement. It means that as you read this sentence you are interacting gravitationally with the rest of the Universe!

Expressed algebraically it is

$$F_{grav} \propto \frac{mM}{r^2}$$

where m and M are the masses of the two interacting bodies and r is the separation of the centres of mass of the bodies.

With the gravitational constant G, the component of the gravitational force acting between the centres of mass of the bodies (the radial component) is

$$F_{grav} = -\frac{GmM}{r^2}$$

The negative sign shows that the force is always attractive (Figure 2).

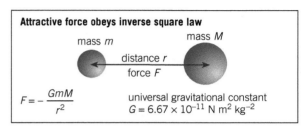

Attractive force obeys inverse square law

mass m mass M

distance r
force F

$F = -\dfrac{GmM}{r^2}$ universal gravitational constant
$G = 6.67 \times 10^{-11}$ N m^2 kg^{-2}

▲ **Figure 2** *Newton's law of gravitation*

 ## Worked example: Martian attraction

The planet Mars orbits the Sun at a distance of 2.3×10^{11} m. Calculate the magnitude of the force between the bodies. (Mass of Mars = 6.4×10^{23} kg, mass of Sun = 2.0×10^{30} kg)

Step 1: To use the equation $F_{grav} = -\dfrac{GmM}{r^2}$ we will assume that the distance given is between the centres of both bodies.

Step 2: magnitude of force $= \dfrac{GmM}{r^2}$

$$= \frac{6.67 \times 10^{-11}\,N\,m^2\,kg^{-2} \times 6.4 \times 10^{23}\,kg \times 2.0 \times 10^{30}\,kg}{(2.3 \times 10^{11}\,m)^2}$$

$$= 1.6 \times 10^{21}\,N \text{ (2 s.f.)}$$

Gravitational fields

We imagine space as 'filled' with a gravitational field. In everyday language we would say that a mass placed anywhere in the field 'feels' a force. Of course, inanimate objects don't feel anything, so it is much better to say that a mass 'is subject to a force in a gravitational field'.

The gravitational field is described by specifying at each point the magnitude and direction of the force on one kilogram. This is g, the **gravitational field strength**, units N kg^{-1}. It describes the effect of the gravitating body of mass M regardless of the mass m placed near it.

$$g = -\frac{GM}{r^2}$$

Radial component of force $= mg = -\dfrac{GmM}{r^2}$.

The gravitational field strength is another example of an inverse-square relationship.

The direction and magnitude of a field can be represented by field lines or by vectors. In the field-line diagram in Figure 3 you can see that the field is radial – the field lines resemble the spokes of a wheel with the Earth at the centre. You can also see that the lines become closer together nearer the surface, representing a stronger field at the surface.

Field lines

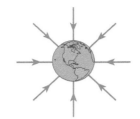

Field direction shown by direction of lines

Field strength shown by closeness of lines

Vectors in space

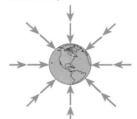

Field direction shown by vector direction

Field strength shown by length of vector

▲ **Figure 3** *Gravitational field lines*

> 🖩 Worked example: The value of g on Earth
>
> The radius of the Earth is 6.4×10^6 m and its mass is 6.0×10^{24} kg. Calculate the gravitational field strength at the surface.
>
> **Step 1:** $g = -\dfrac{GM}{r^2} = -\dfrac{(6.67 \times 10^{-11}\,\text{N}\,\text{m}^2\,\text{kg}^{-2} \times 6.0 \times 10^{24}\,\text{kg})}{(6.4 \times 10^6\,\text{m})^2}$
>
> $\qquad = 9.8\,\text{N}\,\text{kg}^{-1}$ (2 s.f.)

The law in action

We can use Newton's gravitational law to model the field in many situations. Here are just two examples.

The gravitational field inside the Earth

Outside the Earth the field varies in an inverse-square relationship, but how does the strength of the field vary beneath the surface of the Earth? What is the field at the centre? Figure 4 shows a point P within the Earth. The shaded shell shows the part of the Earth above the point. The pull of gravity on P from all parts of this shaded shell cancels out.

The gravitational field strength g_P at P from the unshaded sphere is $g_P = -\dfrac{GM_P}{r_P{}^2}$.

Since mass = density × volume, the mass of the sphere beneath P is $M_P = \rho\dfrac{4}{3}\pi r_P{}^3$, where ρ is the density of the Earth, assumed to be uniform.

Therefore $g_P = -\dfrac{G\rho\dfrac{4}{3}\pi r_P{}^3}{r_P{}^2} = -\dfrac{4}{3}G\rho\pi r_P$

The gravitational field strength inside the Earth is proportional to the distance from the centre. The field strength is zero at the centre.

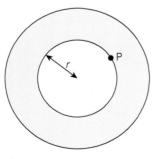

▲ **Figure 4** *The gravitational field at a point P inside the Earth*

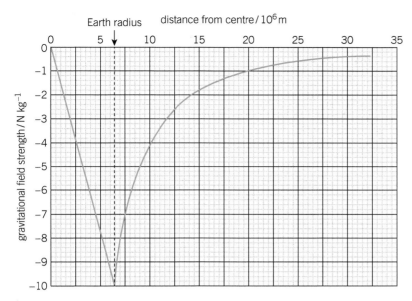

▲ **Figure 5** *Variation of gravitational field strength with distance from the centre of the Earth*

Geostationary satellites

Geostationary satellites, such as those used for satellite TV, appear to stay in the same place in the sky at all times. To achieve this, the satellite orbits the Earth above the equator once a day. We can use Newton's law to calculate the distance from the surface that such a satellite must orbit. If the orbit is too low the satellite will orbit the Earth too quickly. A satellite too far from the Earth will orbit too slowly. The orbital radius required can be found by remembering that the gravitational force acts centripetally. Figure 6 shows you the stages in the calculation.

m = mass of satellite
R = radius of satellite orbit
v = speed in orbit
G = gravitational constant = 6.67×10^{-11} N kg^{-2} m^2
M = mass of Earth = 5.98×10^{24} kg
T = time of orbit = 24 hours = 86400 s

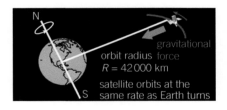

gravitational
orbit radius   force
R = 42000 km

satellite orbits at the
same rate as Earth turns

Calculating the radius of orbit

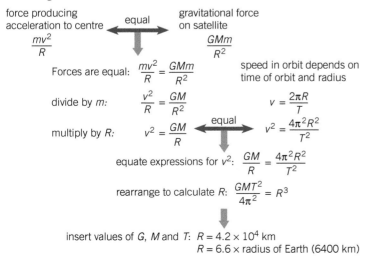

force producing
acceleration to centre ←— equal —→ gravitational force
on satellite
$\dfrac{mv^2}{R}$ $\dfrac{GMm}{R^2}$

Forces are equal: $\dfrac{mv^2}{R} = \dfrac{GMm}{R^2}$ speed in orbit depends on time of orbit and radius

divide by m: $\dfrac{v^2}{R} = \dfrac{GM}{R^2}$ $v = \dfrac{2\pi R}{T}$

multiply by R: $v^2 = \dfrac{GM}{R}$ ←— equal —→ $v^2 = \dfrac{4\pi^2 R^2}{T^2}$

equate expressions for v^2: $\dfrac{GM}{R} = \dfrac{4\pi^2 R^2}{T^2}$

rearrange to calculate R: $\dfrac{GMT^2}{4\pi^2} = R^3$

insert values of G, M and T: $R = 4.2 \times 10^4$ km
$R = 6.6 \times$ radius of Earth (6400 km)

▲ **Figure 6** *Calculation of the orbital radius of a geostationary satellite*

Summary questions

1 Calculate the gravitational pull of a spherical 5 kg mass on a spherical 3 kg mass when their centres are 0.15 m apart. Compare this with the force of the 3 kg mass on the 5 kg mass. *(4 marks)*

2 Newton correctly estimated the density of the Earth as five to six times that of water. Today we know it is $5500 \, \text{kg} \, \text{m}^{-3}$. The volume of a sphere is $\frac{4}{3}\pi r^3$ and the Earth's radius is 6400 km. Show that the mass of the Earth is about 6.0×10^{24} kg, and use the fact that at Earth's surface $g = 9.8 \, \text{N} \, \text{kg}^{-1}$ to show that G is about $6.7 \times 10^{-11} \, \text{N} \, \text{m}^2 \, \text{kg}^{-2}$. *(4 marks)*

3 Describe how the weight of a body would vary en route from the Earth to the Moon. State whether its mass would change. *(3 marks)*

4 You are standing on bathroom scales in a lift. Explain what happens to the reading when the lift is (i) stationary, (ii) starting downwards, (iii) starting upwards, (iv) moving at constant speed. *(4 marks)*

5 Another way of thinking about geostationary satellites is that they must have the same angular velocity ω as the aerial that points towards them. Calculate this angular velocity. *(3 marks)*

6 The Sun, Earth, and Jupiter are sometimes positioned so that all three lie on a straight line, with the Earth and Jupiter on the same side of the Sun. When this happens, what is the ratio of Jupiter's gravitational pull on the Earth to the Sun's gravitational pull? (Masses: Sun 2×10^{30} kg, Earth 6×10^{24} kg, Jupiter 1.9×10^{27} kg. Orbital radii: Earth 1 astronomical unit (AU), Jupiter 5.2 AU.) *(4 marks)*

12.3 Gravitational potential in a uniform field

Specification reference: 5.1.2a(i), 5.1.2a(ii), 5.1.2b(i), 5.1.2b(iv), 5.1.2c(i), 5.1.2c(ii)

The uniform field

Hang a 100 g mass from a spring and measure the extension. You will not observe a change in extension when you move the apparatus to a different part of the room, or take it upstairs. Climb to the top of a tall building and still you will not be able to observe a change of extension.

The extension of the spring is a measure of the weight hanging from it – in other words, the value of the gravitational field strength. Near the surface of the Earth the gravitational field is approximately uniform – it has the same magnitude and direction everywhere. Close up, the Earth is nearly flat and the gravitational field lines are almost parallel (Figure 1).

Gravitational potential and equipotential surfaces

Near the surface of the Earth the change in gravitational potential energy with height, $mg\Delta h$, is always the same for equal changes in height Δh. Lift a 1 kg mass vertically by one metre anywhere and the change in gravitational potential energy = 1 kg × 9.8 N kg^{-1} × 1 m = 9.8 J. Lift the mass a metre higher and the gravitational potential energy will increase by a further 9.8 J. Of course, if you lift a 2 kg mass the same height the change in potential energy will be twice that for the 1 kg mass.

It is sometimes useful to calculate the change in **gravitational potential**, that is, in gravitational energy per kilogram of material, $g\Delta h$, measured in J kg^{-1}. For calculations near the surface of the Earth we set the zero value at sea level. The potential increases by $9.8 \approx 10$ J kg^{-1} for every metre raised.

Equipotential surfaces are surfaces of constant potential. One metre above the ground there is an equipotential surface that is 10 J kg^{-1} greater than that at ground level. You will often use diagrams such as Figure 2, which give two-dimensional representations of the field and show the surfaces as equipotential lines. Note that the spacing between equipotential lines is constant if the potential difference is equal — the separation between the 0 and 10 J kg^{-1} equipotentials (which represents a potential difference of 10 J kg^{-1}) is the same as the separation between the 10 J kg^{-1} and the 20 J kg^{-1} equipotentials. This is a distinguishing characteristic of a uniform field.

Learning outcomes

Describe, explain, and apply:

→ changes in gravitational potential energy and kinetic energy for uniform fields

→ gravitational potential energy change in a uniform field = $mg\Delta h$

→ motion in a uniform gravitational field

→ the terms force, kinetic and potential energy, gravitational field, gravitational potential, equipotential surface

→ diagrams of gravitational fields and the corresponding potential surfaces

→ work done, $\Delta E = F\Delta s$; no work done when the force is perpendicular to the displacement; no work done whilst moving along equipotentials.

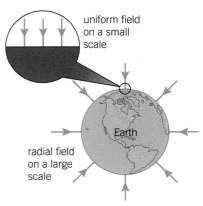

▲ Figure 1 *Near the surface of the Earth the field is approximately uniform*

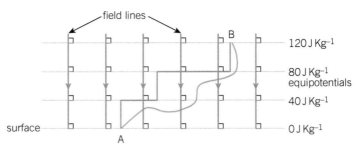

▲ **Figure 2** *Field lines and equipotentials*

Tracking along an equipotential, you can be sure that no gravitational force is acting in that direction because the potential energy is not changing.

Change in energy ΔE when moving Δs along equipotential $= F\Delta s = 0$

∴ force acting along the equipotential $F = 0$

Synoptic link

Change in potential is independent of the route taken in electric fields too, as you will see in Topic 17.1, Uniform electric fields.

Synoptic link

You studied this parabolic motion in Topic 9.4, Projectiles.

For this to be the case, the direction of the gravitational field must always be at right angles to the equipotentials (Figure 2).The change in potential from A to B is independent of the route taken. As long as the change in vertical displacement is the same, so is the change in potential. In Figure 2, the orange line has horizontal sections, in which there is no change in potential, and vertical sections. The change in potential is the same by the orange route and the blue route, $120\,\text{J}\,\text{kg}^{-1}$. This is true for uniform and non-uniform gravitational fields.

Motion in a uniform field

A stone thrown horizontally falls in a parabola. The force (the weight) on the stone is always vertically downwards. There is no force in the horizontal direction.

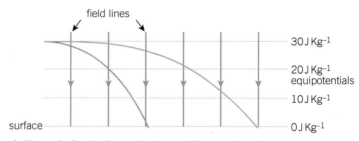

▲ **Figure 3** *Parabolic motion in a uniform gravitational field*

Figure 3 represents the motion of two projectiles projected horizontally near the surface of the Earth. The kinetic energy gained by each particle depends only on the height they fall. Both projectiles will have the same vertical component of velocity when they strike the ground.

Change in gravitational potential energy = change in kinetic energy

$$\Delta mgh = \Delta \tfrac{1}{2}mv^2$$

∴ potential difference $= \Delta gh = \Delta \tfrac{1}{2}v^2$

Worked example: Velocity of a falling object

Calculate the vertical velocity of the two projectiles in Figure 3 as they strike the ground.

Step 1: Read the potential difference for the two projectiles from the equipotential diagram.

The potential difference between the release point and the surface = $30\,\mathrm{J\,kg^{-1}}$.

Step 2: Equate the change in gravitational potential energy to the change in kinetic energy.

$$\Delta gh = \Delta \frac{1}{2}v^2 \qquad \therefore v = \sqrt{2 \times 30\,\mathrm{J\,kg^{-1}}} = 7.7\,\mathrm{m\,s^{-1}}\ (2\ \text{s.f.})$$

Summary questions

1 A ball is thrown vertically upwards at an initial velocity of $20\,\mathrm{m\,s^{-1}}$.
   a Calculate the height of the highest point of the ball's trajectory above the release position. ($g = 9.8\,\mathrm{N\,kg^{-1}}$) (*1 mark*)
   b State the change in gravitational potential between the release position and the highest point of the trajectory. (*1 mark*)

2 Figure 4 shows several cars parked in a multistorey car park. Each floor is 5 m higher than the previous one. (Take $g = 10\,\mathrm{N\,kg^{-1}}$)

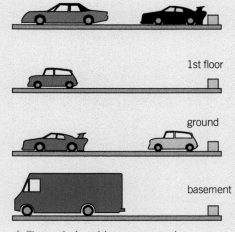

▲ **Figure 4** *A multistorey car park*

   a A family car has a mass of 1 tonne. Calculate how its potential energy changes as it moves from the ground floor to the first floor, to the fourth floor, and to the basement. (*2 marks*)
   b State the gain in potential energy of each kilogram of the car for each of the moves in part (a). (*1 mark*)
   c State the gravitational potential difference between the ground floor and the first floor, between the ground floor and the fourth floor, and between the ground floor and the basement. (*3 marks*)
   d State the gravitational potential difference between the first and the fourth floors; between the first floor and the basement. (*2 marks*)
   e Calculate the potential energy lost by a small car of mass 800 kg in coming down from the fourth to the first floor. (*2 marks*)

3 a Use the relationship field strength $\propto \dfrac{1}{r^2}$ to show that the gravitational field strength 1 km above the surface of the Earth is about $9.8\,\mathrm{N\,kg^{-1}}$. (Radius of Earth = 6.4×10^6 m, gravitational field at surface = $9.8\,\mathrm{N\,kg^{-1}}$) (*3 marks*)
   b Explain why this suggests that the gravitational field can be considered to be approximately uniform near the surface. (*2 marks*)
   c Comment on the spacing of equipotentials separated by equal potential difference in this region. (*1 mark*)

12.4 Gravitational potential in a radial field

Specification reference: 5.1.2a(iii), 5.1.2b(i), 5.1.2b(ii), 5.1.2b(iii), 5.1.2b(iv), 5.1.2c(v), 5.1.2c(vi)

Learning outcomes

Describe, explain, and apply:

→ the gravitational potential of a point mass

→ change in gravitational potential shown as the area under a graph of gravitational field strength against distance, and change in gravitational potential energy shown as the area under a graph of gravitational force versus distance

→ force shown as related to the tangent of a graph of gravitational potential energy versus distance, and field strength shown as related to the tangent of a graph of gravitational potential versus distance

→ diagrams of gravitational fields and corresponding equipotential surfaces

→ gravitational potential energy
$$E_{\text{potential}} \ (\text{or } E_{\text{grav}}) = -\frac{GMm}{r}$$

→ gravitational potential
$$V_{\text{grav}} = \frac{E_{\text{grav}}}{m} = -\frac{GM}{r}$$

→ the terms force, kinetic and potential energy, gravitational field, gravitational potential, equipotential surface.

Gravitational potential wells

Gravitational potential increases as you go up, or away, from Earth. For earthbound calculations it is common to set the zero value at sea level, as we did in Topic 12.3, Gravitational potential in a uniform field. But for space calculations it is easier to set the zero as the potential energy per kilogram at an infinite distance from the planet.

The surface of the Earth is then at the bottom of a potential well. To get a spacecraft far away from the Earth means escaping from the well, which requires a lot of energy. As the potential at infinite distance is set at zero, the potential at any distance less than infinity must be negative. The larger the planet, the more negative the potential at its surface, the deeper the well to escape from.

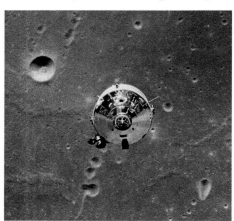

We can map the potential variation around the Earth using data from Apollo 11, the first mission to land humans on the Moon. The craft (Figure 1) returned the astronauts to Earth by 'falling' towards it. The engines were off and the speed continually increased as the craft approached Earth – its kinetic energy increased as its potential energy decreased.

▲ Figure 1 The Apollo 11 command module, the craft that brought astronauts Armstrong, Aldrin, and Collins back to Earth, photographed from the lunar landing module

Suppose Apollo 11 has total energy E_{total}. Then the gravitational potential energy $E_{\text{potential}}$ (or E_{grav}) is just

$$E_{\text{potential}} = E_{\text{total}} - E_{\text{kinetic}}$$

Since the kinetic energy is $\frac{1}{2}mv^2$ and the gravitational potential V_{grav} is $\frac{E_{\text{potential}}}{m}$, dividing this equation by m gives

$$V_{\text{grav}} = \text{constant} - \frac{1}{2}v^2$$

Pairs of observations of speed and distance

Apollo 11 is coasting home downhill with rockets turned off. Distances *r* taken from centre of Earth.

distance /10^6 m	speed /m s^{-1}	$-\frac{1}{2}v^2$ /10^6 J kg^{-1}	10^8 m /r
241.6	1521	−1.16	0.414
209.7	1676	−1.40	0.477
170.9	1915	−1.83	0.585
96.8	2690	−3.62	1.033
56.4	3626	−6.57	1.774
28.4	5201	−13.53	3.518
13.3	7673	−29.44	7.513

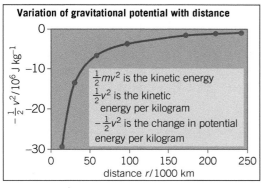
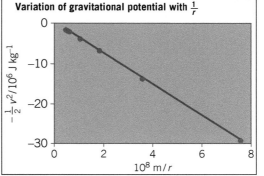

 Figure 2 *Apollo 11 comes back from the Moon*

The data in Figure 2 suggests that

$$V_{grav} \propto -\frac{1}{r}$$

Remember that *r* is the radial distance from the centre of the mass.

You have already met the equations for the gravitational force and field strength for a radial field. The corresponding equations for the gravitational potential energy and the gravitational potential are:

gravitational potential energy $E_{potential} = -\dfrac{GMm}{r}$

gravitational potential $V_{grav} = \dfrac{E_{potential}}{m} = -\dfrac{GM}{r}$

It is important to remember that these equations are for a point mass. We calculate the potential on the surface of a planet by assuming that all the mass of the planet is concentrated at a point at its centre.

Figure 3 shows how the spacing of equipotentials separated by equal potential difference (20 MJ kg^{-1}) increases with distance from a planet. The greater the spacing of equipotentials, the weaker the field.

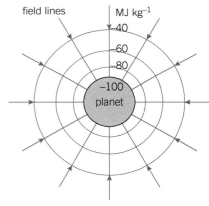

▲ Figure 3 *Equipotentials near a planet*

🖩 Worked example: Earth's gravitational potential

Calculate the gravitational potential at the surface of the Earth.
(Earth radius = 6.4×10^6 m; Earth mass = 6.0×10^{24} kg)

Step 1: $V_{grav} = -\dfrac{GM}{r} = -\dfrac{6.67 \times 10^{-11}\,\text{N m}^2\,\text{kg}^{-2} \times 6.0 \times 10^{24}\,\text{kg}}{6.4 \times 10^6\,\text{m}}$

$= -6.3 \times 10^7\,\text{J kg}^{-1}$ (2 s.f.)

Hint

The four equations for radial fields (force, field strength, potential energy, and potential) all look quite similar. Make sure that you are always clear about which to use – many mistakes have resulted from using the wrong equation.

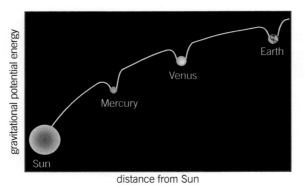

gravitational potential energy

Earth

Venus

Mercury

Sun

distance from Sun

▲ **Figure 4** *Gravitational potential wells in space*

Escape velocity

Picture the gravitational potential in the Solar System and beyond as a large rubber sheet. A ball on it could roll anywhere once set going, but the Sun makes a deep hollow in the middle. Planets must either fall in or move fast enough to go round it. Each planet makes its own hollow, in which moons can orbit and which makes it hard for the inhabitants to climb out into the space beyond (Figure 4).

To escape from Earth a spacecraft must climb out of the Earth's gravitational potential well. The depth of the well is the potential energy per kilogram needed to move a mass from the Earth's surface to a very long distance away. The worked example above shows that the depth of the Earth's potential well is $-6.3 \times 10^7 \, \text{J kg}^{-1}$.

That is a lot of energy for each kilogram of spacecraft. It is needed so that the total energy $E_{\text{kinetic}} + E_{\text{potential}}$ can be equal to or greater than zero, and the spacecraft can coast out of the potential well. That is, in order to escape.

$$E_{\text{kinetic}} + E_{\text{potential}} \geq 0$$

$$\frac{1}{2}mv^2 + -\frac{GMm}{r} \geq 0$$

So, the minimum kinetic energy required to escape is given by

$$\frac{1}{2}mv^2 = \frac{GMm}{r}$$

So, the escape velocity $v_{\text{esc}} = \sqrt{\dfrac{2GM}{r}}$

For Earth, $v_{\text{esc}} = \sqrt{2 \times 6.3 \times 10^7 \, \text{J kg}^{-1}} = 11 \, \text{km s}^{-1}$

Ignoring the effects of air resistance, if you could throw a ball vertically upwards at $11 \, \text{km s}^{-1}$ it would never return!

Of course, rockets do not reach anything like these speeds but still escape from the planet. Rockets are not given a single, explosive push, but have constant thrust from their fuel. For a spacecraft to journey to the outer planets, it must not only escape Earth's gravitational potential well, but also climb up the Sun's gravitational potential well. The way to the outer planets is still uphill. Clever methods can be adopted to find the extra energy. Here are two.

- The rotation of Earth means the launch site itself is moving, so the rocket has kinetic energy even before it is launched.

- The spacecraft can gain kinetic energy through an encounter with another planet, taking energy from that planet's motion, a technique called *gravity assist* or *slingshotting*.

 Extension: Orbital velocity

The speed of a satellite in orbit is $v = \sqrt{\dfrac{GM}{r}}$. If a satellite has a greater velocity than this it will move to a higher orbit. If the velocity is greater than the escape velocity for the height of orbit, it will leave Earth orbit completely.

Consider a 'space plane' designed to take passengers from England to Australia by orbiting at 160 km above the surface. The time for a complete orbit at this height is about 88 minutes, so the plane would need about 45 minutes in space to cross half the globe. To reach a greater velocity at this height it would have to use its rocket motors to produce a force *towards* the centre of the Earth rather than in its direction of motion.

To think about (mass of Earth = 6.0×10^{24} kg):

1 Show that the speed of the ISS (Figure 5) in orbit is about $8\,\text{km s}^{-1}$.
2 Show that the orbital period of the space plane described would be about 88 minutes.
3 Why would this space plane require a force *towards* the centre of the Earth in order to orbit at $v > \sqrt{\dfrac{GM}{r}}$? (Hint: consider centripetal force)
4 What are the arguments against using space planes at a lower altitude to reduce journey times?

▲ **Figure 5** *The International Space Station (ISS) orbits the Earth once every 93 minutes at an average height above the surface of 400 km, and is the source of this picture of Northern Europe, taken on April 2nd 2012 – the bright patch of light in the upper centre is Paris, and the one on the upper right is London*

Synoptic link

You met the idea that energy changes can be calculated from the areas under force–distance graphs in Topic 4.2, Stretching wires and springs, when considering force–extension graphs.

Potential and field strength

The area under a gravitational field–distance graph gives the difference in potential (the energy change per unit mass) for the change in distance chosen (Figure 6(a)). Similarly, the change in energy can be calculated from the area enclosed by a force strength–distance graph (Figure 6(b)).

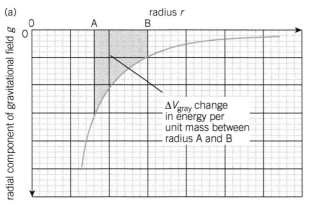

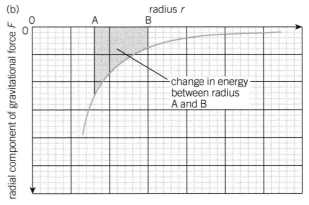

▲ **Figure 6 a** *The difference in potential is the area under the graph of field against radius;* **b** *the difference in energy is the area under the graph of force against radius*

Worked example: Leaving the Moon

On 21st July 1969 the Eagle lander left the lunar surface and docked with the command module orbiting 1.1×10^5 m above the surface. A moment from this journey is dramatically captured in Figure 7.

Figure 8 shows the variation of gravitational field strength near the Moon. Use the graph to calculate the energy required for the lunar module to reach the command module. (Mass of lunar module = 15 000 kg; radius of Moon = 1.74×10^6 m)

▲ **Figure 7** *The Apollo 11 lunar landing module returning to dock with the command capsule*

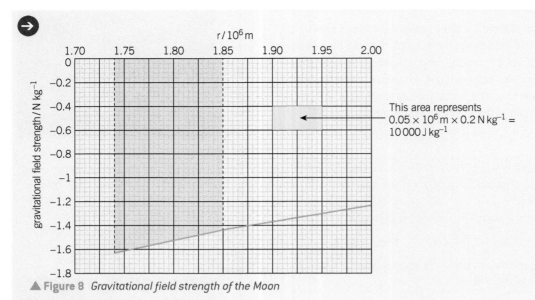

▲ **Figure 8** *Gravitational field strength of the Moon*

Step 1: Identify the height Δr through which the lunar module passes.

The module starts on the surface at $r_0 = 1.74 \times 10^6$ m and rises 1.1×10^5 m to $r = 1.85 \times 10^6$ m.

Step 2: Calculate the change in potential by finding the area under the graph for $r - r_0$, as shown in Figure 8.

The area can be found by a number of methods, including determining the area represented by a graph square and counting the number of squares enclosed by the line and the axis. In our example the shaded rectangle represents a potential difference of $10\,000$ J kg^{-1}. There are about 16.5 such rectangles in the chosen area, giving a potential difference ΔV of $165\,000$ J kg^{-1}.

Step 3: energy required = potential difference ΔV × mass of lunar module m

The energy required to reach the command module = $m\Delta V = 15\,000$ kg × $165\,000$ J kg^{-1}
$$= 2.5 \times 10^9 \text{ J (2 s.f.)}$$

Field and potential gradient

Consider moving a mass m a small vertical distance Δr. Over this distance the potential changes by ΔV. Think, for example, of lifting a book onto a shelf or raising a glass to your mouth.

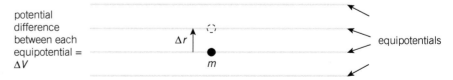

▲ **Figure 9** *A mass m moved upwards through a small distance Δr*

The force exerted to move the mass vertically upwards at constant velocity is equal and opposite to the weight of the mass:

$F = -mg$, where g is the gravitational field strength.

Work done = force × distance = $-mg\Delta r$

We also know that work done = change in gravitational potential energy = $m\Delta V$

$\therefore m\Delta V = -mg\Delta r$

$\therefore -\dfrac{\Delta V}{\Delta r} = g$

The change of potential with radius as the change in radius approaches zero is often referred to as the **potential gradient** and is written as $\dfrac{\mathrm{d}V}{\mathrm{d}r}$.

The radial component of the field at any point in space is equal to the *negative* of the potential gradient at that point. Therefore we can find the value of the radial component of the field strength by finding the gradient of a graph of potential against radius, by drawing the tangent to the curve at the desired point and finding the gradient of the tangent. The graph in Figure 10 shows the variation in potential with radius for the Earth. The gradient at a radius of 2×10^7 m is found to be about $1\,\mathrm{N\,kg^{-1}}$, giving a field strength of $-1\,\mathrm{N\,kg^{-1}}$.

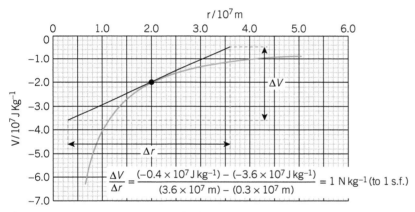

▲ Figure 10 *The field strength for the Earth is the negative of the gradient of a graph of potential against radius*

Similarly, the force on a mass can be found from the negative of the gradient of a graph of potential energy against radius (Figure 11).

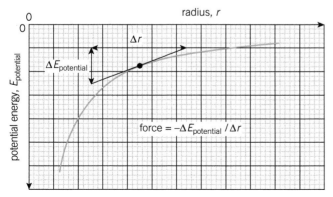

▲ Figure 11 *The force is the negative of the gradient of a graph of potential energy against radius*

Summary questions

1 a Show that the potential energy of a 20 kg mass on the surface of the Earth is about -1.26×10^9 J. *(3 marks)*

   b Calculate the increase in potential energy of the mass when it is raised to a height of 400 km above the surface.
(Mass of Earth $= 6.0 \times 10^{24}$ kg, $G = 6.7 \times 10^{-11}$ N m^2 kg^{-2}, Earth radius $= 6.4 \times 10^3$ km) *(3 marks)*

2 Why should spacecraft be liable to burn up during their descent through the Earth's atmosphere when they did not burn up on their ascent? Hint: compare speeds through the lower atmosphere during ascent and descent. *(3 marks)*

3 The equation $v_{esc} = \sqrt{\dfrac{2GM}{r}}$ can be used to calculate the radius r of a black hole of mass M by setting v_{esc} = velocity of light $= 3.0 \times 10^8$ m s^{-1}. Use the equation to find the radius of a black hole with a mass equal to that of the Earth. *(3 marks)*

4 a Copy and complete Table 1. (Mass of Mars $= 6.4 \times 10^{23}$ kg) *(2 marks)*

▼ **Table 1** *Variation of potential energy with radius for Mars*

Radius /10^6 m	3.4 (surface)	3.9	4.7			14
Potential /10^7 J kg^{-1}	−1.3	−1.1	−0.9	−0.7	−0.5	−0.3

   b Draw a labelled scale diagram showing these equipotentials around Mars. Add eight field lines to your diagram. *(4 marks)*

5 A space probe headed for Mars has escaped Earth. Show that the height of the gravitational potential hill it has yet to climb to reach orbit around Mars is about 3×10^8 J kg^{-1}. (Mass of Sun $= 2 \times 10^{30}$ kg, radius of Earth's orbit $= 150 \times 10^6$ km, radius of Mars's orbit $= 226 \times 10^6$ km) *(4 marks)*

6 Show that the total energy (kinetic + gravitational potential) of a body in orbit of radius r around a mass M is $-\dfrac{GMm}{2r}$. (Hint: remember that the gravitational force is acting centripetally so $\dfrac{GMm}{r^2} = \dfrac{mv^2}{r}$.) *(3 marks)*

Physics in perspective

Standing on the shoulders of giants

In 1676, Isaac Newton suggested in a letter to Robert Hooke that "if I have seen further it is by standing on the shoulders of giants". This has been taken to mean that his insights into science were only possible because he built upon the work of earlier thinkers. Newton's work built upon the astronomical developments of the sixteenth and the seventeenth centuries, including the work of two 'giants', Copernicus (1473–1543) and Kepler (1571–1630).

Copernicus moves the Earth

Five planets are visible with the naked eye: Mercury, Venus, Mars, Jupiter, and Saturn. Careful observations over many months and years show that they move through a fixed band of stars that we now call the constellations of the zodiac. Figure 3 shows the paths of the planets in time lapse against the fixed stars. You can see that some of the planets double-back on themselves on their journey across the sky.

▲ **Figure 1** *The edge of a £2 coin shows the popular phrase used by Newton*

▲ **Figure 2** *Statue of Copernicus in Montreal, Canada*

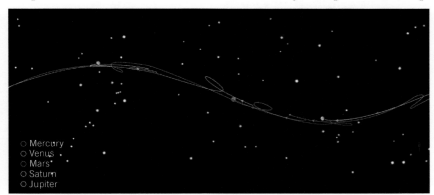

○ Mercury
○ Venus
○ Mars
○ Saturn
○ Jupiter

▲ **Figure 3** *Time-lapse simulation of the apparent movement of the planets through the constellations of the zodiac*

This double-back or retrograde motion was difficult to explain using the 16th century model of the Solar System, which put the Earth at the centre of the stars and planets. Copernicus, a Polish priest, found a simpler explanation, though he did not publish his ideas until he was dying, as he thought his (literally) revolutionary ideas would not be accepted by religious authorities. Copernicus moved the Earth from the centre of the heavens and put it amongst the other planets, which seemed to go against common sense. His model was an improvement, but not perfect — it required the Earth to rotate in a circle within a circle, whose centre was not the Sun — see Figure 4.

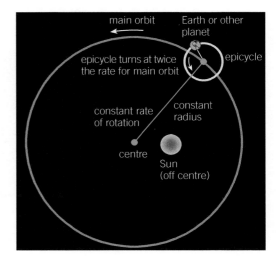

main orbit   Earth or other planet
epicycle turns at twice the rate for main orbit   epicycle
constant rate of rotation   constant radius
centre   Sun (off centre)

▲ **Figure 4** *Because the orbits are really ellipses, in his scheme for planetary motions, Copernicus could only make the orbits circles by putting the Sun off centre and by adding epicycles of his own*

Kepler — the three laws of planetary motion

In 1605, the German-born Johannes Kepler wrote a letter saying "My aim is to show that the heavenly machine is not a kind of divine, living being, but a kind of clockwork." Kepler searched the meticulous data of planetary observations recorded by the Danish astronomer Tycho Brahe to gain an understanding of the movement of the planets around the Sun. He finally discovered a pattern in the data — the planets do not orbit the Sun in circles, but in ellipses with the Sun at one focus. This is Kepler's first law of planetary motion, which greatly simplified the celestial clockwork due to the fact that elliptical orbits remove the need for epicycles or other mathematical guesswork.

▲ **Figure 5** *Statue of Kepler and Brahe in Prague, the Czech Republic*

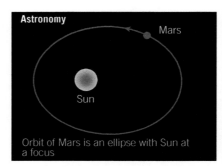 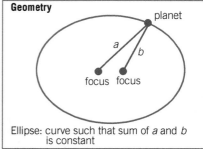

▲ **Figure 6** *Kepler's first law of planetary motion*

Kepler's second law describes how the speed of a planet varies with its distance from the Sun — a line drawn from the planet to the Sun sweeps out equal areas in equal times. As planets move in ellipses they will be nearer to the Sun at some points of their orbit, and further from the Sun in others. The closer they are to the Sun, the faster they move. For example, the Earth is about 5 billion metres nearer the Sun in January than July, and it travels about 3% faster in January than in July.

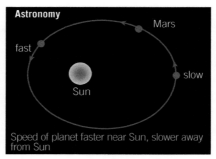 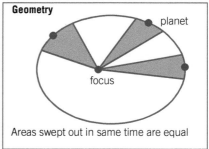

▲ **Figure 7** *Kepler's second law of planetary motion*

Kepler's third law states that the square of the time taken for a planet to orbit the Sun, T, is proportional to the cube of its orbital radius, R. In symbols, $T^2 \propto R^3$

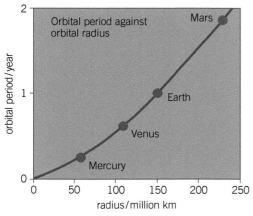

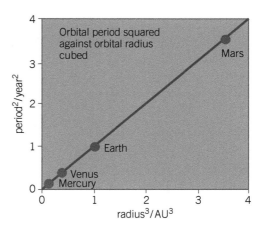

▲ Figure 8 *Kepler's third law of planetary motion*

Kepler reached his laws empirically – he studied Tycho Brahe's data and distilled simple relationships that agreed with the observations. Newton took Kepler's ideas further by showing that the laws of planetary motion can be derived from his theory of universal gravitation. Newton expressed the importance of Kepler, Copernicus and others in his reference to the shoulders of giants. In a similar manner, a little more than two centuries later, Einstein would metaphorically clamber on Newton's shoulders and see still further.

Summary questions

1 Copernicus's model of the solar system seemed to go against common sense. Two 'obvious' problems are given below. For each, give clear explanations to answer the common-sense concerns.
   a If the Earth is moving and rotating on its axis, why don't objects thrown up in the air get left behind?
   b If the Earth sweeps out a very large path around the Sun, why don't the stars look different as we get closer to and farther from them?

2 Kepler's second law states that the speed of a planet increases as it approaches the Sun in its elliptical orbit and decreases as it moves further away. Explain how the total energy of the planet is conserved as it orbits the Sun.

3 Kepler's third law can be derived by showing that the gravitational force on an orbiting body is centripetal. Use the relationship $\frac{mv^2}{R} = \frac{GMm}{R^2}$ to show that $R^3 = \frac{GMT^2}{4\pi^2}$, where T is the time taken for one orbit of the planet (orbital period).

4 Callisto is a satellite of Jupiter discovered by Galileo in 1610. Callisto orbits Jupiter in 16.7 days at a mean distance from the planet of 1.9×10^9 m. Use this data and the equation derived in question 3 to calculate an estimate for the mass of Jupiter.

5 The Earth orbits the Sun at a mean distance of 1.5×10^{11} m. Calculate an estimate for the mass of the Sun.

Practice questions

1 Here are three statements about the gravitational field of the Earth.

   1 The gravitational field of the Earth is an example of a radial field.

   2 The separation of equipotential lines of equal potential difference increases with distance from Earth.

   3 The gravitational field of the Sun exerts a greater force on the Moon than does the gravitational field of the Earth.

   Which of these statements is/are correct?

   A 1, 2 and 3

   B Only 1 and 2

   C Only 2 and 3

   D Only 1   (*1 mark*)

2 Wet clothes are spun in a drum of radius 0.4 m so that they experience an acceleration of $16g$.

   a Calculate the speed of the clothes.
                 (*2 marks*)

   b Calculate the number of revolutions the drum makes in one second.   (*2 marks*)

3 Kepler's third law can be written as

$$R^3 = \frac{GMT^2}{4\pi^2}$$

   where T is the time taken for one orbit of the planet (orbital period), M is the mass of the central body, and R is the orbital radius of the orbiting body. Ganymede is Jupiter's largest moon. Its orbit is circular with a radius of 1.07×10^9 m. The orbital period is seven days, three hours, and 43 minutes. Use this data to calculate the mass of Jupiter.   (*3 marks*)

4 The mass of Mars is 0.11 Earth masses. The radius of Mars is 0.53 Earth radii. The gravitational field strength on the surface of the Earth is $9.8\,\text{N}\,\text{kg}^{-1}$.

   Use only this information and the equation

$$g = -\frac{GM}{r^2}$$

   to calculate the gravitational field strength on the surface of Mars. Show all your working.
                 (*4 marks*)

5 The graph in Figure 1 shows the variation in gravitational field strength near the Earth.

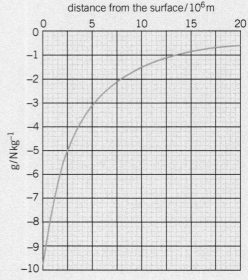

▲ Figure 1

   a The change in gravitational potential between the surface and a distance of 20×10^6 m above the surface is about $4.7 \times 10^7\,\text{J}\,\text{kg}^{-1}$. State how this value can be found from the graph.   (*1 mark*)

   b Calculate the energy required to lift an 800 kg body to 20×10^6 m above the surface of the Earth.   (*2 marks*)

   c Taking four pairs of data points from the graph, show that the data fits the relationship $g \propto -\dfrac{1}{r^2}$ where
     r = distance from centre of the Earth
     = distance from surface + 6.4×10^6 m.
                 (*4 marks*)

6 The surface of the Moon is heavily cratered. Some of the craters are the results of rocks striking the lunar surface.

   a Calculate the potential energy of a 25 kg rock on the surface of the Moon.

     mass of Moon = 7.4×10^{22} kg

     radius of Moon = 1.7×10^6 m   (*2 marks*)

   b State the gain in kinetic energy of a rock that approaches the Moon's surface from a great distance. Explain your reasoning.
                 (*3 marks*)

7 A satellite is in circular orbit around a planet as shown in Figure 2.

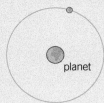

▲ Figure 2

a Copy the diagram and include an arrow showing the direction of the planet's gravitational field on the satellite.

(1 mark)

b Explain why the satellite moves at constant speed even though it is accelerating. *(2 marks)*

c Draw an equipotential between the surface of the planet and the orbit of the satellite. *(1 mark)*

d Here is some data about the planet and the satellite:

mass of satellite = 290 kg

mass of planet = 6.6×10^{23} kg

radius of satellite orbit = 7.0×10^6 m

  (i) Calculate the potential energy of the satellite. *(2 marks)*

  (ii) Calculate the velocity of the satellite in orbit. *(2 marks)*

  (iii) Calculate the kinetic energy of the satellite. *(2 marks)*

  (iv) Calculate the total energy of the satellite. *(2 marks)*

e A second satellite is placed in the same orbit. The second satellite is twice as massive as the first. How do each of the four quantities you calculated in **(d)** compare for the two satellites? You may use further calculations in your answer.

(2 marks)

8 This question is about the motion of Halley's comet.

Figure 3 represents the highly elliptical path of Halley's comet in its 76 year orbit of the Sun.

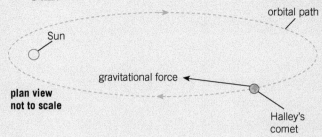

▲ Figure 3

a Figure 3 shows the direction of the gravitational force on the comet.

  (i) Draw an arrow on a copy of Figure 3 to show the component of the gravitational force on the comet which changes its **speed**. Label it **A**.

(1 mark)

  (ii) Draw an arrow on the copy of Figure 3 to show the component of the gravitational force on the comet which changes its **direction**. Label it **C**.

(1 mark)

b At its closest approach to the Sun, the comet is moving at a speed of 54.6 km s⁻¹.

  (i) Show that the **kinetic** energy per unit mass is about 1.5 GJ kg⁻¹. *(2 marks)*

  (ii) The distance from the Sun to the comet is 8.82×10^{10} m at its closest approach. Show that the **total** energy per unit mass is about −20 MJ kg⁻¹.

$G = 6.67 \times 10^{-11}$ N m² kg⁻²

$M_s = 2.00 \times 10^{30}$ kg *(2 marks)*

  (iii) When it is furthest from the Sun, the comet is 5.3×10^{12} m away from the Sun. Calculate the speed of the comet at this distance. *(3 marks)*

OCR Physics B G494 June 2014

13 OUR PLACE IN THE UNIVERSE
13.1 Measuring the Solar System

Specification reference: 5.1.3a(i), 5.1.3a(iii), 5.1.3b(i), 5.1.3c(ii)

Learning outcomes

Describe, explain, and apply:

→ use of radar-type measurements to determine distances within the Solar System; measuring and defining distance in units of time, assuming the relativistic principle of the invariance of the speed of light

→ measurement of distances and relative velocities by radar

→ logarithmic scales of magnitude for quantities such as distance, size, mass, and brightness.

▲ Figure 1 *Ultra Deep Field image from the Hubble Space Telescope*

▶ Figure 2 *Distances can be expressed in units of light travel time*

The size of the Universe

In little more than one human lifetime, astronomers using new and powerful types of telescope have dramatically changed our ideas about the structure and scale of the universe. Figure 1 shows an Ultra Deep Field image captured by the Hubble telescope. Each dot of light is an individual galaxy. This image, taken at infrared wavelengths, includes some of the most distant galaxies visible – some are 13 billion light-years from Earth. The image was captured by aiming the space telescope at a tiny 'empty' patch of sky and letting the image build up over one million seconds. When you look at this picture you see the Universe as it was 13 billion years ago, as that is how long it has taken the light from these very distant galaxies to reach Earth. The current age scale for the Universe is about 13.8 billion years, so this extraordinary image shows us part of the Universe early in its history.

From the around the time of Newton, scientists have had a fair idea of the size of the tiny fraction of the Universe that is the Solar System. It took a few centuries for equipment and ideas to develop to the point where we can capture images like the one above and understand what they show us.

Light-years

A **light-year** is simply the distance light travels through space in one year, approximately 9.5×10^{15} m. This is a useful measure for the vast distances described in astronomy – those distant galaxies in the Figure 1 are

Speed of light
300 000 kilometres per second
300 km per millisecond
300 m per microsecond
300 mm per nanosecond

Light travel time
1 light-second = 300 000 km
1 light-millisecond = 300 km
1 light-microsecond = 300 m
1 light-nanosecond = 300 mm

Time
1 year = 31.5 million seconds

Light-year
1 light-year = 9.4×10^{15} m
= 10^{16} m approximately

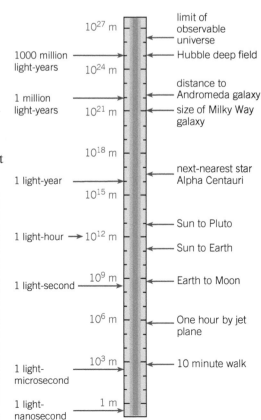

limit of observable universe — 10^{27} m

Hubble deep field

1000 million light-years — 10^{24} m

distance to Andromeda galaxy

1 million light-years — 10^{21} m — size of Milky Way galaxy

10^{18} m

next-nearest star Alpha Centauri

1 light-year — 10^{15} m

1 light-hour → 10^{12} m — Sun to Pluto

Sun to Earth

10^9 m — Earth to Moon

1 light-second

10^6 m — One hour by jet plane

1 light-microsecond — 10^3 m — 10 minute walk

1 light-nanosecond — 1 m

288

about 1.2×10^{26} metres away! A light-year is a distance expressed in light travel time. For shorter distances we can use the 'light-minute'. The Sun is a little under 8.5 light-minutes from Earth, so when you look up at the Sun you are looking 8.5 minutes into the past. Figure 2 shows a range of distances expressed in light travel time. The scale is logarithmic to allow a large range of distances to be shown on the same diagram.

The Hertzsprung–Russell diagram

Another example of the use of logarithmic scales is the **Hertzsprung–Russell diagram** (Figure 3), which plots the luminosity of stars against their temperature.

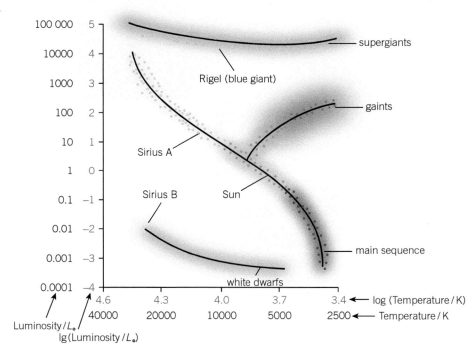

▲ **Figure 3** *Plot of stellar luminosity against surface temperature – note the logarithmic scales for temperature and luminosity, and the reversed scale for temperature. $L_\odot$ is the luminosity of the Sun.*

The top left corner shows blue giant stars. These are the largest and hottest stars. Blue giants burn through their nuclear fuel at a prodigious rate and some will finally become supernovae – outshining entire galaxies in a very brief period of extraordinary energy release. A blue giant such as Rigel (in the foot of the constellation of Orion) is about 66 000 times brighter than the Sun.

Measuring distances in the Solar System

The modern way to measure distances in the solar system is to use **radar** (**ra**dio **d**etection **a**nd **r**anging), just as air-traffic controllers or traffic police do. Venus was successfully targeted for the first time in 1961; a pulse of radio waves was sent out from Earth, and the reflected waves detected returning after a delay. The distance to the planet can be calculated from the delay and the known speed of electromagnetic radiation. With a distance to Venus of the order of 100 million kilometres, the out-and-back delay is several minutes and the reflected signal is very weak. There is no hope of measuring the distance to even the nearest star (not counting the Sun), which is 4.2 light-years away, in this way.

How far away is an asteroid?

Imagine that an asteroid is reported as passing close to Earth. Urgent question – how far away is it and how fast is it approaching or receding from Earth? To get a range, you need a radar pulse to be sent out and reflected back. The delay tells you the time of travel of the radar pulse going out and back – twice the distance to the asteroid. Using the known speed of radar pulses (the speed of light) you can work out the distance. Just multiply the speed of light by half the out-and-back time.

How far away is the Moon?

The distance to the Moon was first estimated by the astronomer Hipparchus in the 2nd century BC. He used trigonometry to obtain a value of about 4×10^8 m. This is close to the modern value of 3.84×10^8 m from the centre of the Earth to the centre of the Moon. This result is obtained using a variation of radar known as **lidar** – light detection and ranging.

Laser light emitted from stations on Earth strikes reflectors set up on the lunar surface by the Apollo astronauts. Although the return signal is incredibly weak, it can be distinguished from the background noise owing to the particular properties of laser light.

Measuring speeds in the Solar System

To get the relative velocity of the asteroid and Earth you can measure the distance again. If the out-and-back time has increased, the asteroid is moving away from Earth and all is well. If it has decreased, the asteroid is getting closer and we might be in danger of collision. The component of velocity along the line of sight is just the change in distance divided by the time between the two measurements (Figure 4).

Assumptions

We have described the calculation of distance and speed from radar measurements as simply as possible. But in fact we made some assumptions. We assumed that the speed of the signal was the same both ways, and that the moment of reflection was just halfway through the time delay of the pulse as observed on Earth.

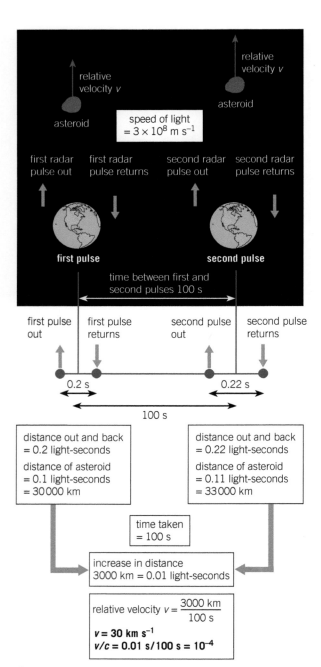

▲ **Figure 4** *Calculating the relative velocity of an asteroid with radar pulses*

Both of these assumptions require that the speed of light is constant no matter how you are moving relative to other objects. That is the basis of the theory of relativity, and this calculation of distance and relative velocity is correct in that theory. Some parts of relativity are quite easy, after all!

Summary questions

1 Calculate the distance in light-seconds between the Earth and Mars when the two planets are at their closest, about 5.5×10^{10} m.   *(2 marks)*

2 Give two reasons why radar is not suitable for measuring distances beyond the Solar System.   *(2 marks)*

3 a Figure 3 shows the luminosity of Sirius A, the brightest star in the northern sky, and the luminosity of its companion, Sirius B, a white dwarf star. Estimate how many times more luminous the star Sirius A is than Sirius B.   *(2 marks)*
   b Suggest why Sirius A is the brightest star in the sky even though it is not the most luminous.   *(1 mark)*
   c Suggest why a logarithmic scale is used for the Hertzsprung–Russell diagram.   *(1 mark)*

4 Table 1 gives the order-of-magnitude mass of a range of objects. Show these examples on a logarithmic scale.   *(4 marks)*

▼ Table 1

Ant	Human	Blue whale	Giant sequoia (tree)	Cruise ship	Comet Halley	Mercury	Sun	Supermassive black hole	Andromeda galaxy
10^{-3} kg	10^{2} kg	10^{5} kg	10^{6} kg	10^{8} kg	10^{14} kg	10^{23} kg	10^{30} kg	10^{37} kg	10^{42} kg

5 A radar pulse makes a return journey to an asteroid in 40.4 s.
   a Calculate the distance to the asteroid at the time of measurement. State any assumptions you make. *(3 marks)*
   b The measurement was repeated 12.0 minutes later. The time for the return journey of the pulse was found to be 40.1 s. Calculate the relative velocity of approach of the asteroid during this time.   *(3 marks)*

Learning outcomes

Describe, explain, and apply:

→ evidence of a 'hot Big Bang' origin of the Universe from cosmological red-shifts (Hubble's law) and cosmic microwave background

→ calculations of distances and ages of astronomical objects.

Synoptic link

If you need to remind yourself of the small-angle approximation, look at Chapter 20, Maths in physics.

Techniques for measuring huge distances

Parallax

Hold your hand out in front of you with your thumb up. Close one eye and observe where your thumb appears in relation to the background. Now observe with your other eye. You will see that the thumb shifts against the background. This is **parallax**. Astronomers use this method to measure the distance to nearby stars. The angular shift (in radians) of the position of the star is the ratio of the baseline to the distance (small-angle approximation). In June the nearby star lines up with stars in region A (Figure 1), whilst 6 months later it is in line with region B. In practice stellar parallax is tiny because the baseline is so small compared with the distance to even the nearest star. In 1838 the known size of the Universe grew when the first parallax of a star was found – a star in the constellation of Cygnus was measured as 10.4 light-years distant, close to the modern value of 11.2 light-years, 700 000 times the distance between the Earth and the Sun. But we can only use parallax for very near stars. How can greater distances be measured?

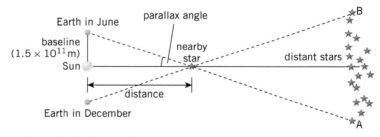

▲ **Figure 1** *Stellar parallax*

Standard candles

Looking at the night sky you will see stars of differing brightness. Why is this? One star may be bigger and brighter than its neighbours or it may just be nearer Earth. How can we tell how bright a star 'really' is? Astronomers call this true brightness the **absolute luminosity** of the star. We need a **standard candle** – a star with a known absolute luminosity. If we know that, then we can calculate the distance to the star by its apparent brightness – a dim star with a high absolute luminosity must be more distant than a star of the same absolute luminosity that appears brighter.

One such standard candle is a type of variable star (a star that changes its brightness over time). In 1912 Henrietta Swan Leavitt (Figure 2) found that a class of variable stars known as Cepheids show a brightness variation that is dependent on their absolute luminosity. Cepheids of greater absolute luminosity vary in brightness over longer periods (Figures 3 and 4).

▲ **Figure 2** *Henrietta Swan Leavitt began work as a volunteer at Harvard College Observatory in 1893, becoming a leader in the then-new science of photographic measurement and ultimately head of the department of photographic stellar photometry*

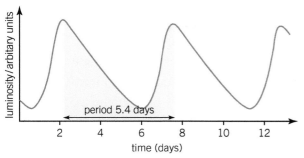

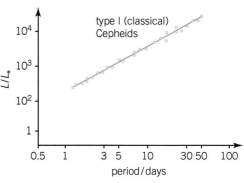

▲ **Figure 3** *Luminosity–time relationship for the star Delta Cephei, the original Cepheid variable*

▲ **Figure 4** *Relative luminosity–period relationship for Cepheids — the longer the variation period, the brighter the star* $\left(\dfrac{L}{L_{\odot}} = \dfrac{luminosity}{luminosity\ of\ Sun}\right)$

Cepheid variables provide a standard candle. Studying the variation in brightness of a Cepheid allows astronomers to find its absolute luminosity. This can be compared to its apparent brightness to reveal its distance. In 1923 Edwin Hubble observed a Cepheid in what was then called the Andromeda Nebula. This is a misty patch of light that some astronomers thought to be a gas cloud within our own galaxy, whilst others thought it was a galaxy beyond our own. The Cepheid observation ended the debate — the misty patch was at least 900 000 light-years distant. The nebula is a galaxy in its own right, called the Andromeda Galaxy. The Universe had grown again.

Measuring speed

As well as the distance to stars and whether there are galaxies outside our own, astronomers were interested in the movement of stars through space. Radar measurements work well for planets, but a different method is required for more distant objects.

Doppler shift

Atoms in a distant star absorb and emit light at particular wavelengths that are accurately known from laboratory measurements. These are the spectral lines of the star. If the star is receding (moving away) from Earth, the wavelengths of its spectral lines increase. If the star is moving towards the Earth, the wavelengths of the spectral lines decrease. Receding stars show **red-shift** because their spectral lines are shifted towards the red (long-wavelength) end of the spectrum. Such a change in wavelength is called a **Doppler shift**.

Suppose that a light source is not moving relative to the observer and emits waves whose peaks are a distance λ (the wavelength) apart and separated by equal intervals of time T, with $\lambda = cT$. If the source is moving away at speed v, the wavelength is stretched out. This is because, between the emission of any two peaks, the source has moved an extra distance vT away. So the wavelength has increased from

$$\lambda = cT$$

by an amount

$$\Delta\lambda = vT$$

> ### Synoptic link
>
> Line spectra were discussed in Topic 7.1, Quantum behaviour.

source and receiver both at rest

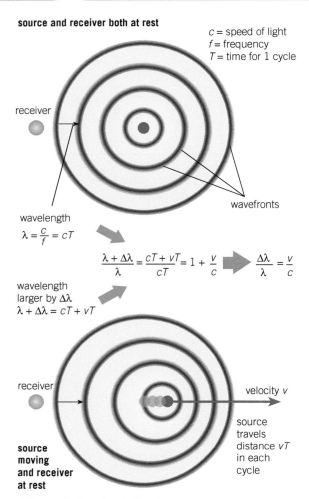

c = speed of light
f = frequency
T = time for 1 cycle

receiver

wavefronts

wavelength
$$\lambda = \frac{c}{f} = cT$$

$$\frac{\lambda + \Delta\lambda}{\lambda} = \frac{cT + vT}{cT} = 1 + \frac{v}{c} \quad \frac{\Delta\lambda}{\lambda} = \frac{v}{c}$$

wavelength
larger by $\Delta\lambda$
$\lambda + \Delta\lambda = cT + vT$

receiver

velocity v

source
travels
distance vT
in each
cycle

source moving and receiver at rest

▲ **Figure 5** *Doppler shift — The wavelength increases when the source travels away from the observer*

The fractional increase in wavelength is therefore

$$\frac{\Delta\lambda}{\lambda} = \frac{v}{c}$$

For speeds v much less than the speed of light c, the equation $\frac{\Delta\lambda}{\lambda} = \frac{v}{c}$ is accurate. However, it is not the full story. Einstein's special theory of relativity modifies the calculation.

Use of the **Doppler effect** (Figure 5) led to one of the most important steps in our understanding of the Universe. Most galaxies were found to be travelling away from us, not approaching. This was our first hint about the expansion of the Universe.

The expanding Universe

Hubble's law

In 1929 the astronomer Edwin Hubble and others used red-shift data to plot the speeds of recession of 24 galaxies against estimates of their distance – in effect, against apparent brightness. The points fell on a more or less straight line — nearly all the galaxies seem to be moving away from us, and the further away they are, the faster they are receding (Figure 6). This is known as **Hubble's law**.

Throughout the 1930s astronomers obtained spectra from galaxies increasingly distant from Earth. Despite the problems of establishing distances, Hubble's law continued to stand. With the red-shift interpreted as a recession velocity, recession velocity v = constant H_0 × distance r.

The constant H_0 is called the Hubble constant.

▶ **Figure 6** *Hubble's law and the age of the Universe*

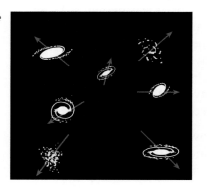

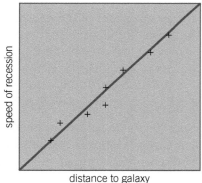

speed of recession

distance to galaxy

speed of recession is directly proportional to distance, $v = H_0 r$

H_0 is the Hubble constant

units of $H_0 = \dfrac{\text{speed}}{\text{distance}} = \dfrac{1}{\text{time}}$

$\dfrac{1}{H_0}$ = Hubble time

units of $\dfrac{1}{H_0}$ = time

If speed v were constant, then $\dfrac{1}{H_0} = \dfrac{r}{v}$ = time since galaxies were close together.

Supported by these observations, between the 1920s and the 1950s an idea developed that the Universe originated in a very hot, very dense state from which it has expanded and cooled. This became known as the **Big Bang** theory. The name stuck, but is misleading. The beginning of the Universe was not, according to present theories, an explosion in an existing empty space, but the creation of an expanding region of space and time.

The bigger the Hubble constant, the faster the Universe must be expanding, and the younger it must be to have reached its present size. So the Hubble constant gives a timescale for the Universe. This time is just the reciprocal, $\frac{1}{H_0}$.

 Worked example: The age of the Universe

In 2013, the Planck Surveyor mission produced data that gave a Hubble constant of $67.1\ \text{km s}^{-1}\ \text{Mpc}^{-1}$. Roughly how old is the Universe? (1 megaparsec (Mpc) = $3.09 \times 10^{22}\ \text{m}$).

Step 1: Convert the value of H_0 to s^{-1}.

$$H_0 = \frac{67\,100\ \text{m s}^{-1}}{3.09 \times 10^{22}\ \text{m}} = 2.17... \times 10^{-18}\ \text{s}^{-1}$$

Step 2: Use the value for H_0 to determine an order-of-magnitude timescale for the Universe.

$$\text{timescale} = \frac{1}{H_0} = \frac{1}{2.17... \times 10^{-18}\ \text{s}^{-1}} = 4.61 \times 10^{17}\ \text{s}\ (2\ \text{s.f.}) \approx 10^{10}\ \text{years}.$$

The 2013 estimate for the age of the Universe is 13.8 ± 0.037 billion years $((1.38 \pm 0.0037) \times 10^{10}\ \text{years})$.

Interpreting the red-shift

We have described Doppler shift in terms of waves from an object moving through space. The very large red-shifts observed from distant galaxies are interpreted differently. They are caused by space itself expanding – the wavelength of light is stretched along with the space through which the light travels. This is known as cosmological red-shift. If the red-shift $z = \dfrac{\text{change in wavelength } \Delta\lambda}{\text{wavelength } \lambda}$, then space has stretched by a factor of $1 + z$ (Figure 7).

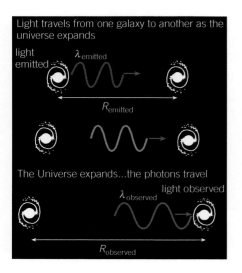

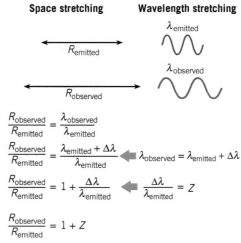

◀ **Figure 7** *The cosmological red-shift (light shifted from blue to red wavelengths) measures the stretching of space*

In the beginning...

there was the Big Bang...

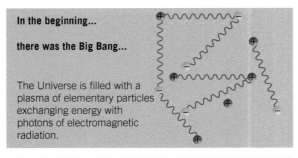

The Universe is filled with a plasma of elementary particles exchanging energy with photons of electromagnetic radiation.

300 000 years after the Big Bang

temperature 3000 K
typical wavelength of radiation, 1 μm

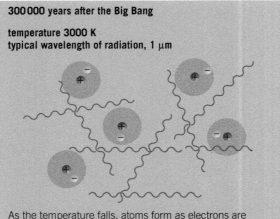

As the temperature falls, atoms form as electrons are held in orbit around nuclei of protons and neutrons.

The Universe becomes transparent to photons, which no longer interact so easily with atoms and so travel unaffected through the Universe.

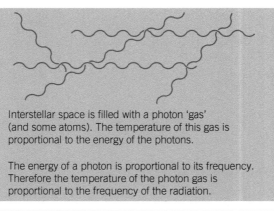

Interstellar space is filled with a photon 'gas' (and some atoms). The temperature of this gas is proportional to the energy of the photons.

The energy of a photon is proportional to its frequency. Therefore the temperature of the photon gas is proportional to the frequency of the radiation.

...14 billion years after the Big Bang

temperature 2.7 K
typical wavelength of radiation, 1 mm

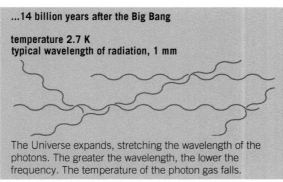

The Universe expands, stretching the wavelength of the photons. The greater the wavelength, the lower the frequency. The temperature of the photon gas falls.

▲ **Figure 8** *Cosmic microwave background radiation*

When we observe objects of red-shift $z = 10$ we are looking at a time when the Universe was about a tenth of its present size. These distant objects show a much younger Universe in which prototype galaxies are forming. The Universe has evolved since then.

Hubble's law provides evidence that the Universe is expanding, but does not give us information about the state of the early Universe. It shows that the Universe is expanding over time, so must have been smaller in the past, but does not suggest that the Universe began in a hot, dense state. The clinching evidence for this picture, the hot Big Bang model, came in 1965.

Cosmic microwave background radiation

In 1965, Arno Penzias and Robert Wilson were calibrating a microwave antenna for use with communication satellites. They found something strange. Wherever the antenna was directed, it detected noise in the signal at microwave wavelengths. Other physicists realised where the microwave background came from – it was the predicted cosmic background radiation that was left over from the hot beginnings of the Universe.

The **cosmic microwave background** has the biggest cosmological red-shift known. It was produced when the Universe became just cool enough for electrons and ions to combine into neutral atoms, emitting photons. That happens at around 3000 K, when the typical wavelength of the photons is around 1 μm. Today these photons are seen, stretched in wavelength, as microwaves with a wavelength of the order of 1 mm – 1000 times longer. The temperature, too, has fallen by a factor of 1000, to just below 3 K (Figure 8). The cosmological expansion $z + 1$ is the same ratio,

$$z + 1 = \frac{\text{radius of the Universe now}}{\text{radius of the Universe then}}.$$

The cosmic microwave background provides convincing evidence that the Universe began in a hot, dense state. Satellites have now produced images of the whole

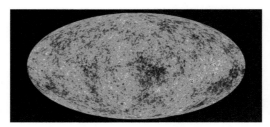

▲ **Figure 9** *Variation in microwave background – in this false-colour image the yellow and red areas indicate denser regions of the early Universe, which later formed the seeds of galaxies*

Universe that show that the microwave background is not perfectly smooth (Figure 9). It has very small differences in temperature. As the microwave background is the red-shifted radiation from the early Universe, this variation must have been present at the early stages of the Universe. It is from these non-uniformities that stars and galaxies are thought to have subsequently formed.

Extension: Mapping the Milky Way

Our galaxy is thought to be a barred spiral shape like the galaxy in Figure 10. But how can we know this when we can't get outside it to capture an image?

The first serious attempt to draw a map of the galaxy was made by the astronomer William Herschel in 1785. He thought that the galaxy was the entire Universe, and wanted to map it. He assumed incorrectly that all stars were of the same absolute brightness and that variation in brightness was due only to distance. With his sister Caroline, Herschel recorded the brightness and position of a vast number of stars, producing the picture in Figure 11.

Herschel also assumed that stars were evenly distributed in space and that his telescope allowed him to see to the very edge of the Universe. He concluded that the directions in which he observed the greatest density of stars must represent the greatest distance to the edge of the Universe.

Nowadays, astronomers observe in different wavelengths, including radio and infrared, so parts of the galaxy in which visible light is obscured by dust and gas clouds can be observed. The European Space Agency satellite Gaia, which began collecting data in 2014, uses parallax to measure the position and distances of stars with extraordinary precision: it can measure angles of relatively bright stars to about 7 microarcseconds, or $2 \times 10^{-9\,\circ}$. It will collect data on about a billion stars over its operational lifetime, 1% of the stars in the galaxy. This will not only give an accurate image of the galaxy, it will help answer many questions, including how galaxies are formed.

▲ **Figure 10** *A barred spiral galaxy visible in the constellation Ophiuchus, taken by the Hubble Space Telescope*

◄ **Figure 11** *Herschel's map of the Milky Way*

To think about:

1. Herschel made his map about 50 years before the distance to a star was measured by parallax. Explain why the debate about whether stars are all equally bright could not be firmly resolved until after distances to a number of stars had been determined.
2. In Herschel's model, why does a great density of stars in a particular direction suggest a great distance to the edge of the galaxy?
3. It is claimed that the angular resolution of the Gaia telescopes is the same angle as that subtended by a single hair at a distance of 1000 km. A hair is about 10^{-5} m wide. Is the claim correct?

Summary questions

1 a State the two pieces of evidence that suggest that the
Universe has expanded from a hot, dense state. (*2 marks*)
   b What does the slight non-uniformity of the cosmic microwave
background suggest about the early Universe? (*2 marks*)

2 Figure 1 shows the parallax angle for a nearby star.
Calculate the parallax angle for a star 11 light-years from Earth
(baseline = 1.5×10^{11} m 1 light-year = 9.5×10^{15} m). (*2 marks*)

3 a In the 1930s one measure of the Hubble constant suggested a value
of $300\ \text{km s}^{-1}\ \text{Mpc}^{-1}$. Use this value to calculate the timescale
of the Universe in years (1 Mpc = 3.1×10^{22} m). (*3 marks*)
   b The age of the Earth is about 4.5 billion years. Why does this suggest
that the 1930s value of the Hubble constant is too high? (*2 marks*)

4 a The radio galaxy 3C324 has a red-shift $z = 1.12$. Show that the
hydrogen spectral line with a wavelength of 21 cm is observed
on Earth at a wavelength of about 44 cm. (*2 marks*)
   b A galaxy in the constellation Hydra has a red-shift $z = 0.203$.
Show that the Universe has expanded by about 20% since
the light from it was emitted. (*2 marks*)

5 Figure 12 shows the history of the expansion of the Universe, from a few
seconds old to now.

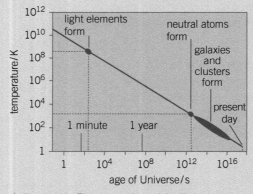

▲ **Figure 12** *The expansion of the Universe*

   a On what kind of scale are the temperature and age plotted? Why was
this necessary? (*2 marks*)
   b Assume that, as the temperature changes, the product λT of
wavelength and temperature is constant. If the photons emitted at
the time neutral atoms formed had a wavelength of about 1 μm, what
would their wavelength have been when the light elements formed? (*2 marks*)

   c By what factor did the Universe expand between these
two times? (*2 marks*)

13.3 Special relativity
Specification reference: 5.1.3a(i), 5.1.3a(ii)

Relative motion

You have often met the term 'at rest', meaning not moving relative to the things around you. As you read this you are probably at rest relative to the furniture in the room (if you are not, you might want to look where you are going!). When you read on a moving train you are at rest relative to the carriage but not at rest relative to the landscape that speeds past you. It seems obvious that you and the train are moving and the track is not. Take care, though. Once again, the common-sense view may not always hold.

Think about observing speeds of objects in the Universe. You know that you sit on a moving platform, the Earth, which orbits the Sun, which moves round the Milky Way galaxy, which moves relative to other galaxies, and so on. Nothing is fixed. All velocities are relative. From this point of view it makes no sense to say which is really moving. We can only detect relative velocity. There is no such thing as 'really moving' (absolute velocity) or 'really at rest'. So now 'being at rest' simply means 'moving with me'. A clock at rest is just one that you carry with you. We will call the time that such a clock records 'wristwatch time', denoted with the Greek letter τ.

Notice that one observer's state of rest may not be the same as another's. They differ by their relative velocity.

These ideas are stated in Einstein's first postulate of **special relativity**, published in 1905. This says that there is nothing special about uniform movement. Travelling at uniform velocity relative to another object changes nothing about physics.

Einstein's first postulate

- Physical behaviour cannot depend on any 'absolute velocity'. Physical laws must take the same form for all observers, no matter what their state of uniform motion in a straight line.

Constant speed of light

The speed of light in a vacuum is defined as $c = 299\,792\,458\,\text{m s}^{-1}$. This is exact. Scientists now define the metre as the distance travelled by light in a vacuum in a time of $\dfrac{1}{299\,792\,458}$ of a second. The decision to define the speed of light as constant was taken for technical reasons to do with precision of measurement. At the same time it agreed with a fundamental shift in thinking in physics introduced by Einstein. This was to regard the speed of light as a fixed conversion of units of distance and time. The idea can be stated as the second postulate of the theory of relativity.

Einstein's second postulate

- The speed of light c is a universal constant. It has the same value, regardless of the motion of the platform from which it is observed. In effect, the translation between distance and time units is the same for everybody.

▲ **Figure 1** *Which is moving, the train or the countryside?*

It doesn't matter how fast you are moving, light will still leave you at $299\,792\,458\,\text{m s}^{-1}$. Imagine that you are on a train that is somehow travelling at $0.5\,c$. If you shine a light from the train you would expect it to travel at c relative to the moving train. This is exactly what happens, but it also travels at c relative to an observer outside the train. The speed of light is constant for all observers.

Space–time diagrams

Space–time diagrams represent objects moving through space and time. These diagrams have the following features:

- Time is conventionally shown on the y-axis.
- Distance is shown on the x-axis, in units of $\frac{x}{c}$ (light-seconds). For simplicity, only one space dimension is shown.
- Every diagram is drawn from the point of view of a given platform. This platform (for example a place on Earth) moves up the time axis as time passes. Observers, clocks, and so on are all carried with the platform. Other platforms are taken to move relative to this one.
- Any other object at rest relative to the platform also moves vertically up the time axis as time passes, staying a constant distance from the platform.

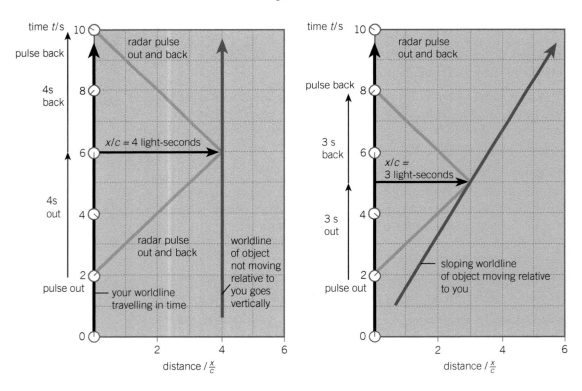

Distance is measured by reflecting a radar pulse, then measuring the time sent and the time received back. The clock used travels with you and the radar equipment (at rest).

Assumptions:
   speed c is constant so reflection occurs halfway through the out-and-back time
   speed c is not affected by the motion of the distant object

▲ **Figure 2** *Space–time diagrams*

- Lines representing paths of objects through space and time are called **worldlines**.

- An object that moves relative to the observer's platform at constant velocity moves along a sloping worldline across the diagram, changing its distance from the observer.

- A light pulse travels at 1 light-second per second, so its worldline travels at 45° across the diagram. Because the speed of light is constant, independent of the observer's motion, this is true for every space–time diagram.

In Figure 2, a radar pulse is released at $t = 2$ s. Its path is at 45° as it travels at 1 light-second every second. In the first diagram it reflects from a stationary object and the pulse returns at $t = 10$ s. The distance to the object is $\frac{x}{c} = 4$ light-seconds. A similar analysis shows that the moving object in the second diagram was at a distance $\frac{x}{c} = 3$ light-seconds when the pulse struck it.

Time dilation and the relativistic factor γ

Clocks moving relative to an observer run slowly as seen by an observer. Each second is stretched or dilated. The greater the relative velocity, the greater the effect. We can think about this using the idea of a **light clock**: a pair of mirrors between which pulses of light bounce back and forth. With this apparatus we can carry out a 'gedankenexperiment', or thought experiment, to explore the consequences of Einstein's postulates.

Imagine that you are sitting on a moving train, watching your light clock. The wristwatch time for one 'tick' is the time taken for a pulse of light to make a return journey between the mirrors, 2τ (Figure 3).

But what will an observer outside the train see when you, the train, and the light clock speed past? The speed of light will be the same for the observer and the passenger on the train, but the observer will see the light beam travelling a greater distance, so the time taken for one tick will increase (Figure 4). Study Figure 4 carefully to make sure that you understand why some distances are given as multiples of t and others are multiples of τ.

Using Pythagoras's theorem,
$$(c\tau)^2 = (ct)^2 - (vt)^2$$
$$\tau^2 = t^2 - \left(\frac{vt}{c}\right)^2 = t^2\left(1 - \frac{v^2}{c^2}\right)$$
$$t = \frac{\tau}{\sqrt{1 - \dfrac{v^2}{c^2}}}$$

The relativistic factor γ is defined as $\gamma = \dfrac{1}{\sqrt{1 - \dfrac{v^2}{c^2}}}$ So
$$t = \gamma\tau$$

Time dilation is a consequence of the speed of light being constant for all observers.

sitting beside the clock

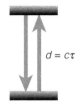

mirrors $d = c\tau$ apart
time out and back
(1 tick) $= 2\tau$

clock records
wristwatch time τ

▲ **Figure 3** *Sitting beside the clock*

clock travelling past you at speed v

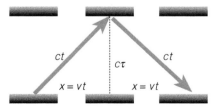

You see the light take a longer path
but the speed is still c,
so the time t is longer.
The moving clock ticks more slowly.

▲ **Figure 4** *Clock travelling past observer at speed v*

Worked example: Half-life dilation

π mesons are unstable particles with mean lifetimes at low speeds of 2.6×10^{-8} s. Calculate the mean lifetime measured by an observer when the π mesons are moving at a relative speed of $2.9 \times 10^8\,\mathrm{m\,s^{-1}}$.

The wristwatch lifetime τ for the π meson is 2.6×10^{-8} s. The observed lifetime t is given by

$$t = \gamma\tau = \frac{\tau}{\sqrt{1 - \dfrac{v^2}{c^2}}}$$

$$t = \frac{2.6 \times 10^{-8}\,\mathrm{s}}{\sqrt{1 - \dfrac{(2.9 \times 10^8)^2\,\mathrm{m^2 s^{-2}}}{(3.0 \times 10^8)^2\,\mathrm{m^2 s^{-2}}}}}$$

$$= 1.0 \times 10^{-7}\,\mathrm{s}\ (2\ \mathrm{s.f.})$$

Time dilation in action

Relativistic time dilation is not merely an idea confined to gedankenexperiments.

● Synchronised, accurate clocks record different time intervals when one is placed on an aircraft that makes a round trip whilst the other remains at the airport.

● The global positioning system has to allow for the effects of relativistic time dilation to pinpoint locations.

● The half-life of particles increases when the particles are accelerated to relativistic speeds. The greater the speed, the greater the time dilation.

One further consequence of time dilation to think about – photons never get old! The wristwatch time of a photon is $\tau = \frac{t}{\gamma}$. For photons, $v = c$ so the relativistic factor is infinite. So, whatever the value of t between the emission and absorption of a photon, even if the photon has travelled across the Universe, the photon sees no time pass by! More carefully stated, it is impossible to travel alongside a photon. This is what started Einstein off thinking about relativity. He realised that if you imagine travelling along with an electromagnetic wave, there can't be any wave, because the electric and magnetic fields would be static. This is what makes light special – it simply has to have the one particular speed that it has.

Summary questions

1   a   Draw a space–time diagram showing an object that is at a distance from an observer of $\frac{x}{c} = 2$ s at time $t = 0$ s. The object is moving at $0.5\,c$ relative to the observer. Show the worldline for the object from $t = 0$ s to $t = 12$ s.   (4 marks)

   b   A pulse of light leaves the observer in (a) at $t = 1$ s. Draw the path of the light on the space–time diagram, showing it reflecting from the object and returning to the observer. (Hint: look at Figure 2 to see an example of such a diagram.)   (2 marks)

   c   Use the diagram in (b) to calculate the distance between the observer and the object when the reflection took place.   (2 marks)

2   A satellite circles Earth, radius 6400 km, every 90 minutes at low altitude. Show that the relativistic time dilation factor $\gamma \approx 1 + 3 \times 10^{-10}$.   (2 marks)

3   Copy and complete Table 1.   (5 marks)
   ▼ Table 1

Velocity of object/velocity of light $\frac{v}{c}$	0.100	0.200	0.300	0.500	0.800			0.999
Relativistic factor γ	1.01	1.02	1.05			2.29	3.20	

4   a   Calculate the gamma factor of a particle moving at $2.95 \times 10^8\,\mathrm{m\,s^{-1}}$ relative to the observer.   (2 marks)

   b   The relativistic factor γ of a particle in a linear accelerator is 2.9. Calculate the speed of the particle.   (2 marks)

5   Atomic clocks in the global positioning system can measure time differences to one part in 10^{12}. Show that, for this precision, corrections for relativistic time dilation start to become needed at speeds of the order of $300\,\mathrm{m\,s^{-1}}$.   (3 marks)

Module 5.1 Creating models

Modelling decay

Modelling radioactive decay

- Exponential decay as a relationship of the form $\frac{dx}{dt} = -kx$, where rate of change is proportional to amount present

- Radioactive decay as a random process with a fixed probability

- The terms: activity, decay constant λ, half-life $T_{1/2}$, probability, and randomness

- Solving equations of the form $\frac{\Delta N}{\Delta t} = -\lambda N$ using iterative numerical or graphical methods

- The equations: $N = N_0 e^{-\lambda t}$ and $T_{1/2} = \ln\frac{2}{\lambda}$

- Practical task – determining the half-life of an isotope such as protactinium

Modelling discharging and charging capacitors

- The equations: $C = \frac{Q}{V}$; $E = \frac{1}{2}QV = \frac{1}{2}CV^2$

- The energy stored on a capacitor as the area below a Q–V graph

- Decay of charge on a capacitor as an example of exponential decay where $\frac{dQ}{dt} = -\frac{Q}{RC}$

- Time constant τ

- The equations $\tau = RC$; $Q = Q_0 e^{\frac{-t}{RC}}$

- Solving equations of the form $\frac{\Delta Q}{\Delta t} = -\frac{Q}{RC}$ iteratively and graphically

- Using the equations $Q = Q_0 e^{\frac{-t}{RC}}$ and $Q = Q_0(1 - e^{\frac{-t}{RC}})$ and corresponding equations for V and I

- Plotting exponential graphs with linear or logarithmic scales

- Practical task – investigating the charging and discharging of a capacitor using meters and data loggers

Modelling oscillations

Simple harmonic motion and circular motion

- The terms: period, frequency, simple harmonic motion

- SHM as oscillation with a restoring force proportional to displacement such that acceleration $\frac{d^2x}{dt^2} = -\frac{k}{x}m$

- Solving equations of the form $\frac{\Delta^2 x}{\Delta t^2} = -\frac{k}{x}m$ by iterative or graphical methods

- SHM of a system where $a = -\omega^2 x$, where $\omega = 2\pi f$, and two solutions are $x = A\sin(\omega t)$ and $x = A\cos(\omega t)$

- The equations $T = 2\pi\sqrt{\frac{m}{k}}$ (mass on a spring) and $T = 2\pi\sqrt{\frac{L}{g}}$ (simple pendulum)

- The equations: $F = kx$, $E = \frac{1}{2}kx^2$, $E_{total} = \frac{1}{2}mv^2 + \frac{1}{2}kx^2$

- Displacement–time, velocity–time, and acceleration–time graphs of simple harmonic oscillators, and their relative phases

- Kinetic and potential energy changes in SHM

- Practical task – measuring the period/frequency of simple harmonic oscillations and relating this to variables such as mass and length

Resonance

- The terms: free and forced oscillations, resonance, damping
- Variation of amplitude of a resonator with driving frequency

- Practical task – qualitative observations of forced and damped oscillations for a range of systems

Gravitational fields

Circular motion

- Motion in a horizontal circle and in a circular gravitational orbit
- Use the equations: $a = \dfrac{v^2}{r}$ and $F = \dfrac{mv^2}{r} = mr\omega^2$ with angular velocity ω in rad s^{-1}

Newton's law of gravitation

- The equations for the gravitational field of a point mass: $F_{grav} = -\dfrac{GmM}{r^2}$ and $g = \dfrac{F_{grav}}{m} = -\dfrac{GM}{r^2}$
- The terms: force, gravitational field

Gravitational potential

- Changes in gravitational potential energy and kinetic energy for uniform fields

- Gravitational potential energy change in a uniform field = $mg\Delta h$
- Motion in a uniform gravitational field
- Understand the meaning of the terms gravitational potential, equipotential surface
- The equation work done $\Delta E = F\Delta s$; no work is done whilst moving along equipotentials
- Sketching and interpreting graphs of: gravitational potential against distance from a point mass; gravitational field strength against distance from a point mass
- Drawing and interpreting diagrams of gravitational fields and corresponding equipotential surfaces
- Equations for the gravitational potential of a point mass: $E_{grav} = -\dfrac{GmM}{r}$ and $V_{grav} = \dfrac{E_{grav}}{m} = -\dfrac{GM}{r}$

Our place in the Universe

Measuring the Solar System

- Radar-type measurements to determine distances and relative velocities within the Solar System; measuring and defining distance in units of time, assuming the relativistic principle of the invariance of the speed of light
- Logarithmic scales of magnitudes of quantities: distance, size, mass, energy, power, and brightness

Measuring the Universe

- Evidence of a 'hot Big Bang' origin of the Universe from cosmological red-shifts and cosmological microwave background
- Hubble's Law to calculate distances and ages of astronomical objects

Special relativity

- Distance in units of time, assuming the relativistic principle of the invariance of the speed of light
- The effect of relativistic time dilation using the relativistic factor $\gamma = \dfrac{1}{\sqrt{1 - \dfrac{v^2}{c^2}}}$

Physics in perspective

The hunt for dark matter

The Big Bang theory has a lot of observational support, including the recession of the galaxies and the cosmic microwave background, but there are still some questions that have yet to be resolved. One question concerns the nature of 'dark matter'.

Dark matter was discovered when astronomers began using Doppler-shift measurements to investigate the rotation of galaxies.

Doppler shift from galaxies

The image from the Hubble space telescope in Figure 1 shows Doppler-shifts across a slice of the galaxy M84. Red-shifted light to the right of the image shows material that is moving away from the Earth. Blue-shifted light on the left is emitted from material that is approaching the Earth.

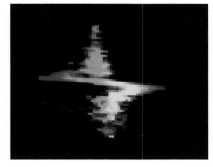

▲ Figure 1 *Red and blue shift from a rotating galactic centre*

Vera Rubin and her colleague Kent Ford measured Doppler shifts of light from the Andromeda galaxy in the late 1960s and early 1970s with incredible results.

Knowing the radius and speed of a star's orbit round a galaxy provides information about the mass of that galaxy. You can find out not only what the mass is but also how it is distributed in the galaxy. For speed v and radius r, the acceleration towards the centre $\frac{v^2}{r}$ measures the gravitational field $\frac{GM}{r^2}$. So the mass $M = \frac{v^2 r}{G}$.

This argument depends on the fact that the gravitational pull anywhere outside a sphere of matter is the same as if the mass were concentrated at the centre, as long as the distribution is spherically symmetrical. Also, inside such a spherical shell of matter there is no resultant gravitational pull from the shell. Most of the visible matter is clustered around the centre – this is where most of the visible light comes from, as you can see from the image of the galaxy NGC2 043 in Figure 2.

▲ Figure 2 *Spiral galaxy NGC 2043 showing visible matter concentrated in the centre of the galaxy*

This concentration of visible matter led Rubin and Ford to expect that matter further from the galactic centre would move more slowly, just as distant planets orbit the Sun at slower speeds than the nearer planets. The observations showed something completely different – the speed of the stars in orbit around the galaxy stayed more or less constant, as shown by the data from the galaxy NGC 2403.

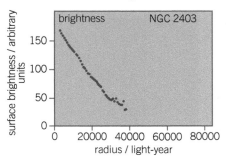

▲ Figure 3 *Brightness and orbital speed data for galaxy NGC 2403*

The left-hand chart shows how the brightness varies with distance from the centre. The brightness drops off quickly away from the galactic centre. This would lead you to expect that the orbital speed would also decrease with distance from the centre.

However, observations showed otherwise — the right-hand chart shows how the orbital speed varies. You can see that the speed reaches a maximum at about a 20 000 light-year radius and then remains constant, an unexpected result which can only be explained by assuming that the mass $M = \dfrac{v^2 r}{G}$ inside radius r must increase in proportion to r.

Although visible stars thin out away from the centre of the galaxy, the unseen mass goes on increasing. Dark matter had been discovered – the NGC 2403 data and similar evidence show that galaxies have haloes of unseen matter filling their outer reaches.

Gravitational lensing

Further evidence for dark matter comes from the phenomenon of gravitational lensing. Light is deflected as it passes a massive object, and the greater the mass the greater the deflection. The image in Figure 4 shows a bright yellow cluster of galaxies. The light from galaxies behind this cluster is smeared out by the gravitational lensing of the cluster — this causes the blue smears you can see in the image. The amount of lensing suggests that the mass distorting the light is far greater than the observed cluster of galaxies. The cluster contains a vast quantity of dark matter. In fact, only about 5% of the Universe is visible matter, with dark matter accounting for about 27% and the rest consisting of 'dark energy'.

▲ **Figure 4** *Light from distant galaxies showing the effects of gravitational lensing*

What is dark matter?

The amount of visible matter in the Universe is in line with theoretical estimates of the amount of hydrogen and helium 'cooked up' in the Big Bang. In other words, dark matter is not just undetected ordinary matter. At the time of writing there are a number of possibilities being considered. These include 'WIMPs' (weakly interacting massive particles) and MACHOs (massive compact halo objects, such as brown dwarf stars or black holes), though it is not certain that MACHOs could account for the amount of dark matter present in the Universe.

Experiments at CERN beginning in 2015 could increase our understanding of WIMPs and may provide evidence for their existence. The nature of dark matter is certainly still an open question, but may be at least partially resolved in the next decade.

Summary questions

Non-relativistic Doppler shift gives the fractional change in wavelength as $\frac{\Delta\lambda}{\lambda} = \frac{v}{c}$, where c is the velocity of light and v is the speed of the source relative to the observer.

1 21 cm radio waves from a hydrogen gas cloud in the Milky Way are observed to be shifted in wavelength by 0.1 mm. Calculate the speed of the cloud relative to the observer. Explain why you cannot say whether the cloud is approaching us or receding from us with the information given.

2 Part of the arm of a galaxy rotates around the galactic centre at a speed of 1000 km s^{-1}, at a distance of 2×10^{20} m.
   a Show that the acceleration towards the centre is about 5×10^{-11} m s^{-2}.
   b Show that the gravitational field strength at this radius is about 5×10^{-11} N kg^{-1}.
   c If the mass of the galaxy inside this radius acts as if it were at the centre, calculate how many solar masses the central region contains. (mass of Sun = 2×10^{30} kg)

3 It has been suggested that the dark matter haloes around galaxies behave like a gas of randomly moving particles. This question explores how such a model may account for the flat rotation curve shown in Figure 3.

Consider a star orbiting at distance r from the galactic centre.

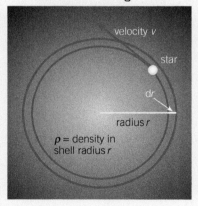

▲ Figure 5

The mass of the shell above the star = $4\pi^2 \rho dr$.

If dark matter behaves like a gas, its density will vary with $\frac{1}{r^2}$.
   a Explain why this density variation means that the mass of each shell of width dr will be the same, whatever the value of r.
   b Explain why the mass inside radius r will increase uniformly with r.
   c By considering the centripetal acceleration of the star, show that the orbital velocity of the star is given by $v = \sqrt{\frac{GM}{r}}$. Explain why a galaxy whose mass increases uniformly with r would give a flat rotation curve, that is, that the orbital velocity of its components is independent of their radius from the centre.

Practice questions

1   Here are some observations about the Universe.

   A   The light from some nearby galaxies has been blue-shifted.

   B   Most radiant matter is found in galaxies.

   C   There is evidence from X-rays that galaxies can have black holes at their centres.

   D   The red-shift of light from most galaxies is proportional to the distance to that galaxy.

   Which statement provides evidence for the 'Big Bang' theory of the formation of the Universe?   *(1 mark)*

2   The star Alpha Centauri is one of the nearest to Earth. It is observed to shift in direction by 4.2×10^{-4} degrees when observed from opposite sides of the Earth's orbit (diameter 3×10^{11} m). Calculate the distance to Alpha Centauri in metres and light-years. (1 year = 3.2×10^{7} s)   *(3 marks)*

3   A nearby galaxy has a small red-shift $z = \frac{\Delta\lambda}{\lambda} = \frac{v}{c} = 0.03$. Calculate the speed at which it is receding from the Earth.   *(1 mark)*

4   A space probe is travelling away from Earth at 30 km s^{-1}. Use the equation $\frac{\Delta\lambda}{\lambda} = \frac{v}{c}$ to show that the Doppler shift of the wavelength of the 1 GHz signals sent from the probe is 0.03 mm.   *(2 marks)*

5   Calculate the relativistic factor γ when $\frac{v}{c} = 0.6$.   *(2 marks)*

6   The Hubble constant is about 70 km s^{-1} Mpc^{-1}. (1 Mpc $\approx 3 \times 10^{19}$ km.)

   a   Show that the Hubble constant has units equivalent to s^{-1}.   *(1 mark)*

   b   Show that this value corresponds to a timescale for the Universe of roughly 1.4×10^{10} years.   *(2 marks)*

7   Figure 1 shows the intensity trace of the spectrum of light from the quasar Q1215 + 333. The peak in the spectrum is the red-shifted ultraviolet line of atomic hydrogen, wavelength 121.6 nm.

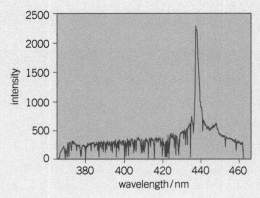

▲ Figure 1

   a   What is the red-shifted wavelength of the line?   *(1 mark)*

   b   Calculate the value of the red shift $z = \frac{\Delta\lambda}{\lambda}$.   *(2 marks)*

   c   By what factor has the Universe expanded since the light was emitted?   *(1 mark)*

   d   What are the photon energies $E = hf$ of the light when it was emitted and when it was received?   *(2 marks)*

   f   What is the ratio of these two energies? How is this connected to the answer to part **(d)**?   *(2 marks)*

8   The Hubble constant is about 210 km s^{-1} per million light-years.

   a   Show that the expansion speed is of the order $\frac{1}{10}c$ for distances of a few hundred million light-years.   *(2 marks)*

   b   Show that, even at this speed, the error in using the non-relativistic Doppler equation $\frac{\Delta\lambda}{\lambda} = \frac{v}{c}$ to find the speed of recession is less than 1%.   *(2 marks)*

9   The half-life of a pion, an unstable particle, is 18 ns when measured at rest in the laboratory.

   a   Calculate the relativistic factor γ for pions moving through the laboratory at a speed of 2.7×10^{8} m s^{-1}.   *(2 marks)*

   b   Calculate the half-life of the pions in part **(a)** as measured by a stationary observer in the laboratory.   *(2 marks)*

10 Unstable particles called muons are formed when cosmic rays enter the atmosphere. The muons travel through the atmosphere at $0.98c$.

a   Show that muons take about 2.7×10^{-5} s to travel through 8.0 km of atmosphere.
(2 marks)

b   Show that, ignoring relativistic factors, about 0.0004% of the muons hitting the top of the atmosphere can be expected to travel 8 km without decaying. The half-life of a muon at rest is 1.5×10^{-6} s.
(3 marks)

c   In fact, the proportion of muons travelling 8 km without decaying is about 8.4%. Confirm that this shows that the observed half-life of the moving muons is about five times that of muons at rest.
(3 marks)

d   The theory of special relativity predicts that the 'wristwatch' time τ for a moving object is given by $\tau = \dfrac{t}{\gamma}$ where t is the time observed by a stationary observer. The relativistic factor γ is given by
$$\gamma = \frac{1}{\sqrt{1 - \dfrac{v^2}{c^2}}}.$$
Calculate γ for particles travelling at $0.98c$ and compare your answer to the value you calculated in part **(c)**.
(4 marks)

11 Figure 2 shows the worldline of a spacecraft which passes the Earth and then returns.

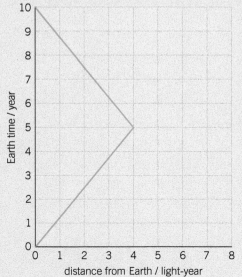

▲ Figure 2

Clocks on the Earth and the spacecraft are zeroed at the instant that the spacecraft passes the Earth.

a   The worldline for the spacecraft is a straight line until $t = 5$ years. What does this tell you about the motion of the spacecraft?
(1 mark)

b   A single pulse of light is sent towards the spacecraft from the Earth when the Earth clock reads $t = 1.0$ years. It reflects off the spacecraft and returns to Earth.

(i)   Why is the worldline for light always at 45° on figure 2?
(1 mark)

(ii) Draw the complete worldline of the pulse of light on a copy of Figure 2.
(2 marks)

c   The arrival of the pulse of light at the spacecraft is the signal for it to turn around and return to the Earth.

(i)   Explain how an observer on Earth can use the times of emission and reception of the pulse to calculate that the spacecraft was 4.0 light-years from the Earth when the pulse reached it.
(2 marks)

(ii) Explain how an observer on the Earth can use the time of emission and return of the pulse to deduce that the spacecraft turned round when the Earth clock read $t = 5.0$ years.
(2 marks)

(iii) Show that the outward speed of the spacecraft relative to the Earth is 2.4×10^8 m s^{-1}.
$c = 3.0 \times 10^8$ m s^{-1}
(1 mark)

d   (i)   Show that the time dilation factor γ for a spacecraft traveling relative to the Earth at velocity $v = 2.4 \times 10^8$ m s^{-1} is about 1.7.
(1 mark)

(ii) Here are some possible times in years for the round trip according to observers on the spacecraft. Write down the correct value.

6.0   8.0   10   17
(1 mark)

OCR Physics B G494 January 2012

MODULE 5.2

Matter

Introduction

The Enlightenment was a period roughly corresponding to Newton's life in the 17th and 18th centuries, in which cultural and intellectual developments in Europe emphasised reason and analysis over tradition and authority. The Royal Society in Britain and the *Académie des Sciences* in France were both founded in the 1660s, and their meetings encouraged development through scientific method and rigorous analysis.

Through the work of scientists such as Priestley, Lavoisier, Dalton, Boyle, Hooke, Newton, Lagrange, and Laplace, huge strides were made in the understanding of the nature of the Universe, on scales both large and small. Newton stated that if he had seen further than others, it was because he stood on the shoulders of giants.

Before the Enlightenment many people thought that all matter consisted of mixtures of four 'elements': earth, air, fire, and water. Carefully observed and measured chemical changes proved that the ancient model was incorrect, which led Dalton to his theory that all elements – using the term in their modern sense – consisted of particles that he called atoms. The ancient Greek philosophers Leucippus and Democritus had suggested that matter consisted of tiny particles 2000 years earlier, but they did not have a clear physical model in mind.

Even so, 150 years after Dalton's death, distinguished Austrian physicist Ernst Mach denied that their existence was proven. Maxwell and Boltzmann went on to show that the behaviour of gases and the nature of temperature could be explained using a simple atomic model combined with Newtonian mechanics.

In **Simple models of matter** you will consider the kinetic model of an ideal gas and learn how to use a mathematical treatment of the movement of gas particles to model and predict gas behaviour. The chapter also examines absolute temperature and internal energy, and the way in which internal energy changes are related to temperature change.

In **The Boltzmann factor** you will relate the energy of particles and the activation energy required for processes to absolute temperature. You will examine how the distribution of particles with different energies depends on temperature, and relate this to the probability of processes occurring, using The Boltzmann factor.

Knowledge and understanding checklist

From your Key Stage 4 or first-year A Level study you should be able to do the following. Work through each point, using your Key Stage 4/first-year A Level notes and the support available on Kerboodle.

☐ Recall that all matter consists of particles: atoms, ions, and molecules.

☐ Calculate power as the rate of energy transfer or work done.

☐ Calculate the pressure of a gas as the force it exerts per unit area on any surface.

☐ Calculate impulse as momentum change and understand its relationship to force.

☐ Recall that temperature is an indicator of particle energy, with higher temperature corresponding to greater particle energy.

☐ Recall that energy levels in atoms define quantum states between which transitions can occur.

Maths skills checklist

Maths is a vitally important aspect of Physics. In this unit, you will need to use the following maths skills. You can find support for these skills on Kerboodle and through MyMaths.

☐ **Using calculators** to solve exponential functions involving the Boltzmann constant.

☐ **Solving algebraic equations** when dealing with the gas laws and the ideal gas equation.

☐ **Interpreting graphs** that illustrate the gas laws and the variation of the Boltzmann factor with energy and temperature.

☐ **Using standard form** and in relating order of magnitude calculations of particle energy to temperature.

MyMaths.co.uk
Bringing Maths Alive

14 SIMPLE MODELS OF MATTER
14.1 The gas laws

Specification reference: 5.2.1a(ii), 5.2.1a(iv), 5.2.1b(i), 5.2.1b(ii), 5.2.1c(ii), 5.2.1d(ii), 5.2.1d(iii)

▲ Figure 1 Air pressure supports a car

Gas pressure

When you pump up car tyres, their volume does not change significantly, but the pressure of the air inside them rises until the rubber walls of the tyres are taut enough to support the weight of the car (Figure 1).

Describing and then explaining the behaviour of gases led to a deeper understanding of the nature of matter, energy, and temperature.

Boyle's law

In 1660, Robert Boyle investigated what he called 'the spring of the air', that is, how hard air pushes back when it is compressed.

 Practical 5.2.1d(iii): Pressure–volume relationship for a gas

Figure 2 is a cross-sectional view of apparatus used to investigate the dependence of gas volume on the pressure that the gas is under. Other arrangements are possible, but all must have a trapped volume of gas whose pressure can be changed and whose volume measured.

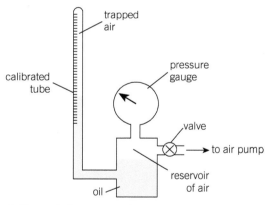

▲ Figure 2 Pressure–volume apparatus

In this apparatus, increasing the pressure of the air in the reservoir forces oil up into the calibrated tube, whilst decreasing the pressure in the reservoir allows some oil out of the calibrated tube. Here, a Bourdon gauge is used to measure the pressure of the air in the reservoir, but other methods, such as a pressure sensor and data logger, could be used. Provided that the temperature of the trapped air is constant and that it can be assumed that the air pressure in the calibrated tube is the same as that in the reservoir, the variation in volume with pressure can be investigated.

In analysing the results it is important to take account of the uncertainty in the measurements — even though the calibrated tube is made from toughened glass, safety considerations should include possible fracture.

In his experiments, Boyle showed that, provided that the temperature and the amount of gas present both remain constant, pressure p and volume V obey a simple relationship — p is inversely proportional to V, that is, pV = constant. This relationship, known as **Boyle's law**, is illustrated in Figure 3.

If the volume of the gas is kept constant but the amount of gas in that volume is increased, then the pressure will increase, so pressure ∝ mass of gas. For example, if the mass of gas in a volume V is doubled, the effect is the same as squeezing the original amount of gas into a volume $\frac{V}{2}$, with the extra gas occupying the other half.

As pV ∝ mass of gas, then $p \propto \frac{mass}{V}$ or p ∝ density ρ.

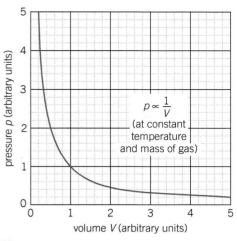
▲ **Figure 3** *Boyle's law*

 Worked example: Deflating a tyre

The pressure in a car tyre is 2.6 bar, where 1 bar is normal atmospheric pressure. Calculate the volume of gas that escapes if the tyre is punctured, assuming the tyre holds its shape. Assume the volume of the tyre is constant at 6.0 litres throughout, and that the temperature of the air does not change.

Step 1: Use pV = constant to calculate the value of the constant. As long as you are consistent you can use litres and bars – no need to convert to m³ and Pa.

constant = 2.6 bar × 6.0 litres = 15.6 bar litre

Step 2: Air will escape until the pressure within the tyre is equal to the external atmospheric pressure. Calculate the total volume of the air in the tyre after it has expanded to a pressure of 1 bar.

pV = 15.6 bar litre

∴ $V = \frac{15.6\,bar\,litre}{p} = \frac{15.6\,bar\,litre}{1\,bar} = 15.6\,litre$

Step 3: Remember that 6.0 litre of air remains in the deflated tyre.

Volume escaping = 15.6 litre − 6.0 litre = 9.6 litre (2 s.f.)

Hint: p and ρ

When writing pressure (p) and density (rho, ρ) make sure they look different.

Particles and moles

Boyle investigated air, but it was soon clear that many other gases identified soon afterwards, including hydrogen and nitrogen, behave in the same way. In 1738 Daniel Bernoulli suggested the reason for this — each gas consists of many very small, moving molecules that exert pressure by rebounding off the walls of their container.

In Figure 4, the weighted piston **EF** is supported by molecules of gas colliding on its underside. The upward force depends on the number of particles colliding, and on their mass and their velocities, as you will see in Topic 14.2, The kinetic model of gases. The model explains Boyle's law — if the volume of the container is halved, the molecules

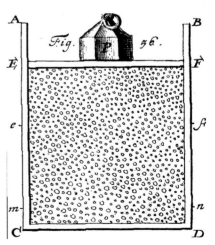
▲ **Figure 4** *Bernoulli's model of gas pressure*

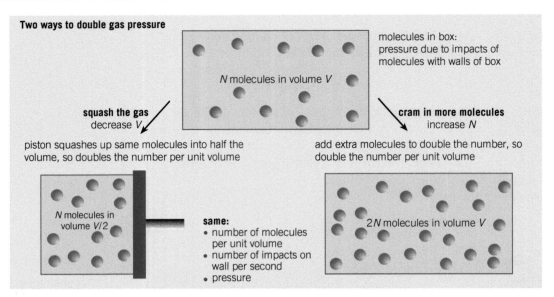

Two ways to double gas pressure

molecules in box:
pressure due to impacts of
molecules with walls of box

N molecules in volume V

squash the gas
decrease V

cram in more molecules
increase N

piston squashes up same molecules into half the
volume, so doubles the number per unit volume

add extra molecules to double the number, so
double the number per unit volume

N molecules in
volume V/2

same:
• number of molecules
  per unit volume
• number of impacts on
  wall per second
• pressure

2N molecules in volume V

▲ Figure 5 *Boyle's law and the number of molecules*

will strike the walls and the piston twice as often, doubling the pressure. The same effect could be produced by doubling the number of molecules in the original volume.

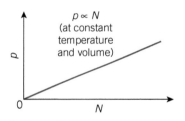

$p \propto N$
(at constant
temperature
and volume)

▲ Figure 6 *The amount law*

If the volume is kept constant and the amount of gas – the number N of molecules – is varied, then the pressure p is directly proportional to N (Figure 6).

Because the number of molecules is a key factor in understanding the behaviour of gases, it is necessary to work with the **amount** of gas in moles instead of the mass of gas in kilograms. A **mole** (symbol mol) of any substance is the amount of substance that contains the same number of particles as there are carbon atoms in 12.0 g of carbon-12. The mass of 1 mol of any substance is called its **molar mass** – the molar mass of helium atoms is $4\,\text{g mol}^{-1}$, that of oxygen atoms is $16\,\text{g mol}^{-1}$, and that of oxygen (O_2) molecules is $32\,\text{g mol}^{-1}$.

The number of particles in a mole is known as the **Avogadro constant**, N_A. The number N of molecules, or ions, or atoms, in a sample of a substance is related to the number of moles n of the substance by

$$N = n \times N_A$$

Hint

The Avogadro constant is a number, rather like 'trillion', but considerably larger: 6.02×10^{23} particles mol^{-1}. The current estimate of the total number of stars in the Universe is about 10^{23} – roughly a sixth of a mole.

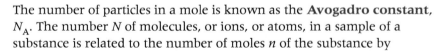 Worked example: Molar mass and the density of air

Calculate the number of molecules of air in a volume of $1.0\,\text{m}^3$. The density of air ρ_{air} is $1.2\,\text{kg m}^{-3}$, and the mean molar mass of air molecules is $29\,\text{g mol}^{-1}$.

Step 1: Calculate the number of moles in a cubic metre of air.

$$n = \frac{\text{mass of sample}}{\text{mass of 1 mol}} = \frac{1.2\,\text{kg}}{29 \times 10^{-3}\,\text{kg}} = 41.3\ldots\,\text{mol}$$

Step 2: Use N_A to calculate the number of molecules N.

$$N = n \times N_A = 41.3...\,\text{mol} \times 6.02 \times 10^{23}\,\text{mol}^{-1} = 2.5 \times 10^{25}$$
$$(2\ \text{s.f.})$$

This is about 250 × the number of stars in the Universe, so $0.004\,\text{m}^3$ of air, which is 4 litres – a small bucketful – contains as many air molecules as the number of stars in the Universe. Molecules are clearly very, very small.

Ideal gases and real gases

When real gases are greatly compressed, Boyle's law no longer holds. This is because the gas molecules interact with each other, exerting attractive van der Waals forces and also reducing the space into which other molecules can move. An **ideal gas** is one in which the molecules do not interact at all, and are so tiny that they occupy negligible volume. All gases behave as ideal gases if the molecules are well separated, but only very small molecules such as helium and hydrogen still behave as ideal gases if the intermolecular separation drops to the sort of spacing indicated in Figure 4.

Pressure, volume, and temperature

Boyle's law applies only if the temperature of the gas is kept constant. The French scientist Jacques Alexandre César Charles investigated how the volume of a gas kept at constant pressure changes when it is heated. He discovered that the volume increases in a linear way with increase in temperature (Figure 7 – note that the symbol θ is used for temperature in °C).

These results were later verified by the chemist Joseph Gay-Lussac, who showed that different gases behave in exactly the same way: for every degree rise in temperature, the increase in volume is $3.66 \times 10^{-3} \times$ the volume of the sample at 0 °C. This relationship is **Charles' law**.

If the volume V of the gas is kept constant and the gas is heated, the increase in pressure p follows the same linear form. This relationship is known as the **pressure law**.

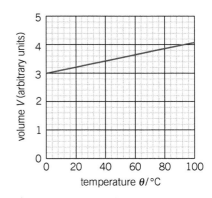

▲ **Figure 7** *Charles' law*

Absolute zero

Charles' law and the pressure law have an important consequence. If the linear graph of Figure 7 is extrapolated backwards, at some temperature the gas will have zero volume or, for the pressure law, zero pressure. This implies that the molecules will have ceased moving altogether. Cooling below this temperature is clearly not possible, so there must be an absolute zero of temperature.

 Worked example: Calculating absolute zero in °C

When any gas is heated at constant pressure, it increases in volume by $3.66 \times 10^{-3}\,V_{0\,°C}$ per degree rise in temperature. Calculate absolute zero in °C.

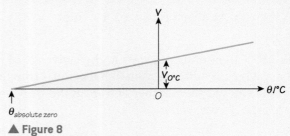

Step 1: Sketch a labelled graph, extrapolated back to absolute zero.

▲ **Figure 8**

Step 2: Write the gradient of the graph in terms of $V_{0°C}$ and $\theta_{\text{absolute zero}}$.

The gradient is the volume increase per degree — it is $3.66 \times 10^{-3} V_{0°C} \,°C^{-1}$.

From the shaded triangle,

$$\text{gradient} = \frac{V_{0°C}}{(0°C - \theta_{\text{absolute zero}})} = 3.66 \times 10^{-3} V_{0°C} \,°C^{-1}$$

Step 3: Solve to find $\theta_{\text{absolute zero}}$.

$$(0°C - \theta_{\text{absolute zero}} \,°C) = \frac{V_{0°C}}{(3.66 \times 10^{-3} V_{0°C} \,°C^{-1})}$$

$$(0°C - \theta_{\text{absolute zero}} \,°C) = \frac{1}{(3.66 \times 10^{-3} \,°C^{-1})} = 273.2...\,°C$$

$$\theta_{\text{absolute zero}} = -273°C \text{ (to 3 s.f.)}$$

The kelvin scale of temperature

In 1848, physicist William Thomson suggested that temperature should be measured not from the melting point of water, but from the absolute zero of temperature. His temperature scale is named after the title he was later given – Baron Kelvin of Largs – and keeps the size of a degree the same as in the Celsius scale by simply adding 273 to the Celsius temperature. There are therefore still 100 degrees between the melting and boiling points of water, but those values are now 273 and 373 kelvin, symbol K (*not* °K). Kelvin (also known as absolute or thermodynamic) temperature has the symbol T, so $T/K = \theta/°C + 273$.

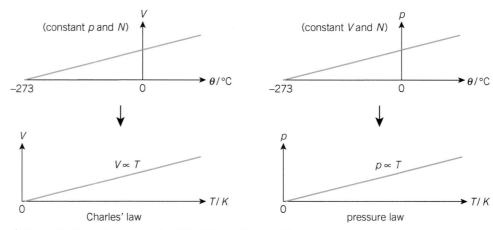

▲ **Figure 9** *The kelvin scale simplifies Charles' law and the pressure law*

The kelvin scale makes the mathematical description of the gas laws very straightforward. Charles' law becomes $V \propto T$ at constant pressure, whilst the pressure law is $p \propto T$ at constant volume.

Practical 5.2.1d(ii): Determining absolute zero

Figure 10 shows a glass capillary tube with a sample of air trapped between the sealed end and a liquid index. The air-filled space in the tube is cylindrical in shape. Measuring the length L of this air column at different temperatures will indicate how the volume of the enclosed gas is changing. The tube can be mounted in a tall glass beaker of water whose temperature can be measured as it is heated over a reasonable range. Care must be taken to ensure that the recorded temperature of the water is close to the temperature of the air inside the capillary tube. Data obtained can be analysed to determine the value of absolute zero.

The ideal gas law

Using the kelvin scale of temperature, the gas laws $pV = $ constant, $p \propto N$, $V \propto T$, and $p \propto T$ can be combined into a single equation,

$$pV \propto NT$$

The constant of proportionality in this equation is the **Boltzmann constant**, k, so $pV = NkT$.

For one mole of gas, $N = N_A$, so this equation has the form $pV = N_A kT = RT$ where the constant R is called the **gas constant** and is given by $R = N_A k$.

Table 1 *The Boltzmann constant and the gas constant in the ideal gas equation*

	For 1 mol	For n mol
Using the gas constant	$pV = RT$	$pV = nRT$
Using the Boltzmann constant	$pV = N_A kT$	$pV = NkT$

Experiments show that the gas constant $R = 8.31\,\text{J mol}^{-1}\text{K}^{-1}$. This allows us to use the Avogadro constant $N_A = 6.02 \times 10^{23}\,\text{mol}^{-1}$ to calculate the value of the Boltzmann constant k.

$$R = N_A k \therefore k = \frac{R}{N_A} = \frac{8.31\,\text{J mol}^{-1}\text{K}^{-1}}{6.02 \times 10^{23}\,\text{mol}^{-1}} = 1.38 \times 10^{-23}\,\text{J K}^{-1}$$

Worked example: Calculation of gas pressures

A helium balloon of volume $0.0050\,\text{m}^3$ contains $0.96\,\text{g}$ of helium at a temperature of 20 °C. Calculate the pressure in the balloon. The molar mass of helium is $4.0\,\text{g mol}^{-1}$.

Step 1: Calculate the number of moles in the balloon.

$$n = \frac{\text{mass of sample}}{\text{mass of 1 mol}} = \frac{0.96\,\text{g}}{4.0\,\text{g mol}^{-1}} = 0.24\,\text{mol}$$

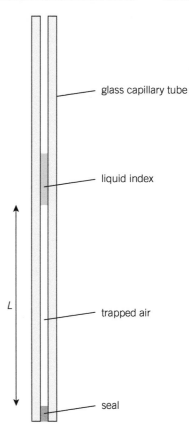

▲ **Figure 10** *Charles' law tube apparatus*
- glass capillary tube
- liquid index
- trapped air
- seal

Synoptic link

In Topic 15.2, The Boltzmann factor, you will see that k relates temperature, in K, to the energy of a particle.

Step 2: Convert the temperature from °C to K.

$$T = \theta + 273 = 20 + 273 = 293\,K$$

Step 3: Use the ideal gas law for n mol.

$pV = nRT$ (you could use $N = nN_A$ and then $pV = NkT$, but $pV = nRT$ is quicker here)

$$p = \frac{nRT}{V} = \frac{0.24\,mol \times 8.31\,J\,mol^{-1}\,K^{-1} \times 293\,K}{0.0050\,m^3}$$
$$= 120\,000\,Pa\ (2\ s.f.)$$

Real gases at low temperatures

Gases like oxygen approximate very well to an ideal gas at room temperature and atmospheric pressure, but at very low temperatures the non-ideal nature of the gas molecules becomes evident. As the oxygen cools and contracts, the molecules become close enough to interact, and at 90 K (−183 °C) the gas condenses into a liquid (Figure 11).

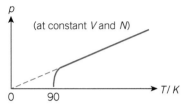

▲ **Figure 11** *The behaviour of oxygen gas becomes non-ideal as it cools to near its boiling point*

Summary questions

1 The largest airship ever built was lifted by 17 000 kg of hydrogen gas. Calculate the number of (a) moles, (b) molecules of hydrogen it contained. Molar mass of hydrogen molecules = 2.0 g mol⁻¹. (*4 marks*)

2 Pressure under water increases with the depth below the surface, increasing by 10 kPa for each 1.0 m of depth. A bubble of volume $9.0 \times 10^{-7}\,m^3$ is released by a diver at a depth of 25 m below the surface. Assuming that its temperature does not change, calculate the volume of the bubble when it reaches the surface, where atmospheric pressure is 100 kPa. (*4 marks*)

3 Air consists of 78% nitrogen, 21% oxygen, and 1% argon by mass (Table 2).
   a Calculate the pressure of 12 kg of air contained in a volume of 11 m³ at a temperature of 25 °C. (*2 marks*)

▼ **Table 2**

Molecule	nitrogen	oxygen	argon
Molar mass /g mol⁻¹	28	32	40

   b Assuming that the total mass of air does not change, calculate the pressure if the volume is compressed to 5.5 m³ and the temperature raised to 56 °C. (*3 marks*)

14.2 The kinetic model of gases

Specification reference: 5.2.1a(ii), 5.2.1a(iv), 5.2.1a(v), 5.2.1a(vi), 5.2.1b(i), 5.2.1c(iii)

Molecular modelling

In the 19th century, many scientists treated the idea of atoms and molecules as an interesting but unproved speculation. James Clerk Maxwell and Ludwig Boltzmann (Figure 1) developed Daniel Bernoulli's explanation of pressure (see Topic 14.1, The gas laws, Figure 4), using the ideas of Newtonian mechanics – the 'clockwork Universe' – to derive a theoretical explanation of the ideal gas equation. Their success is a superb example of how theory works in physics.

▲ **Figure 1** *James Clerk Maxwell (left) and Ludwig Boltzmann*

Assumptions made in the analysis of the kinetic model of an ideal gas

In Topic 14.1, The gas laws, it was stated that:

- an ideal gas comprises molecules that do not interact at all
- the volume occupied by the molecules within the gas is negligible.

In this analysis we add the assumption that:

- all collisions are perfectly elastic, so that no energy is lost in collision.

This last requirement applies not only to collisions between molecules of the gas and its container, but also to collisions between molecules (although, if their volume is negligible, the frequency of intermolecular collisions will be negligible too).

The simplest model: one molecule in a box

Imagine a single molecule of mass m in a box of length x, height y, and depth z moving at a constant speed c in the x-direction (Figure 2). c is used for velocity, rather than v, in the analysis of ideal gases.

Assuming that all collisions are elastic, the molecule will bounce to and fro in the x-direction, with its velocity changing from $+c$ to $-c$ as it rebounds off the right-hand end (Figure 3). Each collision with the right-hand wall of the box exerts an impulse on that wall, equal to the change in momentum, of $+2mc$. In a similar way, each collision with the left-hand wall of the box exerts an impulse on that wall of $-mc - (+mc) = -2mc$. The signs of the two impulses show that the

Learning outcomes

Describe, explain, and apply:

→ temperature as proportional to average energy per particle; average energy $= \dfrac{3}{2}kT \approx kT$ as a useful approximation

→ the random walk of molecules in a gas — distance moved in N steps related to $\sqrt{N}$

→ calculations and estimates of $pV = \dfrac{1}{3}Nm\overline{c^2}$.

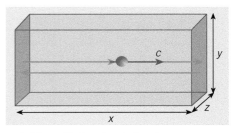

▲ **Figure 2** *A molecule in a box*

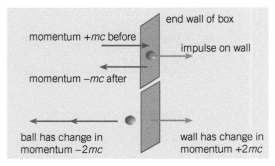

▲ **Figure 3** *Momentum change in one collision*

Synoptic link

You can review impulse in Topic 9.2, Newton's laws of motion and momentum.

forces acting on those walls are outwards – the right-hand wall is pushed in the positive x-direction, and the left-hand wall is pushed in the negative x-direction.

Impulse = $F\Delta t$, so the average force on the right-hand wall is given by $F = \dfrac{\text{impulse}}{\Delta t} = \dfrac{2mc}{\Delta t}$, where Δt is the time between one collision on that wall and the next. The time between two consecutive collisions on the right-hand wall is the time taken for the molecule to travel along the box of length x and back, a distance $2x$, at the steady speed c.

$$c = \frac{2x}{\Delta t} \therefore \Delta t = \frac{2x}{c}$$

so the average force F caused by this single molecule is

$$F = \frac{2mc}{\Delta t} = \frac{2mc}{\left(\dfrac{2x}{c}\right)} = \frac{2mc \times c}{2x} = \frac{mc^2}{x}$$

A more realistic model: N molecules in a box

A typical volume of gas contains rather more than one molecule. If there are N molecules in the box (Figure 4), they will be moving randomly in all possible directions. We can assume that one-third will be moving in the x-direction, one-third will be moving in the y-direction, and one-third will be moving in the z-direction. A more rigorous approach, taking the vector components of velocities in all possible random directions and adding them, gives exactly the same result.

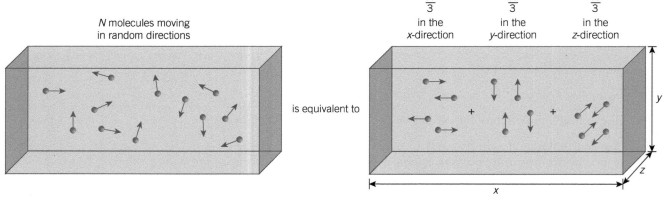

▲ **Figure 4** *Many molecules in the box*

With N molecules in the box, the average force on the right-hand end of the box will be

$$F = \frac{N}{3} \times \frac{mc^2}{x} = \frac{Nmc^2}{3x}$$

We need to know the pressure p on the right-hand end of the box. $p = \dfrac{F}{A}$, and the area A of the right-hand end = $y \times z$, so

$$p = \frac{F}{A} = \frac{Nmc^2}{3x} \div yz = \frac{Nmc^2}{3xyz}$$

As xyz is the volume of the rectangular box, we end up with

$$p = \frac{Nmc^2}{3V} \qquad \text{or} \qquad pV = \frac{1}{3}Nmc^2$$

The full ideal gas model: *N* molecules, with varying speeds, in a box

Molecules not only travel in random directions, they also have, within a range, randomly different speeds. This means that the ideal gas equation $pV = \frac{1}{3}Nmc^2$ should use an average. Note that this is not the average speed. The important measure of the force, even for a single particle, is c^2 and not c $\left(F = \frac{mc^2}{x}\right)$, so it is c^2 that must be averaged.

The mean of a quantity is shown by a bar across the top of the symbol — this means that the expression should contain the mean square $\overline{c^2}$, *not* the square of the mean $(\bar{c})^2$.

 Worked example: Mean square and square of the mean

Three molecules have speeds $200\,\mathrm{m\,s^{-1}}$, $230\,\mathrm{m\,s^{-1}}$, and $250\,\mathrm{m\,s^{-1}}$. Show that the mean square speed $\overline{c^2}$ is not the same as the mean speed squared $(\bar{c})^2$.

Step 1: Square the speeds and find their mean.

$(200\,\mathrm{m\,s^{-1}})^2 = 40\,000\,\mathrm{m^2\,s^{-2}}$, $(230\,\mathrm{m\,s^{-1}})^2 = 52\,900\,\mathrm{m^2\,s^{-2}}$, and $(250\,\mathrm{m\,s^{-1}})^2 = 62\,500\,\mathrm{m^2\,s^{-2}}$

$$\overline{c^2} = \frac{(40\,000\,\mathrm{m^2\,s^{-2}} + 52\,900\,\mathrm{m^2\,s^{-2}} + 62\,500\,\mathrm{m^2\,s^{-2}})}{3}$$

$$= 51\,800\,\mathrm{m^2\,s^{-2}}$$

Step 2: Find and square the mean speed.

$$\bar{c} = \frac{(200\,\mathrm{m\,s^{-1}} + 230\,\mathrm{m\,s^{-1}} + 250\,\mathrm{m\,s^{-1}})}{3} = 226.6...\,\mathrm{m\,s^{-1}}\ (2\ \mathrm{s.f.})$$

$(\bar{c})^2 = (226.6...\,\mathrm{m\,s^{-1}})^2 = 51\,400\,\mathrm{m^2\,s^{-2}}$ (3 s.f.), which is different from $51\,800\,\mathrm{m^2\,s^{-2}}$

This gives us the final form of the theoretical ideal gas equation,

$$pV = \frac{1}{3}Nm\overline{c^2}$$

r.m.s. speed

When quoting molecular speeds, the value you should use is the square root of the mean square speed, $\sqrt{\overline{c^2}}$, which is called the **root mean square speed**. This is usually abbreviated to r.m.s. speed and is sometimes written c_{rms}.

 Worked example: The r.m.s. speed of air molecules at room temperature

Calculate the r.m.s. speed of air molecules, of mean molar mass $29\,\mathrm{g\,mol^{-1}}$, at a temperature of $21\,^\circ\mathrm{C}$.

Step 1: Equate the theoretical ideal gas equation above with the ideal gas law from Topic 14.1, The gas laws.

$pV = \frac{1}{3}Nm\overline{c^2}$ and $pV = NkT$ so $\frac{1}{3}Nm\overline{c^2} = NkT$

Step 2: rearrange to make $\overline{c^2}$ the subject.

$$\frac{1}{3}Nm\overline{c^2} = NkT \therefore \overline{c^2} = \frac{3NkT}{Nm} = \frac{3kT}{m}$$

Step 3: Find absolute temperature T and mass m of 1 molecule.

$$T = 21 + 273\,\text{K} = 294\,\text{K}$$

$$m = \frac{\text{molar mass}}{\text{Avogadro constant } N_A} = \frac{29 \times 10^{-3}\,\text{kg mol}^{-1}}{6.02 \times 10^{23}\,\text{mol}^{-1}}$$

$$= 4.8... \times 10^{-26}\,\text{kg}$$

Step 4: Substitute in values to evaluate the mean square speed $\overline{c^2}$ and then the r.m.s. speed $\sqrt{\overline{c^2}}$.

$$\overline{c^2} = \frac{3kT}{m} = \frac{3 \times 1.38 \times 10^{-23}\,\text{J K}^{-1} \times 294\,\text{K}}{4.8... \times 10^{-26}\,\text{kg}} = 252\,665.6...\,\text{m}^2\text{s}^{-2}$$

$$\sqrt{\overline{c^2}} = 500\,\text{m s}^{-1}\ (\text{2 s.f.})$$

This is a high speed, but not unreasonable. The speed of sound in air is about $340\,\text{m s}^{-1}$, and sound cannot travel faster than the air molecules whose movement passes it on.

Real gases deviate from the ideal gas law

Figure 5 shows a molecule approaching the right-hand wall of the $x \times y \times z$ box.

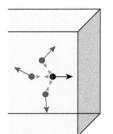

▲ **Figure 5** *Intermolecular forces in a real gas*

Molecules in a real gas can interact with each other. The attractive van der Waals forces between the 'subject' molecule, shown in black, and its nearest neighbours, shown in red, are represented by green arrows, all in pairs as predicted by Newton's third law. The resultant force on the subject molecule will be to the left as it nears the wall, as there are more neighbouring molecules behind it than in front of it. This resultant force will decelerate the subject molecule, so that it strikes the end wall with reduced momentum, which decreases the pressure experienced by the wall.

As van der Waals forces are very short-range forces, any sample of gas with well-separated molecules will display ideal gas behaviour.

Temperature and the energy of molecules

In the theoretical ideal gas equation you can see the term $m\overline{c^2}$, which looks very like the average kinetic energy of a molecule $\left(E_k = \frac{1}{2}mv^2\right)$. This could prompt us to rearrange the right-hand side of the equation as follows:

$$pV = \frac{1}{3}Nm\overline{c^2} = \frac{2}{3}N\left(\frac{1}{2}m\overline{c^2}\right)$$

and, as $pV=NkT$,

$$NkT = \frac{2}{3}N\left(\frac{1}{2}m\overline{c^2}\right) \text{ so}$$

$$\frac{1}{2}m\overline{c^2} = \frac{3}{2}kT$$

For a mixture of gases, like air, this means that the mean kinetic energy of all the molecules is the same at a given temperature.

At room temperature, the nitrogen molecules in air (molar mass $28\,\text{g}\,\text{mol}^{-1}$) will have a higher r.m.s. speed than the argon atoms (molar mass $40\,\text{g}\,\text{mol}^{-1}$) to satisfy $\left(\frac{1}{2}m\overline{c^2}\right) = \frac{3}{2}kT$.

Random walks and diffusion

Although it is assumed that the molecules in an ideal gas occupy negligible volume, they do have a finite size. There are so many molecules in a volume of gas, all whizzing around blindly in random directions, that they are bound to collide frequently. Figure 6 shows a short section of the path of the molecule shown in black.

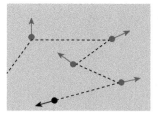

▲ **Figure 6** *Intermolecular collisions in a real gas*

On average, each air molecule travels about 100 nm between collisions, a distance called its mean free path. Under the same conditions of pressure and temperature, a gas such as chlorine, which has bigger molecules, would travel less far between collisions, whilst small helium atoms would travel further.

As each collision changes the direction of the molecules in a random way, the displacement of an individual molecule after many such collisions is hard to predict. Figure 7 shows the same two-dimensional random walk observed for different numbers of 'steps' between collisions. The axes are calibrated in terms of the number of mean free paths this one molecule has travelled. Even the longest path, consisting of 500 steps, would have taken less than a millisecond to complete.

If a perfume bottle is opened the smell will slowly diffuse in all directions. To analyse the movement of the perfume molecules requires statistics. Analysis shows that, if a molecule makes N steps in a random walk, on average it will reach a distance $\sqrt{N}$ steps from its starting point. The mean displacement of all the molecules will be zero, but some molecules will travel a great distance.

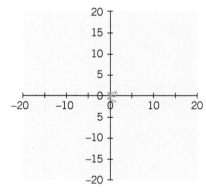

after 10 steps

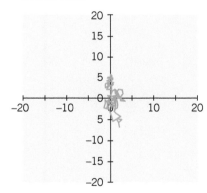

after 100 steps

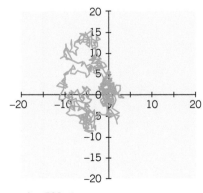

after 500 steps

▲ **Figure 7** *Molecular random walk*

 Worked example: Diffusion of ammonia

A bottle of concentrated ammonia solution is accidentally broken in a lab at 24 °C. Ammonia molecules have a molar mass of 18 g mol^{-1}. Assuming that ammonia molecules moving through air have a mean free path of 1.0×10^{-7} m, calculate the time taken from release for the average ammonia molecule to reach a point 2.0 m from the broken bottle.

Step 1: Calculate the mass m of one ammonia molecule and use it to calculate its r.m.s. speed.

$$m = \frac{\text{molar mass}}{\text{Avogadro constant } N_A} = \frac{18 \times 10^{-3}\,\text{kg}\,\text{mol}^{-1}}{6.02 \times 10^{23}\,\text{mol}^{-1}}$$

$$= 2.9\ldots \times 10^{-26}\,\text{kg}$$

$$T = 24 + 273\,\text{K} = 297\,\text{K}$$

$$\left(\frac{1}{2}m\overline{c^2}\right) = \frac{3}{2}kT \therefore \overline{c^2} = \frac{3kT}{m} = \frac{3kT}{m}$$

$$\overline{c^2} = \frac{3kT}{m} = \frac{3 \times 1.38 \times 10^{-23}\,\text{J}\,\text{K}^{-1} \times 297\,\text{K}}{2.9\ldots \times 10^{-26}\,\text{kg}} = 411\,226.2\,\text{m}^2\,\text{s}^{-2}$$

$$\sqrt{\overline{c^2}} = 641\ldots\,\text{m}\,\text{s}^{-1}$$

Step 2: Use the 'step length' (the mean free path) and the fact that $2.0\,\text{m} = \sqrt{N}$ steps to find the number of steps N.

$$\sqrt{N} \times 1.0 \times 10^{-7}\,\text{m} = 2.0\,\text{m}$$

$$\therefore \sqrt{N} = \frac{2.0\,\text{m}}{1.0 \times 10^{-7}\,\text{m}} = 2.0 \times 10^{7}$$

so $N = (2.0 \times 10^{7})^{2} = 4.0 \times 10^{14}$

Step 3: Calculate the actual total distance travelled by the molecule along its winding path and use the r.m.s. speed to calculate the time this would take.

total distance = N steps = $4.0 \times 10^{14} \times 1.0 \times 10^{-7}\,\text{m} = 4.0 \times 10^{7}\,\text{m}$

$$\sqrt{\overline{c^2}} = \frac{\text{total distance}}{\text{time } t}$$

$$\therefore t = \frac{\text{total distance}}{\sqrt{\overline{c^2}}} = \frac{4.0 \times 10^{7}\,\text{m}}{641\dots\,\text{m}\,\text{s}^{-1}} = 62\,000\,\text{s} \;(2\text{ s.f.})$$

The answer to this worked example, nearly a whole day, is not realistic. In practice you would smell the ammonia (an extremely pungent gas) within a minute. This is because the gas is not moving purely by diffusion. Air currents, from draughts or the sudden movements of the experimenter dropping the bottle, will make the ammonia spread rapidly. To measure the diffusion speed it would be necessary to conduct the experiment in a draught-free environment such as a sealed tube.

Summary questions

1   Five molecules have the speeds given in Table 1. Calculate their r.m.s. speed.   *(3 marks)*

▼ Table 1

Molecule	1	2	3	4	5
Speed /m s^{-1}	240	210	230	220	260

2   Calculate the kinetic energy of a gas molecule at a temperature of 127 °C. Calculate the r.m.s. speed of hydrogen molecules at this temperature (molar mass of hydrogen = 2.0 g mol^{-1}).   *(4 marks)*

3   Explain why a gas can diffuse over 5 cm quite rapidly, but takes much more than 100 times longer to diffuse 5.0 m.   *(2 marks)*

4   The molar mass of helium gas is about 7 times less than the mean molar mass of air. When a person sings a note, the wavelength λ of that note, produced by standing waves in the throat of the singer, is fixed. Use the wave equation $v = f\lambda$ to explain why a person who has breathed in helium gas has a squeaky, high-pitched voice for as long as the gas is in his or her throat.   *(4 marks)*

14.3 Energy in matter

Specification reference: 5.2.1a(i), 5.2.1b(i), 5.2.1c(ii), 5.2.1d(i)

Steam power

The 19th century progress in the understanding of ideal gas behaviour followed the practical development of steam engines that powered the Industrial Revolution. The wish to make steam engines do as much work as possible for the fuel burnt led to a fuller understanding of the nature of **work** and **energy**, the area of physics known as **thermodynamics**.

Heat and work

The Salford brewer and physicist James Joule (Figure 2) investigated the nature of work and heat, comparing the amount of work (force × distance) done by an electric battery consuming a pound of zinc with that from a steam engine consuming a pound of coal. Heat was measured by the rise in temperature it produced in water – one calorie of heat raised the temperature of 1 kg of water by 1 °C. According to the caloric theory accepted at the time, heat was a conserved quantity, but Joule showed that work alone could raise the temperature of water by a consistent amount, so that 1 calorie of heat was always produced by 4200 J of work.

Although it is no longer associated with the discredited caloric theory, the calorie is still popular in non-scientific contexts as a unit for the energy provided by food.

The first law of thermodynamics

The fact that heat and work can produce the same effect led to the concept of **internal energy**, U. It is not possible to state the actual value of the internal energy of a complex system like a steam engine, but *changes* in internal energy ΔU can be related directly to the work W done on the system and the heat Q that it receives. 'Heat' is an ambiguous term that could refer to the energy transferred or to the process of transferring it, so we will call Q the **energy transferred thermally** into the system.

The relationship between these quantities is called the **first law of thermodynamics**.

$$\Delta U = W + Q$$

Note that either or both of W and Q could be negative.

▲ **Figure 1** *Boulton and Watt's 1812 pumping engine at Crofton Pumping Station in Wiltshire is still used to move water into the Kennet and Avon canal*

▲ **Figure 2** *James Joule*

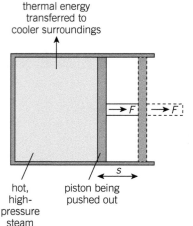

thermal energy
transferred to
cooler surroundings

hot,
high-
pressure
steam

piston being
pushed out

▲ **Figure 3** *Heat engine losing internal
energy thermally and by doing work*

Worked example: Internal energy change in a steam engine

Calculate the internal energy change in the steam engine shown in Figure 3 if 100 J is transferred thermally to the surroundings and a constant force of 400 N pushes the piston out a distance s of 10 cm.

Step 1: Identify whether Q and W are positive or negative.

Both Q and W correspond to energy leaving the steam engine, so both are negative.

Step 2: Calculate Q and W and therefore ΔU.

$$Q = -100\,\text{J and } W = -(Fs) = -400\,\text{N} \times 0.10\,\text{m} = -40\,\text{J}$$

$$\Delta U = W + Q = (-100\,\text{J}) + (-40\,\text{J}) = -140\,\text{J (2 s.f.)}$$

The energy to make things hotter

For most substances there is a simple link between the change in internal energy and the change in temperature. The change in temperature for a given transfer of energy depends on the mass of substance and the kind of substance being made hotter, so that

energy transferred = mass of substance × a constant for the substance × temperature rise.

If the volume of the substance does not change, so no work is done on or by the substance, then the energy transferred will be the change in internal energy ΔU.

$\Delta U = mc\Delta T$ or $\Delta U = mc\Delta\theta$, where c, a constant for the material, is called the **specific thermal capacity** and has the units $\text{J kg}^{-1}\text{K}^{-1}$. The term 'specific' in the name refers to a fixed mass of 1 kg.

Worked example: Boiling a kettle

An electric kettle rated at 2.2 kW contains 1.6 kg of water at 12 °C. Water has a specific thermal capacity of 4200 J kg^{-1} K^{-1}. Calculate the time taken to bring the water to its boiling point (100 °C).

Step 1: Identify the temperature change.

$$\Delta T = \Delta\theta = (100 - 12)\,\text{K} = 88\,\text{K}$$

Step 2: Substitute in values and calculate ΔU.

$$\Delta U = mc\Delta T = 1.6\,\text{kg} \times 4200\,\text{J kg}^{-1}\text{K}^{-1} \times 88\,\text{K} = 591\,400\,\text{J}$$

Step 3: Equate ΔU to the work done by the kettle in the time t.

$$\Delta U = Pt \therefore 591\,400\,\text{J} = 2.2 \times 10^3\,\text{W} \times t$$

$$t = \frac{591\,400\,\text{J}}{2.2 \times 10^3\,\text{W}} = 270\,\text{s (2 s.f.) (= 4.5 minutes)}$$

Note that the equation $\Delta U = mc\Delta T$ cannot be applied if the water is boiling (changing state from liquid to gas). In that situation the temperature does not change, but work is done in breaking intermolecular bonds.

Practical 5.2.1d(i): Determining specific thermal capacities

Figure 4 shows a 12 V electrical immersion heater heating a sample of water. It can readily be adapted for other liquids. For solids such as metals, sample cylinders, drilled with a hole of the same dimension as the heater together with a hole which can take a thermometer, may be used.

In the experiment, the electrical work W done by the heater can be determined by appropriate readings. Care must be taken to minimise Q, the energy transferred thermally into or out of the substance by other means, so that ΔU can be equated to W and so allow c to be calculated.

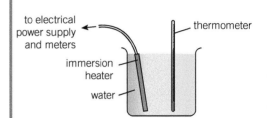

▲ **Figure 4** *Measuring the specific thermal capacity of water*

Extension: The anomalous nature of water

Water molecules are held together in the liquid state by strong hydrogen bonds between hydrogen atoms and oxygen atoms in adjacent molecules. As a consequence, water has a much higher boiling point than might be expected from the position of oxygen in the Periodic Table of the elements. The next element down in the same group is sulfur, which has more massive atoms than oxygen, so hydrogen sulfide, H_2S, might be expected to have a higher boiling point than water, H_2O. However, water boils at 373 K, whilst hydrogen sulfide boils at 213 K. In fact the intermolecular forces in hydrogen sulfide are not hydrogen bonds but only much weaker van der Waals forces.

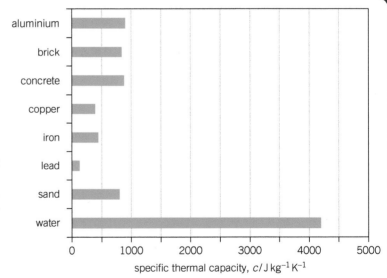

▲ **Figure 5** *Water has a very high specific thermal capacity*

This relatively strong bonding in water is also illustrated in Figure 5, which compares the value of the specific thermal capacity of water with that of a number of other common materials.

This anomalously high value of c has huge environmental significance. The immense mass of water in the seas acts as a huge reservoir of energy. Much of the uncertainty in current projections of the effects of global warming is related to the fact that ocean currents like the Gulf Stream transport vast amounts of energy, so changes in temperature are likely to disturb these flows, with significant feedback effects on temperatures worldwide.

To think about: The specific thermal capacities of metals – the increase in internal energy required to raise the temperature of 1 kg by 1 K – are all different, but their *molar* heat capacities – the increase in internal energy required to raise the temperature of 1 mol (6.02×10^{23} atoms) by 1 K – are very similar. What does this suggest about the bonds between atoms in metals?

Summary questions

1   A steam engine is heated (has thermal energy transferred into it) at a power of 200 W for 10 s. During this time, it does 12 kJ of work. Calculate the change in internal energy of the steam engine.   (*3 marks*)

2   Calculate the molar thermal capacity (the value of ΔU that produces a temperature change of 1 K in 1 mol of the substance) for copper, aluminium, and iron, using the data in Table 1.   (*4 marks*)

▼ Table 1

	Aluminium	Copper	Iron
c / J kg^{-1}K^{-1}	900	390	450
molar mass / kg mol^{-1}	0.027	0.064	0.056

3   The waterfall called Victoria Falls or Mosi-oa-Tunya is 108 m high. By considering the change in gravitational potential energy of 1 kg of water, calculate the temperature increase expected when the water is brought to a halt at the bottom of the waterfall ($g = 9.8$ N kg^{-1}).   (*4 marks*)

4   A swimming pool is 8.0 m wide and 25.0 m long. The bottom slopes uniformly from 1.2 m deep at the shallow end to 3.5 m deep at the deep end (Figure 6). Calculate the output power of the heater required to raise the temperature of the water in the pool from 11 °C to 23 °C over a time of 2 days, and explain why a higher power would be required in practice (density of water $\rho = 1000$ kg m^{-3}).   (*5 marks*)

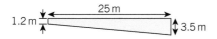

▲ **Figure 6** *Cross-section of a swimming pool*

5   One way to measure the specific thermal capacity of lead is to place a sample of lead consisting of many small spheres of lead shot into a long cardboard tube, having first measured their temperature. The tube is initially held vertically, then inverted a large number of times. Each time it is inverted, the lead falls to the other end of the tube. The tube is then unsealed, the lead poured out, and its temperature re-measured. In an experiment with 100 g of lead in a tube of length 80 cm, inverted 50 times, the temperature of the lead rose by 2.5 °C. The thermometer used had a resolution of ±0.1 °C. Calculate the specific thermal capacity of lead, describing the errors and uncertainties likely to affect the value obtained.   (*5 marks*)

Physics in perspective

Ballooning: the first space race

It was recognised in the 17th century that the atmosphere is like an ocean of air. Things float in an ocean of water, so why not in the air too? Smoke rises, so why not fill a large, light envelope with smoke and let it carry the envelope upwards with it?

The Montgolfier brothers

Brothers Joseph-Michel and Jacques-Etienne Montgolfier first made successful use of this idea in June 1783. A large sphere of cloth and paper was placed over a very smoky fire and held down by eight strong men until it filled, lifted off into the summer sky and was blown away.

The second major Montgolfier balloon flight took up a trio of animal passengers: a sheep, a duck, and a cockerel. On landing they were pronounced fit to eat – not the ideal reward for being the first aeronauts. The first humans to leave the Earth and live to tell the tale were the would-be balloonists François Pilâtre de Rozier and François Laurent Marquis d'Arlandes. This intrepid pair lifted off from Paris on the afternoon of 21 November 1783, and in spite of their balloon fabric catching fire they descended safely 25 minutes after taking off, having travelled eight kilometres.

Hot air balloons rise because the mean density of the balloon is less than that of the surrounding air — the density of air is $1.2\,\text{kg}\,\text{m}^{-3}$ at room temperature and pressure, but at $400\,\text{K}$, close to the maximum temperature that can be achieved without damaging the balloon fabric, the density of air is $1.0\,\text{kg}\,\text{m}^{-3}$. For a balloon of volume $2500\,\text{m}^3$ this means that the air contained has a mass of $500\,\text{kg}$ less than the air which would have occupied its space — it could therefore lift about half a tonne, including all of the balloon materials, fuel, and passengers.

The advent of hydrogen

Within weeks of the Montgolfiers' success, and supported by money from public subscription, Jacques Charles set about conquering the air using the recently discovered and very low-density gas hydrogen. He filled a rather small balloon with hydrogen produced by pouring sulphuric acid over a thousand pounds of iron filings. The balloon lifted off and disappeared over the horizon.

When it landed some distance away, the strange object was attacked and destroyed by angry, frightened villagers. In December 1783, Charles and his co-pilot Nicolas-Louis Robert ascended to a height of about 550 m in a manned balloon. The pioneering use of hydrogen for lift led to this type of balloon being named a Charlière.

▲ **Figure 1** *A postage stamp depicting the Montgolfier brothers*

Although the name Charlière is no longer commonly used for hydrogen balloons, hot-air balloons are called Montgolfières in France to this day.

▲ **Figure 2** *Death of Pilâtre de Rozier*

Soon, a third type of balloon was flown, also in France, by the Montgolfier brothers' first intrepid balloonist Pilâtre de Rozier. The Rozier balloon used a hydrogen envelope inside a hot-air envelope so that less valving and ballasting were needed. It combined the greater lift of Charlière balloons with the better height control of Montgolfières. In 1785, Pilâtre de Rozier was killed whilst attempting to cross the English Channel when the highly flammable hydrogen in his balloon exploded and the balloon came down.

Rozière balloons now use a helium inner envelope, with a surrounding hot air envelope heated by propane. This design was used in the three Breitling Orbiter balloons constructed to circumnavigate the globe. This aim was achieved by Breitling Orbiter 3, a craft the height of the Leaning Tower of Pisa, piloted by Bertrand Piccard and Brian Jones in 1999.

The renaissance of hot air ballooning came in the USA in the early 1960s, under a U.S. Navy contract, but the balloons developed proved to be more valuable for recreation than for military use, and hot air ballooning for sport was reborn.

▲ **Figure 3** *Breitling Orbiter 3 takes off from Château-d'Oex, Switzerland*

Summary questions

1 Hydrogen gas at about 300 K and normal atmospheric pressure has a density of 0.083 kg m^{-3}. Helium, which is considerably more expensive, has double the density of hydrogen. Discuss the advantages and disadvantages of using these two gases in balloons.

2 As a hydrogen weather balloon rises, it moves into air of lower temperature and pressure.
   a Assuming that the balloon envelope is very flexible, explain the changes which occur to the balloon as it rises.
   b As the density of the surrounding air depends on both pressure and temperature, suggest how the force lifting the balloon changes as it moves into cooler, lower-pressure air.

3 A large hot-air balloon can have a volume of 10 000 m^3. Use information from this section to estimate how many passengers could be lifted by this balloon. You can assume that the fabric of the balloon envelope has a mass of 260 kg and that the basket, propane gas tanks, and burners together have a mass of 520 kg.

▲ **Figure 4** *Hot air balloons flying over Turkey*

Practice questions

1 0.50 mol of an ideal gas at 280 K has the p–V graph shown in Figure 1.

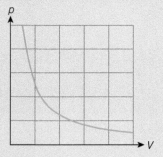

▲ Figure 1

Which of the graphs in Figure 2 is obtained after the temperature of the gas is increased to 560 K whilst the mass of gas remains constant? The dotted curve shows the original graph before the change. *(1 mark)*

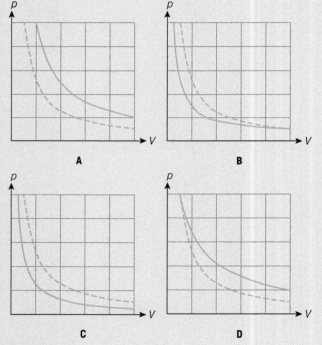

▲ Figure 2

2 Which of the graphs in Figure 2 is obtained when, starting from the first situation in question 1, 0.25 mol of gas escapes while the temperature of the remaining gas remains at 280 K? The dotted curve shows the original graph before the change. *(1 mark)*

3 Three moles of nitrogen gas are at a pressure of 3.0×10^5 Pa and a temperature of 300 K. Calculate the volume of the gas under these conditions. Assume ideal gas behaviour.

State the equation you use in your calculation.
$R = 8.31\,\mathrm{J\,K^{-1}\,mol^{-1}}$ *(3 marks)*
OCR Physics B Paper 2863 June 2010

4 The behaviour of an ideal gas is described by the equation $pV = \frac{1}{3}Nm\overline{c^2}$.

  a State how the pressure of a fixed volume of the gas will change when the speed of each particle doubles. *(1 mark)*

  b The particles in the gas have a range of speeds. Explain why they do not all have the same speed. *(2 marks)*
OCR Physics B Paper 2863 January 2009

5 A mass of 0.70 kg of water is heated using a 1500 W heater.

Calculate the initial rate of change of temperature of the water.

Specific thermal capacity of water = $4200\,\mathrm{J\,kg^{-1}\,K^{-1}}$ *(2 marks)*
OCR Physics B Paper 2863 June 2010

6 The volume of a car interior is $4.2\,\mathrm{m^3}$. In this question, it can be treated as a sealed box.

  a (i) Calculate the number of moles of air of inside the car when the temperature is 11 °C and the pressure 1.0×10^5 Pa.

       Density of air = $1.2\,\mathrm{kg\,m^{-3}}$ in these conditions. Molar mass of air = $0.029\,\mathrm{kg\,mol^{-1}}$. *(3 marks)*

    (ii) A fire extinguisher in the car accidentally releases 1.5 kg of carbon dioxide into the car interior. Calculate the number of moles of carbon dioxide added to the car's interior.

       Mass of a carbon dioxide molecule = 7.3×10^{-26} kg. $N_A = 6.02 \times 10^{23}\,\mathrm{mol^{-1}}$. *(3 marks)*

    (iii) As the carbon dioxide is released from the extinguisher, it cools. Calculate the new pressure inside the car assuming that the final temperature of the gas mixture is −3.0 °C.

       $R = 8.31\,\mathrm{J\,mol^{-1}\,K^{-1}}$ *(2 marks)*

  b Fire extinguishers of this type are not recommended for use in cars. Suggest and explain why they are not suitable. *(2 marks)*

7 This question is about the specific heat capacity of a simple gas in which the molecules each contain only one atom. Argon, of molar mass $0.0399\,\mathrm{kg\,mol^{-1}}$, is a gas of this type.

a **(i)** By applying the gas laws to one mole of argon, show that the mean square speed of argon molecules, $\overline{c^2}$, is given by $\overline{c^2} = \dfrac{3RT}{0.0399}$.

   $R = 8.31\,\mathrm{J\,K^{-1}\,mol^{-1}}$
   $N_A = 6.02 \times 10^{23}\,\mathrm{mol^{-1}}$ *(2 marks)*

   (ii) Show that the mean gain in kinetic energy of an argon atom when the gas temperature is raised from 273 K to 274 K is about $2 \times 10^{-23}\,\mathrm{J}$, and explain why this is a mean value. *(4 marks)*

b Use the answer to **(a)(ii)** to calculate the specific heat capacity of argon. *(2 marks)*

c The calculation in part **(b)** assumes that the volume of the gas is constant. Explain why more energy would be required to raise the temperature of 1 kg of argon gas by 1 °C if the pressure stayed at a constant value of 1 atmosphere, as shown in Figure 3. *(3 marks)*

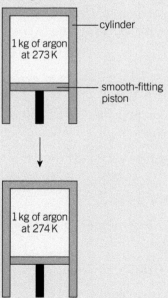

▲ Figure 3

8 This question is about gases under pressure. One mole of helium gas is in a sealed container at 300 K and at a pressure of $2.0 \times 10^5\,\mathrm{Pa}$. The gas is slowly compressed to one third of its original volume without a temperature change. It behaves as an ideal gas.

a Calculate the new pressure of the gas. *(2 marks)*

b Calculate the root mean square speed, c_{rms}, of the molecules at this temperature. (molar mass of helium = $4\,\mathrm{g\,mol^{-1}}$) *(2 marks)*

c Show that the root mean square speed goes up by a factor of $\sqrt{\dfrac{4}{3}}$ when the gas is heated to 400 K. *(2 marks)*

   OCR Physics B Paper 2863 June 2002

9 This question is about the speed of molecules in a gas.
The mass of one mole of nitrogen molecules N_2 is $2.8 \times 10^{-2}\,\mathrm{kg}$.
There are 6.0×10^{23} molecules in one mole of nitrogen.

a Show that the mass of one molecule is about $5 \times 10^{-26}\,\mathrm{kg}$. *(1 mark)*

b The relationship between the kinetic energy of a gas molecule and absolute temperature T is given by
$$\tfrac{1}{2}mv^2 = \tfrac{3}{2}kT$$
Show that the typical speed of nitrogen molecules at 300 K is about $500\,\mathrm{m\,s^{-1}}$.
$k = 1.4 \times 10^{-23}\,\mathrm{J\,K^{-1}}$ *(2 marks)*

c When perfume from an air freshener is released into a room, it gradually diffuses through the volume of the room.

   (i) Assuming the molecules of perfume move at about $500\,\mathrm{m\,s^{-1}}$, show that it would take a molecule about $0.015\,\mathrm{s}$ to travel the length of a room 7.0 m long. *(2 marks)*

   (ii) In fact, it takes very much longer than $0.015\,\mathrm{s}$ for the perfume to travel 7.0 m. Use ideas about diffusion to explain why this is the case.

   A diagram may help your answer. *(3 marks)*

d The perfume molecules are much larger than nitrogen molecules.
Suggest how the rate of diffusion of perfume molecules compares with that of nitrogen molecules. Justify your answer. *(3 marks)*

   OCR Physics B Paper 2863 June 2003

15 THE BOLTZMANN FACTOR

15.1 The ratio $\frac{E}{kT}$

Specification reference: 5.2.2a(ii), 5.2.2c(i)

Learning outcomes

Describe, explain, and apply:

→ calculations and estimates of the ratio of characteristic energies to kT

→ qualitative effects of temperature in processes with an activation energy (for example, changes of state).

▲ **Figure 1** *Cryogenic storage preserves stem cell samples – at the low temperature of liquid nitrogen there is insufficient thermal energy for decomposition reactions*

▲ **Figure 2** *The high temperature of a blowtorch permits the carbonisation reactions responsible for the taste of crème brûlée*

Chilling preserves and heating destroys

To prevent bacterial action destroying organic materials, we store food in freezers, and particularly sensitive biological specimens at the temperature of liquid nitrogen, 77 K, denying bacteria the energy needed to change the tissues (Figure 1). In contrast, cooking changes food in a way that we want, as high temperatures allow chemical processes that alter the food (Figure 2).

Temperature and the energy of particles

In an ideal gas, intermolecular interactions are negligible, so the particles have no potential energy and the mean energy per particle is $\frac{3}{2}kT$. Similar expressions can be found for many other situations, such as vibrating atoms in a solid, or photons emitted by a hot object. In each case, the mean energy per particle is $\frac{3}{2}kT$.

Frequently we need only an order-of-magnitude value for the energy, so we can ignore the small number multiplying kT. This gives us a very useful rule of thumb to use whenever we are dealing with the energy of particles, which could be gas molecules, atoms in a solid lattice, electrons in a current, or photons in a beam of electromagnetic radiation – mean particle energy $E \approx kT$.

 Worked example: An electric grill

The temperature of the heated element in an electric grill is 1500 K. Calculate a typical wavelength of the radiation it emits.

Step 1: Use $E \approx kT$ to calculate the typical energy associated with this temperature.

$E \approx kT = 1.4 \times 10^{-23}\,\text{J}\,\text{K}^{-1} \times 1500\,\text{K} = 2.1 \times 10^{-20}\,\text{J}$

Step 2: Use $E = hf$ to calculate the frequency of a photon with this energy.

$E = hf \therefore f = \dfrac{E}{h} = \dfrac{2.1 \times 10^{-20}\,\text{J}}{6.6 \times 10^{-34}\,\text{J}\,\text{s}} = 3.18... \times 10^{13}\,\text{Hz}$

Step 3: Use $c = f\lambda$ to calculate the wavelength.

$c = f\lambda \therefore \lambda = \dfrac{c}{f} = \dfrac{3.0 \times 10^{8}\,\text{m}\,\text{s}^{-1}}{3.18... \times 10^{13}\,\text{Hz}} = 1 \times 10^{-5}\,\text{m}\ (1\ \text{s.f.})$

$E \approx kT$ is used to give an order of magnitude, not a precise answer, so 1 significant figure is appropriate here. This wavelength (about 0.01 mm) is in the infrared region of the electromagnetic spectrum, which is what you would expect in an electric grill.

Temperature and energy

Parts of the Universe are at temperatures close to absolute zero, whilst in extreme cosmological events such as supernovae temperatures of about 6 GK (six billion degrees) are generated. Figure 3 relates these temperatures to different bond energies and also to the wavelengths of photons with those energies. The energies are expressed in both electronvolts ($1\,\text{eV} = 1.6 \times 10^{-19}\,\text{J}$) and in $\text{kJ}\,\text{mol}^{-1}$ ($1\,\text{kJ}\,\text{mol}^{-1} = 1.7 \times 10^{-21}\,\text{J}$, so $100\,\text{eV} \approx 100\,\text{kJ}\,\text{mol}^{-1}$).

> ### Synoptic link
>
> In Topic 14.1, The gas laws, you learnt about the ideal gas model of gas behaviour, and in Topic 14.2, The kinetic model of gases, you saw that, for an ideal gas, the mean kinetic energy of a molecule is given by
>
> $$E_k = \frac{1}{2}mv^2 = \frac{3}{2}kT.$$

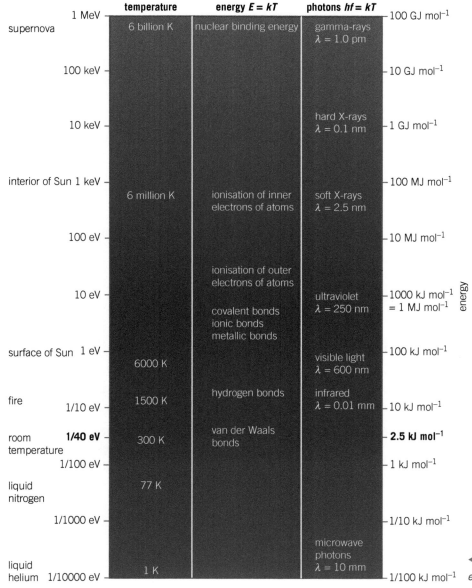

◀ **Figure 3** *Temperature and energy $E \approx kT$*

Processes and activation energies

For many processes, a certain amount of energy is needed to make the process take place. This is the **activation energy** E. In chemistry, it is often necessary to heat reagents in order for a reaction to take place, so that the molecules involved have enough kinetic energy to react. Figure 4 shows a typical example, in which the energy of the products is lower than the energy of the reactants. The energy difference between products and reactants is the energy given out in the exothermic reaction.

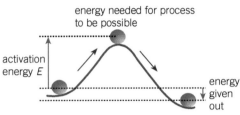

energy needed for process to be possible

activation energy E

energy given out

▲ **Figure 4** *Activation energy and change in energy between reactants and products*

Many other processes, which seem dissimilar at first glance, may be treated in a similar way:

- Water needs thermal energy in order to evaporate.
- Nuclear fusion reactions in the Sun's core can take place only because of the temperature of many million kelvin there.
- The variation of resistance with temperature in thermistors is due to the thermal ionisation of atoms in the semiconductor lattice.

 Worked example: Evaporating water

Experiments show that the energy needed to evaporate 1 kg of water at its boiling point into steam at the same temperature – the specific latent heat of vaporisation of water – is $2.3 \times 10^6 \text{ J kg}^{-1}$. Calculate the energy per molecule required and compare the value with kT at a temperature of 300 K (a warm summer day). The molar mass of water = $0.018 \text{ kg mol}^{-1}$.

Step 1: Calculate the energy required to evaporate 1 mol.

$$E = 2.3 \times 10^6 \text{ J kg}^{-1} \times 0.018 \text{ kg mol}^{-1} = 41\,400 \text{ J mol}^{-1}$$

Step 2: Use the Avogadro constant to calculate the energy E required to evaporate one molecule.

$$E = \frac{41\,400 \text{ J mol}^{-1}}{6.02 \times 10^{23} \text{ molecules mol}^{-1}} = 6.87... \times 10^{-20} \text{ J molecule}^{-1}$$

Step 3: Calculate kT at the temperature given and compare it with E by finding the ratio $\frac{E}{kT}$.

At 300 K, $kT = 1.4 \times 10^{-23} \text{ J K}^{-1} \times 300 \text{ K} = 4.2 \times 10^{-21} \text{ J}$

$$\frac{E}{kT} = \frac{6.87... \times 10^{-20} \text{ J}}{4.2 \times 10^{-21} \text{ J}} = 16 \text{ (2 s.f.)}$$

The worked example seems to indicate that the energy needed to liberate one water molecule from the liquid is 16 times greater than the available energy per particle at room temperature. Why then do wet clothes hung out on a line dry so quickly on a warm day?

The water molecules in the clothes have a range of different energies, and some will have enough energy to escape from the surface. As they do so, the mean energy of the water molecules will drop. The newly cooled water will gain energy by collision with other particles in the warmer environment, giving more water molecules enough energy to escape, and so on until all the water has evaporated.

This is how your clothes dry, and also how perspiration cools you when you are hot.

Things happen between 15*kT* and 30*kT*

The worked example shows that a process requiring $16kT$ happens quite rapidly. In fact, water evaporates at an appreciable rate even just above its freezing point, when evaporation requires $18kT$. If the energy required were more than about $25kT$, very few molecules would evaporate, and the process would be very, very slow. If the energy required were less than about $15kT$, then the process would be so fast that evaporation would rapidly be complete.

Just as the value of kT is a good order-of-magnitude estimate for the energy per particle associated with a particular temperature, so we have a useful rule of thumb for comparing kT with the energy needed for a process to occur. If the energy required is in the range $15kT$ to $30kT$, the process will happen at an appreciable rate. This is illustrated for a range of different processes in Figure 5, where the black diagonal line is the graph of $E = kT$ and the dark blue band above it indicates the range of processes that can proceed at an appreciable rate for each temperature.

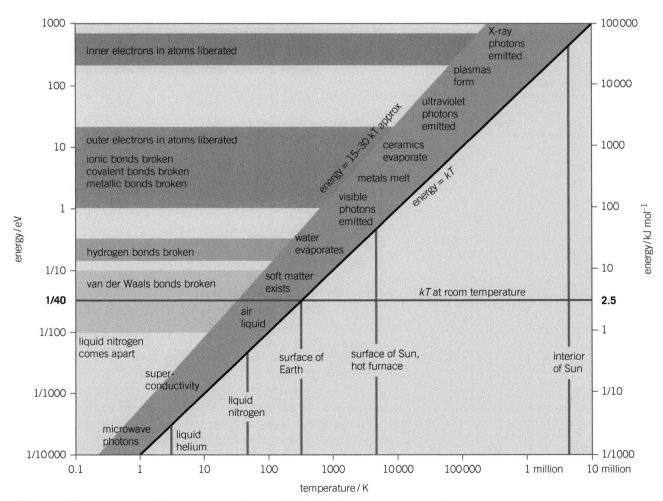

▲ **Figure 5** *Processes requiring energy equivalent to 15 kT – 30 kT happen at an appreciable rate*

Extension: Life at $T = 300$ K

The hydrogen bonds between water molecules are readily formed and broken at temperatures typical on the Earth's surface. At much higher temperatures, the bonds would not form, and water would be found only as a vapour, as it is on Venus. At lower temperatures, as on comets, the bonds would be too strong to break and water would be found as ice.

These same bonds are important to life, as they hold biological molecules in place. For example, in the DNA double helix, the 'letters' making up the genes are safely held in pairs, as shown in Figure 6.

It is essential for the DNA chains to be able to come apart – to be unzipped – so that genetic information can be replicated. But it is also crucial that this doesn't happen too easily, or the DNA molecule would unpeel between replications and might be damaged.

With the help of enzymes that can unzip the DNA easily, the ratio $\frac{E}{kT}$ for hydrogen bonds is about right to achieve this balance at the temperature of the surface of the Earth. At higher temperatures $\frac{E}{kT}$ would be smaller and DNA would

fall apart. The evolution of life as we know it could only occur around 300 K.

To think about: Astronomers seek exoplanets (planets around distant stars) occupying orbits around their star in what is called the 'Goldilocks zone' where water is a liquid. Explain the significance of this zone.

▲ **Figure 6** Image of the DNA double helix with hydrogen bonds holding the two chains together

Summary questions

1 Calculate
   a the mean energy of a gas molecule at 27 °C, *(2 marks)*
   b the mean wavelength of photons emitted by the Universe, treated as an object at 2.7 K. *(2 marks)*

2 The specific latent heat of vaporisation of liquid nitrogen is 200 000 J kg⁻¹. Show that it evaporates freely at 40 K but very slowly, if at all, at 10 K (molar mass of nitrogen = 0.028 kg mol⁻¹). *(5 marks)*

3 Figure 7 shows the distribution of speeds of gas atoms in argon at 25 °C. Argon is a monatomic gas (each separate molecule consists of a single atom). Estimate the mean speed of an argon atom from the graph, and use this value to find the mean kinetic energy of argon atoms.
Use this mean kinetic energy to explain why argon is a gas at 25 °C.
Explain any approximations in these calculations.
(Mass of argon atom = 6.6 × 10⁻²⁶ kg) *(5 marks)*

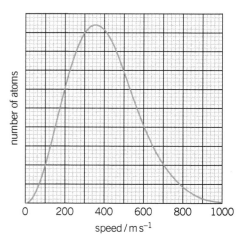

▲ **Figure 7** The distribution of atomic speeds in argon gas at 25 °C

15.2 The Boltzmann factor $e^{-\frac{E}{kT}}$

Specification reference: 5.2.2a(i), 5.2.2a(ii), 5.2.2b(i), 5.2.2c(ii)

Probability and luck

In some games, players roll dice, hoping for sixes. With one die, this happens on average one time in six – a probability of $\frac{1}{6}$ of getting lucky. With two dice, the probability of two sixes is easy to calculate – for every six you roll on die 1, you will get a six on die 2 one-sixth of the time, so two sixes crop up once in every 36 throws. The probability is $\frac{1}{6} \times \frac{1}{6} = \frac{1}{6^2} = \frac{1}{36}$.

Probabilities multiply, they do not add, so the chance of getting four sixes (Figure 1) is $\frac{1}{6} \times \frac{1}{6} \times \frac{1}{6} \times \frac{1}{6} = \frac{1}{6^4} = \frac{1}{1296}$. Promotions sometimes use a dice game to attract public attention or to raise funds. For just £1, you may roll seven dice. If you get seven sixes, you win the big prize: a sports car worth £50 000.

▲ **Figure 1** *Lucky sixes – but what are the chances of throwing seven sixes to win a car?*

Energy exchange by chance

Now consider a situation with many particles, each of which can acquire or lose energy from random interactions with the other particles. The particles could be energy quanta in a hot solid, or molecules in a gas, or reactants in a chemical reaction, or electrons in a semiconductor, or protons in the Sun's core.

To simplify the model, we will assume that all particles can gain or lose an equal-sized quantum E of energy, allowing them to climb up or down the ladder of equally spaced energy levels illustrated in Figure 2.

If a certain fraction f of the particles acquire this energy, or 'get lucky', then the chance of a single particle getting lucky is the same fraction f. (Remember, for dice, $\frac{1}{6}$ of the dice will land with a six upwards, and the chance of rolling a six for a single die is $\frac{1}{6}$.) At any given time, that fraction f of particles have this extra energy. When rolling dice, the probability of getting lucky twice is $\frac{1}{36}$. Here the probability of getting lucky twice, and so having extra energy $2E$, is f^2.

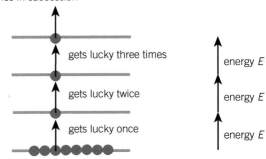

One particle may acquire energy E several times in succession

gets lucky three times — energy E

gets lucky twice — energy E

gets lucky once — energy E

▲ **Figure 2** *Particles can 'get lucky' and acquire energy to climb the ladder of energy levels*

Worked example: How many particles have this energy?

At room temperature, the fraction of particles that can gain an energy of $0.3\,\text{eV}$ is $f = 1.0 \times 10^{-5}$. Calculate the number of particles in 1 mol that have gained energy $0.9\,\text{eV}$ by chance on a ladder of n equally spaced energy levels such as in Figure 2.

Step 1: As $0.9\,\text{eV} = 3 \times 0.3\,\text{eV}$, you must find the fraction of particles that have 'got lucky' three times.

$$\text{fraction} = f^3 = (1.0 \times 10^{-5})^3 = 1.0 \times 10^{-15}$$

Step 2: Calculate the number of particles out of N_A particles that have this energy.

$$\text{number} = 1.0 \times 10^{-15} \times 6.0 \times 10^{23} = 6.0 \times 10^{8}$$

Even though the fraction of particles that gained $0.9\,\text{eV}$ is extremely small, the number of particles in a mole is so enormous that 600 million particles have the required energy.

The Boltzmann factor f

The fraction f used above is called the **Boltzmann factor**, and its mathematical expression can be illustrated by looking at the numbers of particles that climb the energy-level ladder in Figure 3.

The number of particles in level 1 is given by $N_1 = fN_0$, that on level 2 is $N_2 = f^2N_0$, that on level 3 is $N_3 = f^3N_0$, and so on to level X, where $N_X = f^XN_0$. The Boltzmann factor f is the ratio given by

$$f = \frac{\text{number of particles } N_X \text{ on the } X^{\text{th}} \text{ level}}{\text{number of particles } N_{X-1} \text{ on the } (X-1)^{\text{th}} \text{ level}},$$

where the energy difference is E. In the simple example in Figure 3, $f = \frac{1}{2}$.

With many particles, a fraction f acquire energy E at each step of the ladder.

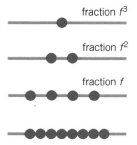

fraction f^3

fraction f^2

fraction f

▲ **Figure 3** *Numbers of particles at different energies*

The way in which the numbers drop by equal ratios for constant increase in energy E indicates an exponential relationship, like the decay of a radioactive nucleus or the discharge of a capacitor. The solution of the radioactive decay equation is $N = N_0e^{-\lambda t}$ and that of the capacitor discharge is $Q = Q_0e^{\frac{-t}{RC}}$. In both situations, the values of the decaying variables decrease with increasing *time*. In Figure 3, the number of particles decreases with increasing *energy* E, implying that $N = N_0e^{-\text{constant} \times E}$. The constant in this equation is $\frac{1}{kT}$, so that

$$N = N_0e^{-\frac{E}{kT}}$$

where the fraction of the particles with the extra energy E, compared with the number not having that energy, is the exponential term $e^{-\frac{E}{kT}}$, which is the Boltzmann factor f.

Worked example: The Boltzmann factor at $15kT$ and at $30kT$

The energy values $15kT$ and $30kT$ have already been used as the range in which processes requiring energy E will proceed at a reasonable rate. Calculate the fractions of particles with this energy in each case.

Step 1: Calculate the term $\frac{-E}{kT}$ in each case.

When $E = 15kT$, $\frac{-E}{kT} = -15$, and when $E = 30kT$, $\frac{-E}{kT} = -30$.

Step 2: Calculate the Boltzmann factor $f = e^{-\frac{E}{kT}}$ for each value of $\frac{-E}{kT}$.

When $E = 15kT$, $f = e^{-15} = 3.1 \times 10^{-7}$ (about 3 in 10 million).
When $E = 30kT$, $f = e^{-30} = 9.4 \times 10^{-14}$ (about 1 in 10000 billion).

Although these fractions are tiny, the actual number of particles involved may be very large, as the worked example 'How many particles have this energy?' shows.

The Boltzmann factor, energy, and temperature

The Boltzmann factor f contains the Boltzmann constant k and two variables — the **activation energy** E needed for a process to take place, and the absolute temperature T. At a constant temperature, kT is constant, so the relationship is $f = e^{-(\text{constant} \times E)}$, mathematically the same as exponential decay of a radioactive sample, where the fraction of nuclei decaying is $f = e^{-\lambda t}$. Figure 4 illustrates this behaviour for two different temperatures. Note that the use of a logarithmic scale converts the exponential decay curve into a straight-line graph, as the value plotted on the y-axis is effectively $\log f$.

The weak intermolecular bonds in liquid nitrogen can be broken by less than 10^{-20} J, so f is close to 1 even at room temperature. The energy of hydrogen bonds is about ten times greater, so processes involving the breaking of hydrogen bonds – the chemistry of life – happen at a significant rate at room temperature (300 K). Breaking stronger ionic and covalent chemical bonds requires about 5×10^{-19} J. At room temperature, f for these processes is very low. Only in a very hot fire, above about 1500 K, is f high enough to break covalent bonds, which is why chemistry only developed after furnaces had been invented.

Figure 5 shows how the Boltzmann factor varies with absolute temperature T, with three different curves for energy E with the values associated with weak van der Waals bonds, with more robust hydrogen bonds, and with strong covalent bonds.

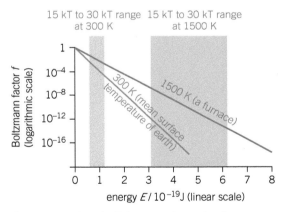

▲ **Figure 4** How the Boltzmann factor f varies with the energy E

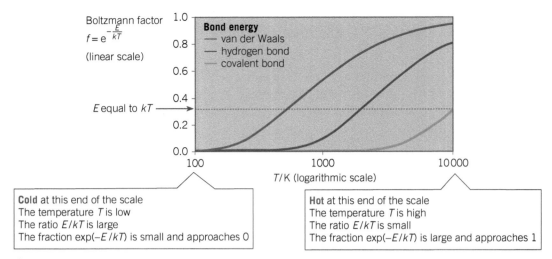

Boltzmann factor $f = e^{-\frac{E}{kT}}$ (linear scale)

E equal to kT

Bond energy
— van der Waals
— hydrogen bond
— covalent bond

T/K (logarithmic scale)

Cold at this end of the scale
The temperature T is low
The ratio E/kT is large
The fraction $\exp(-E/kT)$ is small and approaches 0

Hot at this end of the scale
The temperature T is high
The ratio E/kT is small
The fraction $\exp(-E/kT)$ is large and approaches 1

▲ **Figure 5** How the Boltzmann factor f varies with the temperature T

Viscosity of liquids

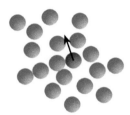

Evaporation of liquids

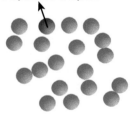

▲ **Figure 6** Random thermal agitation of the surroundings provides the energy for liquids to flow and to evaporate

A range of different activation processes

Many activation processes rely on energy being acquired by chance from the random thermal agitation of the surroundings. In liquids, two different processes take place — liquids flow when molecules in the middle of the liquid gain enough energy to break out of the 'cage' formed by their neighbours, and liquids evaporate when molecules at the surface gain enough energy to break away from their neighbours and leave the liquid (Figure 6). Both of these processes occur more readily at higher temperatures, so as a liquid warms, its viscosity drops, and its rate of evaporation increases. The temperature dependence of viscosity and evaporation are both described by the Boltzmann factor.

The ionisation of atoms in semiconductor crystals is another instance in which the number of atoms with sufficient energy to liberate an electron varies with the temperature as described by the Boltzmann factor.

▦ Worked example: Conductance of thermistors

Thermistors conduct more readily at higher temperatures because a greater fraction of the atoms in the solid acquire the energy E needed to ionise. This means that the number of available conduction electrons at any temperature is given by the Boltzmann factor × total number of atoms in the lattice. As a consequence, the conductivity σ varies according to the equation $\sigma = Xe^{-\frac{E}{kT}}$, where X is a constant. Use the data below to calculate the ionisation energy E of the semiconductor material in the thermistor.

Temperature of thermistor $\theta/°$	25.0	100
Resistance of thermistor R/Ω	10 000	683

Step 1: Calculate the conductance G of the thermistor at each temperature, expressed in K.

At 25°C, $T = (273 + 25)\,K = 298\,K$, and
$$G = \frac{1}{R} = \frac{1}{10\,000\,\Omega} = 1.000 \times 10^{-4}\,S$$
At 100°C, $T = 373\,K$, and $G = \frac{1}{R} = \frac{1}{683\,\Omega} = 1.46\ldots \times 10^{-3}\,S$

Step 2: Rewrite the conductivity equation in terms of conductance, remembering that the length L and cross-sectional area A of the thermistor are constant.

$$G = \frac{\sigma A}{L} = \frac{AX}{L}e^{-\frac{E}{kT}}$$

As A, X, and L are all constants, they can be replaced with a single constant C to give $G = Ce^{-\frac{E}{kT}}$, which can be written $G = C\exp(\frac{-E}{kT})$ for clarity.

Step 3: Take natural logarithms (ln) of this equation. The units of G and C are both siemens(s), which cancel each other out. This step uses two facts about logarithms, namely, $\ln(A \times B) = \ln A + \ln B$, and $\ln(e^x) = x$.

$$\ln G = \ln\{C\exp(\frac{-E}{kT})\} = \ln C + \ln\{\exp(\frac{-E}{kT})\} = \ln C + (\frac{-E}{kT})$$
$$= \ln C - \frac{E}{kT}$$

Step 4: Substitute in the two values of G and T to give a pair of simultaneous equations.

At 25°C, $\ln(1.000 \times 10^{-4}) = \ln C - \dfrac{E}{1.38 \times 10^{-23}\,JK^{-1} \times 298\,K}$
$$\therefore -9.21 = \ln C - \frac{E}{4.11 \times 10^{-21}\,J}$$

At 100°C,

$$\ln(1.46\ldots \times 10^{-3}) = \ln C - \frac{E}{1.38 \times 10^{-23}\,JK^{-1} \times 373\,K}$$
$$\therefore -6.52\ldots = \ln C - \frac{E}{5.15 \times 10^{-21}\,J}$$

Step 5: Subtract one of these equations from the other — it doesn't matter which way this is done, but be careful with signs. This will eliminate the constant C, which we do not need.

$$-9.21 - (-6.52\ldots) = (\ln C - \ln C) + \left\{-\frac{E}{4.11 \times 10^{-21}\,J} - \left(-\frac{E}{5.15 \times 10^{-21}\,J}\right)\right\}$$

$$-2.68\ldots \quad = 0 \quad + E\left\{-\frac{1}{4.11 \times 10^{-21}\,J} + \frac{1}{5.15 \times 10^{-21}\,J}\right\}$$

$$= 0 \quad + E \times (-4.91\ldots \times 10^{19}\,J^{-1})$$

$$\therefore E = \frac{2.68\ldots}{4.88\ldots \times 10^{19}\,J^{-1}} = 5.5 \times 10^{-20}\,J\ (2\ s.f.)$$

Molecules don't care

Throughout this section we have been looking at how the behaviour of matter and energy is governed by the statistics of random distributions. This field of physics, **statistical mechanics**, was developed by Ludwig Boltzmann. Random changes will make the molecules 'try out' all possible arrangements and their energies. No random changes will ever decrease the total number W of these arrangements, so Boltzmann stated the principle that W never decreases, and generally increases. Disordered systems will not, by themselves, become orderly by purely random interactions. This same idea is often expressed in terms of a quantity called entropy, S. Total entropy never decreases and generally increases. This is the **second law of thermodynamics**, which governs the direction of physical and chemical changes. Entropy is related to the number of ways W by the equation $S = k \log W$ (k is, of course, the Boltzmann constant). Boltzmann's tombstone bears this equation (Figure 7).

▲ **Figure 7** *Ludwig Boltzmann's tomb in Vienna*

Summary questions

1 The worked example 'Conductance of thermistors' showed that the energy E needed for an electron to be free for conduction was about 5.5×10^{-20} J. By comparison with the $15kT - 30kT$ energy range at room temperature (about 20 °C, or 293 K), show that you would expect the semiconductor used to conduct electricity at room temperature. *(4 marks)*

2 Figure 8 represents the number of particles at different energies at three different temperatures A, B, and C. The distribution of energies amongst the particles is shown by the number of particles at different levels on the energy level diagram. The levels are all separated by the same energy E in each case. Calculate an estimate of the Boltzmann factor f for each temperature and from this, with no further calculation, compare the three temperatures. *(4 marks)*

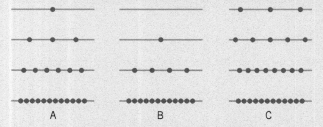

▲ **Figure 8** *Energy-level diagrams for particles at different temperatures*

3 The atmosphere becomes less dense with increasing height above the ground. The number of particles per unit volume n_h at a height h is given by

$$n_h = n_0 \exp\left(-\frac{mgh}{kT}\right)$$

where n_0 is the number of particles per unit volume at ground level.

a Use this relationship to find the height at which the pressure is half the pressure at ground level. You can assume that both gravitational field strength g and temperature T are constant, at 10 N kg^{-1} and 300 K, respectively. Mean molar mass of air = 0.029 kg mol^{-1}. *(3 marks)*

b In reality, both g and T become smaller as you ascend. Without calculation, explain the difference these changes would make to the answer to (a). *(2 marks)*

Module 5.2 Matter

Matter: very simple

The gas laws

- The behaviour of ideal gases
- Graphs indicating the relationships between p, V, N, and T for an ideal gas
- Calculations and estimates using the number of moles n and the Avogadro constant N_A
- Calculations and estimates using $pV = NkT$ where $N = nN_A$ and $Nk = nR$
- Practical task: investigating the pressure–volume relationship for a gas
- Practical task: determining absolute zero

The kinetic model of gases

- Temperature as proportional to average energy per particle; average energy = $\frac{3}{2}kT \approx kT$ as a useful approximation
- Random walk of molecules in a gas: distance gone in N steps related to $\sqrt{N}$
- Calculations and estimates of $pV = \frac{1}{3}Nm\overline{c^2}$

Energy in matter

- Energy transfer producing a change in temperature and the concept of specific thermal capacity c
- Calculations and estimates of temperature and energy change using $\Delta E = mc\Delta\theta$
- Practical task: determining the specific thermal capacity of a substance

Matter: very hot and very cold

The ratio $\frac{E}{kT}$

- Calculations and estimates of the ratio of characteristic energies to kT
- Qualitative effects of temperature in processes with an activation energy (for example, changes of state, thermionic emission, ionisation, conduction in semiconductors, viscous flow)

The Boltzmann factor $e^{-\frac{E}{kT}}$

- Ratios of numbers of particles in quantum states of different energy, at different temperatures (classical approximation only)
- Calculations and estimates of the Boltzmann factor $e^{-\frac{E}{kT}}$
- Sketching and interpreting graphs showing the variation of the Boltzmann factor with energy and temperature

Physics in perspective

Lasers

Lasers, once thought to be an invention without a use, are now found in a vast range of applications, from surgery to home entertainment to surveying. The principle of the laser is a fascinating example of the Boltzmann factor at work and was first explained in 1916 by Albert Einstein.

Einstein's lasers

Einstein showed that there were three possible transitions between atomic energy levels. If an atom is in an excited state, as shown by the electron at the higher energy level E_2 in Figure 1, it can 'fall' to the lower level E_1. Energy is conserved by the emission of a photon of energy $hf = E_2 - E_1$.

The reverse process, absorption, occurs when a photon of exactly the right energy raises an electron to a higher level, as shown in Figure 2.

The third process, stimulated emission, occurs when a photon of the right energy encounters an atom in the excited state — this stimulates the atom into releasing a photon exactly in phase with the incident photon, as shown in Figure 3. This stimulated emission results in the light being amplified, which explains the name 'Laser': Light Amplification by Stimulated Emission of Radiation.

In 1953, the first laser-type device was constructed, which operated at microwave energies and was appropriately called a 'maser'. Throughout the 1950s, efforts were made to develop lasers at ever higher photon energies, and in 1960 Theodore Maiman built the first laser producing visible light.

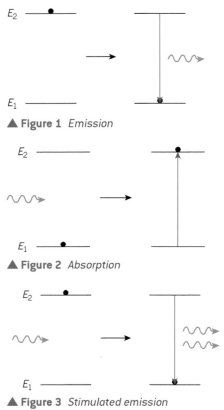

▲ **Figure 1** *Emission*

▲ **Figure 2** *Absorption*

▲ **Figure 3** *Stimulated emission*

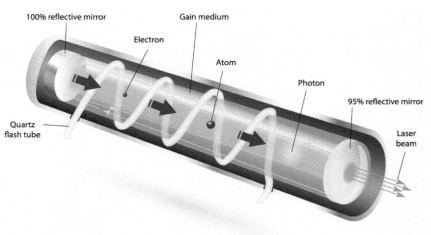

▲ **Figure 4** *Structure of a laser*

A typical laser structure is shown in Figure 4 — the energy input here is the light from a flash tube, but other designs use electric currents. Mirrors are mounted on the two ends of the laser cavity, and these reflect the light to and fro many times through the gain medium in which the stimulated emission happens. Each pass of the light along the tube produces more and more identical photons, all in phase, resulting in an extremely powerful, coherent, and monochromatic beam of light inside the laser.

Even though only a tiny percentage the light in the laser escapes through the mirror, it can still be very powerful — a typical laser used in surgery has an output power of about 100 W, and infrared lasers used as cutting tools in industry operate at several kW.

Although there are many designs of lasers that can emit anything from ultraviolet radiation to infrared, for stimulated emission to occur each needs a number of atoms in a higher energy state that is larger than the number of atoms in a lower energy state. This is not the usual arrangement — for a system in equilibrium at absolute temperature T, the Boltzmann factor f is given by

$$f = \frac{\text{number of atoms at the higher level}}{\text{number of atoms at the lower level}} = e^{-\frac{E}{kT}}$$

where E is the energy difference between the two levels. The Boltzmann factor only approaches 1 as the temperature gets very high, and it certainly cannot exceed 1. This disturbance of the equilibrium conditions, called population inversion, is illustrated in Figure 5.

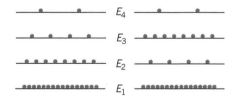

Boltzmann distribution   population inversion

▲ **Figure 5** *Population inversion*

To produce the population inversion shown here, where the numbers in level E_3 exceed those in E_2, a process known as 'pumping' must take place in the gain medium. Different laser designs use different methods to achieve this. In Maiman's first laser design (see Figure 4), the light from the flash tube excites some atoms into a higher state in the gain medium (a synthetic ruby rod in this case) to create the population inversion. On the other hand, in a helium-neon gas laser the pumping mechanism is provided by interatomic collisions.

Far from being 'a solution looking for a problem', as they were once described, lasers are now found everywhere — every supermarket checkout, every DVD player, and every fibre-optic system, not to mention numerous applications in law enforcement, entertainment, the armed forces, and industry.

Summary questions

1 Use ideas from Chapter 7 (in the year 1 course) to calculate the photon energy of a 100 W surgical laser emitting infrared radiation of frequency 2.8×10^{13} Hz.

2 The Boltzmann distribution shown in Figure 5 has equally spaced energy levels. In this simplified example the numbers of atoms at different levels is not realistic. If the difference between adjacent levels, E, is given by $E = 2.8$ eV, show that it corresponds to a temperature of the gain medium of 46 000 K.

3 One way to achieve population inversion in lasers is shown in Figure 6.

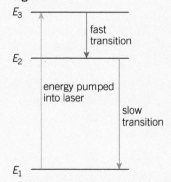

▲ **Figure 6** *A three-level laser*

This method uses a gain medium containing the three levels E_1, E_2, and E_3. The pumping mechanism rapidly raises atoms from level E_1 to E_3, from where they very rapidly drop, without emitting photons, to level E_2. This results in population inversion in levels E_2 and E_1.
Explain why the fast transition from E_3 to E_2 and the slow transition from E_2 to E_1 results in laser action producing photons of energy $(E_2 - E_1)$ and not photons of energy $(E_3 - E_2)$.

Practice questions

1 Here are four temperatures, measured in kelvin.

   A 10^1 K

   B 10^2 K

   C 10^3 K

   D 10^4 K

To the nearest order of magnitude, which temperature corresponds to the energy of a molecule of mass 5.0×10^{-26} kg travelling at a speed of $250\,\text{m s}^{-1}$?   *(1 mark)*

2 To the nearest order of magnitude, which temperature given in question **1** corresponds to the energy of electromagnetic radiation of wavelength 12 μm?   *(1 mark)*

3 The ratio $\dfrac{E}{kT}$ can be used to consider when atoms will evaporate from the surface of a solid, where E is the energy required for an atom to leave the surface of the solid.

   a   State what is indicated by the energy kT.   *(1 mark)*

   b   Aluminium melts at 930 K and boils at 2500 K. Explain how E compares with kT at room temperature.   *(1 mark)*

4 **a**   Show that the particles of a gas at temperature 10 000 K will have an average kinetic energy of the order of 1 eV (1.6×10^{-9} J).

       Boltzmann constant $k = 1.4 \times 10^{-23}\,\text{J K}^{-1}$ *(2 marks)*

   b   Explain why the particles of the gas will not all have a kinetic energy equal to the average value.   *(2 marks)*

       OCR Physics B Paper 2863 January 2003

5 Figure 1 shows the lowest three energy levels of a hydrogen atom with the energy differences between levels E_1, E_2, and E_3 labelled.

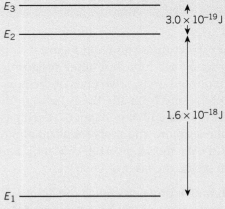

▲ Figure 1

A sample of atomic hydrogen at a temperature of 6000 K has N atoms in the E_1 ground state.

   a   Calculate the number of atoms N_2 at the E_2 level.   *(2 marks)*

   b   Explain why the number of atoms N_3 at the E_3 level is **not** given by the equation $\dfrac{N_3}{N_2} = \dfrac{N_2}{N}$.   *(2 marks)*

6 This question uses the Boltzmann factor to compare the rate of evaporation of ethanol and water at body temperature.

Molar mass of ethanol = 46 g mol^{-1}

Energy required to evaporate 1 kg of ethanol = $8.4 \times 10^5\,\text{J kg}^{-1}$

Avogadro constant, $N_A = 6.0 \times 10^{23}\,\text{mol}^{-1}$

Boltzmann constant, $k = 1.4 \times 10^{-23}\,\text{J K}^{-1}$

   a **(i)**   Show that the energy E required for one molecule of ethanol to evaporate is 6.4×10^{-20} J.   *(2 marks)*

      (ii)   Human body temperature is 310 K. Show that the average energy of a molecule at body temperature is of the order of 4×10^{-21} J.   *(2 marks)*

      (iii) Show that the Boltzmann factor, which estimates the proportion of ethanol molecules with enough energy to vaporise at 310 K, is approximately 3×10^{-7}.   *(3 marks)*

   b   The Boltzmann factor for water at 310 K is approximately 1×10^{-7}.

       Use this information to explain why a drop of an ethanol-based product, such as perfume or aftershave, feels cooler on the skin than a drop of water.   *(3 marks)*

       OCR Physics B Pilot Paper 7733 June 2001

7 This question is about how the density of the air varies with height above the Earth's surface.

At sea level the density of the atmosphere is about $1.2 \, kg \, m^{-3}$.

Consider nitrogen at a temperature of 300 K.

a Show that the average energy of a particle at this temperature is about $4.1 \times 10^{-21} \, J$. *(1 mark)*

b Show that the energy E required to lift a nitrogen molecule of mass $4.6 \times 10^{-26} \, kg$ to a height 3000 m above sea level is about $1.4 \times 10^{-21} \, J$. *(1 mark)*

c Show that the Boltzmann factor, $e^{\frac{E}{kT}}$, for the energy E found in part (b) at a temperature of 300 K is about 0.71. *(2 marks)*

Figure 2 shows a graph of density of air against height above sea level. The atmospheric temperature is assumed to remain constant in this model.

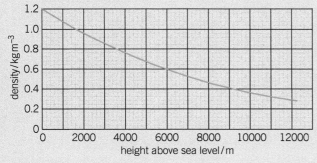

▲ Figure 2

d (i) The graph shows that density decreases with increasing height. Explain how the Boltzmann factor helps to account for this fact. *(2 marks)*

(ii) Propose and carry out a test to decide whether the density falls exponentially with height above sea level. *(3 marks)*

e In fact, atmospheric temperature decreases with height above sea level. Explain why this makes the Boltzmann factor become smaller with height above sea level. *(2 marks)*

OCR Physics B Paper 2863 June 2002

8 This question is about the behaviour of gases at different temperatures.

a Describe how the kinetic theory of gases explains the pressure an ideal gas exerts on the walls of its container. *(3 marks)*

b At room temperature, 288 K, an ideal gas in a cylinder is at a pressure of $5.00 \times 10^5 \, Pa$.

Show that the average energy of a molecule of the gas under these circumstances is about $4 \times 10^{-21} \, J$. *(2 marks)*

c A small proportion of the molecules making up a particular gas are 'dissociated' or split into two or more pieces. The energy E needed to dissociate a molecule is much greater than the average energy of a molecule at room temperature. The fraction of molecules which have energy larger than E is given by the Boltzmann factor, $e^{-\frac{E}{kT}}$.

Calculate the proportion of molecules which are dissociated in the gas at 288 K when $E = 3.4 \times 10^{-20} \, J$. *(2 marks)*

d Figure 3 shows how the Boltzmann factor, $e^{-\frac{E}{kT}}$ varies with temperature T, in kelvin.

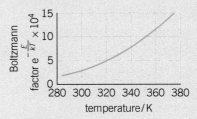

▲ Figure 3

(i) Use the graph to describe how the Boltzmann factor, $e^{-\frac{E}{kT}}$, varies with T over the range shown. *(1 mark)*

(ii) By approximately what factor does the proportion of dissociated molecules increase when there is a modest increase in temperature of the gas from 300 K to 360 K? *(2 marks)*

OCR Physics B Paper 2863 January 2001

MODULE 6.1

Fields

Introduction

The idea of fields was first developed in the early 19th century by Michael Faraday in his attempts to explain the newly discovered phenomena of magnetism and electromagnetism. Without formal education, Faraday invented a powerful visual and mathematical way of describing the way in which magnets and electrically charged objects can have an effect at a distance.

His invention was later elaborated by James Clerk Maxwell, using mathematics well beyond what Faraday could have used. Maxwell's work on thermodynamics and the kinetic theory of gases would have been enough to establish his status as one of the greats of physics. His work on electromagnetism took his reputation to a new level. He was able to sum up all the relationships between electrical and magnetic fields covered in Chapters 15 and 16 of this book in just four (rather complicated) equations.

He used these to show that changes in magnetic fields would produce changing electric fields, and vice versa, resulting in electromagnetic waves that would travel at $3.0 \times 10^8 \, \text{m s}^{-1}$ in free space. In doing this, Maxwell united not only electricity and magnetism, but also optics. His work is a high point of classical physics.

Electromagnetism starts with the idea of the magnetic flux of a magnetic field, and how rate of change of flux leads to electromagnetic induction. This is summed up in the Laws of Faraday and Lenz. Following from this, the applications of these laws to transformers and to generators are discussed, and the chapter finishes with a look at how the force on a current-carrying conductor applies to electric motors.

Chapter 17 starts with **The electric field** and its use by Millikan to find the charge of the electron. This is followed by the force exerted by a magnetic field on moving charges, as used by J J Thomson to find the mass:charge ratio of the electron. The last section deals with Coulomb's Law and the electric field, and electric potential due to point charges.

Knowledge and understanding checklist

From your Key Stage 4 or first-year A Level study you should be able to do the following. Work through each point, using your Key Stage 4/first-year A Level notes and the support available on Kerboodle.

- [ ] Recall that magnetic fields can be represented by flux lines.
- [ ] Understand that current is produced in a circuit by a p.d.
- [ ] Recall that an alternating p.d. can be changed by a transformer.
- [ ] Recall that an electron gains kinetic energy when accelerated across a p.d.

Maths skills checklist

Maths is a vitally important aspect of Physics. In this unit, you will need to use the following maths skills. You can find support for these skills on Kerboodle and through MyMaths.

- [ ] **Changing the subject of equations**, including non-linear equations, such as inverse proportion and inverse square relations of potential and field near a point source
- [ ] **Using sine and cosine relationships** and relating centripetal forces on charges with their magnetic causes.
- [ ] **Finding the gradient of graphs**, such as those of flux linkage when obtaining a value for the induced e.m.f.

MyMaths.co.uk
Bringing Maths Alive

Learning outcomes

Describe, explain, and apply:

→ induced e.m.f. = rate of change of flux linkage

→ $\varepsilon = -\dfrac{d(N\Phi)}{dt}$

→ $\Phi = BA$

→ electromagnetic forces qualitatively, as arising from the tendency of flux lines to contract or from the interaction of induced poles

→ magnetic flux from a coil

→ the terms B-field, magnetic field, flux, flux linkage, flux density, induced e.m.f.

→ diagrams of lines of flux in magnetic circuits, showing continuity of lines of flux.

▲ **Figure 1** *Magnets interact*

▲ **Figure 2** *A magnetic compass*

Magnetism and electricity

It has been known for centuries that steel items that have been in contact with magnetic rocks called lodestones have odd properties. The ends, or poles, of two such magnets attract or repel each other, and a needle of such steel, balanced on a pivot, will align itself so that one end points north.

With the development of electrical circuits in 1800 it became clear that electricity and magnetism are closely related. In fact, the magnetic properties of lodestone, or of permanent magnets like a compass needle, are caused by alignment of circulating currents due to electrons within the atomic structure of the materials.

Fields and flux

Thanks to Michael Faraday, we have two useful concepts which we will use throughout this chapter – **magnetic field** and **magnetic flux**, illustrated in Figure 3.

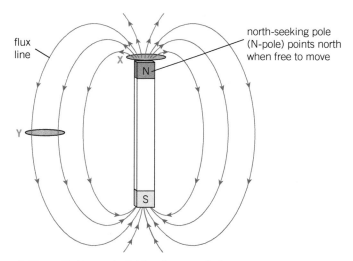

north-seeking pole
(N-pole) points north
when free to move

flux
line

▲ **Figure 3** *Magnetic field and magnetic flux*

From the appearance of the magnetic field pattern created by iron filings scattered near a bar magnet, we can represent the magnetic field by **lines of flux** passing from an N-pole to an S-pole. Note that lines of flux never cross. The word 'flux' ('flow' in Latin) implies that something is flowing, and the arrows on the lines of flux show that this direction is from the N-pole to the S-pole. The N to S direction is arbitrarily chosen, very like the conventional current direction in an electric circuit, and it is helpful to think of flux as analogous to current in an electrical circuit. Unlike electrical current, however, nothing actually flows in a magnetic field.

If you compare the two identical yellow areas **X** and **Y** in Figure 3 you will see that the magnetic field at **X** is stronger than that at **Y**, because the lines of flux are packed more tightly together at **X**. From this we get the name for the measure of the strength of the magnetic field, the **magnetic flux density**, which is given the symbol B. Flux density is also referred to as the **B-field**. B is a vector quantity, as shown by the presence of direction arrows on the lines of flux.

Because of the complicated properties of magnetic materials, it is not clear what happens to the lines of flux inside the magnet. We will often use long, current-carrying coils called **solenoids** as sources of magnetic fields, and these give us the clearer picture shown in Figure 4. You can see that the flux lines form continuous loops. Within the solenoid, the B-field is uniform – this is indicated by parallel, equally spaced lines of flux.

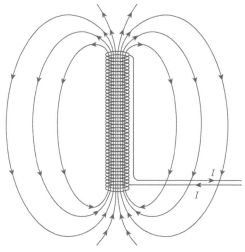
▲ **Figure 4** *Magnetic field in and near a solenoid*

Flux and magnetic forces

When permanent magnets interact, or when a permanent magnet interacts with an electromagnet, the direction of the resultant forces can be deduced from the pattern of the lines of flux between the two interacting bodies. The principle to apply is that the lines of flux will tend to shorten and to straighten.

In Figure 5, the tendency to straighten the lines of flux causes both magnets to rotate, and the tendency to shorten the lines of flux pulls the two magnets together.

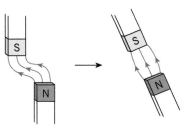
▲ **Figure 5** *Lines of flux tend to shorten and straighten*

Flux Φ and flux density B

Flux Φ and flux density B are related by the equation $\Phi = BA$, with flux Φ (the Greek capital letter phi) measured in **weber** (Wb) and flux density B measured in Wb m^{-2}, provided that the area A, measured in m^2, is perpendicular to the lines of flux.

One complication particular to magnetic fields – it is not present in gravitational or electrical fields – is that the causes and effects of a magnetic field are usually at right angles to the direction of the lines of flux. In two-dimensional diagrams it is therefore often easiest to draw diagrams in which the field is perpendicular to the plane of the diagram, as in Figure 6, with × representing a flux line going into the plane of the paper and · representing a flux line emerging towards the reader. Note that the current that creates the magnetic field is in the plane of the diagram, whilst the B-field is perpendicular to it.

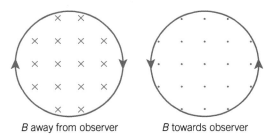
B away from observer B towards observer
▲ **Figure 6** *Magnetic field in a section through a solenoid*

Worked example: Flux through a solenoid

A solenoid of diameter 8.0 cm has a flux density in its centre of $3.5 \times 10^{-3} \, \text{Wb m}^{-2}$. Calculate the flux through the solenoid.

Step 1: Calculate the area A in m^2.

$A = \pi r^2 = \pi \times (4.0 \times 10^{-2} \, \text{m})^2 = \pi \times 1.6 \times 10^{-3} \, \text{m}^2 = 5.02... \times 10^{-3} \, \text{m}^2$

Step 2: Use $\Phi = BA$ to calculate the flux Φ.

$\Phi = BA = 3.5 \times 10^{-3} \, \text{Wb m}^{-2} \times 5.02... \times 10^{-3} \, \text{m}^2$
$= 1.8 \times 10^{-5} \, \text{Wb (2 s.f.)}$

Changing flux Φ and electromagnetic induction

Faraday knew that magnetic fields are created by electric currents, as in a solenoid. He believed in the symmetry of the universe, and so sought for some time to create the opposite phenomenon. He succeeded in showing that a moving magnet near a coil could induce an electric current in that coil, if there were a complete circuit for the current. More generally, any change in magnetic flux in a circuit will result in the **induction** of an e.m.f. across that circuit, and that a faster change of flux results in a greater e.m.f.

The weber and the volt

Faraday's discovery showed that the rate of change of flux in a circuit is proportional to the induced e.m.f. in that circuit.

$$\frac{\Delta \Phi}{\Delta t} \propto \varepsilon$$

As the weber has so far not been defined, but only related to flux density, this gives us a convenient definition for the weber – if the flux in a simple circuit of one coil changes at a rate of $1 \, \text{Wb s}^{-1}$, an e.m.f. of 1 V will be induced. The volt is already defined, so we can say that a weber is the amount of flux that needs to change per second in a circuit to induce an e.m.f. of 1 volt.

Flux and flux linkage

The values of magnetic flux in Wb are often very small for typical magnetic fields, so in real electromagnetic circuits, coils containing many turns are used. This leads to the idea of flux linkage, illustrated in Figure 7. Note that the coil is not a solenoid carrying a current which creates the magnetic field, but rather it is part of a completely separate electrical circuit in which an e.m.f. will be induced.

If the flux though one turn of the coil changes at a rate of (for example) $1 \, \mu\text{Wb s}^{-1}$, then there will be an e.m.f. across that one turn of $1 \, \mu\text{V}$. As the whole coil contains N turns in series, the e.m.f. induced across the whole coil will be $N \times 1 \, \mu\text{Wb s}^{-1}$. By defining a new variable, **flux linkage**, such that flux linkage = number of coils × flux through one coil, we can state that

induced e.m.f. = rate of change of flux linkage, $\varepsilon = \dfrac{\Delta (N\Phi)}{\Delta t}$

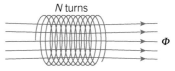

N turns

Φ

▲ **Figure 7** Flux linkage $= N\Phi$

Hint

Volts, webers, and joules: $1 \, \text{V} = 1 \, \text{Wb s}^{-1} = 1 \, \text{J C}^{-1}$

Lenz's law and Faraday's law

Consider the situation where the coil of N turns in Figure 7 forms part of a complete circuit and the flux Φ has just increased from 0 to Φ. The induced e.m.f. caused by this increase in flux will induce a current in the circuit, which will make the coil act as a solenoid, creating a magnetic field. But in which direction will the induced current be?

If the induced current in the solenoid creates flux in the same direction as the original flux, the extra change in flux will cause an increase in induced e.m.f., making the induced current still greater. The increased induced current will increase the flux further, and so increase the induced e.m.f. This positive feedback would continue, with continually increasing induced e.m.f. and current, and of power dissipated in the circuit. This contravenes the principle of conservation of energy, with the slightest induced e.m.f. resulting in ever greater power dissipation.

The logical consequence of this analysis is that the induced current in the solenoid will produce flux in the *opposite* direction to the original flux, reducing the flux increase, and opposing whatever change has caused the increase in flux. This is a consequence of the principle of conservation of energy, and is an illustration of a general principle know as **Lenz's law** – the direction of an induced e.m.f. is always such as to act against the change that causes the induced e.m.f.

As a consequence of Lenz's law, the equation for induced e.m.f. is written:

induced e.m.f. $\varepsilon = -$(rate of change of flux linkage) $= -\dfrac{d(N\Phi)}{dt}$

which is known as **Faraday's law of electromagnetic induction**. The minus sign in this equation is a reminder of Lenz's law – the induced e.m.f. will produce a current which opposes the rate of change of flux responsible for the induction.

> ### Hint
>
> You know that the gradient at one point of a $y-x$ graph can be found by calculating $\dfrac{\Delta y}{\Delta x}$ over a small section of the graph at that point. If the graph is a curve, then $\dfrac{\Delta y}{\Delta x}$ is an approximation to the gradient of the tangent at that point. For smaller values of Δx, the approximation becomes better. As you know if you are studying mathematics, the value that this ratio tends towards as Δx gets smaller and smaller is mathematically represented as $\dfrac{dy}{dx}$, which is the actual gradient of the tangent at this point.

 Worked example: Calculating an induced e.m.f.

The flux through a coil of 50 turns increases from 1.2×10^{-6} Wb to 8.1×10^{-6} Wb in a time of 0.24 s. Calculate the induced e.m.f. across the coil.

Step 1: Calculate the change in flux, $\Delta\Phi$, over this time.

$\Delta\Phi = 8.1 \times 10^{-6}$ Wb $- 1.2 \times 10^{-6}$ Wb $= 6.9 \times 10^{-6}$ Wb

Step 2: Use flux linkage $= N\Phi$ to calculate the change in flux linkage.

$$\begin{aligned} \text{change in flux linkage} &= N \times \Delta\Phi = 50 \times 6.9 \times 10^{-6}\,\text{Wb} \\ &= 3.45 \times 10^{-4}\,\text{Wb} \end{aligned}$$

Step 3: Use Faraday's law to calculate the induced e.m.f, using the approximation $\dfrac{d(N\Phi)}{dt} \approx \dfrac{\Delta(N\Phi)}{\Delta t}$.

$$\begin{aligned} \text{induced e.m.f. } \varepsilon &= -\dfrac{d(N\Phi)}{dt} \approx -\dfrac{\Delta(N\Phi)}{\Delta t} = -\dfrac{3.45 \times 10^{-4}\,\text{Wb}}{0.24\,\text{s}} \\ &= -1.4 \times 10^{-3}\,\text{V (2 s.f.)} \end{aligned}$$

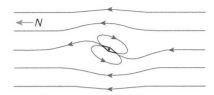

▲ Figure 8 *Compass needle in Earth's magnetic field*

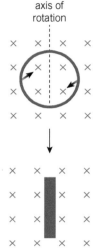

▲ Figure 9 *A gold ring in a magnetic field*

Summary questions

1 Figure 8 shows a magnetic compass (a small bar magnet) placed at an angle to the Earth's magnetic field. The uniform field of the Earth is distorted by the presence of the magnetic compass.
 a Explain how Figure 8 shows that the field is uniform far from the compass, but not uniform near it. (*1 mark*)
 b Explain, in terms of lines of flux shortening and straightening, why the compass rotates until it points north. (*2 marks*)

2 A circular coil containing 250 turns of wire of diameter 4.8 cm is placed perpendicular to a *B*-field of flux density 0.028 Wb m^{-2}. Calculate
 a the flux through the coil, (*1 mark*)
 b the flux linkage in the coil. (*2 marks*)
 c The magnetic field reverses in direction over a time of 0.15 s. Calculate the e.m.f. induced in the coil, assuming that the flux changes uniformly. (*2 marks*)

3 A gold ring has a diameter of 2.0 cm. Treating it as a one-turn coil, its resistance is $1.1 \times 10^{-4}\ \Omega$. It is worn by someone in an MRI scanner with a magnetic field of flux density 2.5 Wb m^{-2}. The wearer moves his hand so that the ring turns through 90°, as shown in Figure 9, in a time of 0.30 s. Calculate
 a the e.m.f. induced in the ring, (*1 mark*)
 b the current in the ring, (*2 marks*)
 c the electrical work done in heating the ring. (*2 marks*)

16.2 Transformers

Specification reference: 6.1.1a(i), 6.1.1a(iv), 6.1.1b(i), 6.1.1c(iii), 6.1.1c(iv), 6.1.1d(iii)

Large electromagnetic machines

Some of the most important electromagnetic machines have no moving parts: **transformers**. Figure 1 shows a power transformer. Although it is very efficient, the small percentage of energy dissipated as heat must be removed from the transformer to prevent it being damaged. The fins protruding from the transformer are filled with oil. Hot oil rising by convection allows heat to escape to the surroundings.

Magnetic circuits

An electrical circuit consists of a potential difference across a continuous conducting loop in which charge will flow (Figure 2).

A magnetic circuit can be constructed in an analogous way, as the lines of flux all form closed loops. The 'magnetic conductor' that forms a complete circuit is shown in Figure 3 as a grey loop. The flux is created by a current I in a solenoid of N turns. The equivalent of p.d. in the magnetic circuit is $I \times N$, so a current of 1 A going around 20 turns is exactly equivalent to a current of 20 A going around one turn. This 'magnetic e.m.f.' will be referred to as **current-turns**.

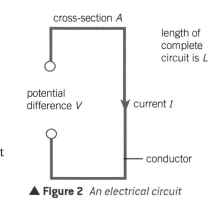

▲ **Figure 2** *An electrical circuit*

The magnetic equivalent of current in the electrical circuit is flux, Φ. A larger value of NI will give a larger value of Φ.

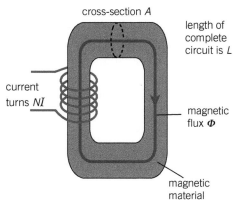

▲ **Figure 3** *A magnetic circuit*

▲ **Figure 1** *An industrial power transformer*

Just as an electrical circuit will give a large current if the conducting loop has high conductance, so the magnetic circuit will give a large flux if the conducting loop has the magnetic equivalent – high **permeance**. This almost invariably means that it will be made of a ferromagnetic material, often iron. In an electrical circuit, current I = conductance $G \times$ p.d. V. In an analogous way, a magnetic circuit has flux = permeance × current-turns. This equation is not as useful as the electrical analogue, as the permeance will depend on the flux, so calculations of flux or permeance will not be expected from you. However, the analogy with electrical circuits does give useful indications of how to build a magnetic machine with as large a flux as possible.

Size is important

In an electrical circuit, the conductance G is given by $G = \sigma\dfrac{A}{L}$, where σ is the conductivity of the material, A its cross-sectional area, and L its length. To get as large a current as possible for a given p.d., the conducting material must have high electrical conductivity, be as short as possible, and have a large cross-sectional area. In exactly the same way, to get as large a magnetic flux as possible for given current-turns, the magnetic material used should have a high magnetic permeability (analogous to conductivity), and should form a circuit with as small a value of L and as large a value of A as possible.

▲ **Figure 4** *A small transformer*

One common type of electromagnetic machine – a small transformer as found in a laboratory power supply – is shown in Figure 4. The length L of the magnetic circuit is made small by making the transformer as compact as possible. The cross-sectional area A is increased by having two possible magnetic circuits in parallel, as shown in the sectional diagram of Figure 5 – the current-turns in the copper coil around the central bar generate flux that divides into the two circuits shown.

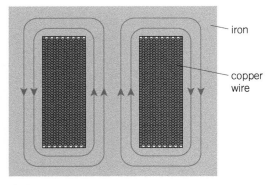

iron

copper wire

▲ **Figure 5** *Section through a transformer*

Note that the transformer has copper (with high electrical conductivity) in the electrical circuit and iron (with high magnetic permeability) in the magnetic circuit.

Air gaps in magnetic circuits

Iron has a magnetic permeability about 200 000 times greater than that of air, so in a complete magnetic circuit like that in Figure 5 little flux will leak out. If there is a gap in the magnetic circuit, as in Figure 6, then the permeance of the entire circuit is very much reduced, resulting in a much smaller flux than a circuit with no gap. This is very like an electrical circuit with a small resistance (the iron) in series with a very large resistance (the air), which would have a very much smaller current than a circuit without the large resistance.

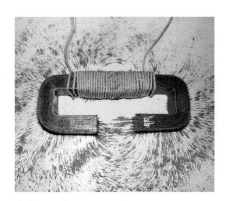

▲ **Figure 6** *A magnetic circuit containing iron and air*

Transformers

The transformer equation

Faraday's law of electromagnetic induction (Topic 16.1) relates induced e.m.f. to the rate of change of flux. In Figure 5 a constant flux direction is shown, but in a transformer an alternating potential difference is connected across the primary coil, resulting in an **alternating current (a.c.)** in that coil. The primary coil generates the flux, so a constantly changing flux is produced.

In the primary circuit, there are two voltages – the alternating p.d. V_P applied to the circuit, and the e.m.f. ε induced in the coil by the changing flux. Although it seems odd that the changing current in the

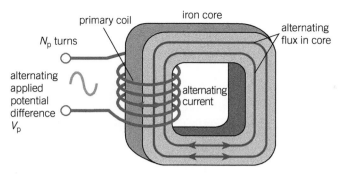

▲ **Figure 7** *The current-turns in the primary coil generate changing flux*

primary coil can produce effects on itself (a bit like lifting yourself up by your own bootlaces), this phenomenon, called **self-inductance**, occurs in any alternating current circuit with magnetic coils in it. By Lenz's law, this self-induced e.m.f will oppose the current that created the changing flux that caused the induction, so the total p.d. across the coil will be $V_P + \varepsilon$, where $\varepsilon = -$ rate of change of flux linkage, from Faraday's law of electromagnetic induction. The total p.d. produces the current I in the primary circuit of electrical resistance R, so that $V_P +$ ($-$ rate of change of flux linkage) $= IR$.

Because the resistance R of the primary circuit is very low, IR is negligibly small, so $V_P -$ rate of change of flux linkage $= 0$, or $V_P =$ rate of change of flux linkage.

However, rate of change of flux linkage $= N_P \times$ rate of change of flux $= N_P \times \dfrac{d\Phi}{dt}$ where N_P is the number of turns in the primary coil,

so $V_P = N_P \times \dfrac{d\Phi}{dt}$, or $\dfrac{d\Phi}{dt} = \dfrac{V_P}{N_P}$.

We can ignore the very small leakage of flux into the air, so the flux Φ is the same in the two coils on the transformer (Figure 8), just as the current is the same throughout a series electrical circuit.

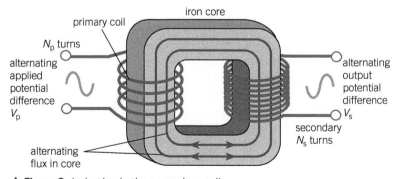

▲ **Figure 8** *Induction in the secondary coil*

The e.m.f. V_S induced in the secondary coil, which has N_S turns, is given by $V_S = -N_S \times \dfrac{d\Phi}{dt}$. As both coils have the same flux through them at all times, the rate of change of flux $\dfrac{d\Phi}{dt}$ is the same.

In the primary coil $\dfrac{d\Phi}{dt} = \dfrac{V_P}{N_P}$, and in the secondary coil $\dfrac{d\Phi}{dt} = -\dfrac{V_S}{N_S}$, resulting in the transformer equation.

$$\dfrac{V_P}{N_P} = -\dfrac{V_S}{N_S} \qquad \text{or} \qquad \dfrac{V_P}{V_S} = -\dfrac{N_P}{N_S}$$

where the minus sign shows that the alternating p.d. across the secondary coil is 180° out of phase with the alternating p.d. across the primary coil (Figure 9). The ratio $\dfrac{N_P}{N_S}$ is called the **turns ratio** of the transformer. If $N_P > N_S$, the transformer is a step-down transformer, because $V_P > V_S$, and if $N_P < N_S$, the transformer is a step-up transformer, as in Figure 9.

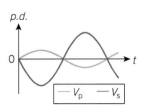

p.d.

0 —→ t

— V_p — V_s

▲ **Figure 9** V_P and V_S are in antiphase

> ### ▦ Worked example: The turns ratio in a transformer
>
> A transformer is constructed to give an output p.d. of 12 V when connected to the 230 V mains supply. It has 690 coils on the primary coil. Calculate the number of turns on the secondary coil.
>
> **Step 1:** Calculate the p.d. ratio.
> $$\dfrac{V_P}{V_S} = \dfrac{230\,\text{V}}{12\,\text{V}} = 19.1...$$
> **Step 2:** The turns ratio is the p.d. ratio – use it to calculate the number of secondary turns. The minus sign can be ignored in calculations of turns ratios, because only the magnitude of the output p.d. is needed, so the phase difference of the p.d.s is not relevant.
>
> This is a step-down transformer, so $N_P > N_S$.
> $$\dfrac{N_P}{N_S} = 19.1... \quad \therefore \quad \dfrac{690}{N_S} = 19.1...$$
> $$Ns = \dfrac{690}{19.1...} = 36$$

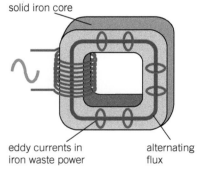

solid iron core

eddy currents in iron waste power   alternating flux

▲ **Figure 10** Eddy currents in a solid iron core

Eddy currents and transformer lamination

Iron is usually the magnetic material in the core of a transformer as it has high magnetic permeance. However, it is a good electrical conductor too. The changing flux within the iron core induces currents in the secondary coil, but it also induces currents inside the core itself (Figure 10). These circulating swirls of electrical current, like eddies of water in a river, are called **eddy currents**. By Lenz's law, the flux produced by eddy currents will oppose the flux created by the primary coil and reduce the efficiency of the transformer. The wasted energy is dissipated as heat within the core by the heating effect of the eddy current.

The loss of energy by eddy currents can be greatly reduced by making the core out of a stack of flat plates, called laminations, instead of a single solid lump. The laminations are separated from each other by thin, electrically insulating layers to minimise eddy currents (Figure 11). A transformer built in this way has few energy losses due to eddy current heating, or heating in the resistance of the wire of the coils, and has a very high efficiency, typically >98%.

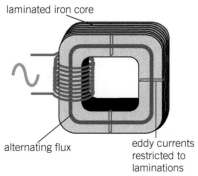

laminated iron core

alternating flux   eddy currents restricted to laminations

▲ **Figure 11** Eddy currents are reduced in a laminated iron core

Currents in a transformer

In an ideal transformer, where energy losses can be ignored, all power delivered into the primary coil is extracted from the secondary coil. This means that, if the voltage increases, the current must decrease:

$$P_P = P_S \therefore I_P V_P = I_S V_S \therefore \frac{I_P}{I_S} = \frac{V_S}{V_P}$$

As $\frac{V_S}{V_P} = \frac{N_S}{N_P}$, this means that $\frac{I_P}{I_S} = \frac{N_S}{N_P}$.

Practical 6.1.1d(iii): Investigating transformers

Figure 12 shows simple laboratory equipment that can be used to investigate transformers. Two closely fitting soft-iron cores, named C-cores on account of their shape, can have coils threaded onto them before they are clipped together (in Figure 12 only the primary coil is shown). The secondary coil used can be a similar pre-wound coil, or you can construct one as part of the investigation. You may use a suitable resistor to complete the secondary circuit.

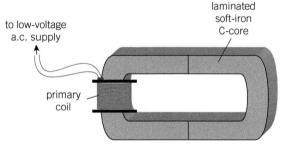

▲ **Figure 12** *Simple transformer (secondary coil not yet added)*

Measurements of p.d. and current can be made in both primary and secondary circuits using appropriate a.c. meters or oscilloscopes. As well as confirming the transformer equations above, the arrangement can be used to investigate:

- the effect of changing the relative positions of the secondary and primary coils
- the effect of decreasing the permeance of the core by introducing sheets of paper between the two C-cores
- the effect of changing the load resistance of the secondary circuit.

Worked example: Input current to a transformer

A step-down transformer of turns ratio 20:1 has an output p.d. of 11 V. The power delivered to the secondary circuit is 0.24 W. Calculate the input current in the transformer.

Step 1: Calculate the current in the secondary circuit.

$$P_S = I_S V_S \therefore I_S = \frac{P_S}{V_S} = \frac{0.24\,\text{W}}{11\,\text{V}} = 0.0218...\,\text{A}$$

Step 2: Use the turns ratio to calculate the primary current.

This is a step-down transformer of turns ratio 20, so $N_P = 20 \times N_S$.

$$\frac{I_P}{I_S} = \frac{N_S}{N_P} = \frac{N_S}{20\,N_S} = \frac{1}{20} \therefore I_P = \frac{I_S}{20} = \frac{0.0218...\,\text{A}}{20} = 0.0011\,\text{A} = 1.1\,\text{mA}$$

(2 s.f.)

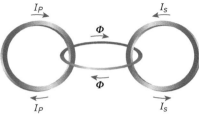

▲ **Figure 13** *Loops of current and flux intersect*

Linking loops of current and flux

Electromagnetic machines contain intersecting loops of current and flux: flux is created by current-turns, where the current loops encircle the flux, and current is created by changing flux, where the flux loops encircle the induced current. Figure 13 illustrates this for a transformer, where the changing primary current I_P creates the changing flux Φ, and the changing flux Φ induces the secondary current I_S.

Summary questions

1 A power station generates an alternating p.d. of 25 000 V. This is stepped up to 400 000 V for transmission through the National Grid.
  a Calculate the turns ratio of the transformer. *(1 mark)*
  b Calculate the current in the secondary coils when the power station is generating 450 MW. *(3 marks)*

2 The core of a transformer is constructed of a stack of vertical laminations, as in Figure 14(a). Explain why a horizontal stack of laminations, as in Figure 14(b), would not be satisfactory. *(3 marks)*

(a)   (b)

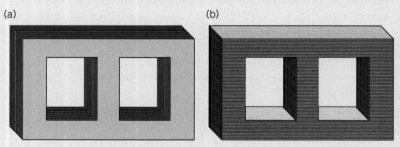

▲ **Figure 14** *Two attempts to make a laminated transformer core*

3 A transformer is 99% efficient when delivering a current of 4.5 A into a load of resistance 14 Ω. Assuming that 20% of the dissipated energy is lost in the copper windings of the primary and secondary coils, calculate the energy dissipated into the iron core per minute. *(4 marks)*

16.3 Generators

Specification reference: 6.1.1a(ii), 6.1.1b(i), 6.1.1b(ii), 6.1.1c(i), 6.1.1d(i)

Generators and dynamos

A transformer is an electromagnetic machine that induces e.m.f. without physical movement – alternating current produces the necessary flux. A **generator** uses motion, often produced by a turbine (Figure 1), to produce the flux changes needed to induce an e.m.f. The output of a generator may be a.c. or d.c. The older term 'dynamo' is usually just an alternative term for generator, but it is sometimes reserved for d.c. generators, with the term 'alternator' used for a.c. generators. To avoid ambiguity, the term 'generator' will be used throughout this chapter.

Moving coils in magnetic fields

In each of the three situations in Figure 2, the flux Φ though the coil changes in the time Δt taken to move the coil, and the change in flux linkage results in an induced e.m.f.

Figure 2(c) can be analysed to give a useful relationship between the velocity of the moving coil and the induced e.m.f.

A coil moving at a constant speed perpendicular to the field

First, let us simplify Figure 2(c) in two ways:

- make the coil rectangular, not circular
- have a uniform magnetic field of flux density B.

Figure 3 shows this coil being moved at a constant velocity v from a position

Learning outcomes

Describe, explain, and apply:

→ the action of a generator in terms of the change of flux linkage produced by relative motion of flux and conductor

→ calculations and estimates using induced e.m.f. = rate of change of flux linkage,
$$\varepsilon = -\frac{d(\Phi N)}{dt}$$

→ observation of induced e.m.f.

(a)

turn the coil so that more flux goes through it

(b)

move the coil into a region where the flux density is larger

(c)

slide the coil so that more flux goes through it

▲ **Figure 2** *Creating changing flux with a constant magnetic field*

▲ **Figure 1** *A d.c. generator inside a wind turbine*

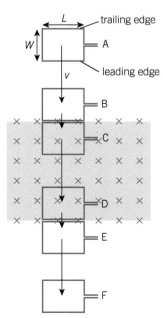

▲ **Figure 3** *A coil moving through a uniform field*

outside the magnetic field on one side, through the magnetic field, and then out of the magnetic field on the other side. The diagram is drawn in the plane of the coil, with the lines of flux passing into the diagram. The movement of the coil is shown in five different stages. At points B and E the coil is on the outside edge of the region of the magnetic field, and at points C and D the coil is on the inside edge of the region.

From A to B there is no induced e.m.f., as there is no flux through the coil at any point, and so there is no change in flux.

From B to C the flux Φ changes from 0 to a maximum value of $B \times$ (total area of coil) = LWB. As the coil is moving at a steady speed, the flux increases at a uniform rate, so there is a constant induced e.m.f. It is not hard to calculate this e.m.f. – the flux rises from 0 to LWB in the time Δt taken for the coil to move a distance W at velocity v.

$$v = \frac{W}{\Delta t} \therefore \Delta t = \frac{W}{v}$$

so the induced e.m.f. ε (ignoring sign) = rate of change of flux linkage

$$= \frac{\mathrm{d}(\Phi N)}{\mathrm{d}t} = \frac{1 \text{ turn} \times LWB}{\Delta t} = \frac{LWB}{\left(\frac{W}{v}\right)} = LWB \times \frac{v}{W} = vLB$$

From C to D the flux stays at its maximum value throughout, so there is no change in flux, which means there is no induced e.m.f.

From D to E the flux changes from a maximum value to 0. As the coil is moving at a steady speed, the flux decreases at a uniform rate, so there is a constant induced e.m.f. $\varepsilon = vLB$. As the flux is decreasing, the e.m.f. is in the opposite direction to that induced between B and C.

From E to F there is no induced e.m.f., just as for the movement from A to B.

These e.m.f. changes are shown in Figure 4.

An alternative way of expressing Faraday's law for coils moving or turning in magnetic fields – a way he used himself – is that the induced e.m.f. is equal to the rate at which lines of flux are being 'cut' by the moving wires. Using that approach, the leading edge of the rectangular coil in Figure 3 starts cutting lines of flux when it leaves position B, and continues to cut the lines at a steady rate, so an e.m.f. is induced until it reaches position C.

Once the coil passes point C, both the leading edge and the trailing edge of the coil are cutting flux lines at the same rate, so an e.m.f. is induced in each. One of these e.m.fs. would result in a clockwise current and the other in an anticlockwise current, so they cancel. After the coil passes point D until it reaches point E, the leading edge is not cutting lines of flux, but the trailing edge is, so an e.m.f. is induced there tending to drive current in the opposite direction to that induced between B and C.

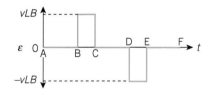

▲ **Figure 4** *e.m.f. induced by the movement in Figure 3*

Worked example: e.m.f. induced across the wings of an aircraft

An airliner of wingspan 60 m flies at 270 m s⁻¹. The flux density of the Earth's magnetic field is 1.7×10^{-4} Wb m⁻², and at the airliner's latitude it acts in a direction at 20° to the vertical. Calculate the e.m.f. induced between the two wingtips of the airliner.

Step 1: To find ε, you will need the value of B at right angles to the moving conductor. Calculate the vertical component of the Earth's B-field. A vector diagram should be drawn first (Figure 5).

$$\frac{B_{vertical}}{B} = \cos(20°) \therefore B_{vertical} = B\cos(20°)$$

$$= 1.7 \times 10^{-4}\,\text{Wb m}^{-2} \times 0.939...$$

$$= 1.59... \times 10^{-4}\,\text{Wb m}^{-2}$$

Step 2: Substitute into $\varepsilon = vLB$.

$$\varepsilon = vLB = 270\,\text{m s}^{-1} \times 60\,\text{m} \times 1.59... \times 10^{-4}\,\text{Wb m}^{-2} = 2.6\,\text{V (2 s.f.)}$$

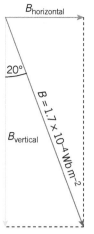

▲ **Figure 5** *Finding the vertical component of the Earth's magnetic field*

 Practical 6.1.1d(i): Induced e.m.f. in a coil with a moving magnet

Figure 2(b) shows that a coil moving towards a bar magnet will have a changing flux linkage, and so an e.m.f. is induced. The same effect can be produced with a stationary coil and a moving magnet. This is easy to investigate – clamp a coil with its axis vertical, and drop a small bar magnet through it (Figure 6).

The induced e.m.f. changes as the magnet enters, falls through, and leaves the coil, and can be recorded as a time-dependent graph using a data logger or a suitable oscilloscope.

The voltage–time graph produced can be analysed and interpreted in terms of the motion of the bar magnet and the shape of its B-field.

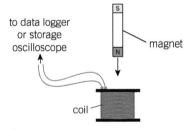

▲ **Figure 6** *Magnet falling through coil*

A rotating coil

Generators usually have a coil rotating in a magnetic field, or else a magnet rotating in a stationary coil, which is equivalent (this is the situation in Figure 2(a)). As the coil rotates, the flux through the coil changes from 0, when the plane of the coil is parallel to B, to a maximum value of $B \times A$, when it is perpendicular to B.

Figure 7 shows a coil of area A rotating in a uniform field. At the time shown in the diagram, the perpendicular to the plane of the coil, shown in green, makes an angle θ to the lines of flux. As in the worked example of the e.m.f. induced in an airliner, the component of B perpendicular to the coil is given by $B\cos\theta$ and so the flux through the coil is $\Phi = B\cos\theta \times A = BA\cos\theta$.

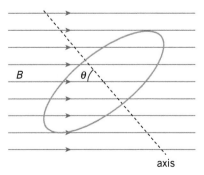

▲ **Figure 7** *Rotating coil in a uniform B-field*

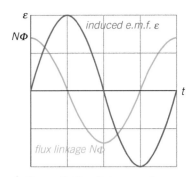

▲ **Figure 8** *Flux linkage and induced e.m.f. in the rotating coil*

If the coil is rotating at a constant angular velocity ω, then $\theta = \omega t$, just as displacement s under constant velocity v is given by $s = vt$.

This means that the flux though the coil is given by $\Phi = BA \cos(\omega t)$, and if the coil has N turns, the flux linkage $= N\Phi = NBA \cos(\omega t)$. Through one entire rotation, the flux linkage changes as shown in Figure 8.

The induced e.m.f. = −(the rate of change of flux linkage), $-\dfrac{\mathrm{d}(\Phi N)}{\mathrm{d}t}$, which equals −(the gradient of the $N\Phi$–t graph). Note that:

- ε has its largest values when the $N\Phi$–t graph is steepest, as $N\Phi$ is changing fastest
- ε is 0 when the $N\Phi$–t graph is a maximum or minimum, as the gradient is zero at those points
- increasing $N\Phi$ produces negative ε, whilst decreasing $N\Phi$ produces positive ε.

Real generators

Real generators, like real transformers, are magnetic machines engineered to have as large a flux as possible. This can be done in two ways – by increasing the permeance of the magnetic circuit, and by using an electromagnet instead of a permanent magnet. Figure 9 shows one design of a generator incorporating these improvements. Note also that, instead of a coil rotating inside a stationary magnet, this has a rotating electromagnet (the **rotor**) spinning inside fixed coils wound on the stationary iron core (the **stator**).

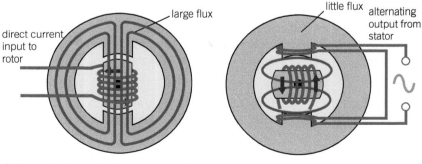

▲ **Figure 9** *A practical generator design*

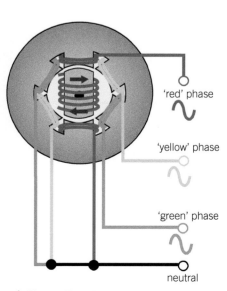

▲ **Figure 10** *A three-phase generator*

The large generators found in power stations make a further improvement on this design: they have three sets of stator coils (Figure 10).

This design effectively combines three generators in one. The three secondary coils are 120° apart and so produce three alternating e.m.f.s that are out of phase by 120°. Besides the improved efficiency of the generator, the availability of different phases of alternating voltage allows the construction of more powerful electromagnetic machines, such as three-phase electric motors.

Summary questions

1 The flux linkage in the stator coil in a generator varies as shown in Figure 11.

  a On a copy of this graph, sketch a graph to show how the induced e.m.f. changes. (*1 mark*)

  b Explain how the e.m.f. would be different if the rotor speed in the generator were doubled. (*3 marks*)

▲ **Figure 11** *Variation of flux linkage in a stator coil of a generator*

2 Bicycles can be fitted with lights powered by dynamos constructed as in Figure 12.

  a State when the flux through the coils is greatest, and when it is smallest, as the magnet turns. (*2 marks*)

  b The magnet rotates 10 times a second, generating an e.m.f. of maximum value 3.0 V. Assuming that the time taken for the flux to change to produce this e.m.f. is 25 ms, estimate the largest flux through the coils if the coils have 500 turns in all. (*3 marks*)

3 A generator with a 200-turn coil as a stator rotates at a frequency of 50 Hz. The maximum flux through the stator is 0.024 Wb. Using the relationship between flux linkage and induced e.m.f. shown in Figure 8, estimate the maximum e.m.f. produced by the generator. (*4 marks*)

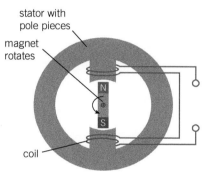

▲ **Figure 12** *A bicycle dynamo*

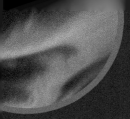

16.4 Flux changes and force

Specification reference: 6.1.1a(iii), 6.1.1c(i), 6.1.1c(ii), 6.1.1d(ii)

Learning outcomes

Describe, explain, and apply:

→ the action of a motor in terms of the tendency of flux lines to contract

→ calculations and estimates involving $F = ILB$ for a straight current-carrying conductor in a uniform magnetic field

→ use of a current balance to determine the uniform magnetic flux density between the poles of a permanent magnet

→ calculations and estimates using induced e.m.f. = rate of change of flux linkage, $\varepsilon = -\dfrac{d(\Phi N)}{dt}$.

Motors big and small

Along with transformers and generators, motors are a type of electromagnetic machine. They range in size from tiny, low-power precision motors used in DVD drives (Figure 1) to large motors used in heavy industry (Figure 2). The same underlying physics applies to all motors.

The catapult force

To understand motor action, consider the effect of putting a straight current-carrying conductor into a magnetic field. A wire carrying a current generates its own field, seen in Figure 3 from two perpendicular directions, (a) and (b). In Figure 3(a) the lines of flux above the wire are emerging from the plane of the paper and those below the wire are going into the plane. In Figure 3(b) the current is going into the plane of the paper. The flux lines form loops around the wire in Figure 3(b). These loops become further and further apart further away from the wire, showing that the B-field is becoming weaker. This is also shown by the increased separation of the B-field markers in Figure 3(a).

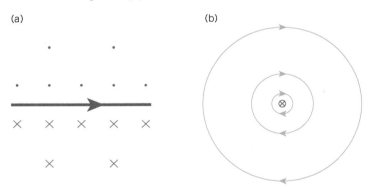

▲ **Figure 3** *B-field near a straight current-carrying wire: (a) B in the plane parallel to the wire; (b) B in the plane perpendicular to the wire*

If this current-carrying wire is placed in a uniform magnetic field, as in Figure 4, the field due to the wire adds vectorially to the uniform field to give a distorted field that is stronger above the wire and weaker below. Because lines of flux tend to shorten and straighten, a downward force results – the catapult force, shown as a green arrow. (The different shades of blue used for lines of flux here are not important – they are introduced so you can use Figure 4 as a key to interpret the three-dimensional representation in Figure 5.)

A motor contains more than a straight wire. In the simplest design, the rotating part of the motor, called the **armature**, consists of a coil.

▲ **Figure 1** *A motor 1 cm across from a DVD drive – note the six tiny electromagnets*

▲ **Figure 2** *Large motors used for pumping*

Figure 5 shows how a rectangular one-turn coil is affected by catapult forces. The side **AB** of the coil is identical to the situation in Figure 4, and so has a downward catapult force. On the opposite side, **CD**, the current is in the opposite direction, and so the force pushes upwards. The currents in the other two sides, **BC** and **DA**, are parallel to the lines of flux, and so no force acts on them. The two forces produce a torque that rotates the coil anticlockwise.

When the rotating coil reaches a vertical position, a switching mechanism – a **commutator** – reverses the connections to the coil. The coil's inertia carries it through the vertical position, and the torque produced by the newly reversed current makes the coil continue to rotate anticlockwise.

▲ **Figure 4** *The catapult force*

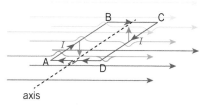

▲ **Figure 5** *Rotation of an armature coil*

Real motors are more complex than this simple model. The obvious improvements to the armature are to use a multi-turn coil, with an iron core to increase the permeance of the magnetic circuit. Other designs, such as the motor in a washing machine, omit commutators entirely by using an alternating B-field that continually moves ahead of the armature, stretching the lines of flux, so that the shortening and straightening lines of flux pull the armature round.

Teslas and webers per square metre

The magnitude F of the catapult force depends on the electric current I, the length L of wire in the uniform field, and the flux density B. Increasing any one of these factors increases F in proportion, so the equation relating them is $F \propto ILB$.

If we introduce a new unit of B-field, the **tesla** (T), defined as the force per unit current per unit length of conductor perpendicular to B, then

$$F = ILB$$

where B is in T, and $1\,\text{T} = \dfrac{1\,\text{N}}{(1\,\text{A} \times 1\,\text{m})} = 1\,\text{N}\,\text{A}^{-1}\,\text{m}^{-1}$.

It seems confusing to have two separate units for measurement of B-field strength, but fortunately they can be shown to be identical. Figure 6 shows two smooth parallel conducting rails with a metal bar resting on them. There is a uniform B-field perpendicular to this arrangement. The metal bar is being pulled at a constant velocity v as shown by the green arrow.

As was shown in Topic 16.3, Generators (in the section titled 'A coil moving at a constant speed perpendicular to the field'), the e.m.f. induced between the ends of a conductor of length L moving at a velocity v perpendicular to a field of flux density B is given by $\varepsilon = vLB$, where B is the flux density in Wb m^{-2}.

The circuit is completed off to the left of the diagram, so a current I flows. The electrical power generated by the moving rod cutting through lines of flux is given by power = current × e.m.f., $P = I\varepsilon$, so that $P = I(vLB)$.

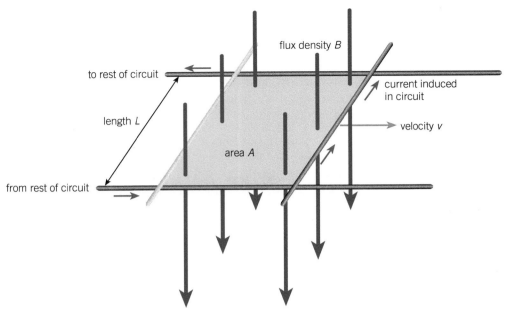

▲ **Figure 6** *Combining F = ILB and ε = vLB*

If electrical power is being generated, conservation of energy tells us that it must be produced by the force F needed to move the rod at the velocity v. The rod is moving at a constant speed, and the rails are smooth, so there are no external forces acting on the rod. There cannot be a resultant force acting on the rod, or it would accelerate or decelerate. There are two equal and opposite forces acting on the sliding rod, from Newton's first law:

- the force exerted by whatever is pulling the rod to the right at the constant velocity v
- the catapult force pulling the rod to the left.

Mechanical power delivered = electrical power generated, so equating mechanical power $P = Fv$ to the electrical power $P = I(vLB)$ gives $Fv = IvLB \therefore F = ILB$.

This analysis started with an equation using B in Wb m^{-2}, and it results in the equation used to define the tesla, so clearly $B\,/\,\text{Wb m}^{-2} = B\,/\,\text{T}$.

Notice that in this simple example the moving rod is a generator, because it is creating an e.m.f., and it is also a motor, because the catapult force is pushing it to the right. This is an important observation – every motor is also a generator, and every generator is also a motor.

> **Hint**
>
> You can use either T or Wb m^{-2} as the unit for flux density – they are exactly equivalent.

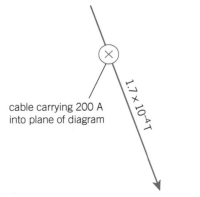

cable carrying 200 A into plane of diagram

$1.7 \times 10^{-4}\,\text{T}$

▲ **Figure 7** *A power cable in the Earth's magnetic field*

🖩 Worked example: Force on a power line

A horizontal power cable of length 60 m is carrying a current of 200 A d.c. The magnetic field of the Earth is $1.7 \times 10^{-4}\,\text{T}$. Figure 7 shows a sectional diagram through the power line, indicating the direction of the B-field. Calculate the force acting on the power cable. On a sketch of Figure 7, indicate the line along which this force acts.

➡

Step 1: Substitute values into $F = ILB$ and calculate the force.

$F = ILB = 200\,\text{A} \times 60\,\text{m} \times 1.7 \times 10^{-4}\,\text{T} = 2.0\,\text{N}$ (2 s.f.)

Step 2: Remembering that F, B, and I are all mutually perpendicular, determine the line along which F acts. (You are *not* required to give the direction in which F acts along this line.)

I is into the diagram. As F is perpendicular to I, it must be in the plane of the diagram. F is also perpendicular to B, so it acts along the line sketched in Figure 8.

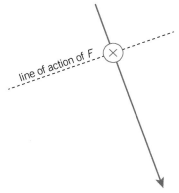

▲ **Figure 8** *Line of action of the force on a power cable*

Practical 6.1.1d(ii): Using a current balance to measure *B*-fields

Figure 9 shows a stiff wire bent into right angles, firmly clamped in place, with a length L of the wire midway between the facing poles of a pair of ceramic magnets. The wire is connected to a d.c. power supply and ammeter. The digital balance can be tared (zeroed) once the magnets are in position for convenience of reading, but this is not essential.

By Newton's third law, an upward force on the current-carrying wire from the magnets will cause an equal and opposite downward force on the magnets from the current-carrying wire. This force can be deduced from the mass reading on the balance. Changing the current through the stiff wire will change the force F = ILB, and a suitable graph will enable you to obtain a value for B. It is necessary to be careful with the positioning and

orientation of the stiff wire. The choice of the length L may also be critical, as it is unlikely that the B-field is uniform over the whole volume between the two poles.

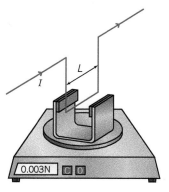

▲ **Figure 9** *A current balance*

Motors as generators

It was observed above that every generator is a motor, and every motor is a generator. If you rotate a generator by hand, as soon as you use the generator to deliver current to a load you immediately feel resistance as the generator, acting as a motor, pushes against you. In an exactly analogous way, once a motor starts spinning, it generates an e.m.f., called **back e.m.f.** By Lenz's law, this opposes the p.d. driving the motor, and limits the speed of the motor (Figure 10). This is similar to the way that air resistance affects the speed of a free-falling object, except that in this case the armature coil is rotating, so angular velocity ω stands in place of linear velocity v, and torque on the armature corresponds to force on the falling object.

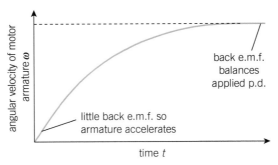

▲ **Figure 10** *Back e.m.f. opposes motor rotation*

These two examples – the opposing force produced in a generator, and the back e.m.f. generated in a motor – seem as if nature is just being awkward, but they are both examples of the coupling between the electrical and magnetic circuit loops. The force acting on the generator coils is how mechanical work is transferred into the electrical circuit, and the back e.m.f. acting on the motor transfers the electrical work into mechanical work done on the motor load.

➕ Extension: Brushless motors

In a motor, either the current-carrying coil rotates, or the electromagnet does. Either way, how can the rotor be part of an electrical circuit without the cables tangling immediately? One solution is the use of sliding contacts, called brushes, which may make contact with a commutator, as mentioned above. Another system uses electromagnetic induction to induce currents inside the rotor, producing a magnetic field that interacts with the field from the surrounding stator coils.

A rotor of this kind, which has no electrical contacts to the outside world, is called a squirrel cage, although 'hamster wheel' might be a better name. It consists of a solid cylinder of laminated iron in which are embedded copper conductors (Figure 11).

This rotor is mounted inside an iron stator carrying three pairs of coils, each carrying a.c. that is 120° out of phase with the adjacent pair of coils. The alternating magnetic fields produced by the stator coils induce eddy currents in the copper conductors of the squirrel cage. These currents make the rotor magnetic and make it turn (Figure 12).

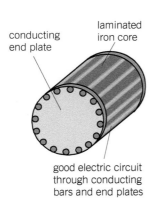

conducting end plate

laminated iron core

good electric circuit through conducting bars and end plates

▲ **Figure 11** *A squirrel-cage rotor*

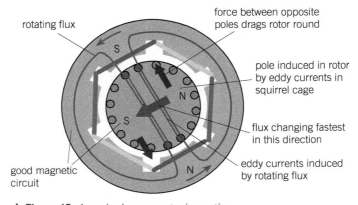

rotating flux

force between opposite poles drags rotor round

pole induced in rotor by eddy currents in squirrel cage

flux changing fastest in this direction

eddy currents induced by rotating flux

good magnetic circuit

▲ **Figure 12** *A squirrel-cage motor in section*

The phase difference between adjacent stator coils means that the magnetic flux appears to rotate around the motor. The squirrel-cage rotor turns more slowly than the magnetic flux, so it rotates constantly to try to catch up with the moving magnetic field.

The squirrel-cage motor is a simple and robust design – the magnetic and electric circuits are good, and there is no need for easily worn brushes.

To think about:

> Brushless motors of this sort are often called induction motors. Explain why this term is appropriate.

Summary questions

1 A wire carrying a 4.2 A current is placed between the pole-pieces of a magnet so a 0.26 T B-field is perpendicular to the wire. The length of wire in the magnetic field is 5.0 cm. Calculate the force acting on the wire.
   (3 marks)

2 Two parallel vertical wires, mounted a short distance apart, carry identical current. Use Figure 3 to explain why there is a force on each wire, and predict the direction in which these forces act.   *(4 marks)*

3 A d.c. motor has a 50-turn coil measuring 4.0 cm × 6.5 cm in a 0.12 T magnetic field, as shown in Figure 13.
   a Calculate the force on each of the long sides of the coil when there is a current of 0.34 A in the coil.   *(2 marks)*
   b Using the approximation that the flux through the coil changes uniformly as it rotates through 90° from the horizontal position shown in Figure 13, calculate the back e.m.f. in this motor when the coil is rotating at a frequency of 12 Hz.   *(3 marks)*

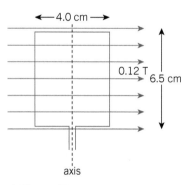
◄═ 4.0 cm ═► 0.12 T 6.5 cm axis

▲ **Figure 13**

Eddy current and regenerative braking

In a typical train journey, a large moving mass needs to be brought to a stop quite frequently, and then it needs to be accelerated once more to its cruising speed. The large kinetic energy of the moving mass must be dissipated safely each time it brakes, but there are two big issues with braking.

Dissipation and damage

The first issue with braking is that the method used to dissipate the kinetic energy can result in physical damage. This is equally true in cars, where the same method is used – friction. Brake pads on the moving vehicles are pushed against metal discs or drums attached to the rotating wheels, and the kinetic energy is transferred into the rotating metal parts. These get hot, and the increased internal energy is lost by conduction and radiation.

The large frictional forces involved in slowing down a large moving mass inevitably result in the brake pads and metal parts against which they rub wearing away. As a result, both parts need replacement at some point.

An alternative method of transferring kinetic energy of the moving vehicle into the internal energy of the brake disc uses eddy current braking. When the train needs to brake, an electromagnet is used to produce a strong magnetic field perpendicular to a metal disc rotating with the wheel axle. The rotating disc moves relative to this magnetic field and so has an e.m.f. induced in it along the radius of the disc, resulting in loops of electric current called eddy currents – see Figure 1.

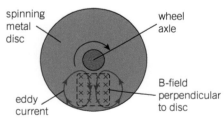

spinning metal disc
wheel axle
eddy current
B-field perpendicular to disc

▲ **Figure 1** *Eddy current braking*

By Lenz's Law, these electric currents interact with the magnetic field to produce a force ($F = ILB$) that opposes the motion of the wheel. This acts to reduce the angular velocity ω of the axle and decelerates the train. But where does the energy go? The eddy currents heat the disc by joule heating proportional to I^2R, so as before the dissipated energy increases the internal energy of the disc. However, this differs from the use of frictional brakes in one important aspect – there is no contact between the disc and the electromagnet, so neither is worn away by frictional forces.

Wasted energy

Although eddy current braking offers an improvement on the use of friction, there is a second issue which needs addressing in braking systems – they are wasteful of energy. All of the kinetic energy of the moving train is dissipated as heat every time it stops. This can be improved upon by the use of regenerative braking.

In regenerative braking, the force which decelerates the vehicle does not dissipate the kinetic energy. The work done as it brings the vehicle to rest transfers the energy in such a way that a substantial fraction of it can subsequently be returned to the vehicle as kinetic energy. In some systems, the rotational kinetic energy of the vehicle axles is transferred to a massive flywheel inside the vehicle. The spinning flywheel acts as an energy store while the vehicle slows down and stops, and this rotational kinetic energy can subsequently be used to accelerate the vehicle once more.

With electric and hybrid petrol-electric cars, and for electric trains, another opportunity presents itself. The principle here is similar to that of the Dinorwig Power Station, which uses its generators as motors in times of low energy demand to pump water uphill into a gravitational potential energy store. This enables the power station to generate electricity in times of high demand by running the water back down. In the vehicle, the motor is used as a generator to recharge the battery when the vehicle is braking.

Synoptic link

The Dinorwig power station was covered in Topic 9.5, Work and power.

▲ **Figure 2** *Regenerative braking means this electric car needs charging less frequently*

Summary questions

1 Vehicles that use eddy current or regenerative braking also need traditional brakes that act by friction on the wheels. Suggest why these are necessary.

2 Table 1 shows properties of some common metals. Discuss which of these would be best used as the spinning metal disc in an eddy current braking system.

▼ Table 1

metal	aluminium	copper	iron
density / $kg\,m^{-3}$	2700	8900	7900
electrical conductivity / $S\,m^{-1}$	3.5×10^7	6.0×10^7	1.0×10^7
cost / £ ton^{-1}*	600	3000	100

*approximate scrap metal prices February 2015

3 An electric car has regenerative braking that allows kinetic energy to be stored in a battery and then returned to kinetic energy with an efficiency of 50%. Discuss how this will affect the range of an electric car in
  a long journeys along motorways,
  b typical use in a town or city.
  You should think about the effect of frequently turning corners and driving on hilly terrain.

Practice questions

1   An ideal transformer with 600 primary turns and 120 secondary turns has a lamp attached to the secondary coil. The primary coil carries a current of 0.24 A, and the p.d. across the secondary coil is 3.0 V.

Which of the following statements is/are true?

1   The power of the lamp is 3.6 W

2   The conductance of the lamp is 0.40 S

3   An e.m.f. of 15 V is induced in the primary coil

A   1, 2, and 3

B   Only 1 and 2

C   Only 2 and 3

D   Only 1   (*1 mark*)

2   A d.c. motor is connected to a 12 V power supply. The rate of rotation varies with time as shown in Figure 1, becoming constant after the time *T*.

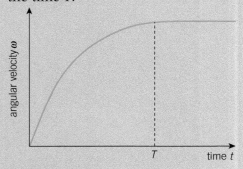

▲ Figure 1

Which of the following statements is/are true?

1   The current in the coil is greatest at the start

2   The induced e.m.f. in the coil decreases after time *T*

3   The magnetic force on the coil increases up to time *T*

A   1, 2, and 3

B   Only 1 and 2

C   Only 2 and 3

D   Only 1   (*1 mark*)

3   A straight wire of resistance 0.25 Ω has a p.d. of 1.8 V across it. A pair of magnets are placed each side of the wire as shown in Figure 2, with a magnetic flux density between them of 1.7×10^4 T.

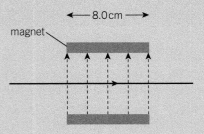

▲ Figure 2

Calculate the force acting on the wire. (*3 marks*)

4   An iron rod has 400 turns of wire coiled around it. There is a current in the coil. The rod has a cross-sectional area of $1.25 \times 10^{-5}\,m^2$.

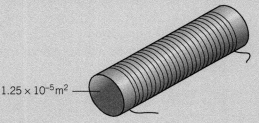

$1.25 \times 10^{-5}\,m^2$

▲ Figure 3

The **flux linkage** of the coil is 4×10^{-4} Wb.

a   Calculate the flux linking **one** turn of the coil.   (*1 mark*)

b   Calculate the flux density in the iron.   (*2 marks*)

OCR Physics B Paper 2864 June 2002

5   This question is about the e.m.f. induced in a coil of wire.

Figures 4a and 4b show different views of a simple seismometer (earthquake detector).

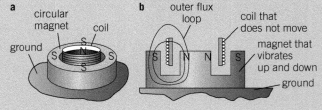

▲ Figure 4

The circular magnet sits on the ground and vibrates in a vertical direction when there is an earthquake. The cylindrical coil, which does not move, sits between the poles of the magnet. The movement of the magnet induces an e.m.f. in the coil.

a   Figure 4b shows two loops of flux linking the electric and magnetic circuits.

(i) On a copy of Figure 4b, Mark with an X the point on the outer flux line where the flux density is **lowest**.
(1 mark)

(ii) On the copy of Figure 4b, sketch **one** other flux loop. *(1 mark)*

b When the magnet is not moving, the flux linkage of the coil is 1.2×10^{-3} Wb turns.

To test the seismometer, the coil was completely removed from the field of the magnet in a time of 0.25 s.

Calculate the average e.m.f. across the coil during this time. *(2 marks)*

c The graph of Figure 5 shows how the flux linkage of the coil varies with time during an earthquake.

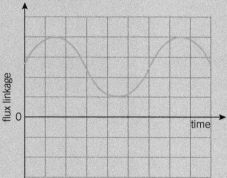

▲ Figure 5

(i) On a copy of Figure 5, sketch a graph to show how the e.m.f. across the coil changes with time. *(3 marks)*

(ii) In order to increase the amplitude of the e.m.f. across the coil during an earthquake, the construction of the seismometer is to be modified.

Describe **two** modifications that would increase the e.m.f.

Explain how they work. *(3 marks)*

OCR Physics B Paper 2864 January 2004 Q10

6 This question is about the changing magnetic fields in transformers.

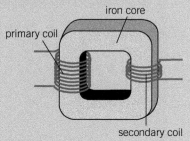

▲ Figure 6

An iron core is wound with primary and secondary coils of insulated copper wire to make a transformer, as shown in Figure 6.

a On a copy of Figure 6, sketch two complete loops of magnetic flux which pass through the secondary coil, when there is a current in the primary coil.
(2 marks)

b The ends of the secondary coil are now connected to an oscilloscope to obtain the e.m.f.–time graph in Figure 7.

On a copy of Figure 7, sketch the variation with time of the magnetic flux in the secondary coil. *(3 marks)*

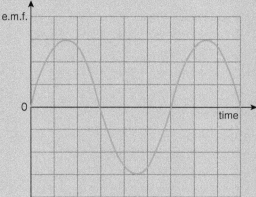

▲ Figure 7

c For an ideal transformer, the magnetic flux in the secondary coil is the same as the magnetic flux in the primary coil.

Use this to explain why the quantity

$$\frac{\text{e.m.f. across the coil}}{\text{turns of wire in the coil}}$$

Has the same value for both primary and secondary coils in an ideal transformer.
(2 marks)

d In a real transformer, eddy currents in the iron core will alter the flux in the two coils.

(i) Explain why eddy currents are set up in the core and suggest why this alters the flux in the core. *(2 marks)*

(ii) State and explain how the core should be constructed so as to reduce eddy currents. *(2 marks)*

OCR Physics B Paper 2864 January 2005

17 THE ELECTRIC FIELD
17.1 Uniform electric fields

Specification reference: 6.1.2a(i), 6.1.2a(iv), 6.1.2a(vi), 6.1.2b(i), 6.1.2b(iv), 6.1.2c(ii)

Hint

Gravity only pulls, but electric fields can pull or push. The similar equations for electrical and gravitational fields are helpful, but electrical charges can be negative or positive, so electric forces may be 'down-field' or 'up-field'. If a negative charge had been used in Figure 3, the force would have been upwards.

Synoptic link

In Topic 3.1, Current, p.d., and electrical power, in the section on p.d., charges moving in an electrical circuit were compared with objects moving downhill through a gravitational potential difference. You looked at gravitational fields in more detail in Topic 12.3, Gravitational potential in a uniform field. On this basis, the comparison can now be developed.

Particle accelerators

What is the connection between a dental X-ray machine and CERN's Large Hadron Collider (Figures 1 and 2)? Although they are very different in scale, both use potential differences to accelerate charged particles.

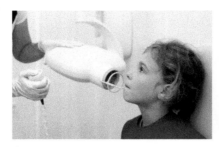

▲ **Figure 1** Dental X-ray machine

▲ **Figure 2** Part of CERN's Large Hadron Collider

Uniform fields

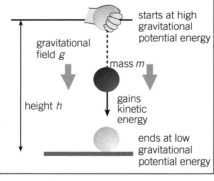

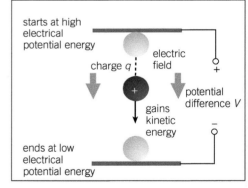

▲ **Figure 3** Gravitational and electrical energy changes

The analogy between electric and gravitational fields is extremely useful. For gravitational fields, the field strength $g = \dfrac{F_{grav}}{m}$, so the weight $F_{grav} = mg$. In the same way, the electric field strength $E = \dfrac{F_{electric}}{q}$ so the electric force $F_{electric} = qE$.

For the uniform fields in Figure 3, the force is constant, so the work done by the field on the object is given by force × distance. In the gravitational case this is just the gravitation potential energy, $W = (mg) \times h = mgh$. In the electrical case the distance moved is d, so $W = (qE) \times d = qEd$. The electric potential difference is given by $V = \dfrac{W}{q}$, so $W = Vq$. Together, these equations give $qEd = Vq \therefore E = \dfrac{V}{d}$.

 ## Worked example: Electric field strength in a spark plug

The spark plug in a car's petrol engine ignites fuel by applying a voltage of 30 000 V across a gap of 1.3 mm to create a spark. Calculate the strength of the electric field in the spark gap.

Step 1: Convert the measurements to SI base units and calculate E.

$$E = \frac{V}{d} = \frac{30\,000\,\text{V}}{1.3\,\text{mm}} = \frac{30\,000\,\text{V}}{1.3 \times 10^{-3}\,\text{m}} = 2.30\ldots \times 10^{7}\,\text{V m}^{-1}$$

Step 2: Choose the correct units and round the answer to match the number of significant figures in the data.

From the equation $E = \dfrac{F_{\text{electric}}}{q}$, the units of E are N C^{-1}. They are the same as V m^{-1}, which arise naturally out of this calculation. You can use either, so $E = 2.3 \times 10^{7}\,\text{N C}^{-1}$ $= 2.3 \times 10^{7}\,\text{V m}^{-1}$ (2 s.f.).

▲ **Figure 4** *The powerful E-field in a spark plug creates a spark that ignites the petrol–air mixture*

Equipotentials in electric fields

As in a uniform gravitational field, **equipotential surfaces** linking points of equal potential form equally spaced planes in a uniform electric field. Figure 5 illustrates the equal spacing and also shows how the field becomes non-uniform near the edges of the two charged plates.

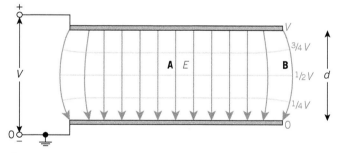

▲ **Figure 5** *Uniform electric field*

> ### Synoptic link
>
> Equipotential surfaces are perpendicular to the field lines for g-fields, just as for E-fields (Topic 12.3, Gravitational potential in a uniform field).

The equipotential surfaces are perpendicular to the field lines. Note that $E = \dfrac{V}{d}$ can only be applied where the E-field lines are straight. Where the field lines curve, as at the edges of Figure 5, the equation that applies is the more general one: $E = -\dfrac{\mathrm{d}V}{\mathrm{d}r}$, where r is distance measured along the field line. The minus sign in the equation shows that the direction of the E-field is in the opposite direction from the increase in V. The minus sign is not usually written when using the uniform field equation $E = \dfrac{V}{d}$, as this invariably uses parallel charged plates and E is in the direction from the positive plate to the negative one.

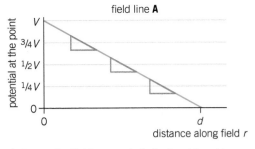
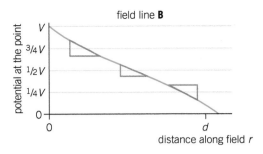

▲ **Figure 6** *Field strength E obtained from V–r graph*

The variations of potential along field lines **A** (in the central region) and **B** (at the edge) in Figure 5 are depicted in Figure 6, where the distance r is measured along each field line. Tangents are drawn on each graph where each field line crosses the equipotentials to obtain E at the point by calculating the gradient $\frac{\mathrm{d}V}{\mathrm{d}r}$ at those points. Each of the gradient triangles, drawn in the colour of the field lines in Figure 5, has the same-sized base so that the vertical height of each triangle indicates the size of the gradient. The gradient along line **A** is constant, as the field along field line **A** is $E = \frac{V}{d}$ everywhere. Along field line **B**, the gradient varies. Calculations of the gradient give $E = 0.96\frac{V}{d}$ at the $\frac{3}{4}V$ and $\frac{1}{4}V$ equipotentials and $E = 0.58\frac{V}{d}$ at the $\frac{1}{2}V$ equipotential. These estimates are confirmed by the spacing of field lines in Figure 5, where field line **A** is equidistant from its neighbours along its whole length, whilst field line **B** is further from its neighbours near the centre.

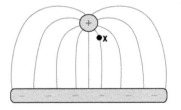

▲ **Figure 7** *The electric field pattern between a positively charged sphere and a negatively charged plate*

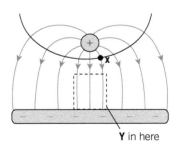

▲ **Figure 8** *Answer to Worked example: Sketching equipotentials*

> ### 🖩 Worked example: Sketching equipotentials
>
> For the electric field pattern in Figure 7: (a) use arrowheads to indicate the direction of the field; (b) sketch an equipotential curve through the point marked **X**; (c) write **Y** in a region where the electric field is approximately uniform.
>
> **Step 1**: Identify the direction from the fact that electric field (the direction of the force on a positive charge) goes from + to −.
>
> **Step 2**: Draw a free-hand curve through **X** perpendicular to each field line it crosses.
>
> **Step 3**: Mark **Y** in a region where the field lines are (roughly) straight and equally spaced.
>
> The completed sketch could be that shown in Figure 8.
>
> (In an examination, precise right angles would not be expected in the equipotential sketch.)

Millikan's oil-drop experiment

In 1897, J J Thomson identified the electron as a fundamental particle and measured the ratio of its charge to its mass. However, as no-one knew the charge of an electron, he could not deduce its mass. American physicists Robert Millikan and his student Harvey Fletcher tackled this problem, and in 1913 Millikan published his Nobel-prize-winning paper on the charge of the electron. The arrangement used by Millikan and Fletcher is shown, in simplified form, in Figure 9.

Oil is squirted through a hole in a metal plate and falls as a spray of tiny droplets in view of a microscope. The squirting charges many of the drops positively or negatively. Applying a p.d. across the two metal plates results in a uniform field in the gap. The direction and magnitude of the p.d. is then adjusted so that a particular drop is held suspended under the equal and opposing effects of the uniform gravitational and electric fields between the plates (Figure 10).

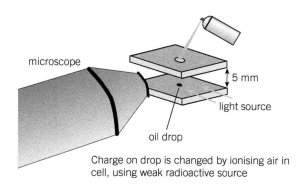

Charge on drop is changed by ionising air in cell, using weak radioactive source

▲ **Figure 9** *Millikan's experimental set-up*

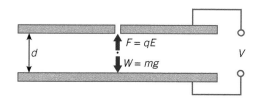

▲ **Figure 10** *Forces on a stationary oil drop*

When the p.d. needed to hold the drop stationary has been measured, the power supply is switched off, allowing the drop to fall. Because the drop is so tiny, it rapidly reaches its terminal velocity in air. By timing the falling drop over a marked distance, the known viscosity of air can be used to calculate the radius of the drop. As the density of the oil is known, the mass of the drop can be calculated.

From Newton's first law, Figure 10 shows that $qE = mg$.

As $E = \frac{V}{d}$, $qE = \frac{qV}{d} = mg$ ∴ $q = \frac{mgd}{V}$.

Millikan and Fletcher obtained thousands of values of charge in their experiments and found that all of them were simple multiples of the same value: 1.6×10^{-19} C. No value less than this was obtained. The conclusion is inescapable – the charge of the electron, symbol e, is -1.6×10^{-19} C. As hydrogen atoms are neutral, this means that the charge of the proton is $+1.6 \times 10^{-19}$ C.

 Worked example: Calculating the charge on an oil drop

In a modern version of Millikan's experiment, an oil drop of mass 5.20×10^{-15} kg was held stationary by a p.d. of 470 V between two plates 4.42 mm apart. Calculate the charge on the oil drop, using $g = 9.81$ N kg^{-1}.

Step 1: Calculate the weight of the drop.

$$W = mg = 5.20 \times 10^{-15} \text{kg} \times 9.81 \text{N kg}^{-1} = 5.1012 \times 10^{-14} \text{N}$$

Step 2: Calculate the electric field strength E.

$$E = \frac{V}{d} = \frac{470 \text{V}}{4.42 \times 10^{-3} \text{m}} = 106\,334.8... \text{V m}^{-1}$$

Step 3: Equate forces on the drop and solve to find q.

$$F_{\text{electric}} = W \text{ so } F_{\text{electric}} = qE = 5.101 \times 10^{-14}\,\text{N}$$

$$\therefore q = \frac{5.1012 \times 10^{-14}\,\text{N}}{106\,334.8\ldots\text{V}\,\text{m}^{-1}} = 4.80 \times 10^{-19}\,\text{C} \text{ (3 s.f.)}$$

This drop carried a charge of $3e$.

Summary questions

1 Two metal plates 5.0 cm apart are connected to a 500 V supply.
   a Draw a diagram with solid lines to represent the electric field between and just outside the plates, with arrows to indicate the direction of the field. (*1 mark*)
   b Add and label dotted lines to represent lines of equipotential at 100 V intervals. (*2 marks*)
   c Calculate the electric field strength between the plates. (*2 marks*)

2 Damp air will ionise and conduct electricity in electric fields greater than about $5 \times 10^5\,\text{V}\,\text{m}^{-1}$. Lightning strikes the earth from a thundercloud 200 m above the ground. The bottom of the cloud carries a charge of 20 C. Calculate
   a the p.d. between the ground and the cloud, (*1 mark*)
   b the energy dissipated if the lightning discharges the cloud base. (*2 marks*)

3 In a repeat of Millikan's experiment, the following values of charge on the oil drops were obtained. The unit used for charge was not the coulomb, but an older unit called the franklin.

Charge /franklin	9.59×10^{-10}	2.40×10^{-9}	4.32×10^{-9}	2.88×10^{-9}	3.36×10^{-9}

   Use the data to deduce the electronic charge in franklin and to calculate the value of 1 franklin in C. (*5 marks*)

17.2 Deflecting charged beams

Specification reference: 6.1.2a(v), 6.1.2b(i),
6.1.2c(ii), 6.1.2c(iv), 6.2.1c(i)

Identifying the electron

In Topic 17.1, Uniform electric fields, it was mentioned that J J Thomson (Figure 1) had measured the mass:charge ratio for the electron. How did he do this? He had suspected that the newly discovered 'cathode rays' were particle beams, as they carry electrical charge. He knew that moving charges could be affected by both electric and magnetic fields, and he used this physics elegantly to give the result that earned him the Nobel Prize.

▲ **Figure 1** *The English physicist J J Thomson, who discovered the electron*

Learning outcomes

Describe, explain, and apply:

→ the term electronvolt

→ the force on a moving charged particle due to a uniform magnetic field

→ calculations and estimates for motion of a charged particle in a uniform magnetic field, $F = qvB$

→ calculations and estimates involving $E = \dfrac{V}{d}$ for a uniform electric field.

Deflecting beams of charged particles in uniform electric fields

In Millikan's oil-drop experiment (Topic 17.1, Uniform electric fields), charged oil drops were brought to a stop by applying an E-field in the direction in which they are falling. Figure 2 shows what happens if the E-field is perpendicular to the particle velocity instead.

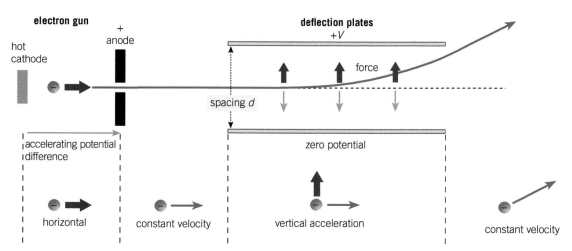

▲ **Figure 2** *Deflecting an electron beam with an electric field*

Electrons 'boil off' the hot cathode and are accelerated by the electron gun, gaining kinetic energy equal to the potential energy drop between cathode and anode, $\frac{1}{2}mv^2 = eV_{gun}$, where e is the electronic charge. This relationship crops up so often in physics that it led to a convenient unit of energy, the **electronvolt** (eV). This is defined as the work done on or by an electron when it moves across a potential difference of 1 volt. From this definition, $1\,eV = 1.6 \times 10^{-19}\,C \times 1\,V = 1.6 \times 10^{-19}\,J$, so you convert an energy value in eV into joules by multiplying by $1.6 \times 10^{-19}\,J\,eV^{-1}$.

Having left the electron gun, the electrons move with a constant horizontal component of velocity as there is no longer any horizontal force acting on them. In the vertical direction, though, there is a constant vertical force – the weight of the electrons – pulling them downwards.

 Worked example: Electrons falling in a simple deflection tube

A typical electron deflection tube used in schools has an accelerating voltage of up to 5000 V and is 30 cm long horizontally. Calculate the vertical distance fallen by the electrons whilst they travel across the tube when the accelerating voltage is 1000 V ($1\,eV = 1.6 \times 10^{-19}\,J$, $m = 9.1 \times 10^{-31}\,kg$).

Step 1: Calculate the horizontal velocity component of the electrons.

$$\frac{1}{2}mv^2 = 1000\,eV = 1.6 \times 10^{-19}\,J\,eV^{-1} \times 1000\,V = 1.6 \times 10^{-16}\,J$$

$$\therefore mv^2 = 2 \times 1.6 \times 10^{-16}\,J = 3.2 \times 10^{-16}\,J$$

$$\therefore v^2 = \frac{3.2 \times 10^{-16}\,J}{m} = \frac{3.2 \times 10^{-16}\,J}{9.1 \times 10^{-31}\,kg} = 3.51... \times 10^{14}\,m^2\,s^{-2}$$

$$v = \sqrt{3.51... \times 10^{14}\,m^2\,s^{-2}} = 1.87... \times 10^7\,m\,s^{-1}$$

Step 2: Calculate the time taken for an electron to cross the apparatus, a distance of 30 cm = 0.30 m.

The horizontal component of velocity is constant, so $v = \frac{s}{t}$

$$\therefore t = \frac{s}{v} = \frac{0.30\,m}{1.87... \times 10^7\,m\,s^{-1}} = 1.59... \times 10^{-8}\,s$$

Step 3: Use the appropriate *suvat* equation to calculate the distance fallen under gravity in this time.

$s = ut + \frac{1}{2}at^2 \therefore s = \frac{1}{2}at^2$ as there is no initial vertical component of velocity

$$s = 0.5 \times 9.81\,m\,s^{-2} \times (1.59... \times 10^{-8}\,s)^2$$
$$= 1.3 \times 10^{-15}\,m\,(2\,s.f.)$$

The tiny distance fallen by the electron in the worked example (about the diameter of a hydrogen nucleus) means that the effects of gravity can be ignored in these electron tubes. At higher accelerating p.d.s the distance fallen would be even less.

Once between the deflection plates, the electrons are in a uniform electric field of strength $E = \frac{V}{d}$. In Figure 2 the electric field E is vertically downwards, but the electrons have negative charge so that the force acting on each is $F_{electric} = (-e)E = -eE$ in the opposite direction to E — upwards. This upward force means that the vertical component of the electrons' velocity, initially ~0, increases as they pass through the plates. It is useful to compare the path between the plates with that of a projectile thrown horizontally in a constant gravitational field — this trajectory is a parabola, and so is the path taken by the electrons. Imagine Figure 2 inverted (or turn the book upside down) and you will see the similarity.

Once the electrons leave the uniform E-field between the deflection plates, they continue in a straight line.

 Worked example: Electron deflection in a simple deflection tube

In the electron deflection tube of the previous worked example, the deflection plates are 8.0 cm long in the horizontal direction and separated by a vertical distance $d = 5.0$ cm (see Figure 2). The electron gun voltage is 1000 V and a p.d. of 400 V is applied between the deflection plates. Calculate the vertical displacement of electrons as they emerge from the deflection plates.

Step 1: Use the horizontal velocity component of the electrons, v_{horiz}, from the previous example to calculate the time for which the electrons are in the vertical E-field.

$$v_{horiz} = \frac{s}{t} \therefore t = \frac{s}{v_{horiz}} = \frac{0.080\,\text{m}}{1.88 \times 10^7\,\text{m s}^{-1}} = 4.25... \times 10^{-9}\,\text{s}$$

Step 2: Calculate the field strength E and use this to find the force acting on the electrons.

$$E = \frac{V}{d} = \frac{400\,\text{V}}{0.050\,\text{m}} = 8000\,\text{N C}^{-1}$$

$$F_{vert} = eE = 1.6 \times 10^{-19}\,\text{C} \times 8000\,\text{N C}^{-1} = 1.28 \times 10^{-15}\,\text{N}$$

Step 3: Calculate the vertical acceleration of each electron and therefore the vertical displacement whilst the electrons are between the plates.

$$F_{vert} = ma_{vert} \therefore a_{vert} = \frac{F_{vert}}{m} = \frac{1.28 \times 10^{-15}\,\text{N}}{9.1 \times 10^{-31}\,\text{kg}}$$
$$= 1.40... \times 10^{15}\,\text{m s}^{-2}$$

(This is about $10^{14}\,g$, so the deflection will be considerably more than in the previous worked example!)

$$s_{vert} = \frac{1}{2}a_{vert}t^2 = 0.5 \times 1.40... \times 10^{15}\,\text{m s}^{-2} \times (4.25... \times 10^{-9}\,\text{s})^2$$
$$= 0.013\,\text{m} = 1.3\,\text{cm (2 s.f.)}$$

Moving charged particles in uniform magnetic fields

Particle constrained in a wire

If a wire carrying a current cuts across a magnetic field, a force acts on it. The charges making up the current are moving, so each electron in the wire is being pushed to one side, and the sum of all these forces is the resultant force on the wire.

Figure 3 shows a length L of wire in a magnetic field. In this wire, the current I is due to N charges in the wire, moving together at a mean

Synoptic link

Topic 16.4, Flux changes and force, showed that the force F on a wire of length L carrying a current I perpendicular to a magnetic field B is given by $F = ILB$, with F being perpendicular to both the wire and the magnetic field.

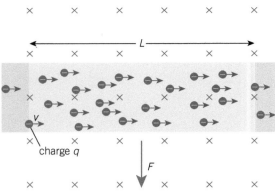

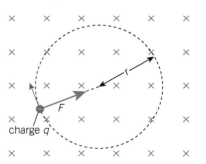

▲ **Figure 3** *Force on moving charges in a magnetic field*

drift velocity v. As the resultant force F_{total} on the wire is ILB, the force on one electron must be $F_{charge} = \dfrac{F_{total}}{N}$.

The N charges in the length L of the wire in the B-field move with velocity v, so the time taken for all N to move out of this length is the time taken for the very last one (the labelled charge in Figure 3) to move that distance L. This is the time t, where $v = \dfrac{L}{t}$, so $t = \dfrac{L}{v}$.

The total charge Q flowing out of the length L in time t is $Q = Nq$, where q is the charge of one of the moving particles (electrons in this example). By definition, $Q = It$, so $Nq = It = I\dfrac{L}{v}$. As the total force on the wire is $F_{total} = ILB$, we can write $F_{total} = NqvB$. This means that the force on each electron, $F_{charge} = \dfrac{F_{total}}{N} = \dfrac{NqvB}{N} = qvB$.

Particles in beams

The force in Figure 3 is perpendicular to the wire and the field, and makes the whole wire move. If the charges q are all moving independently, not constrained by the wire, then their movement will be different, as each independent charge experiences a force at right angles to its velocity.

When a freely moving body travelling at a steady speed experiences a constant force at right angles to its velocity, then it will move in a circle, with the centripetal force $F = \dfrac{mv^2}{r}$, where r is the radius of the circle. In planetary orbits, the centripetal force is caused by gravitational attraction. In particle accelerators, it is the force $F = qvB$.

Figure 4 sums up the effect of a uniform magnetic field on moving charged particles — the magnetic force $F = qvB$ causes the centripetal motion described by $F = \dfrac{mv^2}{r}$, where r is the radius of the circular orbit produced. The circular motion of charged particles in uniform magnetic fields can be used to analyse their properties, as the next worked example illustrates.

▲ **Figure 4** *Centripetal motion due to qvB*

⊞ **Worked example: Finding the speed of a subatomic particle**

In a cloud-chamber photograph, a curved path is visible. The curve is an arc of a circle of radius 3.5 cm, and the path of the particle is perpendicular to a uniform magnetic field of flux density 1.2 T. Assuming that the particle is a proton (charge of $+1.6 \times 10^{-19}$ C, mass 1.7×10^{-27} kg), calculate its speed.

Step 1: Equate the magnetic force to the centripetal force and rearrange to make v the subject.

$$qvB = \frac{mv^2}{r} \therefore qB = \frac{mv}{r} \therefore v = \frac{qBr}{m}$$

Step 2: Substitute the data into the equation.

$$v = \frac{qBr}{m} = \frac{1.6 \times 10^{-19}\,\text{C} \times 1.2\,\text{T} \times 3.5 \times 10^{-2}\,\text{m}}{1.7 \times 10^{-27}\,\text{kg}}$$
$$= 4.0 \times 10^6\,\text{m s}^{-1}\ (2\ \text{s.f.})$$

Electric and magnetic fields in particle accelerators

Large particle accelerators use electric fields to accelerate the charged particles in several stages, with each electrode pair giving the particles another push. To achieve the required kinetic energy, the protons in CERN's Large Hadron Collider must be accelerated for about 20 minutes, during which they will travel a distance comparable with the circumference of the Earth's orbit. This can only be done in a circular accelerator, with the protons going round very many times. The very strong magnetic fields ($B > 8\,T$) produced by superconducting electromagnets are used to make the high-speed protons follow a circular path.

Synoptic link

You will learn more about particle accelerators in Topic 18.2, Accelerating charges and electron scattering.

Summary questions

1 A proton travelling at $1.5 \times 10^5\,m\,s^{-1}$ passes into a region of uniform magnetic field $B = 0.36\,T$ in a direction perpendicular to the B-field (mass of proton = $1.7 \times 10^{-27}\,kg$, charge of proton = $+e = +1.6 \times 10^{-19}\,C$). Calculate
   a the magnetic force acting on the proton, *(2 marks)*
   b the radius of the circular path followed by the proton. *(3 marks)*

2 Figure 5 is a cloud-chamber photograph showing the tracks of a proton and an alpha particle (consisting of two protons and two neutral neutrons of the same mass as protons) in a uniform magnetic field. The two particles both have the same initial speed of $1.8 \times 10^7\,m\,s^{-1}$. Explain
   a how the image shows that each particle is losing kinetic energy as it travels along its path, *(2 marks)*
   b why the initial radius of curvature of the alpha particle is greater than that of the proton. *(2 marks)*

3 Figure 6 shows an electron beam passing into a region where a magnetic field B is at right angles to an electric field E. The field strengths are adjusted until the electric force upwards exactly balances the magnetic force downwards. The p.d. across the electron gun that produces the electron beam is the same as the p.d. V between the two plates. Show that,
   a the speed v of the electrons leaving the electron gun is
   $$v = \sqrt{\frac{2eV}{m}},$$
   (2 marks)
   b the ratio $\dfrac{e}{m}$ is given by $\dfrac{e}{m} = \dfrac{eV}{2d^2B^2}$. *(3 marks)*

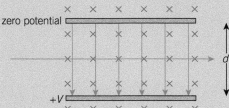

▲ **Figure 6** *Electron beam passing through electric and magnetic fields*

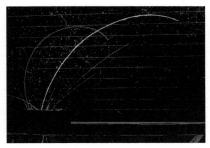

▲ **Figure 5** *False-colour cloud-chamber photograph*

17.3 Charged spheres

Specification reference: 6.1.2a(ii), 6.1.2a(iii), 6.1.2b(i), 6.1.2b(ii), 6.1.2b(iii), 6.1.2c(i), 6.1.2c(iii)

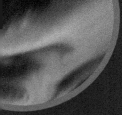

Learning outcomes

Describe, explain, and apply:

→ the fact that a spherically symmetrical distribution of charges is equivalent to a point with the same total charge at its centre

→ the electric field of a charged object, and the force on a charge in an electric field; inverse-square law for a point charge, $\frac{1}{r^2}$ relationship

→ electrical potential energy and electric potential due to a point charge, $\frac{1}{r}$ relationship

→ difference in electric potential as the area under a section of the curve of a graph of electric field versus distance

→ electric field strength related to the tangent to the graph of electric potential versus distance

→ changes in electrical potential energy as the area under a section of the curve of a graph of electric force versus distance

→ electric force related to the tangent to the graph of electrical potential energy against distance

→ electrical potential energy = $k\frac{qQ}{r}$ and $V_{electric} = k\frac{Q}{r}$

→ for radial components from a point charge, $F_{electric} = k\frac{qQ}{r^2}$ and $E = \frac{F_{electric}}{q} = k\frac{Q}{r^2}$, where $k = \frac{1}{4\pi\varepsilon_0}$.

A hair-raising experience

Although the head of the student in Figure 1 is not a perfect sphere, it is certainly charged. The high voltage generated by the van de Graaff machine has transferred charge to her with the result that the surface of her body is covered with positive electrical charges. The mutual repulsion of these charges makes her fine, straight hair stand out radially from her head, following the electrical field lines (with gravity pulling it down a little).

▲ **Figure 1** *Field lines of hair*

Coulomb's law

In the 1780s Charles-Augustin de Coulomb (Figure 2) investigated the pair of equal and opposite forces between two electrically charged conducting spheres at varying distances. He established that the force between two charges q and Q is directly proportional to each of the charges, and inversely proportional to the square of the distance r between the centres of the two spheres. Writing the constant of proportionality as k, **Coulomb's law** is

▲ **Figure 2** *The French physicist Charles-Augustin de Coulomb*

$$F_{electric} = k\frac{qQ}{r^2}$$

If the charges are opposite in sign (one positive and one negative), the negative value for force indicates that the force is attractive (Figure 3). For two positive or two negative charges, the force is repulsive.

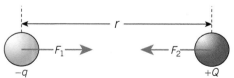

r is the position of q measured from Q
F_1 is the force on q due to Q
F_2 is the force on Q due to q
$F_1 = \frac{kqQ}{r^2} = F_2$

▲ **Figure 3** *Opposite charges attract*

As electric field strength $E = \frac{F_{electric}}{q}$, Coulomb's law leads to

$E = \frac{F_{electric}}{q} = k\frac{qQ}{r^2} \times \frac{1}{q} = k\frac{Q}{r^2}$, where Q is the source of the electric field.

Worked example: The electric field and force inside a hydrogen atom

A hydrogen atom has a proton (charge $+1.6 \times 10^{-19}$ C) and an electron (charge -1.6×10^{-19} C) at a mean distance of 110 pm from the proton. Calculate the electric field at the electron's position and the force acting on the electron.

Step 1: Convert r into m and calculate E, using $k = 8.98 \times 10^9 \, \text{N m}^2 \, \text{C}^{-2}$.

$$r = 110 \times 10^{-12} \, \text{m} = 1.1 \times 10^{-10} \, \text{m}$$

$$E = k\frac{Q}{r^2} = 8.98 \times 10^9 \, \text{N m}^2 \, \text{C}^{-2} \times \frac{1.6 \times 10^{-19} \, \text{C}}{(1.1 \times 10^{-10} \, \text{m})^2}$$

$$= \frac{1.44 \times 10^{-9} \, \text{N m}^2 \, \text{C}^{-1}}{1.21 \times 10^{-20} \, \text{m}^2} = 1.19... \times 10^{11} \, \text{N C}^{-1}$$

Step 2: Calculate F_{electric} from E.

$$E = \frac{F_{\text{electric}}}{q}$$

$$\therefore F_{\text{electric}} = qE$$

$$= -1.6 \times 10^{-19} \, \text{C} \times 1.19... \times 10^{11} \, \text{N C}^{-1}$$

$$= -1.9 \times 10^{-8} \, \text{N} \quad (2 \text{ s.f.})$$

Inverse-square laws, flux, and point sources

The similarity between electric fields and gravitational fields was mentioned in Topic 17.1, Uniform electric fields. Here you see that the force between two charges is similar to the gravitational force between two masses m and M. Inverse-square laws are not restricted to these two cases. Whenever a quantity spreads out from a point, if that quantity is conserved, then it will follow this same rule. Another example is the intensity of light. Figure 4 shows a cone of light spreading to illuminate an area A at a distance r. If that same cone of light travels to a surface twice as far away, the area illuminated is $4A$. Spreading the same energy over a surface area four times larger means that the luminous intensity (the power per unit area received by the surface) is reduced four times. Trebling the distance would reduce the intensity $3^2 = 9$ times.

For electric and gravitational fields, the spacing between lines indicates the field strength. If there are N field lines spreading outwards in all directions from a point charge Q, at a distance r from the central charge they will meet the surface of a sphere of radius r (Figure 5).

The sphere of radius r has surface area $4\pi r^2$, so the density of field lines on the surface, and therefore the field strength, is $\frac{N}{4\pi r^2}$. The number of field lines N is proportional to Q, so the field strength $\propto \frac{Q}{r^2}$.

Experiments with 'point sources' always use small (or not-so-small) charged spheres, but a charged sphere is exactly equivalent to a point source at its centre bearing the same charge. Look at Figure 5 and imagine the spherical surface of the charge shrinking, whilst keeping the same number of field lines spreading out.

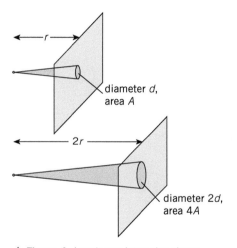

▲ **Figure 4** *Luminous intensity obeys an inverse-square law*

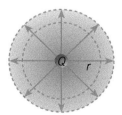

▲ **Figure 5** *Conservation of field lines around a point charge Q*

Synoptic link

Look back at Topic 12.4, Gravitational potential in a radial field, to remind yourself about the gravitational potential for an inverse-square gravitational field.

Electric potential near a charged sphere

The gravitational potential at a radial distance r from the centre of a mass M, defined as $V_{grav} = -\dfrac{\text{gravitational potential energy}}{m}$, is given by $V_{grav} = -\dfrac{GM}{r}$ for an inverse-square gravitational field $g = -\dfrac{GM}{r^2}$. In exactly the same way, at a radial distance r from the centre of a charge Q, where the electric field is $E = k\dfrac{Q}{r^2}$, the electric potential is $V = k\dfrac{Q}{r}$. Note that the minus sign in the gravitational case indicates the attractive nature of the force. The electrical case has no minus sign because the force is attractive only if one of the two charges is negative, in which case their product qQ supplies the negative sign.

Gravitational and electric fields and potential

The analogy between these two fields can be used when considering how the field–distance graph leads to potential. Figure 6 compares the variation in the gravitational field with radial distance r from a mass M and the variation in the electric field with radial distance r from a charge $-Q$. A negative charge was chosen, as field strength = force per unit (positive) charge, and making Q negative means that, like the gravitational case, the force on a positive charge q will be attractive, and the field lines will head inwards.

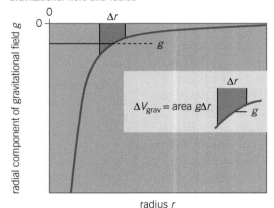

Gravitational field and radius

ΔV_{grav} = area $g\Delta r$

Electric field and radius

ΔV_{elec} = area $E\Delta r$

▲ **Figure 6** *The field graph gives potential*

Study tip

To reinforce this work on gradients and areas, it is worth comparing the situation here with displacement and velocity, where displacement is found from an area under the velocity–time graph (an integral) and velocity is found from the gradient of a tangent to the displacement–time graph (a differential).

In each graph a distance Δr is indicated. The potential difference ΔV across this distance, that is, the work done in moving a unit mass/charge radially across Δr, is equal to (– force per unit mass or charge) multiplied by the distance moved, that is, (–field strength) × Δr. The minus sign is needed as the field points back towards the centre of the mass or charge whilst the potential increases as r increases, which is the opposite direction. In each graph the mean value of the field strength in the interval is labelled. The area between the curve and the r-axis, shown shaded, is equal to this potential difference.

The value of field is similarly related to the potential graph, as shown in the pair of graphs in Figure 7.

Gravitational potential and radius

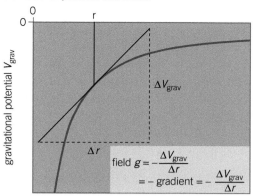

Electrical potential and radius

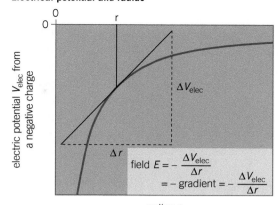

▲ **Figure 7** *The potential graph gives field*

Field strength (g or E) at a distance r from the centre $= -\dfrac{\mathrm{d}V}{\mathrm{d}r}$, which is the gradient of a tangent to the graph at that position.

🖩 Worked example: Ionising a hydrogen atom

Using the data from the previous worked example, calculate the potential difference ΔV between the position 110 pm from a proton and a second position a great distance from the proton. Use this potential difference to find the energy needed to ionise the atom.

Step 1: Calculate the potential V_{elec} 110 pm from the proton.

$$V_{\text{elec}} = k\frac{Q}{r} = 8.98 \times 10^9\,\text{N}\,\text{m}^2\,\text{C}^{-2} \times \frac{1.6 \times 10^{-19}\,\text{C}}{1.1 \times 10^{-10}\,\text{m}} = 13.0\ldots\,\text{V}$$

Step 2: Calculate ΔV, using the fact that at a great distance $r \approx \infty$ from the proton, $k\dfrac{Q}{r}$ will be vanishingly small.

$V_{\text{elec}} = 0$ far from the proton so $\Delta V = $ final V – initial V
$= 0 - 13.0\ldots\,\text{V} = -13.0\ldots\,\text{V}$

Step 3: Use ΔV and the electron charge to calculate the work done using $W = $ p.d. $\times$ charge.

$W = \Delta V e = (-13.0\ldots\,\text{V}) \times (-1.6 \times 10^{-19}\,\text{C})$
$= 2.1 \times 10^{-18}\,\text{J}$ (2 s.f.)

Hint

The difference in electric potential ΔV_{elec} between two points is exactly the same quantity as the potential difference (p.d.) between two points in electrical circuits.

Electric force and potential energy

In Figures 6 and 7 electric field E and electric potential V are related — in the E–r graph, the area is the change in potential V, whilst in the V–r graph, the tangent gives the field E. We can construct similar plots of electric force F_{electric} versus r and of electrical potential energy versus r.

From Coulomb's law, $F_{\text{electric}} = k\dfrac{qQ}{r^2}$, and because electrical potential energy $= qV = k\dfrac{qQ}{r}$ (from Topic 17.1, Uniform electric fields), it follows that the former graph can be used to calculate changes in electrical potential energy from the area, and the latter will have a tangent that represents the electric force.

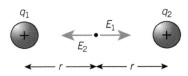

▲ **Figure 8** *Field between two positive charges*

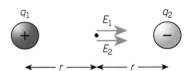

▲ **Figure 9** *Field between a positive charge and a negative charge*

Vectors and scalars

Electric field E, like electric force F_{elec}, is a vector quantity, whilst electric potential V, like electrical potential energy qV, is a scalar quantity. This distinction is important when considering the interaction of two electric fields, such as in Figure 8, where two positive charges are placed a distance $2r$ apart. Midway between the two charges, the electric fields are in opposite directions. If $q_1 = q_2$, then $E_1 = E_2$ and the resultant electric field E is zero. Potential is a scalar, so the resultant potential V at the midpoint is obtained by adding the two separate potentials to give $V = \dfrac{kq_1}{r} + \dfrac{kq_2}{r}$.

If q_2 is replaced by a negative charge, as in Figure 9, the situation becomes different. Now the two electric fields are in the same direction, so E_1 and E_2 will add. However, the sign of q_2 is negative, so the resultant potential is $V = \dfrac{kq_1}{r} + \dfrac{k(-q_2)}{r} = \dfrac{kq_1}{r} - \dfrac{kq_2}{r}$. For the case where q_1 and q_2 are the same size, then $V = 0$.

Summary questions

1 Two small metal spheres, each carrying a charge of 2.5 µC, are mounted on insulating supports a distance of 0.15 m apart. Calculate
 a the force each charge exerts on the other, *(2 marks)*
 b the potential energy stored in the system. *(2 marks)*

2 The graphs of Figure 10 show how electric field E and potential V vary with distance r from a point charge.
 a Use the $E-r$ graph to find the potential difference ΔV in moving from 10 cm to 15 cm from the point charge, and use the $V-r$ graph to confirm this value. *(3 marks)*
 b Use the $V-r$ graph to find the electric field E at a distance of 10 cm from the point charge, and use the $E-r$ graph to confirm this value. *(6 marks)*

3 Figure 11 shows four point charges arranged in a square of side 25 cm.

 Calculate: (a) the electric field E; (b) the potential V at the point X in the centre of the square. *(5 marks)*

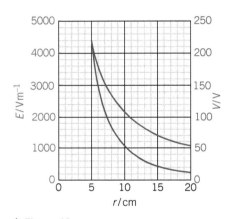

▲ **Figure 10**

▲ **Figure 11**

Magnetic fields

Faraday and Lenz:

- Calculations and estimates using induced e.m.f. = rate of change of flux linked,
$$\varepsilon = -\frac{\mathrm{d}(N\Phi)}{\mathrm{d}t}$$

- Describing and explaining electromagnetic forces qualitatively as arising from tendency of flux lines to contract, or from the interaction of induced poles

- Magnetic flux from a coil

- Appropriate use of the terms: B-field, flux, flux linkage, induced e.m.f.

- Sketching and interpreting diagrams of lines of flux in magnetic circuits, showing their continuity

Transformers:

- Simple linked electric and magnetic circuits

- Flux produced by current turns

- The need for large conductance and permeance in an electromagnetic machine and the effect on those factors of increasing the dimensions

- The effect of iron and air gap in a magnetic circuit

- The effect of eddy currents in an electromagnetic machine

- The action of a transformer

- Calculations and estimates for an ideal transformer involving $\frac{V_1}{V_2} = \frac{N_1}{N_2}$ and $\frac{I_2}{I_1} = \frac{N_1}{N_2}$

- Practical task: investigating transformers

Generators:

- The action of a generator in terms of the change of flux linked, produced by relative motion of flux and conductor

- Practical task: investigating induced e.m.f. in a coil with a moving magnet

Flux changes and force:

- The action of a motor in terms of the tendency of flux lines to contract

- Calculations and estimates involving $F = ILB$ for a straight current-carrying conductor, where current and velocity are perpendicular to a uniform magnetic field

- Practical task: using a current balance to measure B-fields

Charge and field

Uniform electric fields:

- Uniform electric field $E = \dfrac{V}{d}$
- Evidence for discreteness of charge on electron (e.g. Millikan oil drop experiment)
- Calculations and estimates involving $E = -\dfrac{dV}{dr}$ giving $E = \dfrac{V}{d}$ for a uniform field
- Diagrams of electric fields and the corresponding equipotential surfaces

Deflecting charged beams:

- The force on a moving charged particle due to a uniform magnetic field
- Calculations and estimates involving motion of a charged particle in a uniform magnetic field involving $F = qvB$ and also $E = \dfrac{V}{d}$ for a uniform electric field

Charged spheres:

- The electric field of a charged object, and the force on a charge in an electric field; inverse square law for a point charge; $\dfrac{1}{r^2}$ relationship

- Electrical potential energy and electric potential due to a point charge; $\dfrac{1}{r}$ relationship
- Appropriate use of the fact that a spherically symmetrical distribution of charges is equivalent to a point, with the same total charge, at its centre
- Sketching and interpreting graphs of electric field versus distance and identifying electric potential as area under curve
- Sketching and interpreting graphs of electric potential V, and potential energy against distance, and relating electric field to tangent to the $V-r$ graph from $E_{electric} = -\dfrac{dV}{dr}$.
- Make calculations and estimates involving, for radial components from a point charge, $F_{electric} = k\dfrac{qQ}{r^2}$ and $E = \dfrac{F_{electric}}{q} = k\dfrac{Q}{r^2}$ where $k = \dfrac{1}{4\pi\varepsilon_0}$

Physics in perspective

Mass spectrometers

After J J Thomson measured the mass:charge ratio for the electron, he applied the same principle to atomic nuclei. In 1913 he used the first mass spectrometer to show that neon consists of at least two different forms with different masses, which we now recognise as the isotopes $^{20}_{10}$Ne and $^{22}_{10}$Ne.

Over the next three decades the technique was refined to identify the multitude of isotopes you will meet in chapter 19, and mass spectrometry was one of the methods used to separate the isotope uranium-235 used in the first nuclear weapons. Modern applications of mass spectrometers involve much larger charged particles than nuclei – they are routinely used in chemical analysis, as shown in Figure 1.

▲ **Figure 1** *Mass spectrometer used in protein analysis*

How does it work?

The process used in a typical mass spectrometer is shown as a flow chart in Figure 2.

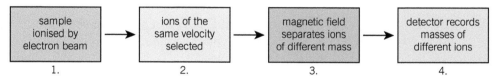

▲ **Figure 2** *Mass spectrometer process*

The process uses electric and magnetic fields at a number of different stages.

1   The electron beam used to create ions is produced by an electric field, accelerating electrons across a potential difference. The energy needed depends on the ions that it is intended to create. Electrons of low kinetic energy will dislodge electrons from molecules – for a sample containing ethanol this could result in ions such as $C_2H_5OH^+$. A more energetic electron beam will break the target molecules apart, resulting in ionised fragments such as $C_2H_5^+$ and OH^+.

2   The resulting mix of ions, both positive and negative, then goes into a velocity selector stage. In this region both an electric and magnetic field, at right angles to each other and to the velocity of the ions, ensure that all ions travelling straight through in an undeviated path have the same velocity.

3   The positive ions, all with identical velocity v, then pass into a region with a uniform magnetic field. The magnetic force qvB on each ion makes it travel in an arc of a circle of radius r, where the centripetal force is provided by the magnetic force, so $F = qvB = \dfrac{mv^2}{r}$.

Providing that the ions all have the same charge +e, then B, v, and q are all constant so that $v \propto r$. More massive ions travel along less sharply-curved paths whilst less massive ions follow paths that are much more curved, as shown in Figure 3.

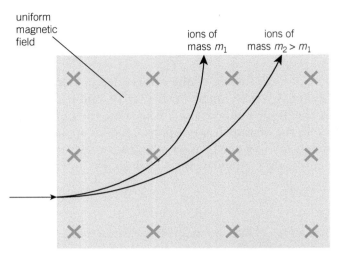

▲ **Figure 3** *The paths of ions depend on mass*

4   In early mass spectrometers, the detection of ions following different paths was done with photographic film – where ions hit the film, the film became blackened, with the intensity of the blackening increasing where more ions struck the film. Modern mass spectrometers, however, use electronic detectors that can be moved to scan across the different ion paths. More usually, the ion detectors can be fixed and the strength of the magnetic field in region 3 (see Figure 2) is varied from low to high values so that, as the paths increase in curvature, more massive ions are detected. As the electronic detectors are effectively counting ions, the detector output can be used to give the proportions of different ions present more accurately than attempting to measure the blackening of photographic film.

Mass spectrometers out of this world

Early mass spectrometers were massive constructions, but modern devices have been miniaturised for use in space missions. These devices operate on different principles from those shown in Figure 2 – the selection of ions is done by oscillating electric fields, resulting in all ions with the 'wrong' mass:charge ratio having unstable paths and being absorbed by electrodes alongside the ion path. The mass spectrometer on the Mars rover 'Curiosity' (see Figure 4) is of this sort, and has been used to analyse gases released from Martian rocks when they are heated.

▲ **Figure 4** *Artwork of the Curiosity rover on Mars*

Summary questions

1 The element chlorine consists of two isotopes: three-quarters of the atoms are $^{35}_{17}Cl$ and one-quarter are $^{37}_{17}Cl$. Explain why
   a the mean molar mass of chlorine is 35.5 g,
   b the mass spectrum of Cl_2^+ ions produced from chlorine gas has peaks at three different mass:charge ratios.

2 The velocity selection stage of a traditional mass spectrometer (Figure 2) has ions of a certain velocity v passing through without deviation. Draw a diagram indicating both the E-field and the B-field and use it to explain why any ions with velocity $v = \dfrac{E}{B}$ will pass through without deviation.

3 Suggest why mass spectrometers used in space missions need to be as small as possible.

4 It has been believed for some years that all the water on Earth arrived some time after its creation, from comets crashing into its surface. In 2014, the European Space Agency's Rosetta mission reached a comet. Rosetta's mass spectrometer measured the isotopes found in a cloud of water vapour on the comet, and found that the different isotopes of hydrogen and oxygen were in proportions quite different from those found in water on Earth. Explain the significance of this result.

Practice questions

1   Which of the following is a possible unit for electrical field strength?

    A Cm^{-1} **B** JC^{-1} **C** NV^{-1} **D** Vm^{-1}
                                              (*1 mark*)

2   Figure 1 shows two isolated point charges.

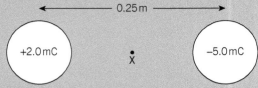

▲ Figure 1

    Which of the following is the correct expression for the electric potential at point **X**, exactly midway between the two charges? ($k = \dfrac{1}{4\pi\varepsilon_0}$)

    A $(-0.056\,C\,m^{-1}) \times k$

    B $(-0.024\,C\,m^{-1}) \times k$

    C $(+0.024\,C\,m^{-1}) \times k$

    D $(+0.056\,C\,m^{-1}) \times k$ (*1 mark*)

3   An ion of mass $5.3 \times 10^{-26}\,kg$ and charge $+1.6 \times 10^{-19}\,C$ enters a uniform magnetic field of flux density $0.018\,Wb\,m^{-2}$ at a velocity of $250\,ms^{-1}$ parallel to the field.

    a   Calculate the magnetic force acting on the ion.   (*1 mark*)

    b   Calculate the radius of the circular path it will follow.   (*2 marks*)

4   Figure 2 shows some equipotential lines around an electricity power cable at +500 V.

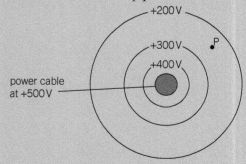

▲ Figure 2

    a   What feature of the diagram shows that the electric field strength is strongest next to the power cable?   (*1 mark*)

    b   On a copy of Figure 2, draw a line to represent the direction of the electric field at point P.   (*2 marks*)

         OCR Physics B Paper 2864 June 2003

5   This question is about using electric fields to control the motion of charged particles.

    Figure 3 shows a machine for printing labels on sacks of flour.

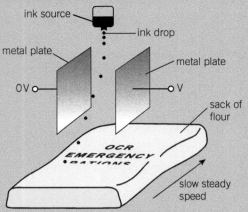

▲ Figure 3

    Drops of ink are charged as they leave the ink source. They fall between a pair of metal plates, onto a sack which is on a conveyor belt moving at a steady speed. By varying the voltage V of the right-hand metal plate, the point of impact of each drop on the sack can be controlled.

    a   Each ink drop has a charge of +0.80 nC.

        Calculate the number of electrons which need to be removed from the drop as it leaves the source.

        $e = 1.60 \times 10^{-19}\,C$ (*2 marks*)

    b   **(i)**   On Figure 3, the positively charged drops are pushed to the left.

            State and explain the sign of the potential V required on the right-hand plate.   (*1 mark*)

      (ii)   On a copy of Figure 4, draw **three** lines to represent the uniform electric field between the parallel plates.

▲ Figure 4   (*3 marks*)

    c   **(i)**   The force F on a charge Q in an electric field E is given by

            $F = EQ$.

Use this to show that the potential V of the right-hand plate of Figure 4 is given by

$$V = \frac{Fd}{Q}$$

where d is the separation between the plates. *(2 marks)*

(ii) Use the data listed below to calculate the value, in volts, of the potential V.

$F = 3.6\,\mu\text{N}$

$d = 150\,\text{mm}$

$Q = 0.80\,\text{nC}$ *(2 marks)*

OCR Physics B Paper 2864 January 2004

6 This question is about controlling electrons in an oscilloscope.

Details of the accelerating plates of the oscilloscope are shown in Figure 5.

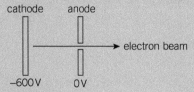

▲ **Figure 5**

Electrons from the cathode are accelerated towards the anode. The electrons travel in a vacuum.

a Electrons leave the cathode at a potential of $-600\,\text{V}$ with negligible kinetic energy. They are accelerated towards the anode at $0\,\text{V}$.

(i) On a copy of Figure 5, sketch an equipotential line for $-200\,\text{V}$ between the anode and the cathode. *(2 marks)*

(ii) Show that the kinetic energy gained by the electrons passing from cathode to anode is about $1 \times 10^{-16}\,\text{J}$.

$e = 1.6 \times 10^{-19}\,\text{C}$ *(2 marks)*

b The electrons approach a pair of deflecting plates on their way towards the screen, as shown in Figure 6.

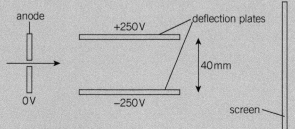

▲ **Figure 6**

(i) On a copy of Figure 6, sketch five lines to represent the electric field in the space between the deflecting plates. *(2 marks)*

(ii) State the potential difference between the deflecting plates. *(1 mark)*

(iii) Calculate the electric field strength between the deflecting plates. Include the unit in your answer. *(3 marks)*

OCR Physics B Paper 2864 January 2003

7 This question is about the motion of charged particles.

a Electrons enter an accelerator with negligible speed, and leave with a speed of $1.8 \times 10^7\,\text{m s}^{-1}$. By calculating the kinetic energy of the electrons, show that they are accelerated by falling through a potential difference of about $900\,\text{V}$.

$e = 1.6 \times 10^{-19}\,\text{C}$

$m_e = 9.1 \times 10^{-31}\,\text{kg}$ *(4 marks)*

b The electrons enter a region of uniform magnetic field, as shown in Figure 7. The electrons are moving in a vacuum.

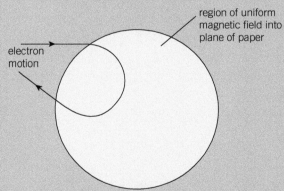

▲ **Figure 7**

The magnetic field of strength B causes each electron of mass m, velocity v and charge e to move in a circle of radius r. By considering the centripetal force on each electron of kinetic energy E_k, show that

(i) $r = \dfrac{mv}{Be}$ *(1 mark)*

(ii) $r = \dfrac{\sqrt{2E_k m}}{Be}$ *(2 marks)*

OCR Physics B Paper 2864 January 2008

MODULE 6.2
Fundamental particles

Introduction

This part of the course introduces you to the structure and binding of atoms and nuclei. We also consider practical uses of radioactivity, and the risks associated with using the atom. Particle physics was born at the turn of the 20th century and rapidly developed, but it is still advancing today.

The upgrade of the 27 km CERN beam to energies of 13 TeV in 2015 has begun to produce vast quantities of data that will allow particle physicists to refine and revise their models and answer some of the questions concerning dark matter that astronomers and cosmologists are working on.

Looking inside the atom begins by considering the scattering experiments of Ernest Rutherford's team and the discovery of the nucleus. You will see that much of particle physics relies on scattering particles at higher and higher energies, and that relativistic effects must be considered when particles are accelerated to great energies – the protons in the beam at CERN can move at a speed that is a fraction of a percent lower than the speed of light.

Experiments in the second half of the 20th century led to new ideas about protons and neutrons, which are now known to be composite particles made up of simpler particles called quarks. You will be introduced to the concept of antimatter before the chapter ends with a return to the larger atomic scale and a consideration of the wave behaviour of electrons within atoms.

Using the atom opens by describing the nature and effects of ionising radiation and looking at the damage they can do when absorbed by tissue. You will be introduced to ways of calculating absorbed dose and a method of estimating the risks of absorbing different forms of ionising radiation.

The second part of the chapter focuses on the binding energy of nuclei and how changes in binding energy per nucleon drive nuclear decays and explain the energy released in fission and fusion. You will use Einstein's 1905 equation $E = mc^2$ to calculate values for the energy released in such processes.

Knowledge and understanding checklist

From your Key Stage 4 or first-year A Level study you should be able to do the following. Work through each point, using your Key Stage 4/first-year A Level notes and the support available on Kerboodle.

☐ Recall the nuclear model of the atom.

☐ Recall the differences in numbers of protons and neutrons related to masses and identities of nuclei.

☐ Understand and use equations representing radioactive decays.

☐ Understand ionisation, as well as absorption and emission of radiation, in terms of changes in atomic energy levels.

☐ Relate nuclear decays to changes in proton number, nucleon number, and charge.

☐ Describe nuclear fission and fusion.

☐ Calculate energy gained by charges accelerated through a potential difference.

Maths skills checklist

Maths is a vitally important aspect of Physics. In this unit, you will need to use the following maths skills. You can find support for these skills on Kerboodle and through MyMaths.

☐ **Changing the subject of equations** such as $E = mc^2$ and the relativistic factor $\gamma = \dfrac{1}{\sqrt{1 - \dfrac{v^2}{c^2}}}$. You will also need to use the equations of exponential decay when considering the activity of radioisotopes used in nuclear medicine and the absorption of gamma rays.

☐ **Interpreting logarithmic plots**, for example to find the decay constant of a radionuclide. You will also interpret decay curves to find half-life.

☐ **Understanding simple probability** in the context of radioactive decay. You will need to make calculations using the percentage risk of contracting illness from exposure to radiation.

MyMaths.co.uk
Bringing Maths Alive

18 LOOKING INSIDE THE ATOM
18.1 Probing the atom with alpha particles

Specification reference: 6.2.1a(ii), 6.2.1b(i), 6.2.1b(ii), 6.2.1c(ii)

Synoptic link

The discovery of radioactivity was described in Topic 10.1, Radioactive decay and half-life.

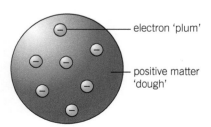

▲ Figure 1 *Plum-pudding model of the atom*

electron 'plum'

positive matter 'dough'

Inside the atom

When radioactivity was first investigated at the end of the 19th century, not all physicists accepted the existence of atoms. The atomic theory gained more support over the decade spanning the turn of the 20th century, and one of Einstein's 1905 papers gave a mathematical argument for discrete particles of matter – atoms.

In 1897 the Cambridge physicist J J Thomson performed experiments that showed that 'cathode rays' were discrete particles. Thomson pictured the atom as like a 'plum pudding' with the newly discovered negative particles – electrons – embedded throughout a positive sphere (Figure 1).

You can't simply look inside an atom to see if it looks like a plum pudding. Instead, we investigate the structure of atoms with scattering experiments. Particles are accelerated to high energy and directed at a target. Arrays of detectors surrounding the target track and identify the particles created and scattered in the collision.

Alpha-particle scattering

Hans Geiger and Ernest Marsden carried out the first scattering experiment under the direction of Ernest Rutherford in 1909. They fired alpha particles at a sheet of gold foil (Figure 2). The pattern of scattering produced led to Rutherford's model of the atom – a tiny, dense, positive nucleus and distant orbiting electrons.

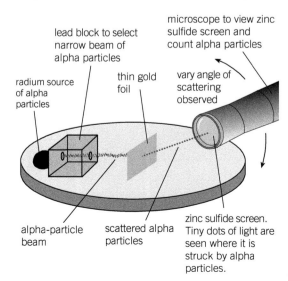

lead block to select narrow beam of alpha particles

microscope to view zinc sulfide screen and count alpha particles

radium source of alpha particles

thin gold foil

vary angle of scattering observed

alpha-particle beam

scattered alpha particles

zinc sulfide screen. Tiny dots of light are seen where it is struck by alpha particles.

◀ Figure 2 *Rutherford's scattering experiment*

Alpha particles were known to be helium atoms that had lost their electrons and therefore were positively charged. It was also known that alpha particles have a lot of energy, typically 5 MeV.

It was expected that these particles would shoot through the thin gold foil with barely any deflection, but Geiger and Marsden got a surprise. A few of the alpha particles bounced right back from the foil, being scattered by angles greater than 90°.

Rutherford reasoned that:

- the core of the atom must be massive (on an atomic scale) to deflect alpha particles through large angles

- the core of the gold atoms and the alpha particles must both be very small (much smaller than an atom) or more alpha particles would 'hit' and bounce back – in fact, only around one alpha particle in 10 000 bounces back

- the alpha particles are deflected because of electrical repulsion between the positively charged alpha particle and a positively charged core in the gold atoms.

Rutherford concluded that all atoms consist of a very small, massive, positively charged nucleus, surrounded at great distances by electrons (Figure 3). The alpha particle was simply the nucleus of helium.

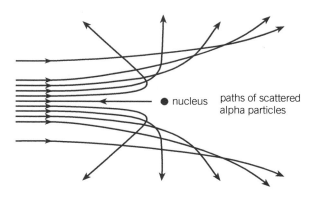
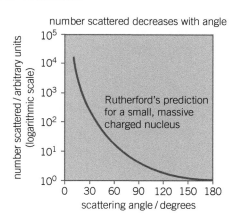

▲ **Figure 3** *Rutherford's picture of alpha-particle scattering*

Several tests of this picture were possible (Figure 4):

- If the alpha particles were slowed down, would more be deflected at greater angles, since the nucleus should now more easily turn them back? Yes, they were.

- Would nuclei of smaller electric charge scatter alpha particles less strongly, as expected? Yes, they did.

- Would the pattern of numbers of alpha particles scattered at different angles fit the pattern expected from an inverse-square law for electrical repulsion? Yes, the predicted pattern was a good fit.

The really big surprise was how small the nucleus of an atom turns out to be, concentrating very nearly the whole mass of the atom in a tiny

Synoptic link

This analysis uses ideas about the electrical potential of a point charge that you met in Chapter 17, Charge and field.

core. The large energy of alpha particles means that they can get fairly close to a nucleus before being turned back. For 5 MeV alpha particles scattered by gold nuclei, the distance of closest head-on approach turns out to be as little as 4 to 5×10^{-14} m. So the nucleus must be even smaller than that.

Estimating the size of the nucleus

Figure 5 shows how the potential energy of an alpha particle increases as it makes a head-on approach to a positive nucleus – in this case a gold nucleus. At its closest approach its kinetic energy is zero, and its electrical potential energy is equal to its initial kinetic energy at a large distance from the nucleus.

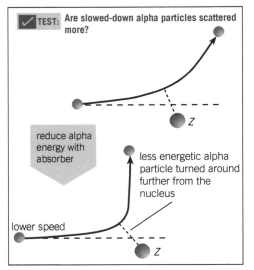

TEST: Are slowed-down alpha particles scattered more?

reduce alpha energy with absorber

less energetic alpha particle turned around further from the nucleus

lower speed

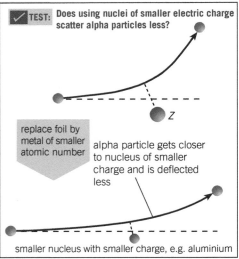

TEST: Does using nuclei of smaller electric charge scatter alpha particles less?

replace foil by metal of smaller atomic number

alpha particle gets closer to nucleus of smaller charge and is deflected less

smaller nucleus with smaller charge, e.g. aluminium

▲ **Figure 4** *Careful investigation of alpha-particle scattering supported the nuclear model*

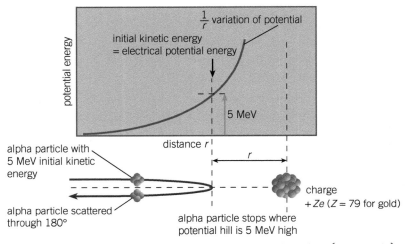

alpha particle with 5 MeV initial kinetic energy

alpha particle scattered through 180°

alpha particle stops where potential hill is 5 MeV high

charge $+Ze$ ($Z = 79$ for gold)

▲ **Figure 5** *Head-on approach of an alpha particle to a gold nucleus (not to scale)*

This energy change is analogous to a ball rolling up a curved track. At the bottom it has zero potential energy and all its energy is kinetic energy. When it stops it has gravitational potential energy equal to its original kinetic energy. It then rolls back down. Similarly, the alpha particle climbs an electrical potential 'hill' as it approaches the gold nucleus.

The electrical potential energy of a charge q at distance r from a charge Q is given by

$$\text{energy} = \frac{kQq}{r}, \text{ where } k = \frac{1}{4\pi\varepsilon_0} = 8.98 \times 10^9 \,\text{N m}^2\,\text{C}^{-2}$$

In this case q and Q are the charges of the alpha particle and the gold nucleus, respectively. We can therefore calculate r for an estimate of the size of a gold nucleus.

Worked example: Estimating the size of a gold nucleus

Use the data above to calculate the distance of closest approach r for the alpha particle, and from this put a limit on the size of the gold nucleus.

Step 1: Figure 5 shows that the electrical potential energy at r is $5.0\,\text{MeV} = 5.0 \times 10^6 \times 1.6 \times 10^{-19}\,\text{J}$
$$= 8.0 \times 10^{-13}\,\text{J}$$

Step 2: Calculate the charges on the alpha particle q and the gold nucleus Q.

$q = +2e = 2 \times 1.6 \times 10^{-19}\,\text{C} = 3.2 \times 10^{-19}\,\text{C}$

$Q = +79e = 1.264 \times 10^{-17}\,\text{C}$ (retaining all the significant figures in this intermediate value)

Step 3: Now use these values to find r.

Energy $= \dfrac{kQq}{r}$

$\therefore 8.0 \times 10^{-13}\,\text{J} = \dfrac{8.98 \times 10^9\,\text{N}\,\text{m}^2\,\text{C}^{-2} \times 3.2 \times 10^{-19}\,\text{C} \times 1.264 \times 10^{-17}\,\text{C}}{r}$

$\therefore r = \dfrac{8.98 \times 10^9\,\text{N}\,\text{m}^2\,\text{C}^{-2} \times 3.2 \times 10^{-19}\,\text{C} \times 1.264 \times 10^{-17}\,\text{C}}{8.0 \times 10^{-13}\,\text{J}} = 4.5 \times 10^{-14}\,\text{m (2 s.f.)}$

The nucleus of a gold atom must have a smaller radius than $4.5 \times 10^{-14}\,\text{m}$.

> **Hint**
>
> Remember that the charge on a particle is not the same as the charge number. For example, the charge on an alpha particle is not $+2$, it is $+2e = 3.2 \times 10^{-19}\,\text{C}$.

Summary questions

1 The radius of an atom is of the order of $10^{-10}\,\text{m}$. The radius of the nucleus is of the order of $10^{-14}\,\text{m}$.
   a Calculate the ratio of the volume of the atom to the volume of the nucleus. *(2 marks)*
   b The relative size of the nucleus to the atom has been compared to a pinhead in a medium-sized house. Assuming that a pinhead has a radius of about 1 mm, state whether the comparison is reasonable. Support your answer with calculations. *(3 marks)*

2 Figure 6 shows the path of an alpha particle near a nucleus. Copy the diagram.
   a Add a path representing an alpha particle with less energy. The path starts from the same point and in the same direction as the original path. *(2 marks)*
   b Explain the differences and similarities in the shape of the paths. *(4 marks)*

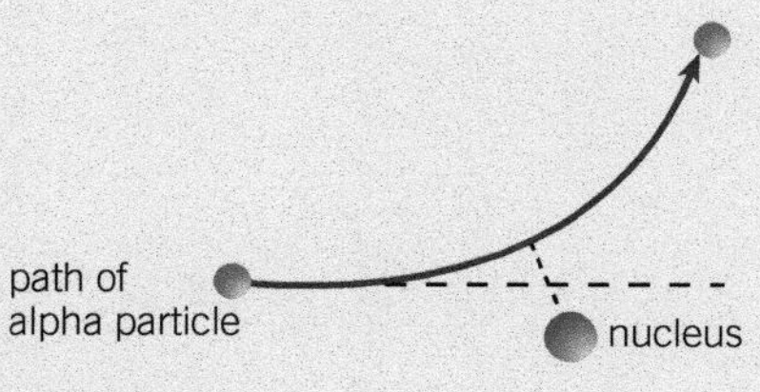

path of alpha particle

nucleus

▲ Figure 6

3 An alpha particle makes a head-on approach to a plutonium nucleus. The distance between the centres of the two particles is $9.0 \times 10^{-15}\,\text{m}$ at closest approach. Calculate the electrical potential energy of the alpha particle at this distance (charge on alpha particle $= +3.2 \times 10^{-19}\,\text{C}$, charge on plutonium nucleus $= +1.3 \times 10^{-17}\,\text{C}$, $k = 8.98 \times 10^9\,\text{N}\,\text{m}^2\,\text{C}^{-2}$). *(2 marks)*

4 Show that the distance of closest approach of a 6 MeV alpha particle to a nucleus of iron is about $1.2 \times 10^{-14}\,\text{m}$. An iron nucleus has 26 protons. *(2 marks)*

5 A 5 MeV alpha particle and a proton accelerated to 5 MeV are both scattered by the same massive nucleus. State which gets closer to the nucleus in a head-on collision, and explain your answer. *(3 marks)*

18.2 Accelerating charges and electron scattering

Specification reference: 6.2.1a(i), 6.2.1a(vii), 6.2.1c(iii), 6.2.1c(iv)

Synoptic link

This analysis uses ideas about uniform electric fields that you met in Chapter 17, Charge and field.

Scattering electrons

The alpha-particle scattering experiments of Rutherford, Geiger, and Marsden gave a maximum value for the radius of the nucleus – it could be no bigger than the closest approach of the alpha particle to the centre of charge (Topic 18.1, Probing the atom with alpha particles).

Electrons are scattered by the nucleus in the same way as alpha particles, with only a small proportion scattered through large angles. Of course, in this case the force between the scattered particle and the nucleus is attractive. For these scattering experiments, electrons must have more energy than those released in beta decay. Suitable high-energy electrons are produced using particle accelerators.

Accelerating charges

In a dental X-ray tube electrons are 'boiled off' a heated wire and accelerate towards a positive target. X-rays are produced as the electrons crash into the target (Figure 1).

Getting an electron out of an atom

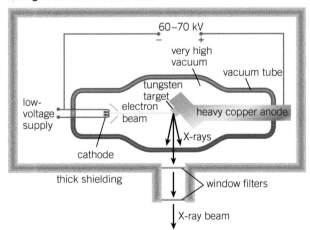

▶ **Figure 1** A high p.d. provides energy to accelerate electrons in a dental X-ray machine

Synoptic link

As you saw in Topic 17.2, Deflecting charged beams, the energy gained by a particle of charge equal in magnitude to an electron moving through a p.d. of one volt is 1 electronvolt (eV), where 1 eV = 1.6 × 10⁻¹⁹ J.

The electrons gain kinetic energy $\frac{1}{2}mv^2$ as they pass through the electric potential difference V, losing electrical potential energy $= qV$, where q is the charge on the electron.

loss of electrical potential energy of particle = gain of kinetic energy of particle

$$qV = \frac{1}{2}mv^2$$

$$\therefore v = \sqrt{\frac{2qV}{m}}$$

Worked example: Electron and proton velocities

a Calculate the velocity of an electron when it is accelerated through a p.d. of 1000 V. The mass of an electron is 9.1×10^{-31} kg, and the magnitude of its charge is 1.6×10^{-19} C.

$$v = \sqrt{\frac{2qV}{m}} = \sqrt{\frac{2 \times 1.6 \times 10^{-19}\,C \times 1000\,V}{9.1 \times 10^{-31}\,kg}} = 1.9 \times 10^7\,m\,s^{-1} \text{ (2 s.f.)}$$

b Calculate the velocity of a proton accelerated through the same potential difference. The mass of a proton is 1.7×10^{-27} kg.

$$v = \sqrt{\frac{2qV}{m}} = \sqrt{\frac{2 \times 1.6 \times 10^{-19}\,C \times 1000\,V}{1.7 \times 10^{-27}\,kg}} = 4.3 \times 10^5\,m\,s^{-1} \text{ (2 s.f.)}$$

Both particles gain the same kinetic energy. The velocity of the proton is smaller than that of the electron because the proton has greater mass.

The equation $v = \sqrt{\frac{2qV}{m}}$ suggests that there is no limit to the velocity a particle can reach as long as a large enough potential difference is available. Indeed, the equation suggests that particles should be able to travel faster than the speed of light if the accelerating potential difference is sufficient. You already know that this is impossible – no material particle can travel as fast as light.

Einstein's equation $E_{rest} = mc^2$

The equations we have used for momentum ($p = mv$) and kinetic energy are accurate enough at low speeds, but not at high speeds. For higher speeds we use equations published by Einstein in his 1905 paper on special relativity (Figure 2). Einstein showed that momentum is given by

$p = \gamma mv$ where γ is the relativistic factor, $\gamma = \dfrac{1}{\sqrt{1 - \dfrac{v^2}{c^2}}}$.

When $v \ll c$, $\gamma \approx 1$ and Newton's momentum relationship $p = mv$ works well.

Synoptic link

For a reminder about the physics of the speed of light and the relativistic factor γ, review Topic 13.3, Special relativity.

 Worked example: Relativistic velocity

A particle has a relativistic factor $\gamma = 2$. Calculate the velocity of the particle.

Step 1: Rearrange the equation for the relativistic factor.

$$2 = \frac{1}{\sqrt{1 - \dfrac{v^2}{c^2}}} \quad \therefore \sqrt{1 - \frac{v^2}{c^2}} = \frac{1}{2}$$

$$\therefore 1 - \frac{v^2}{c^2} = \left(\frac{1}{2}\right)^2 = \frac{1}{4}$$

$$\therefore \frac{3}{4} = \frac{v^2}{c^2}$$

Step 2: Use c to calculate v.

$$\therefore v = c\sqrt{\frac{3}{4}}$$

$$= 3.0 \times 10^8\,m\,s^{-1} \times 0.86...$$

$$= 2.6 \times 10\,m\,s^{-1} \text{ (2 s.f.)}$$

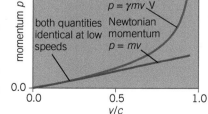

Problem:
time t depends on relative motion because of time dilation

Newton's definition of momentum
$p = mv$
$p = m\dfrac{\Delta x}{\Delta t}$

Einstein's solution:
Replace Δt by $\Delta \tau$, the change in wristwatch time τ, which does not depend on relative motion

from time dilation:
$\gamma = \dfrac{1}{\sqrt{1 - \dfrac{v^2}{c^2}}}$
$\Delta t = \gamma \Delta \tau$
substitute for $\Delta \tau$

Einstein's new definition of momentum
$p = m\dfrac{\Delta x}{\Delta \tau}$
$p = \gamma m\dfrac{\Delta x}{\Delta t}$

$\dfrac{\Delta x}{\Delta t} = v$

$p = \gamma mv$

relativistic momentum

$p = \gamma mv$

▲ **Figure 2** *Einstein redefines momentum*

Hint

The relativistic factor is never less than 1.

$$\gamma = \frac{\text{total energy}}{\text{rest energy}}$$

$$= \frac{(\text{kinetic energy} + \text{rest energy})}{\text{rest energy}}$$

$$= 1 + \frac{\text{kinetic energy}}{\text{rest energy}}$$

Worked example: Rest energy

The mass of an electron is 9.1×10^{-31} kg. Calculate its rest energy in joules and MeV.

Step 1: rest energy $= mc^2$

$= 9.1 \times 10^{-31}$ kg $\times$ $(3.0 \times 10^8 \, \text{m s}^{-1})^2$

$= 8.19 \times 10^{-14}$ J

Step 2: $1 \, \text{eV} = 1.6 \times 10^{-19}$ J, so rest energy in eV

$= 8.19 \times 10^{-14} \, \text{J} / 1.6 \times 10^{-19} \, \text{J}$

$= 5.1 \times 10^5 \, \text{eV} = 0.51 \, \text{MeV}$
(2 s.f.)

▲ **Figure 3** *The ATLAS detector, one of the six detector experiments at the LHC*

Next, Einstein showed that the energy of a free particle must be thought of as having two parts – kinetic energy and what is now called rest energy. The fundamental quantity is the particle's total energy.

total energy = kinetic energy + rest energy
= γmc^2 where m is the particle mass.

When the particle is at rest $\gamma = 1$. The particle has total energy = rest energy $E_{\text{rest}} = mc^2$

$$\frac{\text{total energy}}{\text{rest energy}} = \frac{\gamma mc^2}{mc^2} = \gamma$$

This is a useful result as it gives us another way of calculating the relativistic factor γ.

At high speeds, $\gamma \gg 1$, $v \approx c$

When $v \approx c$, momentum $p = \gamma mv \approx \gamma mc$

As $E_{\text{total}} = \gamma mc^2$, for high speeds, $p \approx E_{\text{total}} / c$

Particle physicists often express the masses of particles in energy units, frequently electronvolts. An electron has a mass of about 0.5 MeV in these units, for example. A proton has a mass of about 1 GeV (2000 times greater). With both sides of the equation in the same units, you can simply think of the mass as being the rest energy, $E_{\text{rest}} = m$.

Relativity in action – the Large Hadron Collider

The Large Hadron Collider (LHC) at CERN in Geneva is one of the world's foremost particle accelerators (Figure 3). In one experiment, protons are accelerated to kinetic energies of 7×10^{12} eV (7000 GeV).

The rest energy of a proton is about 1 GeV.

$$\gamma = \frac{\text{total energy}}{\text{rest energy}} = \frac{7000 \, \text{GeV} + 1 \, \text{GeV}}{1 \, \text{GeV}} \approx 7000$$

We can use $\gamma = \dfrac{1}{\sqrt{1 - \dfrac{v^2}{c^2}}}$ to show that a particle with a relativistic

factor of 7000 has a speed of $0.999\,999\,990\,c$. However much kinetic energy the accelerated particles gain, they will never reach the speed of light. The real point of a particle accelerator is not the final speed of the particles but the energy and momentum that they carry.

The linear accelerator

How can electrons (and other particles) be accelerated to such huge energies? One method is to use a linear accelerator.

In a linear accelerator, bunches of electrons (or other charged particles) are accelerated between a pair of electrodes. They travel on to another pair where they are accelerated again, and so on (Figure 4). The energy gained by the particles is limited only by the length of the accelerator. The trick is to time the changing p.d. between each pair of electrodes so that, exactly as the electrons arrive at the gap, the p.d. is in the right direction to accelerate them. As the electrons go faster, they have to be allowed to travel farther between gaps so that the time from gap to gap is constant. But when the electrons have reached nearly the speed of light, the gap spacings need to be made equal. Relativity determines the design of the machine.

The accelerating field

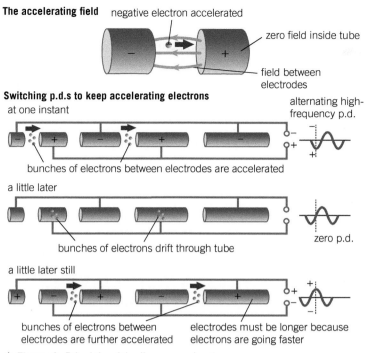

▲ **Figure 4** *Principle of the linear accelerator*

Electron scattering

We have said that electrons scatter off the nucleus in a similar manner to alpha particles. However, as Figure 5 shows, there is a kink in the scattering curve. Remember that electrons can be thought of as having a wavelength given by de Broglie's relation, $\lambda = \dfrac{h}{p}$, where p is the momentum and h is the Planck constant.

Synoptic link

This section uses ideas about diffraction and quantum behaviour you met in Chapters 6 and 7, Wave behaviour and Quantum behaviour.

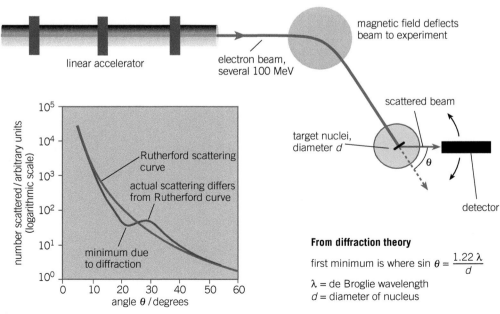

From diffraction theory

first minimum is where $\sin \theta = \dfrac{1.22\,\lambda}{d}$

λ = de Broglie wavelength
d = diameter of nucleus

▲ **Figure 5** *Diffraction of electron beams by nuclei*

Electrons scattered by a small spherical nucleus behave much like photons diffracted by a small disc. The diffraction pattern is a series of concentric rings round a strong central maximum. The size of the nucleus can be found from the size of the rings. The first diffraction minimum is at angle θ, where $\sin\theta \approx \dfrac{1.22\,\lambda}{\text{nuclear diameter}}$. The diffraction pattern is superimposed on the Rutherford scattering curve – the kink in Figure 5 is the first diffraction minimum.

To get a small enough wavelength for diffraction and therefore to measure the nucleus, the electrons must be accelerated to several hundred megaelectronvolts (MeV), much more than their rest energy mc^2 of 0.5 MeV. The relativistic calculation of the momentum and so of the wavelength is then very simple. To a good approximation, $p \approx \dfrac{E}{c}$ and so, as $\lambda = \dfrac{h}{p}$, $\lambda \approx \dfrac{hc}{E}$.

 Worked example: de Broglie wavelength for diffraction

a Calculate the de Broglie wavelength of an electron of energy 300 MeV (1 eV = 1.6 × 10⁻¹⁹ J).

Step 1: $\lambda \approx \dfrac{hc}{E} = \dfrac{6.6 \times 10^{-34}\,\text{J s} \times 3.0 \times 10^{8}\,\text{m s}^{-1}}{300 \times 10^{6} \times 1.6 \times 10^{-19}\,\text{J}}$

$\qquad = 4.1 \times 10^{-15}\,\text{m (to 2 s.f.)}$

b 300 MeV electrons give a diffraction minimum at 20° when scattered from target nuclei. Use the equation $\sin\theta \approx \dfrac{1.22\,\lambda}{\text{nuclear diameter}}$ to give an estimate for the nuclear diameter.

Step 1: $\sin 20° = \sin\theta = \dfrac{1.22 \times 4.1 \times 10^{-15}\,\text{m}}{\text{nuclear diameter}}$

nuclear diameter $= \dfrac{5.0 \times 10^{-15}\,\text{m}}{\sin 20°} = 1.5 \times 10^{-14}\,\text{m (2 s.f.)}$

Summary questions

1 An electron is accelerated through a p.d. of 500 V.
   a Calculate the kinetic energy of the accelerated electron. *(2 marks)*
   b Calculate its speed (mass = 9.1 × 10⁻³¹ kg). *(2 marks)*
   c State why it is not necessary to take relativistic effects into account in this calculation. *(1 mark)*

2 Calculate the de Broglie wavelength of
   a an electron accelerated through 1000 V, *(2 marks)*
   b an alpha particle with 5 MeV of kinetic energy, *(2 marks)*
   c an electron with energy 10 GeV, and state why you must use the relativistic approximation $p = \dfrac{E}{c}$. *(3 marks)*

3 An electron and a proton are both accelerated in a vacuum between the same pair of electrodes.
   a Explain why their kinetic energies are the same but their directions of travel are opposite. *(2 marks)*
   b Show that the speed of the electron is about 40 times larger than the speed of the proton if relativistic effects are not important. *(2 marks)*

4 The rest energy of a proton is about 1 GeV.
   a Show that the relativistic factor $\gamma = 30$ for a proton accelerated to a kinetic energy of 29 GeV. *(2 marks)*
   b Calculate the speed of the proton. *(2 marks)*

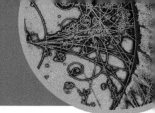

18.3 Inside the nucleus

Specification reference: 6.2.1a(v), 6.2.1a(vi), 6.2.2b(i)

The structure of nuclei

You have seen how the size of the nucleus has been estimated from alpha-particle and electron scattering experiments. You know that the nucleus is positively charged and that, for a neutral atom, this charge is balanced by negatively charged electrons.

In 1917 Ernest Rutherford performed experiments that showed that when a nitrogen nucleus is struck by an alpha particle a hydrogen nucleus can be released. The hydrogen nucleus was found to have a positive charge of the same magnitude as the negatively charged electron. In 1920 Rutherford gave the hydrogen nucleus the name **proton**, from the Greek for 'first'. Protons were identified as constituents of all atomic nuclei.

The number of protons, and so the charge on the nucleus, is equal to the atomic number Z of the element in the Periodic Table. The mass of an ion can be found by bending a beam of the ions in electric and magnetic fields. The mass of the ion nucleus is found by subtracting the tiny mass of its electrons. For all ions except hydrogen, nuclear masses are always greater than the values expected from the number of protons.

The extra mass can be explained if the nucleus contains two kinds of particle – **nucleons** – of approximately equal mass – electrically charged protons, and uncharged **neutrons** (because they are neutral), which were discovered in 1932 (Figure 1). Neutrons add mass without adding charge, and are the reason that atoms of the same element often have exactly the same chemical properties but slightly different mass. These **isotopes** have the same **proton number** but different numbers of neutrons, and so they have a different total **nucleon number**. The number of protons decides the positive charge on the nucleus, and therefore the number of negative electrons in a neutral atom. The number of electrons determines the chemical properties of the atom. But a small difference in the number of neutrons makes a difference to the mass, and may make an isotope radioactive. For example, carbon-12 (with 6 protons and 6 neutrons in its nucleus) is stable, whereas carbon-14 (with 6 protons and 8 neutrons) is unstable, with a half-life of about 5700 years.

Learning outcomes

Describe, explain, and apply:

→ the terms proton, neutron, nucleon, quark, gluon, nucleon number, proton number, isotope

→ a simple model of the quark structure of protons and neutrons

→ conservation of mass/energy and charge in balanced nuclear equations.

Synoptic link

See Topic 17.2, Deflecting charged beams, for a reminder of the ways that magnetic and electric fields affect beams of charged particles.

Inside the nucleus

mass number A = number of nucleons
$\qquad = Z + N$

chemical symbol

$${}^{A}_{Z}\text{X}$$

atomic number Z = number of protons

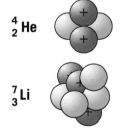

	atomic number Z	mass number A	neutron number N
${}^{4}_{2}$He	2	4	2
${}^{7}_{3}$Li	3	7	4

proton ⊕   neutron ○

▲ **Figure 1** *Protons add mass and charge to the nucleus, neutrons simply add mass*

test

Worked example: Uranium nuclei

a The nucleon number of a uranium isotope is 238. The proton number is 92. How many neutrons are in the nucleus?

Step 1: number of neutrons = nucleon number – proton number
$$= 238 - 92 = 146$$

b Uranium-238 decays by alpha emission. Complete the equation for the decay.

$$^{238}_{92}U \rightarrow ^{A}_{Z}Th + ^{4}_{2}\alpha$$

Step 1: Balance the 'top line' of the equation (nucleon number) and the 'bottom line' (proton number).

$$A = 238 - 4 = 234, Z = 92 - 2 = 90$$

$$^{238}_{92}U \rightarrow ^{234}_{90}Th + ^{4}_{2}\alpha$$

Note that an alpha particle can be represented as $^{4}_{2}\alpha$ or $^{4}_{2}He$.

Quark list

	electric charge		electric charge
● up	$+\frac{2}{3}e$	○ anti-up	$-\frac{2}{3}e$
● down	$-\frac{1}{3}e$	○ anti-down	$+\frac{1}{3}e$

Proton p = uud

total charge = $1e$
$u + u + d = p$
$+\frac{2}{3}e \ +\frac{2}{3}e \ -\frac{1}{3}e = 1e$

Neutron n = udd

total charge = 0
$u + d + d = n$
$+\frac{2}{3}e \ -\frac{1}{3}e \ -\frac{1}{3}e = 0$

Antiproton $\bar{p} = \bar{u}\bar{u}\bar{d}$

total charge = $-1e$
$\bar{u} + \bar{u} + \bar{d} = \bar{p}$
$-\frac{2}{3}e \ -\frac{2}{3}e \ +\frac{1}{3}e = -1e$

Antineutron $\bar{n} = \bar{u}\bar{d}\bar{d}$

total charge = 0
$\bar{u} + \bar{d} + \bar{d} = \bar{n}$
$-\frac{2}{3}e \ +\frac{1}{3}e \ +\frac{1}{3}e = 0$

Deeper still – inside the nucleon

In the early 1960s, improved accelerator designs enabled the energies of the particles used in scattering experiments to be increased. Many new kinds of particles were produced and named, like Δ (delta) and Σ (sigma), to join those already known, such as π⁻ and K mesons. Physicists look for simple models, and a particle 'zoo' of many fundamental particles is not a simple model. Surely all these new particles couldn't be truly fundamental? A more elegant idea is that protons, neutrons, and other species in the particle zoo are combinations of more fundamental components, named **quarks**. Quarks have charges of either $\frac{1}{3}$ or $\frac{2}{3}$ of the magnitude of the charge on an electron (e), and can be positively or negatively charged.

Protons and neutrons are composed of two types of quarks, the up quark, u (charge = $+\frac{2}{3}e$), and the down quark, d (charge = $-\frac{1}{3}e$). The 'recipe' for a proton is uud, giving a charge of $+\frac{2}{3}e + \frac{2}{3}e + (-\frac{1}{3})e = +1e$. A neutron is ddu.

Antiparticles are represented with a line above the usual symbol, so an antiproton is represented by $\bar{p}$. The quark combination for an antiproton is $\bar{u}\,\bar{u}\,\bar{d}$, spoken 'anti-up, anti-up, anti-down'.

As well as the combinations of quarks needed to produce protons, neutrons, and their antiparticles (Figure 2), others are possible (Figure 3), accounting for particles such as the Δ⁻ (quark combination ddd) and Δ⁺⁺ (quark combination uuu).

◄**Figure 2** *The nucleon family is built from two kinds of quark*

Delta particles

$\Delta^- = ddd$

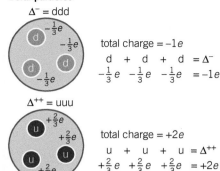

total charge $= -1e$

$$d + d + d = \Delta^-$$
$$-\tfrac{1}{3}e \quad -\tfrac{1}{3}e \quad -\tfrac{1}{3}e \quad = -1e$$

$\Delta^{++} = uuu$

total charge $= +2e$

$$u + u + u = \Delta^{++}$$
$$+\tfrac{2}{3}e \quad +\tfrac{2}{3}e \quad +\tfrac{2}{3}e \quad = +2e$$

... and their antiparticles $\overline{ddd}$ and $\overline{uuu}$

Mesons

$\pi^+ = u\bar{d}$

total charge $= +1$

$$u + \bar{d} = \pi^+$$
$$+\tfrac{2}{3}e \quad +\tfrac{1}{3}e = 1e$$

$\pi^- = d\bar{u}$

total charge $= -1$

$$d + \bar{u} = \pi^-$$
$$-\tfrac{1}{3}e \quad -\tfrac{2}{3}e = -1e$$

▲ **Figure 3** *Quarks explain other particles*

Hunting the quark

You may wonder whether these patterns of quarks are just number games or whether protons and neutrons really contain smaller objects. There is evidence to suggest that nucleons are indeed composed of three particles.

Electrons accelerated to energies of 20 GeV can map the charge distribution inside a nucleon. Make sure you are clear about that – the internal structure of a single nucleon (proton or neutron) is probed by the electrons. When a high-energy electron collides with a quark, the deflection of the electron is not the only outcome. In addition, particles are created from the energy of the interaction. A jet of new particles emerges that roughly follows the path of the quark as it is given a huge kick by the electron. Many of these new particles are mesons – quark–antiquark pairs.

The American physicist Richard Feynman realised that relativistic effects make this complicated situation much simpler. To the accelerated electron, the proton looks like a particle approaching it at nearly the speed of light. Time dilation slows down the motion of the quarks so that they seem almost stationary. They become sitting targets. This makes it easier to deduce the pattern of electric charges in the nucleon from the pattern of scattering. It was shown that the scattering was consistent with the existence of three particles inside a neutron or proton. From the dependence of the scattering on the charge on the scattering particle, it was clear that quarks do indeed have charges which are a fraction of the fundamental unit e.

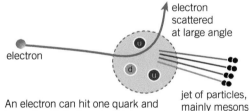

electron scattered at large angle

electron

jet of particles, mainly mesons

An electron can hit one quark and be scattered. Exchange of high-energy photons leads to the creation of a jet of particles and antiparticles

▲ **Figure 4** *At high energies individual quarks scatter electrons*

Gluons

Something very strong must hold the quarks together inside a particle. Quarks attract one another by exchanging particles unimaginatively called **gluons**. They act as the glue that keeps protons and neutrons in one piece and they also lead to the further forces that make nucleons attract one another, keeping the nucleus together. This gluon interaction is known as the **strong interaction**.

Extension: Nuclear matter

The radii of different nuclei have been measured by electron scattering (Topic 18.1, Probing the atom with alpha particles). The graph of nuclear volume against nucleon number is a straight line (Figure 5), so a nucleon takes up the same volume however many nucleons there are in the nucleus.

The radius r of a nucleus is proportional to the cube root of its volume. As volume is proportional to mass number A, we can write $r \propto \sqrt[3]{A}$

$\therefore r = r_0 \sqrt[3]{A}$.

The value of r_0 (close to 10^{-15} m) is a good measure of the radius of an individual nucleon.

The 'nuclear matter' that makes up a nucleus is astonishingly dense. This matter is found in neutron stars. When some massive stars run out of nuclear fuel they collapse and explode, outshining galaxies for a few days. Such *supernovae* leave behind a core of neutrons a few kilometres across, like a gigantic atomic nucleus. Rapidly spinning neutron stars can be detected by their radio emissions and are called *pulsars*. The Crab Nebula (Figure 6) has a pulsar at its centre.

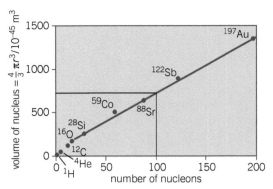

▲ **Figure 5** *Nuclear volume plotted against number of nucleons: double the number of nucleons and the volume roughly doubles*

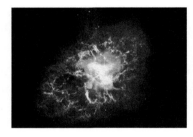

◀**Figure 6** *The Crab Nebula is the remnant of a supernova that was observed on Earth in 1085*

To think about:

1 Use data from the graph to show that the density of nuclear matter is about 2×10^{17} kg m^{-3} (nucleon mass = 1.7×10^{-27} kg).

2 Calculate the approximate mass of a matchbox full of nuclear matter.

Summary questions

1 a Identify particle X in the following equation.
$^{3}_{1}H + {}^{2}_{1}H \rightarrow {}^{4}_{2}He + X$ *(1 mark)*

 b Complete the following equation representing the decay of polonium-216.
$^{216}_{84}Po \rightarrow {}^{...}_{...}He + {}^{...}_{...}Pb$ *(2 marks)*

 c State how many neutrons there are in a nucleus of uranium-235, $^{235}_{92}U$. *(1 mark)*

2 How many quarks are there in an alpha particle? *(2 marks)*

3 The decay of a neutron to a proton can be thought of as the change of one down quark in the combination udd to an up quark, producing the combination uud. Show that the values of the quark charges require charge $-e$ to be carried away in this process. *(2 marks)*

4 Describe how up quarks of charge $+\frac{2}{3}e$ and down quarks of charge $-\frac{1}{3}e$ can be used to construct the following:
 a a proton, charge $+1e$; b a neutron, charge 0; c a Δ^{++} particle, charge $+2e$; d a Δ^{-} particle, charge $-1e$. *(4 marks)*

18.4 Creation and annihilation

Specification reference: 6.2.1a(vi), 6.2.1b(i), 6.2.1c(iii)

Matter and antimatter

The idea of antimatter has long been used in science fiction. You might think it is an invention of imaginative writers, but scientists, technologists, and medical physicists use and experiment with antimatter on a routine basis. Indeed, a well-known method of medical imaging, positron emission tomography (PET scanning), depends upon matter and antimatter particles annihilating one another to produce gamma-ray photons.

In 1931, Paul Dirac predicted the existence of antimatter, and the **positron** – the **antiparticle** of the electron – was discovered the next year. By 1956, the antiproton and antineutron had also been discovered. For each subatomic particle, there is an antiparticle, although some neutral particles, such as the photon, can be their own antiparticles. A particle and its antiparticle have the same mass, but opposite charge.

Classification of particles

Particles and antiparticles are classified into two groups:

- **Leptons**: fundamental particles that interact through a nuclear force called the weak interaction. The weak interaction is responsible for beta decay. Electrons are leptons.

- **Hadrons**: composite particles (made up of quarks) that interact through the strong interaction. The strong interaction holds nuclei together. Protons and neutrons are hadrons.

In terms of particle physics, the composition of the world is breathtakingly simple. The ordinary matter around you is made from just one lepton (the electron) and two different quarks (up and down) in protons and neutrons. These three, together with the **neutrino** that goes with the electron, explain almost all the visible matter in the Universe (Table 1).

▼ **Table 1** *The world around you*

Lepton	Electric charge	Rest energy /MeV
electron	−1e	0.511
positron	+1e	0.511
ν_e electron-neutrino	0	very near 0
$\bar{\nu}_e$ anti-electron-neutrino	0	very near 0

Quark	Electric charge	Rest energy /MeV
d down	−1/3e	8
u up	+2/3e	4
$\bar{d}$ anti-down	+1/3e	8
$\bar{v}$ anti-up	−2/3e	4

Annihilation

Positrons are emitted from nuclei in a process called β⁺ decay. A positron cannot exist long in our world of normal matter. Within about a millimetre of travel it encounters the outer electron of an atom and the two particles annihilate one another. They vanish.

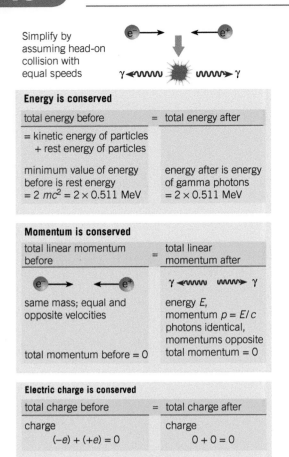

Simplify by assuming head-on collision with equal speeds

Energy is conserved

total energy before	= total energy after
= kinetic energy of particles + rest energy of particles	
minimum value of energy before is rest energy = 2 mc^2 = 2 × 0.511 MeV	energy after is energy of gamma photons = 2 × 0.511 MeV

Momentum is conserved

total linear momentum before	=	total linear momentum after
same mass; equal and opposite velocities		energy E, momentum $p = E/c$ photons identical, momentums opposite
total momentum before = 0		total momentum = 0

Electric charge is conserved

total charge before	= total charge after
charge $(-e) + (+e) = 0$	charge $0 + 0 = 0$

▲ **Figure 1** *Energy, momentum, and electric charge are always conserved in electron–positron annihilation*

Matter is destroyed, but the energy remains and is carried away by a pair of gamma-ray photons travelling in opposite directions.

Although particles disappear in an annihilation event, some quantities are conserved:

- electric charge
- linear and angular momentum
- total energy (including rest energy $E_{rest} = mc^2$)
- lepton number.

Figure 1 analyses an electron–positron annihilation. We have simplified the event by assuming that the particle–antiparticle pair meet head-on with zero kinetic energy. This means that the energy of the gamma photons produced is simply the rest energy of the particle–antiparticle pair. The lepton number before and after the event is zero – an electron has a **lepton number** of +1 and a positron (antielectron) a lepton number of −1.

Pair creation

The reverse of annihilation is creation. Pair production is the creation of a positron and an electron by a gamma-ray photon of sufficient energy near a massive nucleus. Such a gamma-ray photon must have energy equal to at least $2mc^2$, the total rest energy of an electron and a positron, which comes to 1.022 MeV. As with annihilation, creation always involves a particle–antiparticle pair.

A 1.022 MeV gamma photon has enough energy to create an electron–positron pair, but it cannot create such a pair in free space. This is because the photon has momentum $p = \dfrac{E}{c}$ and momentum must be conserved. However, pair creation can occur near a nucleus, which then carries away some momentum so that both energy and momentum are conserved. The lepton number and electric charge before and after the event are zero, so these quantities have also been conserved as required (Figure 2).

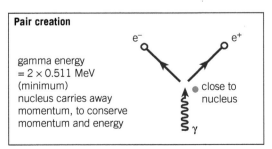

Pair creation

gamma energy = 2 × 0.511 MeV (minimum) nucleus carries away momentum, to conserve momentum and energy

▲ **Figure 2** *Pair creation*

Worked example: Creating matter

What is the minimum energy of a gamma ray required to create a proton–antiproton pair? Give the energy in joules and GeV (mass of proton = 1.67×10^{-27} kg).

Step 1: The minimum energy required will equal the rest energy of the two particles created.

Rest energy of particles in joules
= $2 \times 1.67 \times 10^{-27}$ kg $\times$ $(3.00 \times 10^8 \, \mathrm{m\,s^{-1}})^2 = 3.006 \times 10^{-10}$ J.

Step 2: $1 \, \mathrm{GeV} = 10^9 \times 1.60 \times 10^{-19} \, \mathrm{J} = 1.60 \times 10^{-10} \, \mathrm{J}$

Energy in GeV $= \dfrac{3.006 \times 10^{-10} \, \mathrm{J}}{1.60 \times 10^{-10} \, \mathrm{J}} = 1.88 \, \mathrm{GeV}$ (3 s.f.)

Creation of a proton–antiproton pair needs much more energy than creation of a positron and an electron, as the rest masses are each 938 MeV – such events occur in particle accelerators.

Figure 3 shows a false-colour image of tracks recorded in a bubble chamber. Two gamma-ray photons have entered from the top of the picture and each has created an electron–positron pair. The electron tracks are coloured green, the positron tracks red. A magnetic field deflects the particles in curved paths. In the event just below centre, the incoming photon has given all its momentum to the electron–positron pair. The tracks of this pair are only slightly curved, showing that this pair are moving at high speeds. In the event at the top most of the energy and momentum of the incoming photon is given to an electron knocked out of the atom that produces the long green track with little curvature. An electron-positron pair is also produced — their tracks are heavily spiraled because the particles lose energy to the atoms in the bubble chamber and slow down.

▲ **Figure 3** *Pair production in a bubble chamber*

Beta decay and conservation laws

A nucleus that has too many neutrons can decay by emitting an electron. This is beta-minus (β^-) decay. When β^- decay was first investigated it seemed to violate several conservation laws. Consider the decay of strontium-90 to yttrium-90 by beta-minus decay. A first attempt at an equation for the decay is

$$^{90}_{38}\mathrm{Sr} \rightarrow {}^{90}_{39}\mathrm{Y} + {}^{0}_{-1}\mathrm{e}$$

The lower number (atomic number or proton number) is unchanged by the decay ($38 = 39 + -1$), showing that electric charge has been conserved. The upper number (nucleon number) is also unchanged when a neutron changes to a proton. But there is something wrong – there is an electron (a lepton) on the right-hand side of the equation and no balancing lepton on the left-hand side. Lepton number has not been conserved.

There is a further problem with our simple equation. The difference in rest energy of the strontium and yttrium nuclei is 0.546 MeV, so every beta particle released should have the same energy, 0.546 MeV (Figure 4). This is not the case. Beta particles are emitted with range of energies, as shown in Figure 5.

Beta decay of strontium-90

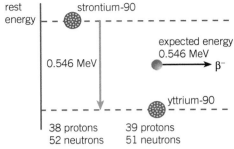

▲ **Figure 4** *Expected energy of beta particles released in the decay of strontium-90*

Energy spectrum of beta decay of strontium-90

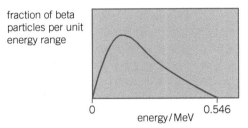

▲ **Figure 5** *Energy spectrum of the beta decay of strontium-90*

This result gave physicists cause for concern. Where does the energy go if it is not taken away by the beta particles? It seemed that beta decay might violate the conservation of energy.

In 1930, Wolfgang Pauli suggested that another particle was emitted in the reaction, carrying off the extra energy. This particle would have to be somewhat odd, with no charge, a very tiny mass, and extremely weak interactions with matter (or else it would be detected). The particle was a new type of lepton, which Enrico Fermi named **neutrino** in 1934 (meaning 'little neutral one'), given the symbol ν. Because it has no charge and contains no nucleons, in nuclear equations it is represented as $_{0}^{0}\nu$. It was 26 years before the neutrino was detected. It interacts so weakly with matter that most neutrinos would pass through a light-year of lead without interacting with anything – billions of neutrinos are passing through you as you read this sentence!

When strontium decays into yttrium, a beta particle is emitted along with an antineutrino $_{0}^{0}\bar{\nu}$. Lepton number is therefore conserved, as the lepton number of the antineutrino is −1. The 'missing' energy is the energy carried away by the antineutrino. The full equation for the beta decay of strontium is

$$_{38}^{90}\text{Sr} \rightarrow {}_{39}^{90}\text{Y} + {}_{-1}^{0}\text{e} + {}_{0}^{0}\bar{\nu}$$

with conserved particle properties:

- nucleon number: $90 = 90 + 0 + 0$
- electric charge: $38 = 39 - 1 + 0$
- lepton number: $0 = 0 + 1 - 1$.

Beta decay of strontium-90, including antineutrino emission

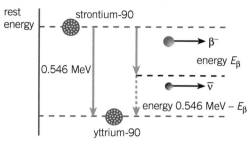

▲ **Figure 6** *Antineutrinos carry away the 'missing' energy in the beta decay of strontium-90*

Summary questions

1   State which of the following particles are leptons: protons, electrons, positrons, antiprotons, neutrinos, neutrons, antineutrinos, antineutrons.
(2 marks)

2   Explain why neutrinos are so difficult to detect.   *(2 marks)*

3   State the conservation laws that apply to the alpha decay in the equation below, and say how they are satisfied.

$$^{238}_{92}U \rightarrow \,^{234}_{90}Th + \,^{4}_{2}\alpha$$
(2 marks)

4   Free neutrons are unstable and decay to protons (this is the underlying process in beta decay). Explain why the equation below cannot be fully correct for neutron decay, and write the correct equation.

$$^{1}_{0}n \rightarrow \,^{1}_{1}p + \,^{0}_{-1}e$$
(2 marks)

5   Explain how pair production of electrons and positrons conserves:
a electric charge; **b** energy; **c** lepton number.   *(3 marks)*

6   Calculate the minimum energy of a gamma ray required to create:
a an electron–positron pair; **b** a muon–antimuon pair. Give your answers in joules and MeV. (Mass of electron $= 9.1 \times 10^{-31}$ kg; mass of muon $= 207 \times$ mass of electron)   *(6 marks)*

18.5 Electrons in atoms

Specification reference: 6.2.1a(iii), 6.2.1a(iv), 6.2.1b(i), 6.2.1b(iii)

Synoptic link

You met the idea of energy levels and emission spectra in Topic 7.1, Quantum behaviour.

Atomic energy levels

Every yellow sodium streetlight or orange neon sign gives evidence that electrons in atoms have discrete energy levels. Their light comes in sharp, discrete spectral lines. Each line is light carried by photons of discrete frequency, and so of discrete energy $E = hf$. The energies of the photons correspond to the difference in energy between two of the possible energy levels of electrons in the atom (Figure 1). Emission spectra show that each atom has a fixed pattern of energy levels (Figure 2).

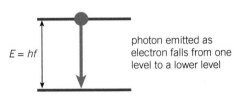

$E = hf$

photon emitted as electron falls from one level to a lower level

▲ Figure 1 Photon emission

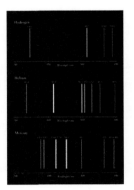

▲ Figure 2 Emission spectra of hydrogen, helium, and mercury

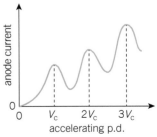

heated filament   grid   anode

sensitive ammeter

V

▲ Figure 3 Franck and Hertz apparatus

anode current

0 V_c $2V_c$ $3V_c$

accelerating p.d.

▲ Figure 4 Variation of anode current with accelerating p.d. in the Franck and Hertz apparatus

Franck and Hertz experiment 1914

This experiment, which led to the Nobel Prize in Physics in 1925, gives further evidence of discrete energy levels in atoms. The filament on the left of Figure 3 is heated, and electrons 'boil off' in a process known as thermionic emission. The electrons are accelerated towards a wire grid, which is held at a positive potential. The anode is at a slightly lower potential than the grid. Electrons accelerate from the filament, pass through the grid and are decelerated in the short distance between the grid and the anode. Only those electrons with sufficient kinetic energy will reach the anode from the grid. For example, if the anode potential is 0.2 V less than the grid potential only electrons passing through the grid with a kinetic energy of 0.2 eV or greater will reach the anode. The current registered on the sensitive ammeter is a measure of the number of electrons reaching the anode each second. The tube contains gas at a low pressure so the accelerated electrons collide with the gas atoms as they travel along the tube.

Figure 4 shows the shape of the line obtained as the accelerating p.d. is increased. At first an increase in accelerating p.d. increases the current, but at a critical potential difference V_c the current begins to fall before rising once again. This is repeated at $2V_c$, $3V_c$, and higher integer multiples of V_c, which are not shown.

The shape of the graph is explained as follows:

- At low accelerating p.d.s the electrons from the filament make elastic collisions with the gas atoms and 'bounce off' the gas atoms without loss of energy.

- Increasing the p.d. increases the current registered because more electrons reach the anode each second.

- At p.d. V_c, the accelerated electrons have sufficient energy to knock an electron in the gas atom to a higher energy level. This means that some of the energy of the accelerated electron is given to the gas atom in an inelastic collision. The electrons involved in inelastic collisions do not have sufficient energy to reach the anode, so the current falls.

- Drops in current are also observed at $2V_c$ and $3V_c$, because at these points the accelerated electrons can transfer energy to two or three gas atoms, respectively.

- The drops in current show that inelastic collisions only happen at specific electron energies. This shows that there are specific energy levels in the atom – the electrons in the atom are limited to fixed energy levels.

 Worked example: Photon energy

In a Franck–Hertz experiment with mercury vapour there is a drop in anode current at an accelerating p.d. of 4.90 V. This energy represents the difference between two energy levels in the mercury atom. Calculate the wavelength of ultraviolet radiation that is emitted when the electron in the atom falls back to the lower level.

Step 1: Calculate the energy of an electron accelerated through 4.9 V.

$$\text{Energy} = qV = 1.60 \times 10^{-19}\,\text{C} \times 4.90\,\text{V} = 7.84 \times 10^{-19}\,\text{J}$$
$$= 4.9\,\text{eV}$$

Step 2: $E = hf \therefore \lambda = \dfrac{hc}{E} = \dfrac{6.63 \times 10^{-34}\,\text{J s} \times 3.00 \times 10^{8}\,\text{m s}^{-1}}{7.84 \times 10^{-19}\,\text{J}}$
$$= 2.54 \times 10^{-7}\,\text{m (3 s.f.)}$$

Synoptic link

The explanation for the discrete energy levels of atoms takes ideas about standing waves from Topic 6.1, Superposition of waves, and ideas about electron waves from Topic 7.3, Electron diffraction.

Electron standing waves

Why should electrons only be 'allowed' to occupy certain, fixed energies? Remember that electrons have de Broglie wavelengths associated with them. An atom can be thought of as a kind of pocket, or box, in which electrons can be trapped. The positively charged nucleus creates a potential energy well in which negative electrons can be bound, unable to escape (Figure 5).

If electrons are constrained, trapped in some sort of box, electron standing waves can form. These are analogous to the standing waves that form along a guitar string when it is

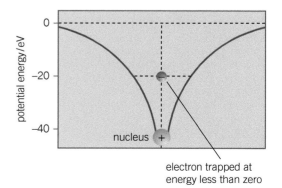

▲ **Figure 5** An electron trapped in a potential well

Synoptic link

The electrical potential well that holds the electron is similar to the way that you are trapped by Earth's gravitational potential well (Topic 12.4 Gravitational potential in a radial field).

plucked. Just as the wavelengths that form on a guitar string must fit into length of the string, the possible de Broglie wavelengths of the trapped electron are limited by the size of the box.

The de Broglie relationship tells us that $p = \dfrac{h}{\lambda}$.

As the energy of the electron is quite small, we can use the non-relativistic expression

$$E_k = \frac{1}{2}mv^2 = \frac{p^2}{2m}$$

$$\therefore E_k = \frac{h^2}{2m\lambda^2}$$

The smaller the 'box' trapping the electron, the smaller the wavelength and so the larger the momentum and kinetic energy. The shape and size of the box determines which wavelengths are allowed, which in turn selects the allowed energy values. An electron in an atom can only have one of these allowed values. No in-between states exist. The permitted energies are therefore the energy levels of electrons in atoms. Electrons in atoms occupy specific energy levels because the electrons are associated with de Broglie standing waves.

A simplified model of the hydrogen atom

Let's apply the model of electrons trapped in an atomic box to hydrogen, the simplest element, with a single electron trapped in the box. Our model is obviously imperfect; instead of the curving slopes of the potential energy well of the nucleus represented in Figure 5 we are considering hard walls a fixed distance apart. But it has one correct, crucial feature – the electron waves are trapped in a definite

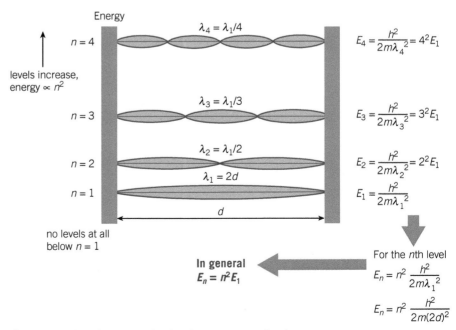

▲ **Figure 6** *Standing waves lead to discrete energy levels*

space. In this model (Figure 6), the energy of the electrons depends only on the width of the box and the number n of half-wave loops in the standing wave. This produces a number of allowed energy levels, labelled by a quantum number n starting at the ground state, the lowest level. The model suggests that the hydrogen atom has levels of energy increasing in proportion to the square of the quantum number n. This is not the case. In fact the electron energy levels in hydrogen increase as $\frac{1}{n^2}$.

In 1926, the Austrian physicist Erwin Schrödinger found a way to calculate the shapes and energies of the electron standing waves in the hydrogen atom. The energy levels of hydrogen came out as

$$E_n = -\frac{13.6\,\text{eV}}{n^2}.$$

This result agrees with the observed emission lines of hydrogen. Figure 7 shows the energy levels of hydrogen. The visible lines (the Balmer series) are produced when electrons drop to the $n = 2$ level. The bigger drops to the $n = 1$ level produce photons of higher energy (in the ultraviolet region).

As n increases, the energies get closer and closer together, and the energy E_n gets closer and closer to zero. If $E > 0$ the electron has positive total energy and breaks free of the atom – the atom ionises. The ionisation energy is just the energy needed to get an electron from the lowest level at $-13.6\,\text{eV}$ up to zero energy, when it is free. So the ionisation energy of hydrogen is $13.6\,\text{eV}$.

> **Hint**
>
> If you are studying chemistry, you may have used the notation 1s, 2s, 2p, and so on to denote electron orbitals. The principal quantum number n is this number (e.g., for 2s, $n = 2$).

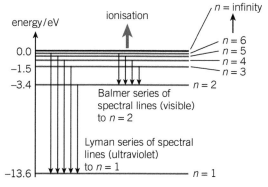

▲ **Figure 7** As n increases, E_n gets closer to zero

Why aren't atoms smaller?

You have seen that the nucleus is tiny compared to the volume of an atom. So why aren't atoms smaller?

The electron in a hydrogen atom must be bound to the nucleus. Its total energy must therefore be negative. The attraction of the nucleus gives it negative potential energy. But if the electron is boxed into a closed space, the standing waves require it to have momentum, and therefore kinetic energy. The kinetic energy gets bigger the smaller the box. Adding the positive kinetic energy to the negative potential energy always brings the total energy up nearer towards zero.

> **Synoptic link**
>
> The explanation for the minimum size of an atom takes ideas about potential energy of a point charge from Chapter 17, Charge and field.

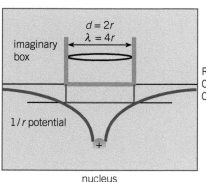

Replace $1/r$ potential by a box of width $d = 2r$
Calculate kinetic energy for waves $\lambda = 2d = 4r$
Calculate potential energy at r

standing wave $\lambda/2 = d$
momentum $p = h/\lambda$
kinetic energy $= p^2/2m$

◄ **Figure 8** How small can a hydrogen atom be?

For an atomic 'box' of width $d = 2r$ (Figure 8) the electric potential energy is given by

$$E_p = -\frac{e^2}{4\pi\varepsilon_0 r}$$

The kinetic energy of a de Broglie wave trapped in an atom of radius $r = \frac{\lambda}{4}$ is given by

$$E_k = \frac{h^2}{2m\lambda^2} = \frac{h^2}{32mr^2}$$

As we squeeze the atom, reducing its radius r, the potential energy will become more negative. For example, halve the radius and the magnitude of the potential energy will double. However, the kinetic energy of the particle ($\propto \frac{1}{r^2}$) will increase by a factor of four. The magnitude of the kinetic energy will equal the magnitude of the potential energy at the minimum possible atomic radius.

At this radius, $E_k + E_p = 0$. If the kinetic energy is of greater magnitude than the potential energy, the electron will escape. Squeeze a hydrogen atom too hard and its electron will pop out.

 Worked example: The minimum radius of the hydrogen atom

Calculate an estimate for the radius of a hydrogen atom. Assume that the kinetic and potential energies of the electron are of equal magnitude at this radius (mass of electron = 9.1×10^{-31} kg, Planck constant = 6.6×10^{-34} J s, electron charge = -1.6×10^{-19} C, $\varepsilon_0 = 8.85 \times 10^{-12}$ C V^{-1} m^{-1}).

Step 1: Equate the magnitudes of the kinetic and potential energies of the electron.

$$E_k = E_p \qquad \text{so} \qquad \frac{h^2}{32mr^2} = \frac{e^2}{4\pi\varepsilon_0 r}$$

$$\therefore r = \frac{h^2 4\pi\varepsilon_0}{32me^2}$$

$$= \frac{(6.6 \times 10^{-34}\,\text{J s})^2 \times 4\pi \times 8.85 \times 10^{-12}\,\text{C V}^{-1}\text{m}^{-1}}{32 \times 9.1 \times 10^{-31}\,\text{kg} \times (-1.6 \times 10^{-19}\,\text{C})^2}$$

$$= 6.5 \times 10^{-11}\,\text{m} \ (2 \text{ s.f.})$$

This estimate gives an atomic diameter in the order of 10^{-10} m, which agrees with other order-of-magnitude calculations.

Summary questions

1 The spectrum of light from sodium vapour has a bright yellow line at wavelength 590 nm. Show that sodium atoms have a pair of energy levels differing in energy by about 2.1 eV. *(3 marks)*

2 a The energy level of hydrogen in its ground state = −13.6 eV. Calculate the ionisation energy of hydrogen in J. *(2 marks)*

 b Use the equation $E_n = -\dfrac{13.6\,\text{eV}}{n^2}$ to find the energy of the $n = 2$ level in hydrogen. *(1 mark)*

 c Calculate the energy required to ionise a hydrogen atom from the $n = 2$ level. Give your answer in J and eV. *(4 marks)*

3 The units in the calculation of the minimum atomic radius in the worked example are given as $\dfrac{J^2\,s^2 \times C\,V^{-1}\,m^{-1}}{kg \times C^2}$. Show that these units simplify to metres (m). *(3 marks)*

4 Carotene is a straight molecule about 2 nm long (Figure 9). Certain delocalised electrons effectively occupy the whole length of the molecule. Treating their standing waves as similar to those on a stretched string, calculate,

 a the electron wavelengths for the lowest energy state ($n = 1$) and the next state ($n = 2$), *(3 marks)*

 b the difference in kinetic energy of an electron in the $n = 1$ and $n = 2$ states, *(3 marks)*

 c the frequency of a photon emitted in a transition from the $n = 2$ to $n = 1$ state, *(1 mark)*

 d the wavelength of the photon in (c). *(1 mark)*

 e Give a reason why carotene molecules can absorb light in the visible part of the spectrum. *(1 mark)*

Carotene molecule $C_{40}H_{56}$

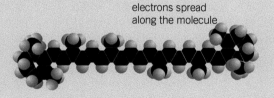

electrons spread along the molecule

Analogy with guitar string

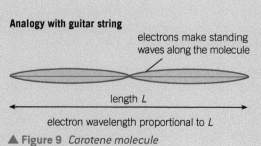

electrons make standing waves along the molecule

length L

electron wavelength proportional to L

▲ **Figure 9** *Carotene molecule*

Physics in perspective

Using antimatter to image the brain

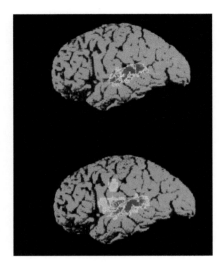

▲ **Figure 1** *PET scan of a brain responding to words*

Figure 1 shows a positron-emission tomography (PET) scan image of a brain. The activity of different areas of the brain is measured by the blood flow to those areas, where red- and yellow-coloured areas represent high blood flow. In the upper image the patient is listening to words – the auditory region of the brain is busy, as shown by the red and yellow areas.

In the lower image, the patient is listening to and repeating words. Another part of the brain is active – the higher yellow area. This is the area involved in speech.

PET scans are helping to advance our understanding of the changes to brain function caused by conditions such as Alzheimer's disease, epilepsy, and schizophrenia, as well as the working of the healthy brain.

Labelling blood

In Figure 1, the oxygen carried by blood to the brain is 'labelled' with a small amount of the radioactive isotope oxygen–15. The half-life of oxygen–15 is about 2 minutes, and it decays by emitting a positron, a process known as beta-plus decay.

$$^{15}_{8}O \rightarrow {}^{15}_{7}N + {}^{0}_{+1}e + {}^{0}_{0}\nu$$

Once created, the positron does not get far. Within a millimetre or so it encounters an ordinary electron in an atom and the two particles annihilate one another. The energy released by the annihilation is carried away by a pair of gamma-ray photons travelling in opposite directions, and detectors around the patient pick up these photons.

From many such events, a computer-generated map of the activity in your brain can be built up. More gamma photons from a brain region shows more oxygen in that region, which indicates greater activity.

Detecting the photons

Figure 2 shows the head of a patient surrounded by a bank of scintillators linked to photomultipliers. In the scintillation crystal, each gamma photon creates many thousand visible photons in a burst of light (scintillation). These photons enter the photomultipliers, leading to a cascade of electrons that give a detectable electrical pulse.

Pairs of gamma photons produced through electron-positron annihilation arrive at the detectors almost simultaneously, which allows the computer to calculate the area where the annihilation took place. For example, in Figure 2, the annihilation must have occurred along the line AB. This is a 'line of response' (LOR). Many positrons are emitted from each small region of the brain that has taken up the oxygen–15, and there will be many such LORs for each region. Calculating where the LORs cross allows the site of the positron emissions to be found.

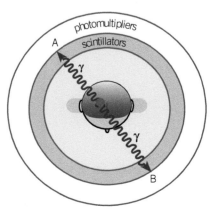

▲ **Figure 2** *Schematic diagram of gamma detectors in a PET scanner*

Time of flight imaging

If the timing system is precise and accurate enough, the time difference between detection of each photon in a pair can be found. This is known as time of flight imaging. This will improve the resolution of the image and make imaging possible with smaller amounts of radioisotope. Recent improvements in timing resolution have led to renewed interest in its development.

Producing oxygen–15

To get hold of a radioisotope with a half-life of two minutes, the hospital makes it on-site, as close to the scanner as possible. Deuterium ions (^2_1H, called deuterons) are accelerated in a cyclotron (a circular accelerator) and fired at nitrogen, producing the reaction shown.

$$^2_1H + {}^{14}_7N \rightarrow {}^{15}_8O + {}^1_0n$$

The oxygen–15 is incorporated into water which is readily taken up by the body. It's a race against time to produce the oxygen–15, synthesise it into water, and inject it into the patient. Positron emission tomography is an excellent example of a theoretical idea (antimatter) being used to further our knowledge in useful areas such as neuroscience and medicine.

▲ **Figure 3** *A medical cyclotron*

Summary questions

1 This question is about electron-positron annihilation. Assume particles have no kinetic energy when they meet. The mass of an electron is 9.1×10^{-31} kg.
   a Explain why the energy of each of the pair of gamma ray photons produced in the electron-positron annihilation is about 0.51 MeV. You should include calculations in your answer.
   b Explain why at least two photons must be produced in the annihilation and why the two photons will always move away from each other at 180°.

2 Oxygen–15 has a half-life of 124 s.
   a Calculate the time taken for the activity of a sample of oxygen–15 to fall to ten percent of its original value.
   b Assume it takes ten minutes after production to inject oxygen–15 into the patient. Calculate the activity of the oxygen–15 as a percentage of its original activity when it is injected into the patient.
   c State how the emission of a neutrino conserves lepton number in the positron decay.
   d Using a tracer with a short half-life means a smaller amount can be used compared to one with a longer half-life. Explain why. Suggest one other advantage to using a tracer with a short half-life.

3 The wavelength of visible light is of the order of 10^{-7} m. Estimate by calculation the maximum number of visible photons a gamma photon of energy 0.51 MeV could produce in a scintillation crystal.

4 A medical cyclotron accelerates deuterons to energies of 9 MeV. Show that this energy is sufficient to bring the deuteron ion to within 10^{-14} m of the centre of a nitrogen nucleus. What is the importance of this result?

Practice questions

1 Here is a list of particles:

   A electron   **B** proton

   C positron   **D** neutrino

Choose from the list:

  a The particle which is not fundamental.

  b The particle with the highest rest energy.

  c The particle with a neutral anti-particle.

  d The particle with the lowest rest energy.

                             (*4 marks*)

2 An electron, mass 9.11×10^{-31} kg, is accelerated to half the speed of light ($0.50c$).

  a Calculate the relativistic factor, γ. (*2 marks*)

  b Calculate the momentum of the particle.

                             (*2 marks*)

3 The rest energy of an electron is 0.51 MeV. Calculate the p.d. through which an electron must be accelerated to reach a relativistic factor of 2.0. (*2 marks*)

4 Calculate the de Broglie wavelength of a proton of kinetic energy 2.2×10^{-15} J. (*2 marks*)

5 A 5.4 MeV alpha particle makes a head-on approach to a uranium–238 nucleus (proton number = 92).

  a Calculate the distance of closest approach.

                             (*2 marks*)

  b How, if at all, would your answer change if the alpha particle were approaching a uranium–235 nucleus? (*1 mark*)

6 **a** Calculate the relativistic factor γ of a proton accelerated to 7 TeV (7×10^{12} eV).

      Rest energy of proton = 0.96 GeV (*2 marks*)

  b Calculate the speed of such a proton. Comment on your answer. (*2 marks*)

7 The equation represents beta-minus decay:

$$^{1}_{0}n \rightarrow \, ^{1}_{1}p + \, ^{0}_{-1}e + \, ^{0}_{0}\bar{\nu}$$

Calculate the energy released in the decay, assuming the neutrino has negligible mass.

mass of electron = 0.0009×10^{-27} kg

mass of proton = 1.6726×10^{-27} kg

mass of neutron = 1.6749×10^{-27} kg   (*2 marks*)

8 The radius of a nucleus is given by the relationship

$$\text{radius } r = r_0 \, A^{1/3}$$

where A is the nucleon number of the nucleus and r_0 is constant. The radius of a helium atom is 1.9×10^{-15} m. Calculate the radius of a uranium–238 nucleus. (*2 marks*)

9 Figure 1 shows the first four energy levels of a hydrogen atom. A hydrogen atom has energy −13.6 eV in its ground state.

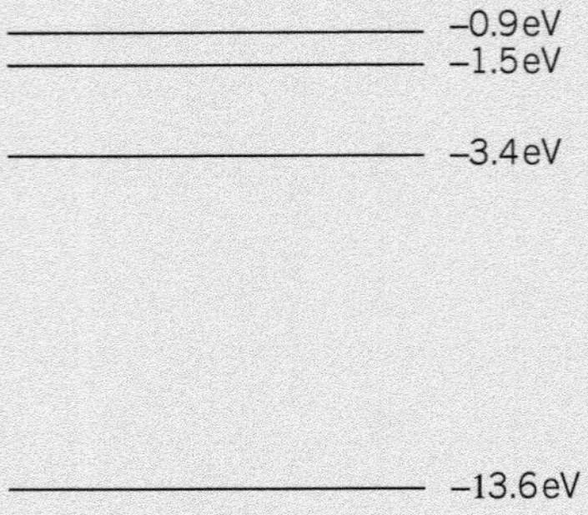

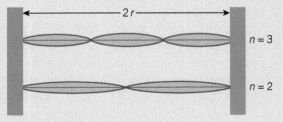

                          −0.9 eV

                          −1.5 eV

                          −3.4 eV

                          −13.6 eV

▲ **Figure 1**

  a Copy Figure 1 and show all the possible transitions between energy levels.

                             (*1 mark*)

  b Calculate the lowest frequency of photon emitted from the transitions you have shown in part **(a)**. (*2 marks*)

  c Calculate the highest frequency of photon emitted from the transitions. (*2 marks*)

  d An electron of energy 11.5 eV has an inelastic collision with a hydrogen atom in the ground state. Explain why the electron has energy 1.3 eV after the collision. (*2 marks*)

10 The hydrogen atom can be modelled simply as an electron standing wave in a box of length $2r$, where r is the radius of the atom (Figure 2).

                    2*r*

                                         n = 3

                                         n = 2

▲ **Figure 2**

  a Use the equations $E_k = \frac{1}{2}mv^2$ and $\lambda = \frac{h}{mv}$ to show that $E_k = \dfrac{h^2}{2m\lambda^2}$ (*1 mark*)

  b Estimate by calculation the kinetic energy of an electron at the $n = 2$ and $n = 3$ levels.

      Order-of-magnitude radius of hydrogen atom = 10^{-10} m. (*4 marks*)

  c Explain why the electron remains bound to the hydrogen nucleus. (*2 marks*)

11 This question is about the decay of strontium–90, a beta-emitter commonly used in schools. A school source has an activity of 7.0×10^4 Bq.

a (i) Show that the minimum number of strontium–90 nuclei in the source is about 9×10^{13}.
Half-life of strontium = 9.2×19^8 s
(2 marks)

(ii) Suggest why this is a minimum figure. *(1 mark)*

Strontium–90 decays into yttrium–90 by releasing an electron and an anti-neutrino. Figure 3 shows the energy spectrum of electrons released in this decay.

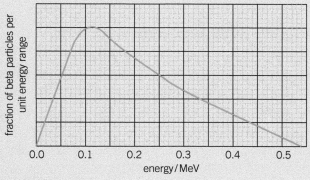

▲ Figure 3

b (i) Use the graph to estimate the most common energy of a beta particle released in the decay. *(1 mark)*

(ii) Each strontium–90 nucleus releases just over 0.5 MeV of energy when it decays. Explain how the graph suggests that particles other than beta particles are also emitted in the decay. *(3 marks)*

c A beta particle released from strontium–90 has a kinetic energy of 0.45 MeV. Calculate the relativistic factor γ for a beta particle with this kinetic energy and use this value to find the speed of a 0.45 MeV beta particle.
Rest energy of electron = 0.511 MeV
$c = 3.0 \times 10^8$ m s^{-1} *(4 marks)*

OCR Physics B Paper G495 June 2011

12 This question is about the scattering of alpha particles by nuclei as shown in Figure 4.

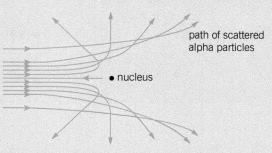

path of scattered alpha particles

nucleus

▲ Figure 4

Alpha particles of energy 5.0 MeV are fired at a thin gold foil. It is observed that 4 in every 100 000 alpha particles are scattered through large angles and 'bounce back' from the foil. It is assumed that these particles have made collisions with individual, stationary gold nuclei.

a State and explain any changes in the proportion of alpha particles which bounce back when the following changes are made to the experiment.

(i) A **thicker** gold foil is irradiated with 5 MeV alpha particles. *(2 marks)*

(ii) The original gold foil is irradiated with alpha particles of **higher** energy than 5 MeV. *(2 marks)*

b When a 5 MeV alpha particle makes a direct collision with a nucleus, the alpha particle is momentarily stationary.

(i) State the electrical potential energy of such an alpha particle when it is at its closest to the gold nucleus. *(1 mark)*

(ii) Convert the value in **(i)** into joules.
$e = 1.6 \times 10^{-19}$ C *(1 mark)*

(iii) Use the value in **(ii)** to calculate the distance of closest approach of the particle to the nucleus.
charge on alpha particle = $+2e$
charge on gold nucleus = $+79e$
$k = \dfrac{1}{4\pi\varepsilon_0} = 9.0 \times 10^9$ N m^{-2} C^{-2} *(2 marks)*

c Assuming that the distance calculated in **b(iii)** equals the radius of the nucleus, the experiment suggests that the ratio $\dfrac{\text{nuclear radius}}{\text{atomic radius}}$ is about 6×10^5.
Use the value of the ratio to estimate the density of a gold nucleus. State one other assumption that you make.
Density of gold = 1.9×10^4 kg m^{-3} *(3 marks)*

OCR Physics B Paper G495 February 2011

19 USING THE ATOM
19.1 Ionising radiation
Specification reference: 6.2.2a(i), 6.2.2d(i)

Synoptic link

Look back to Topic 18.5, Electrons in atoms, for more about ionising atoms by providing enough energy to lift an electron out of the potential well created by the nucleus.

Humans use ionising radiation

Ionising radiation can both treat and cause cancer. This stark statement demonstrates the energetic nature of ionising radiation. It has many medical uses, from treatment to imaging and diagnosis to sterilising instruments. Energy from the decay of radioisotopes can be used in thermoelectric generators – the New Horizons probe to Pluto is powered by a generator containing 11 kg of plutonium oxide pellets. Scaled-down versions of similar generators have even been implanted in humans to power pacemakers.

Properties of ionising nuclear radiations

Ionising radiation produces ions in the atoms of the materials it passes through. There are many kinds of ionising radiations including cosmic rays, ultraviolet light, alpha particles (α), beta particles (β), and gamma rays (γ).

It takes about 14 eV to ionise a nitrogen atom by removing an electron. Photons of visible light have energies around 2 eV, so visible light is not ionising radiation. Ultraviolet photons have greater energy and can ionise atoms and break molecular bonds, as sunburn demonstrates. Nuclear radiations from radioactive decay can have energies of a few MeV per particle, so all of them (α, β, and γ) have enough energy to ionise 100 000 atoms or so.

- Alpha radiation is strongly ionising. Alpha particles (helium nuclei) are not very penetrating. Each ionisation transfers energy from the alpha particle, so it rapidly loses energy. Therefore an alpha particle has a range of only a few centimetres in air. It soon captures two electrons and becomes a neutral helium atom. Most of the helium found on Earth has been produced like this.

- Beta radiation is weakly ionising – electrons and positrons produce fewer ions per metre of travel than alpha particles. In consequence, a beta particle travels further than an alpha particle of the same initial energy – its range is of the order of 1 metre.

- Gamma radiation (gamma photon) is far less ionising than the charged nuclear radiations. In air, gamma radiation follows the inverse square-law. For example, the intensity falls by a factor of four when the distance from the source doubles.

Cloud-chamber tracks

A cloud chamber reveals the path of particles that pass through it. The chamber contains supersaturated vapour. When a particle of ionising radiation passes through the vapour it produces ions, which act as centres on which the vapour condenses as a vapour trail like that of a jet aircraft, marking the path of the particle. Figure 1 shows alpha particle tracks. These are straight and thick, and have a definite range. Their thickness suggests many ionisations per metre. They are straight because alpha particles are not deflected by the atoms they ionise. The definite range shows that alpha particles are released with specific energies, unlike beta particles.

Figure 2, recorded in 1932 by the American physicist Carl Anderson, is the image that proved the existence of the positron. The track, curving upwards to the left, is weak because the positron, like the electron, is only weakly ionising. The track curves because there is a uniform magnetic field acting into the paper. The direction of the curve shows that the particle is positive. The curvature is greater after the particle passes the horizontal white line, which is a lead plate. The degree to which the particle was slowed by the lead proved that the particle has the same mass as an electron.

Absorption of nuclear radiations

The effect of air in absorbing nuclear radiations has been stated above. Denser materials reduce the range of ionising particles because there are more atoms to interact with per metre of path.

▲ Figure 1 *Tracks of alpha particles*

Practical 6.2.2d(i): Studying the absorption of α, β, and γ radiation

The apparatus shown in Figure 3 can be used to investigate the absorption of nuclear radiations.

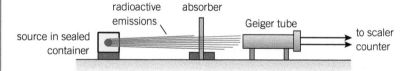

▲ **Figure 3** *Apparatus for investigating absorption of radiation by different materials*

Take a reading of the background count rate (with no source present). This reading is subtracted from all readings taken with the source present, to establish the count rate from the source (the corrected count). The degree of absorption can be established by measuring the count rate on the scaler counter with no absorber present, inserting an absorber, and repeating the process. The Geiger tube will need to be close to the source for alpha radiation as alpha particles are readily absorbed by air.

Changing the thickness of a particular absorber (aluminium, say) allows you to investigate the relationship between absorber thickness and count rate.

▲ **Figure 2** *The discovery image of the positron*

- Alpha particles are stopped by a few sheets of paper or a thin metal foil.
- Beta particles are stopped by a few millimetres of aluminium.
- Gamma rays are not completely absorbed by a few millimetres of lead.

A careful measurement of the absorption of gamma rays in lead shows that, to a good approximation, a given thickness of material reduces the number of gamma-ray photons by a constant fraction – the intensity of the radiation decreases exponentially with thickness. Just as half-life is defined for exponential radioactive decay, so for exponential decrease of intensity with thickness you can define the **half-thickness**. This is simply the thickness of absorber needed to halve the number of photons that, on average, get through.

If I_0 is the intensity with no absorber present and I is the intensity with an absorber of thickness x present,
$I = I_0 e^{-\mu x}$, where μ is the absorption coefficient (m^{-1}).

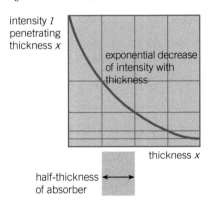

intensity I penetrating thickness x

exponential decrease of intensity with thickness

thickness x

half-thickness of absorber

Absorption by successive thicknesses

intensity reduced to one half by each block

▲ **Figure 4** *Exponential decrease of gamma ray intensity with thickness of absorber*

Synoptic link

The equation $I = I_0 e^{-\mu x}$ is of the same form as the decay equation $N = N_0 e^{-\lambda t}$ in Topic 10.2, Another way of looking at radioactive decay.

Just as we can find half-life from the decay constant, we can find half-thickness from the absorption coefficient.

Half-thickness $x_{1/2} = \dfrac{\ln 2}{\mu}$

Effects of ionising radiation on living tissue

Ionising radiation damages cells and their DNA. The more energy transferred from the radiation to the cells, the more damage. However, there is no simple relationship between, say, the energy of the particles emitted by a source and the potential damage caused to the cells the particles travel through. For example, rapidly dividing cells are most at risk – testes and ovaries are roughly ten times more sensitive to radiation than skin and bone surfaces.

Summary questions

1 The energy required to ionise a nitrogen atom is about 14 eV. Calculate the maximum number of nitrogen atoms an alpha particle of energy 3.0 MeV can ionise. *(2 marks)*

2 Calculate the longest wavelength of an ultraviolet photon that could ionise an atom with an ionisation energy of 10 eV. *(3 marks)*

3 0.5 m of concrete reduces the intensity of a gamma ray source by half.
   a Calculate the absorption coefficient of the concrete. *(2 marks)*
   b Calculate the thickness of concrete required to reduce the intensity to 5% of its original value. *(3 marks)*

4 Using your own research, write a description of how one of the following works, including diagrams and images if helpful:
   - home smoke detectors that use alpha radiation
   - sterilisation of medical equipment or of foodstuffs by gamma radiation
   - power generation in spacecraft using radioactive decay
   - use of radioactive tracers in medical diagnosis
   - a medical use of X-rays. *(10 marks)*

19.2 Effects of radiation on tissue

Specification reference: 6.2.2b(i), 6.2.2c(i), 6.2.2c(ii), 6.2.2c(iii)

Learning outcomes

Describe, explain, and apply:

→ the terms absorbed dose, effective dose, risk

→ absorbed dose in gray (Gy) = energy deposited per unit mass

→ effective dose in sievert (Sv) = absorbed dose in gray × quality factor

→ activity of a sample of radioactive material (related to half-life and decay constant).

Synoptic link

You have made calculations of half-life and the decay constant in Topics 10.1, Radioactive decay and half-life, and 10.2, Another way of looking at radioactive decay.

Radiation dose

A little knowledge is a dangerous thing. For much of the first half of the 20th century, a little knowledge is all that physicists and doctors had about the effects of ionising radiations on living matter. Physicists knew how to count particles and so how to measure radiation levels in becquerels. In the 1930s Hal Gray, one of the first medical physicists, realised that as well as the amount of radiation, what matters is the energy absorbed by the tissue. The harm is done by the energy of the radiation and the ionisation damage it inflicts on cells.

The **absorbed dose** – the number of joules absorbed per kilogram of tissue – is now measured in grays (Gy), where $1\,\text{Gy} = 1\,\text{J}\,\text{kg}^{-1}$. When a patient is exposed to ionising radiation during treatment or investigation you should not assume that the radiation is absorbed (and energy transferred) over the whole body. The treatment is often targeted at a much smaller mass of tissue.

 Worked example: Calculating dose

A source emitting beta particles with an average energy of $9 \times 10^{-14}\,\text{J}$ per particle is absorbed by a thyroid gland of mass $40\,\text{g}$. The initial activity of the source is $200\,\text{kBq}$. The decay constant λ of the source is $1.0 \times 10^{-6}\,\text{s}^{-1}$. Calculate the dose received by the gland over 2 days. What assumption is made in the calculation?

Step 1: First estimate by calculation the number of decays in the period, using the equation $N = N_0 e^{-\lambda t}$.

The initial number of source nuclei can be found from activity $= \lambda N$.

$200 \times 10^3\,\text{s}^{-1} = 1.0 \times 10^{-6}\,\text{s}^{-1} \times N_0$

$N_0 = 2 \times 10^{11}$ nuclei

The number remaining after 2 days ($1.728 \times 10^5\,\text{s}$) is
$N_0 e^{-\lambda t} = 2 \times 10^{11} \times e^{-0.1728} = 1.682... \times 10^{11}$

Number decayed $= N_0 - N = 2 \times 10^{11} - 1.682... \times 10^{11}$
$= 3.17... \times 10^{10}$

Therefore, the number of beta particles emitted over two days is $3.17... \times 10^{10}$.

Step 2: Energy absorbed = number of particles × energy per particle

$= 3.17... \times 10^{10} \times 9 \times 10^{-14}\,\text{J} = 2.85... \times 10^{-3}\,\text{J}$

Step 3: Dose in gray $= \dfrac{\text{energy absorbed}}{\text{mass of absorbing tissue}} = \dfrac{2.85... \times 10^{-3}\,\text{J}}{40 \times 10^{-3}\,\text{kg}}$
$= 0.07\,\text{Gy}$ (1 s.f.)

The calculation assumes that all the energy of the beta particles is transferred to the thyroid – in other words, that all the beta particles emitted are absorbed by the thyroid.

However, the energy absorbed isn't the whole story. The amount of damage also depends on the type of radiation and the type of tissue exposed. Alpha radiation is highly ionising, beta and gamma radiation less so. The gray – energy absorbed per kilogram – is not a good enough unit for predicting possible consequences of exposure to radiation. So the sievert (Sv), named after the Swedish radiologist Rolf Sievert, is used. The various components of the absorbed dose in grays are multiplied by quality factors depending on the type of radiation and the type of tissue to give the **effective dose** in sieverts. The *dose equivalent* takes into account the quality factor of the radiation (see Figure 1) but does not consider the type of tissue irradiated.

Radiation quality factors		
radiation	factor	dose equivalent of 1 gray/Sv
alpha	20	20
beta	1	1
gamma	1	1
X-rays	1	1
neutrons	10	10

▲ **Figure 1** *Radiation quality factors for estimation of effective dose and dose equivalent – the round numbers show that they are not precisely understood*

Worked example: Dose equivalent

An individual breathes in a small amount of particulates that emit alpha particles of energy 9.0×10^{-13} J. These remain in the lungs for 6 months, decaying with a constant activity of 200 Bq. Calculate the dose equivalent received during this time, assuming that the particulates are evenly spread in the lungs (mass of lungs = 2.0 kg, 6 months = 1.6×10^7 s).

Step 1: Calculate the energy absorbed over the period.

energy absorbed = energy per decay × activity × time

$= 9.0 \times 10^{-13}$ J × 200 Bq × 1.6×10^7 s = 2.88×10^{-3} J

Step 2: dose in gray = energy absorbed per kg = $\dfrac{2.88 \times 10^{-3} \text{ J}}{2.0}$

$= 1.44 \times 10^{-3}$ Gy

Step 3: From the table in Figure 1, the quality factor of alpha radiation is 20.

dose equivalent = dose in gray × quality factor

$= 1.44 \times 10^{-3}$ Gy × 20 = 0.029 Sv (2 s.f.)

Radiation and risk

What is a large dose of radiation? Each year we all get a dose equivalent of a little less than 2000 μSv (microsievert) from naturally occurring background radiation. Perhaps a large dose is one that adds significantly to this background level. From Figure 2 you can see that all artificial sources taken together account for 22% of the effective dose. Taken as a whole, artificial sources add significantly to the background. However, individual artificial sources of radiation, such as nuclear power, only add a relatively small amount to the total dose equivalent.

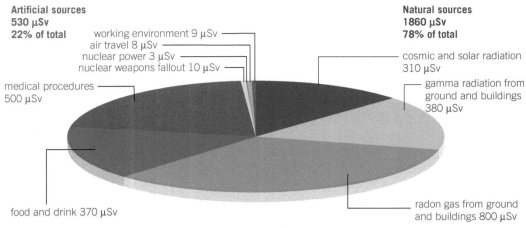

▲ **Figure 2** *Average whole-body effective dose per year (Europe)*

Another way of thinking about dose is to consider the risk for a given equivalent dose. The latest guidelines from the International Commission on Radiological Protection suggest that the incidence of cancer due to ionising radiation is about 5% per sievert.

Medical procedures contribute about 500 µSv per year – is this a large dose? It is a lot smaller than the dose from natural sources, but what about the risk of developing cancer from such a dose?

risk = equivalent dose × incidence per sievert = $500 \times 10^{-6} \times 5\%$
    = 0.0025%

This shows that the likelihood of developing cancer from medical procedures is small for any one individual. However, for a population of, say, 60 million, this suggests that around 1500 people will develop cancer from medical procedures each year.

Working on the simple hypothesis that there is a linear relationship between the dose and the effect (for example, twice the dose implies twice the risk), it is clear that we should do all that we can to minimise radiation dose. This is particularly important for those working with ionising radiation. This principle is known as ALARA, an acronym for As Low As Reasonably Achievable.

Another way of thinking about risk

We can assess risk in everyday life using the simple expression,
risk = probability of an event occurring × consequence if it does

Humans are not very good at handling probabilities or risks. For example, the probability of a failure of a nuclear power plant in a year might be 0.02%. This sounds fairly safe, but it suggests that with several hundred reactors around the world we should expect a serious failure every decade or so. That sounds much more worrying, and of course the consequences of a failure are considerable.

People also get used to risks. For example, in Britain about 2000 people are killed in car accidents each year, but we continue to use cars quite happily. The chance of an accident whilst skiing is relatively high, but many families choose skiing holidays. New kinds of risk, which people do not understand or feel unable to control, are more worrying. This may be one reason for public anxiety about risks from radiation.

Allowed doses

The maximum allowed radiation exposure for a member of the public from artificial sources is 1000 μSv per year for frequent exposure. This means that the authorities accept a risk of 50 cancers per million. Is that too much? To compare, a risk of 50 deaths per million is associated with smoking 75 cigarettes a year or cycling 2 km to work each day. What level of risk do you find acceptable?

Summary questions

1 Calculate how many beta particles of average energy 8×10^{-14} J must be absorbed in 0.2 kg of tissue to give a dose of 0.5 mGy. (*1 mark*)

2 Calculate the number of alpha particles of energy 5 MeV that must be absorbed in 0.3 kg of tissue to give a dose equivalent of 0.5 mSv (quality factor of alpha radiation = 20). (*2 marks*)

3 An alpha emitter has a half-life of 138 days. The energy of the alpha particles emitted is 8.5×10^{-13} J. A sample of initial activity 2×10^3 Bq is absorbed by 1.5 kg of tissue. Calculate an estimate of the dose equivalent delivered to the tissue in one week (quality factor of alpha radiation = 20). (*3 marks*)

19.3 Stability and decay

Specification reference: 6.2.2a(ii), 6.2.2b(i), 6.2.2c(iv)

Learning outcomes

Describe, explain, and apply:

→ stability and decay of nuclei in terms of binding energy; transformation of a nucleus on emission of radiation; qualitative variation of binding energy with proton and neutron number (nuclear valley)

→ the terms binding energy, atomic mass unit

→ energy changes from nuclear transformations, $E_{rest} = mc^2$.

Stability

Why are some nuclei stable and other unstable? The answer lies in the balance between the number of protons and neutrons. Stable isotopes of light (low nucleon number) elements tend to have equal numbers of protons and neutrons. For example, carbon–12 (6 protons, 6 neutrons) is stable whilst carbon–14 (6 protons, 8 neutrons) is radioactive, decaying by beta decay into stable nitrogen–14 (7 protons, 7 neutrons).

Figure 1 shows a plot of proton number against neutron number. Stable isotopes of the light elements lie along the $N = Z$ line, but stable heavier isotopes have more neutrons than protons in their nuclei.

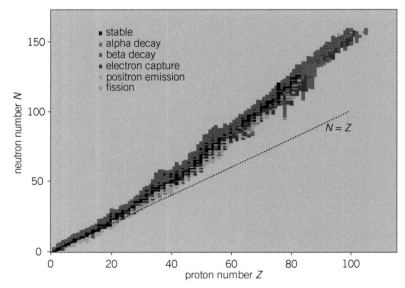

▲ **Figure 1** *Stable and unstable nuclei: the balance of numbers of protons and neutrons*

A strong attraction acts between nucleons, that is, between protons and protons, neutrons and neutrons, and neutrons and protons. This is the **strong nuclear force** and, as its name suggests, it overcomes the electrical repulsion of the protons. As the proton number increases, so does the potential energy of their mutual electrical repulsion. The extra neutrons 'dilute' the protons and provide further nuclear attraction to overcome the electrical repulsion. Consequently, stable heavier nuclei have more neutrons than protons.

The Pauli exclusion principle

Particles like electrons, protons, and neutrons, which belong to a class of particles called fermions, have a special property. The amplitudes for two identical particles arriving at exactly the same point in space–time subtract, that is, they add up with opposite phase and the phasor arrows point in opposite directions. As a result, the total amplitude is zero, so such particles can never come together in exactly the same state. The rule that no two identical fermions can share the same quantum state is the **Pauli exclusion principle**.

Synoptic link

Phasors were covered in Topic 7.2, Quantum behaviour and probability.

This principle gives a further reason for nuclei having roughly equal numbers of protons and neutrons – a proton and a neutron can share the same state, but two protons or two neutrons cannot. This is also why matter is hard – however heavily you press down on an object you cannot squeeze the particles in it into the same quantum states.

Photons belong to a different class of particles, called bosons, which do not follow the exclusion principle. They do the opposite. The amplitudes for two identical photons at the same point in space–time add up with the same phase, and the amplitude doubles. The laser is the best known application of this behaviour.

Binding energy

Stability is a question of energy. If energy is needed to pull apart the protons and neutrons in a nucleus, the combination must be stable. The more energy needed, the more stable the nucleus. The energy of a nucleus must therefore be less than the energy of its protons and neutrons taken separately.

We calculate the rest energy of a nucleus from $E_{rest} = mc^2$, where m is the mass of the nucleus, that is, the mass of the atom minus the mass of its electrons. The individual rest energies of all the protons and neutrons are then subtracted from the rest energy of the nucleus. The resulting value is the **binding energy** of the nucleus. The more negative this value is, the more stable the nucleus.

The atomic mass unit

The masses of atoms and nuclei are often expressed in the **atomic mass unit** because their masses are far smaller than 1 kg.

One atomic mass unit (u) is defined as the mass of a carbon–12 *atom*, that is, the carbon–12 nucleus and the six electrons bound to the atom. $1\,u = 1.660\,56 \times 10^{-27}\,kg$.

The binding energy of a carbon–12 nucleus

To calculate the binding energy of the carbon–12 nucleus, we must first calculate its mass:

- mass of a carbon atom = 12.0000 u
- mass of one electron = 0.000 549 u
- mass of 6 electrons = 6 × 0.000 549 u = 0.003 29 u.

Therefore, mass of the carbon–12 nucleus = 12.0000 u – 0.003 29 u
$$= 11.996\,71\,u$$

Now we calculate the mass of six unbound protons and neutrons:

- mass of one proton = 1.007 28 u
- mass of one neutron = 1.008 66 u
- mass of six protons and six neutrons = 6 × (1.007 28 u + 1.008 66 u)
$$= 12.0956\,u$$

Difference in mass = 11.9967 u – 12.0956 u = –0.0989 u
$$= -1.643 \times 10^{-28}\,kg$$

This difference in mass is sometimes called the **mass defect**.

> **Hint**
>
> Calculations of binding energies require more significant figures than most of the calculations we have performed.

Hint

Make sure that you use the mass of the *nucleus* in binding energy calculations. If you are given the mass of the atom you must subtract the mass of its electrons to find the mass of the nucleus.

The binding energy in mass units is -1.643×10^{-28} kg. Converted into joules, $\Delta E_{rest} = \Delta mc^2 = -1.643 \times 10^{-28}$ kg $\times (3.0 \times 10^8 \, \text{m s}^{-1})^2$
$$= -1.479 \times 10^{-11} \, \text{J}$$

$$\text{Binding energy in eV} = \frac{-1.479 \times 10^{-11} \, \text{J}}{1.6 \times 10^{-19} \, \text{J eV}^{-1}} = -92.42 \, \text{MeV}$$

Binding energy per nucleon

The value we have calculated above, -1.479×10^{-11} J, is the binding energy for the carbon–12 nucleus. This is the energy required to separate all 12 nucleons from one another. The binding energy per nucleon is a measure of how strongly each individual nucleon is bound. For carbon–12 this comes out as -1.479×10^{-11} J/12 $= -1.233 \times 10^{-12}$ J (or about -7.7 MeV) per nucleon.

The nuclear valley

Unstable nuclei have less-negative binding energies per nucleon than stable nuclei. Spontaneous radioactive decay happens because nucleons tend towards lower-energy (more-negative) states. Figure 2 shows a plot of the way binding energy varies with proton and neutron number. This diagram is worth spending some time over. The plot looks like a deep, narrow valley. The binding energy is negative everywhere in the valley. The energy is zero up on the high flat plains above the valley, where nuclei have been taken apart into free protons and neutrons.

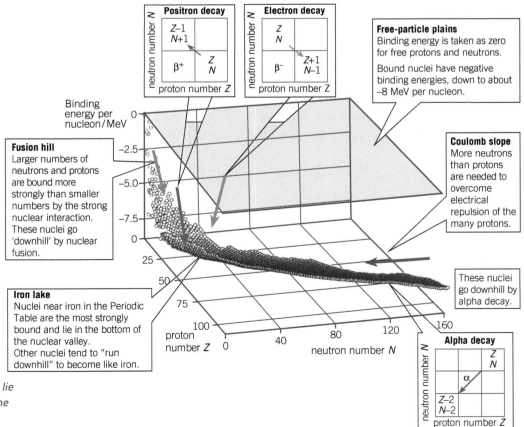

Pauli cliffs
The sides of the valley rise steeply. The Pauli exclusion principle keeps numbers of protons and neutrons approximately equal. Binding is less strong if either is in excess.

Positron decay

Z–1 N+1	
β⁺	Z N

Electron decay

Z N	
β⁻	Z+1 N–1

Free-particle plains
Binding energy is taken as zero for free protons and neutrons.

Bound nuclei have negative binding energies, down to about −8 MeV per nucleon.

Fusion hill
Larger numbers of neutrons and protons are bound more strongly than smaller numbers by the strong nuclear interaction. These nuclei go 'downhill' by nuclear fusion.

Coulomb slope
More neutrons than protons are needed to overcome electrical repulsion of the many protons.

Iron lake
Nuclei near iron in the Periodic Table are the most strongly bound and lie in the bottom of the nuclear valley. Other nuclei tend to "run downhill" to become like iron.

These nuclei go downhill by alpha decay.

Alpha decay

	Z N
α	
Z–2 N–2	

▶ **Figure 2** *Stable nuclei lie along a narrow band in the nuclear valley*

Beginning with the lightest nuclei, the valley descends steeply as extra neutrons and protons are added and the strong nuclear force takes effect. This is called the 'fusion hill' – you can picture the nuclei in stars moving down the hill as they combine into heavier elements in nuclear fusion reactions (Topic 19.4, Fission and fusion). The lowest point, that is, the strongest binding, is around the element iron, $_{26}^{56}$Fe, with a binding energy of $-8.8\,$MeV per nucleon. Beyond the 'iron lake', as nuclei get larger the valley floor rises gently and the binding per nucleon becomes weaker. This rise, the 'Coulomb slope', occurs because of the reduction in strength of binding caused by the growing electrical Coulomb repulsion of the protons. Nuclei can fall down this slope, back towards iron. Many do so by alpha decay, losing two protons and two neutrons in one go. Some undergo fission, breaking into smaller pieces (Topic 19.4, Fission and fusion).

The valley has steep side walls of unstable nuclei with a slight excess of protons or neutrons. We have called these walls the 'Pauli cliffs'. The Pauli exclusion principle means that extra neutrons or protons must have states of higher energy because they cannot all occupy the lowest energy state, so the energy cliffs rise on both sides of the valley. Nuclei fall down the sides of the valley by β^- or β^+ decay, emitting electrons or positrons.

Nuclei, like atoms, can exist in excited states of higher energy – the nucleus has risen above the floor of its part of the valley. Like atoms, nuclei emit photons when they fall from an excited state to a lower one – in this metaphor, when they drop back to the valley floor. These are very high-energy gamma photons, with energy in the range 0.1–$5\,$MeV or more. This is the origin of the gamma rays emitted by radioactive materials.

Decay processes and decay chains

The changes undergone by nuclei in alpha, beta, or gamma decays are shown in Figure 3. Such changes are accompanied by release of energy, as the resulting nucleus has a more negative binding energy per nucleon than the starting nucleus.

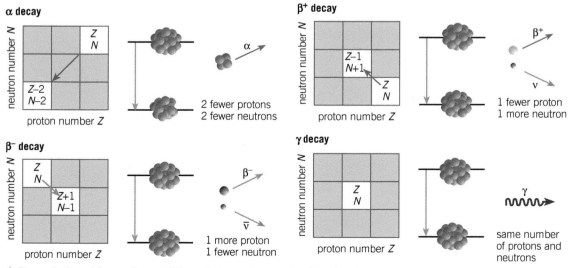

▲ **Figure 3** *N and Z are affected differently by each type of radioactive decay process*

Worked example: Radium decay

Radium–224 undergoes alpha decay, producing radon–220 as shown. Calculate the energy released in the decay (mass of radium nucleus = 223.971 91 u; mass of radon nucleus = 219.9642 u; mass of helium nucleus = 4.001 50 u; 1 u = 1.660 56 × 10^{-27} kg).

$$^{224}_{88}\text{Ra} \rightarrow \, ^{220}_{86}\text{Rn} + \, ^4_2\text{He}$$

Step 1: Calculate the mass defect Δm.

$$\Delta m = 223.971\,91\,u - 219.9642\,u - 4.001\,50\,u = 0.006\,21\,u$$
$$= 1.031... \times 10^{-29}\,\text{kg}$$

Step 2: Energy released = $\Delta m c^2$.

$$\text{energy released} = 1.031... \times 10^{-29}\,\text{kg} \times 9.00 \times 10^{16}\,\text{m}^2\text{s}^{-2}$$
$$= 9.28 \times 10^{-13}\,\text{J (3 s.f.) (about 5.8 MeV)}$$

Summary questions

1. Bismuth–212 ($^{212}_{83}$Bi) decays in two ways – by emitting an alpha particle or by emitting an electron. Write symbols for the resulting nuclei.
   The nearby elements are thallium (Tl), Z = 81; lead (Pb), Z = 82; polonium (Po), Z = 84; astatine (At), Z = 85. *(2 marks)*

2. A decay chain starting at uranium–238 ($^{238}_{92}$U) takes six steps. The decay processes, in order, are α, β⁻, β⁻, α, α, α. Find the nucleon number and proton number of the isotope at the end of the chain. *(2 marks)*

3. The mass of the deuterium nucleus ($^2_1$H) consisting of one proton and one neutron, is 3.344 × 10^{-27} kg. Calculate the binding energy per nucleon (proton mass = 1.007 28 u; neutron mass = 1.008 66 u; 1 u = 1.660 56 × 10^{-27} kg). *(3 marks)*

4. The mass of the radium–224 atom is 224.020 22 u. Using data from question 3 and below, calculate the binding energy per nucleon of radium–224 (electron mass = 0.000 549 u). *(4 marks)*

Unstable nuclei can decay to produce further unstable nuclei. In this manner, decay chains are formed in which an unstable nucleus goes through successive decays until it forms a stable nucleus. An individual nucleus can take millions of years to arrive at a stable form if some of the steps in the chain have long half-lives. Figure 4 shows a decay chain starting at thorium–232 and ending at stable lead–208.

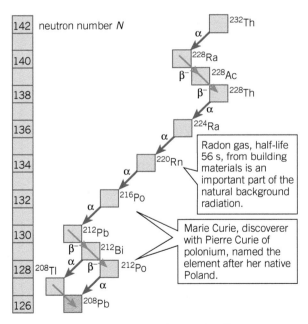

▲ **Figure 4** *Decay chain from thorium–232*

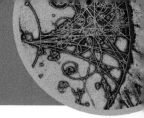

Nuclear fission

Energy is always in the news. Oil prices affect the world's economy, new methods of extracting gas such as fracking raise questions about environmental impact, whilst nuclear power plants are discussed in reference to the need for a safe and carbon-neutral energy supply. Sooner or later a replacement for oil and gas must be found. Decisions made in the next decades will decide the part nuclear power has to play after oil.

Nuclear fission is the splitting of heavy nuclei to form lighter nuclei. This process releases energy. Fission reactions in the world's first reactor, built in a University of Chicago squash court, were first recorded in December 1942.

A plot of nuclear binding energy per nucleon against nucleon number (Figure 2) shows that, on the right-hand side of the plot, large nuclei have less-negative binding energies per nucleon than the smaller nuclei. The binding energy per nucleon begins to rise again for nucleon numbers smaller than 56 (the 'iron lake' point of lowest energy and greatest stability discussed in Topic 19.3, Stability and decay).

The drop from the largest nuclei at the top of the slope to the iron lake at the bottom is about 1 MeV per nucleon. Fission exploits this energy difference by breaking a large nucleus into smaller pieces. The nucleus behaves rather like a liquid drop. A neutron captured by a uranium–235 nucleus can set the drop-like nucleus oscillating, sometimes into a dumbbell-like mode (Figure 3). If its shape gets close to that of two drops, it may split in two. The reason is that the two smaller droplets together have less energy than the original larger nucleus – they are nearer the valley bottom.

▲ **Figure 1** *Sizewell B nuclear power station in Suffolk – the fission reactor is housed in the domed building*

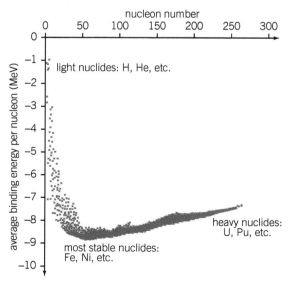

▲ **Figure 2** *Nuclear binding energy per nucleon against nucleon number*

443

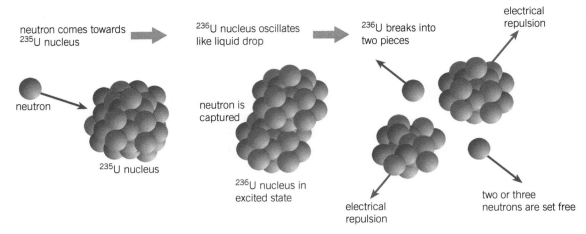

▲ **Figure 3** *Nuclear fission of uranium–235*

 Worked example: Uranium–235 fission energy

A possible fission reaction for uranium–235 is $^{235}_{92}U \rightarrow {}^{146}_{57}La + {}^{87}_{35}Br + 2{}^{1}_{0}n$. Binding energies per nucleon are: uranium–235 = –7.6 MeV, lanthanum–146 = –8.2 MeV, bromine–87 = –8.6 MeV.

Calculate the energy released when one uranium–235 nucleus undergoes fission in this manner, and the energy released per nucleon.

Step 1: Energy released = binding energy of nucleus before fission – binding energy of nuclei after fission

$= (-7.6\,\text{MeV} \times 235) - ((-8.2\,\text{MeV} \times 146) + (-8.6\,\text{MeV} \times 87))$

$= -1786\,\text{MeV} - (-1945.4\,\text{MeV})$

$= 159.4\,\text{MeV} = 160\,\text{MeV}\ (2\ \text{s.f.}).$

Step 2: Calculate the energy release per nucleon.

$159.4\,\text{MeV}/235 = 0.68\,\text{MeV}\ (2\ \text{s.f.}).$

Figure 4 shows another possible fission process for uranium–235 and the energy released.

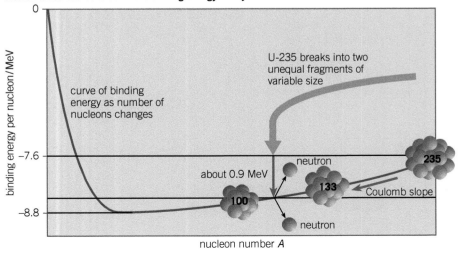

▲ **Figure 4** *Fission takes nucleons down the binding-energy valley*

Chain reactions and power from fission

A **chain reaction** is one in which the products of one reaction go on to start one or more further reactions. Chain reactions in nuclear fission depend on the neutrons released from one fission event being absorbed by other nuclei to trigger further fission. The fissile uranium–235 (U-235) isotope makes up less than 1% of naturally occurring uranium, which is mostly uranium–238 (U-238). So most neutrons from fission events are absorbed by U-238 nuclei without causing fission, or escape from the reactor without interacting with nuclei.

The number of fission events can be increased by using a **moderator**, a material (often water) used to slow down the neutrons, as slower neutrons are more likely to be captured by nuclei. In addition, since the volume-to-surface area ratio of the fuel is larger for a large mass than a smaller mass in the same shape, fewer neutrons will escape from the larger mass of fuel, so the number of new fission reactions will increase. A reaction 'goes critical' when it becomes self-sustaining, that is, when each fission event triggers, on average, one other event.

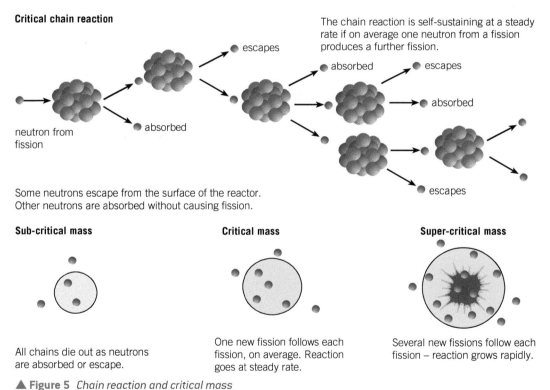

Critical chain reaction

The chain reaction is self-sustaining at a steady rate if on average one neutron from a fission produces a further fission.

escapes
absorbed
escapes
absorbed
neutron from fission
absorbed
escapes

Some neutrons escape from the surface of the reactor. Other neutrons are absorbed without causing fission.

Sub-critical mass

All chains die out as neutrons are absorbed or escape.

Critical mass

One new fission follows each fission, on average. Reaction goes at steady rate.

Super-critical mass

Several new fissions follow each fission – reaction grows rapidly.

▲ **Figure 5** *Chain reaction and critical mass*

If too many neutrons produce further fission events the chain reaction can grow exponentially. To avoid this, reactors use control rods, which absorb neutrons. Inserting control rods between the rods containing the fissile material (fuel assemblies) will slow down the reaction. The further the control rods are lowered into the fuel, the more neutrons are absorbed. Figure 6 shows a schematic diagram of a reactor that uses pressurised water as a moderator and as a coolant. Energy from fissions is transferred to the pressurised water in the cooling circuit. This heats the water to produce steam to drive the generator.

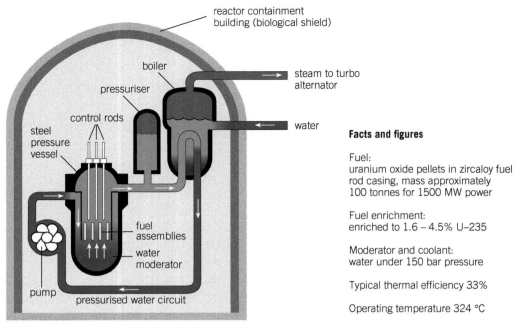

▲ **Figure 6** *The design of a pressurised water reactor*

Facts and figures

Fuel:
uranium oxide pellets in zircaloy fuel
rod casing, mass approximately
100 tonnes for 1500 MW power

Fuel enrichment:
enriched to 1.6 – 4.5% U–235

Moderator and coolant:
water under 150 bar pressure

Typical thermal efficiency 33%

Operating temperature 324 °C

How much fuel does a reactor use?

The answer to this obviously depends on the output power of the reactor. Let's consider a reactor that produces an electrical power output of 800 MW.

- Assuming an overall efficiency of 33%, the power required by the reactor is 2400 MW.
- Each fission provides about 3.2×10^{-11} J (200 MeV).
- Number of fissions required per second = $2400 \times 10^6/3.2 \times 10^{-11}$ = 7.5×10^{19} fissions per second
- Mass required per second = mass of nucleus (mass of a uranium–235 nucleus is about 4×10^{-25} kg) × fissions per second = 4×10^{-25} kg × 7.5×10^{19} s^{-1} = 3.0×10^{-5} kg s^{-1}
- mass of uranium–235 used each year (3.2×10^7 seconds) = 3×10^{-5} kg s^{-1} × 3.2×10^7 s = 960 kg

Enriched uranium fuel is roughly 3% uranium–235, so about 32 000 kg (32 tonnes) of fuel is used in a year.

Nuclear fusion

Fusion is the reaction that powers the stars. At very high temperatures small nuclei collide and fuse. As they do so energy is released – they fall down the initial steep slope of the valley of nuclear stability. But it isn't easy to fuse protons, because very high energies are needed to get them close enough to each other against the electric potential barrier. Also, at the very moment of fusion, one of these protons must decay to a neutron so that together they can form deuterium. This is why nuclear fusion in the Sun (Figure 7) is a very slow process.

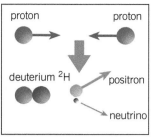

Two protons fuse, converting one
to a neutron, to form deuterium ^{2}H.

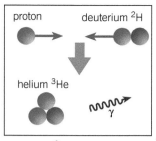

Deuterium ^{2}H captures
another proton to form ^{3}He.

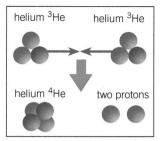

Two ^{3}He nuclei fuse, giving ^{4}He
and freeing two protons.

▲ **Figure 7** *Nucelar fusion in the Sun, a three-stage process*

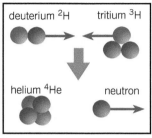

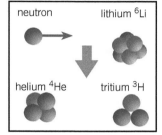

Deuterium and tritium are heated to a very high temperature. Neutrons from their
fusion then fuse with lithium in a 'blanket' around the hot gases. Tritium is renewed.

▲ **Figure 8** *Fusion in reactors on Earth, a two-stage process*

Designs for a fusion reactor on Earth use a different, two-stage
fusion reaction (Figure 8). Overcoming the electrical repulsion of
the two fusing nuclei requires great energies and therefore very
high temperatures. A very hot plasma (a state hot enough for all the
electrons to be stripped from the atoms) of tritium and deuterium is
confined by a very strong magnetic field. An electrical discharge heats
the plasma to the necessary millions of degrees. The big problem is
keeping the plasma hot enough and compressed enough for long
enough for fusion to happen. One way is to surround the plasma with
a blanket of lithium. The lithium captures neutrons from the fusion
of deuterium and tritium. This produces helium and regenerates the
tritium. The lithium blanket also carries the cooling water, which takes
energy away to raise steam. Another way uses laser pulses to compress
pellets of deuterium and tritium.

Although fusion has been achieved, sustained power production is still
some time away. The Joint European Torus (JET) experimental reactor
in Oxfordshire succeeded in producing 10 MW of fusion power for a
period of 0.5 s. The International Thermonuclear Experimental Reactor
(ITER) in France is designed to produce ten times more power from
fusion than it consumes. At the time of writing the reactor is under
construction. The plasma (shown in pink in Figure 9) is contained by
strong magnetic fields. The scale of the project is mind-boggling. For
example, the core shown is three times the weight of the Eiffel Tower,
and 100 000 km of niobium–tin superconducting strands are used in
the magnets.

▲ **Figure 9** *Schematic diagram of the
core of the ITER*

Extension: Inertial confinement fusion

Inertial confinement uses deuterium–tritium pellets roughly 1 mm across. Many high-intensity laser beams strike the pellet simultaneously. The outer layers of the pellet evaporate and form an expanding plasma.

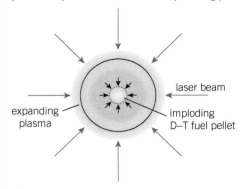

▲ **Figure 10** *Laser beams create a plasma that compresses the interior of the deuterium–tritium (D–T) pellet*

The expansion of the plasma exerts a compressive force on the pellet, increasing the density and temperature of the core. The density reaches about 10^6 kg m^{-3}, not high enough to start fusion itself, but the shock waves formed travel through the fuel and further compress the core. The resulting temperature and density are high enough to allow the nuclei to overcome the Coulomb energy barrier so that fusion can take place.

To think about:

1 Tritium is an isotope of hydrogen with two neutrons. Deuterium is a hydrogen isotope with one neutron. Why does the presence of neutrons lower the energy barrier for fusion to occur?

2 Why does an expanding plasma compress the fuel pellet?

3 Why does this form of fusion require both high temperatures and high densities?

Summary questions

1 The mass of the $^{235}_{92}$U nucleus is 3.902×10^{-25} kg. Calculate the binding energy per nucleon ($m_p = 1.672\,649 \times 10^{-27}$ kg, $m_n = 1.674\,954 \times 10^{-27}$ kg). *(2 marks)*

2 The fission energy of U-235 is about 1 MeV per nucleon. Show that about 5×10^{12} J is released if all the U-235 nuclei in 1 kg of nuclear fuel containing 5% U-235 undergo fission. *(3 marks)*

3 Calculate the energy released in the fusion reaction $^2_1H + ^2_1H \rightarrow ^3_1H + ^1_1H$. Give your answer in J and MeV (masses: $^2_1H = 2.014\,102$ u; $^3_1H = 3.016\,049$ u; $^1_1H = 1.007\,825$ u; 1 u = $1.660\,56 \times 10^{-27}$ kg). *(3 marks)*

Module 6.2 Summary

Investigating the atom

Probing the atom with alpha particles

- Evidence of a small, massive nucleus from scattering; paths of scattered particles
- Kinetic and potential energy of a scattered charged particle

Accelerating charges and electron scattering

- Particle accelerators to generate high-energy beams
- Relativistic calculations for particles
- The equations: $E_{rest} = mc^2$ and $E_{total} = \gamma E_{rest}$

Inside the nucleus

- The terms: scattering, proton, neutron, nucleon, quark, gluon, hadron, nucleon number, proton number, isotope
- Quark structure of protons and neutrons

Creation and annihilation

- The terms: electron, positron, neutrino, lepton, hadron, antiparticle, lepton number
- Conservation of mass/energy, charge and lepton number in balanced nuclear equations

Electrons in atoms

- The term: energy level
- Discrete energy levels in atoms
- The model of the atom; quantum behaviour of electrons in a confined space
- Electron standing waves in simple models of an atom

Using the atom

Ionising radiation

- Nature and effects of ionising radiations; penetrating power; effects on living tissue
- Practical task – studying the absorption of α, β, and γ radiation by appropriate materials

Effects of radiation on tissue

- The terms: absorbed dose, effective dose, dose equivalent, risk
- The equations: absorbed dose (gray) = energy deposited per unit mass; effective dose (sievert) = absorbed dose (gray) × quality factor
- Calculating the activity of a sample of radioactive material

Stability and decay

- Stability and decay of nuclei in terms of binding energy; transformation of a nucleus by emission of radiation; qualitative variation of binding energy with proton and neutron number
- The terms: binding energy, atomic mass unit
- Energy changes from nuclear transformations: $E_{rest} = mc^2$

Fission and fusion

- The terms: nuclear fission, chain reaction, nuclear fusion, nuclear power generation
- Plots of binding energy per nucleon of nuclei against nucleon number

Fission – past and future

▲ **Figure 1** *Lise Meitner and Otto Hahn, who worked together for nearly thirty years*

▲ **Figure 2** *Enrico Fermi (left) working on the first man-made nuclear reactor in 1942*

In 1939, in exile in Copenhagen from Nazi Germany, the Austrian physicist Lise Meitner and her nephew Otto Frisch had an idea that changed the world. They realised that results obtained by the German chemist Otto Hahn meant that the uranium–235 nucleus could split in two. As the energy released in this nuclear fission was much larger than that obtained from simple radioactive decay, this was a discovery with far-reaching consequences.

The Italian navigator

Events moved fast. The US government started the Manhattan Project to develop a nuclear bomb and physicist Enrico Fermi – another exile, this time from fascist Italy – set about building the world's first man-made nuclear reactor in a squash court at the University of Chicago.

Fissile uranium–235 (U–235) makes up less than 1% of naturally occurring uranium, which is mostly uranium–238 (U–238), an isotope that does not undergo fission. Most neutrons from the fission of U–235 are absorbed by U–238 nuclei, or escape from the surface of the reactor.

Fermi's team constructed a 'pile' of uranium fuel bricks separated by graphite blocks, with the graphite acting as a moderator to slow down the neutrons released in the fission of U–235. Slowing the neutrons down greatly increases the chance of their capture by a U–235 nucleus, leading to further fission events.

As an object increases in size, its volume increases at a greater rate to its surface area. So as the 'pile' of uranium bricks grew, a smaller and smaller proportion of the neutrons produced escaped through the surface, and a larger proportion could cause new fissions. The fifty-first layer was enough to make the reaction self-sustaining. Figure 3 shows the read-out from the detecting instruments at the moment that the pile went critical at 3:36 pm on 2nd December 1942. This success was announced through the coded sentence 'The Italian navigator has landed in the New World'.

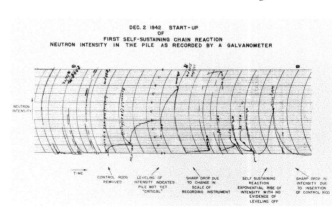

▲ **Figure 3** *Record of the first nuclear chain reaction*

Fermi's reactor had control rods made of a material that strongly absorbs neutrons, so that he could stop the chain reaction if it grew too big. The trace in Figure 3 shows the rate of neutron release by the pile. It rises when the control rods are removed, but levels off – the reaction is sub-critical at this point. There is a drop at the centre of the image because the scale of the recording instrument was changed. Towards the right end of the line you can see that the neutron intensity is rising without levelling off – self-sustaining nuclear fission had been achieved. A control rod was quickly inserted to ensure that the reaction did not grow out of control.

Nature gets there first

Building the reactor was a huge technological achievement, so it may have seemed a far-fetched idea when, in 1972, the French physicist Francis Perrin suggested that there may have been a "natural fission reactor" in the Oklo deposit of uranium ore in Gabon. After all, Fermi and his team had needed specially enriched uranium, control rods, a moderator and much more besides. What was the evidence?

Perrin noticed that a sample from the Oklo ore deposit seemed a bit low in the isotope uranium–235. This was a problem for the French Atomic Energy Commission, whose uranium handling facilities must keep careful track of all the uranium isotopes handled to be sure that none are diverted for use in weapons. More analysis on ore from the same part of the mine showed it was all short on U–235, with around 200 kg "missing" – enough for several nuclear bombs. Where had the missing U–235 gone?

A few billion years ago the concentration of U–235 would have been higher, because the half-life of U–235 (7×10^8 years) is smaller than that of U–238 (4.5×10^9 years). When the Oklo deposit was formed 1.8 billion years ago, the concentration of U–235 would have been not far from the 3% needed to sustain a chain reaction. Perhaps the Oklo deposit had been the site of a natural nuclear reactor using up some of the U–235 present. Evidence for this idea came from the presence of fission products in the region, including substantial excesses of the isotopes neodymium–143 and ruthenium–99.

Today's picture is that about 2 billion years ago there were 16 natural nuclear reactors in the Oklo region, releasing 15 000 MW-years of energy over hundreds of thousands of years. The average power was, however, quite small – up to 100 kW, or enough to raise the temperature of the ore by a few hundred degrees Celsius. Groundwater is thought to have acted as a neutron moderator, leading to a stable chain reaction.

Thorium – a fuel for fission's future?

Identified resources of uranium amount to about 5.5 million tonnes, with roughly another 10 million tonnes in currently unidentified locations. Nuclear power currently uses about 70 000 tonnes per year. This suggests that uranium can fuel the current number of reactors for about 220 years. However, it is likely that the use of nuclear power will increase in the future and alternative fuels are being considered. One such fuel is thorium ($^{232}_{90}$Th), a metal found in small amounts in many rocks. Thorium is not a fissile material but it is fertile – it can produce fissile uranium–233 by absorbing a neutron and then undergoing two beta decays:

$$^1_0n + {}^{232}_{90}Th \rightarrow {}^{233}_{90}Th \xrightarrow{\beta^-} {}^{233}_{91}Pa \xrightarrow{\beta^-} {}^{233}_{92}U$$

Uranium–233 undergoes fission, releasing two or three neutrons. The neutrons captured by uranium will lead to further fission events, producing a self-sustaining chain reaction in the correct conditions. Some neutrons will be captured by the thorium, leading to the production of more uranium fuel.

Summary questions

1 a Explain why a moderating material is required in a uranium fission reactor.
   b Explain the use of control rods in a fission reactor.
   c Explain why a smaller proportion of neutrons escaped from Fermi's nuclear pile when its size was increased.

2 Use the concept of binding energy to explain how energy is released in nuclear fission.

3 Sustained fission requires a $\frac{\text{uranium–235}}{\text{uranium–238}}$ ratio of about 0.03 (3%).
   By calculating the reduction through decay of the two uranium isotopes over the period from the formation of the Oklo deposits to the present day, show that a deposit which could sustain fission at the time of the deposit would now show a $\frac{\text{uranium–235}}{\text{uranium–238}}$ ratio of about 0.007.

4 a Show that, at present rate of energy production, uranium reserves will last for about 220 years.
   b Write out the complete decay equation for $^{233}_{90}$Th beta decay and explain how the equation shows that proton number, nucleon number and lepton number have been conserved.
   c About 200 MeV are released from each uranium–233 fission. Assuming that all the uranium–233 is produced from thorium–232, calculate the mass of thorium required per second to produce a reactor power of 1400 MW. (mass of thorium–232 atom $\approx 4 \times 10^{-25}$ kg)

Practice questions

1   Which of the following statements about nuclear fission is correct?

    A   The total mass of the products of a fission reaction is greater than the mass of the nucleus undergoing fission.

    B   The average binding energy per nucleon of the products of the fission reaction is more negative than the binding energy per nucleon of the nucleus undergoing fission.

    C   Energy is released in fission when low-mass nuclei join to make more massive nuclei.

    D   Uranium is the only element to undergo fission.   *(1 mark)*

2   The Sun is powered by the fusion of hydrogen into helium. One stage of the process is:

$$^{1}_{1}H + ^{1}_{1}H \rightarrow ^{2}_{1}H + \underset{......}{\overset{......}{?}} + ^{0}_{0}\nu$$

    a   Copy and complete the equation and identify the particle '?'.   *(2 marks)*

    b   State how the release of the neutrino conserves lepton number in the fusion reaction.   *(1 mark)*

    c   The Sun releases energy at a rate of 4×10^{26} W. Calculate the mass lost by the Sun per second.   *(2 marks)*

3   This question is about nuclear fission. A uranium–235 nucleus can capture a neutron to form uranium–236. This nucleus undergoes fission. One possible reaction is:

$$^{236}_{92}U \rightarrow ^{90}_{36}Kr + ^{144}_{56}Ba + 2^{1}_{0}n$$

    a   Calculate the energy in joules released in the reaction.

      Data: U–236 binding energy per nucleon = −7.6 MeV

      Kr–90 binding energy per nucleon = −8.7 MeV

      Ba–144 binding energy per nucleon = −8.3 MeV   *(3 marks)*

    If conditions are correct a chain reaction can develop.

    b   (i)   State what is meant by the term 'chain reaction'.   *(1 mark)*

      (ii)   Explain why a certain mass of uranium–235 is required to sustain a chain reaction, and the need for a 'moderator' in a fission reactor.   *(3 marks)*

4   Protons accelerated to energies of 180 MeV can be used to irradiate tissues in medical procedures. Calculate the number of protons required to deliver a dose equivalent of 100 mSv to a cluster of cells of mass 2.5×10^{-4} kg.

    Quality factor of protons = 10   *(3 marks)*

5   The maximum dose equivalent allowed for individuals, such as dentists, who work with X-rays is 2.0×10^{-2} Sv per year.

    a   Calculate the risk of a dentist developing cancer due to exposure to X-rays at the maximum allowed dose equivalent over a 40-year career. The risk of developing cancer due to radiation exposure is about 5% per sievert.   *(1 mark)*

    b   When using X-ray machines, dental workers distance themselves from the patient before taking the X-ray. Explain why this reduces the risk for the dentist.   *(3 marks)*

6   The mass of the helium nucleus $^{4}_{2}He$ is 4.0015 u.

    a   Show that the mass of the nucleus is about 0.75% less than the sum of the mass of its original components.   *(1 mark)*

    b   Calculate the binding energy per nucleon of the helium helium.

      proton mass = 1.007 28 u

      neutron mass = 1.008 66 u

      1 u = $1.660 56 \times 10^{-27}$ kg   *(3 marks)*

7   This question is about the radiation dose received from a variety of sources.

    The typical dose equivalent from natural sources is about 2000 μSv per year. The risk of developing cancer is about 5% per sievert.

    a   Taking the population of Britain as 60 million, estimate by calculation the number of cancer cases due to natural sources in a year.   *(2 marks)*

b Perform a similar calculation for typical dose equivalent of $500\,\mu Sv$ per year from medical and dental procedures. *(2 marks)*

c Nuclear power gives an average effective dose of $3\,\mu Sv$ per year. Should we be concerned about cancer if more of Britain's energy is provided by nuclear power in the future? Include calculations and comparisons with your answers to **(a)** and **(b)** in your response. *(4 marks)*

8 This question is about beta emissions within the body. Much of the beta radiation comes from the decay of potassium–40.

a The half-life of potassium–40 is about 1.3×10^9 years. A $65\,kg$ body contains about 3.6×10^{20} atoms of potassium–40. Show that there will be approximately 6000 decays in the body per second. *(3 marks)*

b The energy deposited in the body from each decay is about $4 \times 10^{-14}\,J$. Calculate the annual dose equivalent in sieverts over one year for a $65\,kg$ person.

Quality factor of beta radiation = 1 *(3 marks)*

c The risk of contracting cancer from radiation is given as 5% per sievert. Use this and your answer to **(b)** to calculate an estimate of the number of cancers contracted in a population of 60 million individuals over the course of a year. *(3 marks)*

d A student suggests that the risk of contracting cancer is not affected by the mass of the individual. Comment on this suggestion, explaining why you agree or disagree. *(3 marks)*

9 This question is about a nuclear fusion reaction that may be able to provide a future energy source. The symbol equation for the first stage of the reaction is given below:

$$^2_1H + {}^3_1H \rightarrow {}^4_2He + {}^1_0n$$

The word equation for the reaction is:

Deuterium nucleus + tritium nucleus → helium nucleus + neutron.

a **(i)** Here is some data for the reaction:

Mass of deuterium nucleus = 2.0135 u
Mass of tritium nucleus = 3.0155 u
Mass of helium nucleus = 4.0015 u
Mass of neutron = 1.0087 u

Show that the total mass of the particles decreases by about $3 \times 10^{-29}\,kg$ in each fusion reaction. $1\,u = 1.66 \times 10^{-27}\,kg$ *(2 marks)*

(ii) Calculate the energy released in one reaction. $c = 3.0 \times 10^8\,m\,s^{-1}$ *(2 marks)*

b The separation of the deuterium and tritium nuclei must be about $10^{-14}\,m$ for fusion to occur. Show that the work done in bringing the two nuclei to within $10^{-14}\,m$ is about $2 \times 10^{-14}\,J$.

electric force constant $\dfrac{1}{4\pi\varepsilon_0} = 9.0 \times 10^9\,N\,m^2\,C^{-2}$

$e = 1.6 \times 10^{-19}\,C$ *(2 marks)*

c In a working fusion reactor, a tritium-deuterium plasma is kept at high temperatures to give the nuclei the energy required for fusion to occur.

(i) Use $E \approx kT$ to estimate the temperature required of the tritium and deuterium plasma for fusion to occur.

Minimum energy of nuclei required for fusion = $1 \times 10^{-14}\,J$
Boltzmann constant $k = 1.4 \times 10^{-23}\,J\,K^{-1}$ *(1 mark)*

(ii) Explain why this fusion reaction may occur when the plasma is at a lower temperature. *(2 marks)*

OCR Physics B Paper G495 June 2013

Section A — Multiple Choice

1 The upper frequency limit of human hearing is about 20 kHz.

Which value gives the minimum rate needed to accurately sample sound of this frequency?

(1 mark)

  A > 10 kHz   **B** > 20 kHz

  C > 40 kHz   **D** > 80 kHz

2 'Conductive putty' is a malleable material which conducts electricity. A piece of this putty is rolled into the cylinder shown in Figure 1. The conductance between the ends of the cylinder is measured as 0.18 S.

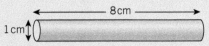

▲ Figure 1

This piece of putty is then rolled out further to make a uniform cylinder 12 cm long. What will be the conductance between the ends of this new cylinder? *(1 mark)*

  A 0.08 S   **B** 0.12 S

  C 0.27 S   **D** 0.41 S

3 The circuit in Figure 2 contains three identical lamps, each marked *6.0 V, 0.2 A*.

'Normal brightness' is the brightness of one bulb with a current of 0.2 A.

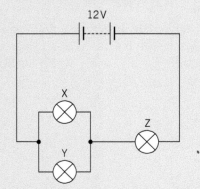

▲ Figure 2

Which of the following statements about the bulbs in Figure 2 are correct?

  1   Bulbs X and Y have the same brightness

  2   The p.d. across bulb Y is more than 6 V

3 Bulb Z is dimmer than normal

  A 1, 2, and 3 are correct

  B Only 1 and 2 are correct

  C Only 2 and 3 are correct

  D Only 1 is correct   *(1 mark)*

4 Light passing through a narrow gap diffracts. Which of the following changes would decrease the amount by which the light diffracts?

  A Decreasing the intensity of the light

  B Decreasing the width of the gap

  C Decreasing the wavelength of the light

  D Decreasing the amplitude of the light

(1 mark)

5 A converging lens of power +7.0 D focuses an image at a distance of 15 cm from the lens. Which of the values below is the object distance? *(1 mark)*

  A −0.14 m   **B** −3.0 m

  C −0.33 m   **D** −0.08 m

6 Which of the following units are the correct units for stress? *(1 mark)*

  A $N m^{-3}$ **B** $J m^{-3}$

  C $J m^{-1}$ **D** $N m^{-1}$

7 The work function of potassium is 3.5×10^{-19} J. What is the maximum speed of the emitted photoelectrons when the potassium is illuminated by light of wavelength 3.9×10^{-7} m? *(1 mark)*

Electron mass = 9.1×10^{-31} kg

Planck constant = 6.63×10^{-34} J s

  A $3.5 \times 10^{11} m s^{-1}$ **B** $5.9 \times 10^{5} m s^{-1}$

  C $4.2 \times 10^{5} m s^{-1}$ **D** $2.9 \times 10^{5} m s^{-1}$

8 Which of the following values is the best estimate of the de Broglie wavelength of an electron travelling at ten percent of the speed of light? *(1 mark)*

Electron mass = 9.1×10^{-31} kg

Planck constant = 6.63×10^{-34} J s

  A 2.4×10^{-12} m   **B** 4.8×10^{-12} m

  C 2.4×10^{-11} m   **D** 4.8×10^{-11} m

9 A metal spring extends elastically by 0.17 m. The energy stored by the spring at this extension is 0.87 J. What is the force constant k of the spring? *(1 mark)*

A $60\,\mathrm{N\,m^{-1}}$ **B** $30\,\mathrm{N\,m^{-1}}$

C $10\,\mathrm{N\,m^{-1}}$ **D** $5\,\mathrm{N\,m^{-1}}$

10 Graphs **A** to **D** describe four different straight-line trips, each in two parts. In each graph, the *x*-axis variable is time.

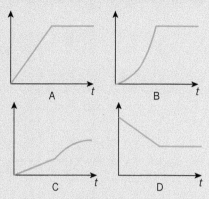

Assuming the *y*-axis variable is displacement, which graph shows deceleration over one part of the trip? *(1 mark)*

11 Figure 3 shows two soft balls of equal mass rolling towards each other on a horizontal, frictionless surface.

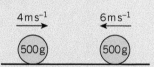

▲ Figure 3

What is the magnitude of the total momentum of the two balls after the collision? *(1 mark)*

A $1\,\mathrm{N\,s}$ **B** $5\,\mathrm{N\,s}$

C $1000\,\mathrm{N\,s}$ **D** $5000\,\mathrm{N\,s}$

12 What is the maximum possible total kinetic energy of the two balls in Figure 3 after the collision, assuming they stick together? *(1 mark)*

A $0.5\,\mathrm{J}$ **B** $1\,\mathrm{J}$

C $9\,\mathrm{J}$ **D** $25\,\mathrm{J}$

Section B

1 Figure 5 shows an oscilloscope waveform of a pure sound wave.

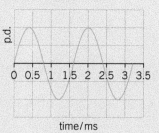

▲ Figure 4

Calculate the frequency of the note. Give the units. *(3 marks)*

2 A data projector has an illuminated display that is 48 mm wide. A lens projects an image of the display on to a screen. The width of the image is 1.7 m.

a Calculate the magnification of the image. *(1 mark)*

b The distance between the illuminated display and the lens in 8 cm. Calculate the distance from the lens to the screen. *(1 mark)*

c Calculate the power of the lens. *(2 marks)*

3 Light is partially polarised when it reflects from a sheet of glass. A student investigates this phenomenon using a polarising filter, as shown in Figure 5.

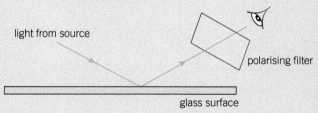

▲ Figure 5

a Describe how the student could use the filter to observe the polarisation of light. *(2 marks)*

b State the observation that would make the student conclude that the light was only partially polarised. *(1 mark)*

4 Figure 5 represents light waves incident on a lens. The lens focuses the waves at point *F*.

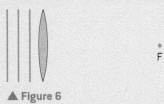

▲ Figure 6

a Copy Figure 4 and draw the three waves between the lens and the point of focus.
(2 marks)

b What does Figure 6 indicate about the distance between the light source and the lens? Explain your answer. *(2 marks)*

5 A lens has power +15 D.

The curvature of light waves incident on the lens is −3 D.

a Calculate the curvature of the waves as they pass out of the lens. *(1 mark)*

b Calculate the distance from the lens at which the light is focused. *(1 mark)*

6 Figure 7 shows the result of plane wavefronts passing through a converging lens.

The lens is replaced with one of the same shape and dimensions, made from a material with a higher refractive index.

focal point

▲ Figure 7

a Copy Figure 7, then draw another diagram to show the effect of using a lens with *higher* refractive index on the wavefronts to the right of the lens.
(2 marks)

b State with a reason whether the power of the new lens in part **(a)** is larger, smaller, or the same as that of the original lens.
(1 mark)

OCR Physics B Paper 2860 January 2009

7 A memory cell has an area of 1×10^{-14} m^2. It stores 1 bit of information.

The total area of memory cells on a microchip is 6×10^{-5} m^2.

Calculate the number of bytes of information the chip can store. *(2 marks)*

8 An analogue signal has a voltage variation of 6.7 V. It is sampled with 1024 voltage levels.

a Calculate the resolution of each sample. *(1 mark)*

b Calculate the number of bits required for 1024 levels. *(1 mark)*

c The noise in the signal gives a random voltage variation of ± 0.1 V. Calculate the number of bits worth using to sample the signal. *(3 marks)*

9 Figure 8 shows light entering glass from air. The angles of incidence and refraction are given.

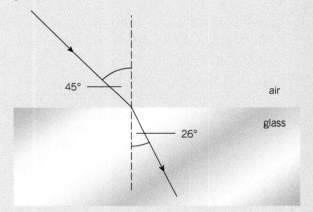

45°

air

26°

glass

▲ Figure 8

a Use Figure 8 to calculate the refractive index of the glass. *(2 marks)*

b The velocity of light in air = 3.00×10^8 m s^{-1}. Calculate the velocity in the glass. *(2 marks)*

c Light passes from the glass into air. The angle of incidence in the glass is 35°. Calculate the angle of refraction in the air. *(2 marks)*

10 A variable resistor is connected to a battery of e.m.f. ε and and internal resistance r via an ammeter. The p.d. V across the terminals of the battery is measured for different values of current I to give the graph in Figure 9.

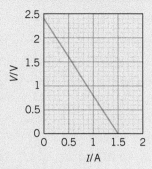

▲ Figure 9

a Explain why the graph shows that the e.m.f. of the battery is 2.4 V. *(1 mark)*

b Use the graph to calculate the internal resistance r of the battery. *(2 marks)*

11 Table 1 displays the electrical conductivity of four different materials at room temperature.

▼ Table 1

material	gold	lithium	pure silicon	doped silicon
conductivity / S m^{-1}	4.9×10^7	1.2×10^7	4.7×10^{-4}	22

a Gold and silicon are both metals. In metals, each atom provides one free electron to the cloud responsible for conduction. Suggest and explain two reasons for the difference in resistivity between gold and silicon. *(2 marks)*

b The doped silicon has one atom in every ten million replaced by a boron atom. Suggest why this tiny proportion of foreign atoms increases the conductivity so significantly. *(2 marks)*

12 A student makes the following measurements to find the tensile stress in a wire:

tension in wire = 147 ± 1 N
cross-sectional area = (0.86 ± 0.10) × 10^{-6} m^2

a Calculate the value of the tensile stress in the wire. *(1 mark)*

b Calculate the largest possible value of the stress, given the uncertainties in the data. *(2 marks)*

c The student wishes to reduce the uncertainty in the value of the stress.

State which measurement you would choose to improve to achieve this.

Explain your choice. *(1 mark)*

OCR Physics B Paper G491 June 2010

13 Metals can show plastic behaviour under stress.

a Explain what is meant by the terms plastic behaviour and stress. *(2 marks)*

b Use what you know about the microscopic structure of metals to describe how they can show plastic behaviour. *(2 marks)*

14 A light-emitting diode emits photons of energy 3.7×10^{-19} J.

a Calculate the wavelength of the radiation emitted. *(2 marks)*

b The output power of the diode is 40 mW. Calculate the number of photons emitted each second. *(2 marks)*

15 Figure 10 shows some energy levels in an atom. Calculate the frequency of the photon emitted when an electron falls from level A to level C. *(2 marks)*

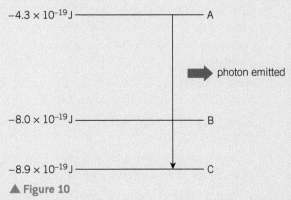

▲ Figure 10

16 A photon travels from X to Y. Three possible paths for the photon are shown in Figure 11.

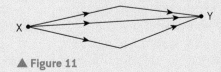

▲ Figure 11

The phasors at Y for the three paths are shown in Figure 12. Each has the same amplitude A.

▲ Figure 12

a   Draw a diagram of the phasor arrows adding tip-to-tail and show the resultant phasor. *(2 marks)*

b   Show by calculation that the amplitude of the resultant phasor is between $2A$ and $2.5A$. *(2 marks)*

17 A standing wave is formed on a stretched wire, as shown in Figure 13.

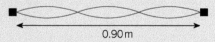

0.90 m

▲ Figure 13

a   State the wavelength of the standing wave on the wire. *(1 mark)*

b   The frequency of vibration of the wire is 360 Hz. Calculate the speed of transverse waves on the wire. *(2 marks)*

c   On a copy of the diagram, label the positions of a displacement node and a displacement antinode with the letters N and A. *(1 mark)*

18 A stone is dropped down a well. The time for it to reach the water is measured to be 1.6 s.

a   Calculate the depth of the well, assuming that $g = 9.8 \, \text{m s}^{-2}$. *(2 marks)*

b   The timing has an uncertainty of ±0.1 s. Calculate the uncertainty in your calculated depth in part **(a)**. *(2 marks)*

c   The speed of sound in air is 340 m s^{-1}. Use your answer to **(a)** to calculate the time taken for the sound of the splash to reach the person timing the fall, and comment on how the systematic error will affect the estimate of the depth obtained in **(a)**. *(2 marks)*

19 A car travels a distance s in a time t with constant acceleration a. In this time, the velocity of the car increases from an initial velocity u to a final velocity v.

The equations below model the motion.

$$s = \frac{(u + v)t}{2} \qquad \text{equation 1}$$
$$v = u + at \qquad \text{equation 2}$$

a   Rearrange each of these equations to make t the subject of the equation. *(2 marks)*

b   Equate the two expressions for t and hence show that $v^2 = u^2 + 2as$. *(1 mark)*

20 On a test drive on a straight, horizontal track, a high-performance car accelerates from 0 to 60 mph (27 m s^{-1}) in 6.2 s. The mass of the car and driver is 1400 kg.

a   Calculate the mean resultant accelerating force over the 6.2 s. *(2 marks)*

b   Draw a labelled sectional diagram showing all forces acting on the car during this acceleration. Label the forces with appropriate descriptions, but without values. *(3 marks)*

c   The car continues to accelerate to its top speed of 162 mph (72 m s^{-1}). Assuming that the accelerating force between the wheels and the road is the same as in part **(a)**, calculate the power dissipated when travelling at this speed. *(3 marks)*

21 This question is about stretching polythene.

a   A long narrow sample strip of polythene is cut from a shopping bag. It stretches elastically up to a strain of 0.082 at a stress of 14 MPa. This is the elastic limit of the material.

(i)   Calculate the Young modulus of the polythene and state the unit. *(3 marks)*

(ii) The cross-sectional area of the sample is $1.9 \times 10^{-7} \, \text{m}^2$.

Calculate the force applied to the sample to produce a stress of 14 MPa. *(2 marks)*

b Figure 14 shows the stress against strain graph for the sample to its breaking point.

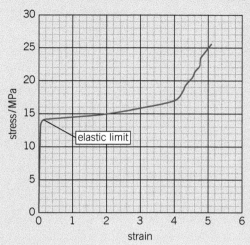

▲ Figure 14

(i) Describe the behaviour of the sample as it is stretched from the elastic limit to its breaking point. *(2 marks)*

(ii) Use Figure 14 to calculate the extension of the sample at the breaking point. The original length of the sample is 15 cm. *(3 marks)*

c Suggest and explain what is happening to the **long chain molecules** in the sample between the elastic limit and the break point as stress is increased slowly.

You may wish to use labelled diagrams.

(4 marks)

OCR Physics B Paper G491 June 2012

22 This question is about electron diffraction.

The equation for single slit diffraction is $\lambda = b \sin\theta$, where b is the width of the slit and θ is the angle to the first minimum.

a (i) Show that an electron accelerated through a potential difference of 900 V will gain kinetic energy of about 1.4×10^{-16} J. *(1 mark)*

(ii) The kinetic energy of an electron is related to its momentum in the equation

$$\text{kinetic energy} = \frac{\text{momentum}^2}{2m}$$

where m is the mass of an electron $= 9.11 \times 10^{-31}$ kg.

Calculate the momentum mv of the electron. *(2 marks)*

(iii) Use your answer from (a)(ii) to calculate the angle to the first minimum when electrons of energy 1.4×10^{-16} J pass through a gap of width 4×10^{-9} m *(3 marks)*

b State how the position of the first minimum will change when the electrons are accelerated through a greater potential difference. Explain your answer. *(3 marks)*

23 A photocell generates electric current by the photoelectric effect.

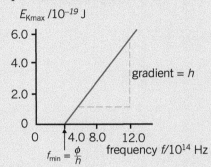

▲ Figure 15

Light of a wavelength 4.6×10^{-7} m is incident on a metal plate called the photocathode.

a Show that the energy of a photon of this light is about 4×10^{-19} J. *(2 marks)*

About 3×10^{17} photons strike the photocathode every second. A current of 1.3 mA is detected.

b (i) Calculate the number of electrons released from the photocathode each second. *(2 marks)*

(ii) Suggest and explain why the number of electrons released each second is smaller than the number of photons incident on the photocathode each second. *(2 marks)*

c When red light of wavelength 6×10^{-7} m is incident on the photocathode there is no current detected, even though the number of photons striking the photocathode each second is greater than 3×10^{17}. Explain this observation. *(2 marks)*

Paper 2 Practice questions (AS)

Section A

1 This question is about the operation of a gas-filled pixel in a plasma TV screen. A plasma is a conducting ionised gas. It is formed by a high voltage pulse across a pair of electrodes in the pixel as shown in Figure 1.

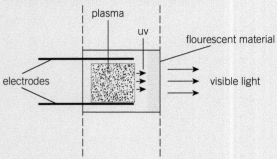

▲ **Figure 1** *Schematic diagram of a pixel in plasma display*

a Describe what is meant by an ionised gas.
   (2 marks)

b Plasma emits UV radiation at a frequency of 2.9×10^{15} Hz.

   Calculate the wavelength of this radiation.

   Speed of light = $3.0 \times 10^8 \, \text{m s}^{-1}$ *(1 mark)*

c Gas atoms can be ionised by collision with fast-moving electrons. The p.d. between the electrodes provides energy for these electrons.

   Calculate the energy gained by an electron of charge 1.6×10^{-19} C when it passes through a p.d. of 240 V. *(2 marks)*

d Once started by a high voltage pulse the plasma in a pixel can be maintained at a lower voltage. The plasma can be ended by switching off the voltage. Figure 2 shows how the current in the gas in a pixel changes as the p.d. is raised to 290 V and lowered back to 0 V.

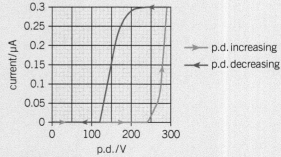

▲ **Figure 2**

(i) **1** State the voltage at which ionisation starts.

   2 State the voltage at which ionisation stops. *(2 marks)*

(ii) There are 6.2×10^6 pixels in the display. When emitting visible light pixels operate at 180 V.

   Use data from Figure 2 to calculate the total operating power of the display with all the pixels on.
   (3 marks)

   OCR Physics B Paper G491 Jan 2011

2 This question is about the materials from which cutting tools such as drill bits are made.

a **(i)** Metals have a polycrystalline structure.

   Explain the term *polycrystalline* as applied to the structure of a metal.

   You may wish to use labelled diagrams in your answer. *(2 marks)*

(ii) Drill bits can be made from steel alloy. Figure 3 shows the microstructures of pure iron metal and a steel alloy.

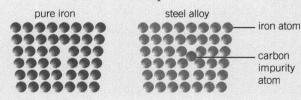

▲ **Figure 3**

   Steel alloy containing carbon is less ductile than pure iron.

   State the meaning of the term *ductile* and describe how Figure 3 can be used to help explain why steel is less ductile than iron. *(3 marks)*

b **(i)** Diamond is much harder than steel. This gives a diamond-coated steel drill bit an advantage over a steel one.

   1 State what is meant by *hardness*.

   2 Explain the advantage. *(2 marks)*

(ii) The atoms in steel have metallic bonding and in diamond the atoms have covalent bonding. Describe these types of bonding. Use your description to explain the difference in hardness between steel and diamond.
You may wish to use labelled diagrams in your answer. *(4 marks)*

OCR Physics B Paper G491 Jan 2013

3 This question is about *relative* and *resultant* velocities.

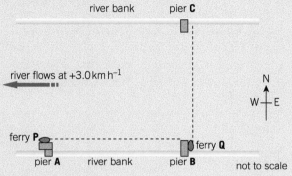

river flows at $+3.0\,\text{km}\,\text{h}^{-1}$

▲ **Figure 4**

Figure 4 shows part of a wide river on which there are three piers. The river flows from east to west at a constant velocity of $+3.0\,\text{km}\,\text{h}^{-1}$ as shown.

a Ferry **P** travels from pier **A** to pier **B**, and then back again. The ferry travels at a speed of $5.0\,\text{km}\,\text{h}^{-1}$ through still water.

(i) Calculate the velocity of the ferry relative to the river bank as it sails.

1 From **A** to **B**

2 From **B** to **A** *(2 marks)*

(ii) Piers **A** and **B** are 2.0 km apart.
Show that the total sailing time for a return journey for ferry **P**, sailing from pier **A** to **B** and back again to **A**, is 1.25 hours.
Ignore the time taken for the boat to turn around at pier **B**. *(2 marks)*

b There is another pier **C** directly across the river from pier **B**, as shown in Figure 4.
A second ferry **Q** travels between piers **B** and **C** which are 2.0 km apart. This ferry also travels at a speed of $5.0\,\text{km}\,\text{h}^{-1}$ through still water.

(i) By scale drawing, or some other method of your choosing, show that the ferry **Q** must sail in a direction 37 degrees east of north in order to travel due north across the river, from pier **B** to pier **C**. *(2 marks)*

(ii) Show that the resultant velocity of this ferry relative to the river bank is $4.0\,\text{km}\,\text{h}^{-1}$ due north. *(2 marks)*

c Ferry **Q** sets off from pier **B** on an outward bound journey to **C** at the same time as ferry **P** sets off from pier **A** towards pier **B**.
Show that the bearing of ferry **Q** from ferry **P** is about 27 degrees east of north, when **Q** just reaches pier **C**. *(2 marks)*

OCR Physics B Paper 2861 June 2004

4 Figure 7 shows the graph of force against extension for a metal wire **A**.

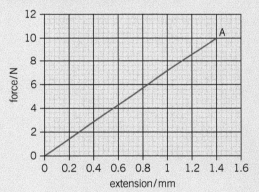

▲ **Figure 7**

a (i) Draw on a copy of Figure 5 the graph you would expect for a wire of the same material and diameter as **A**, but of **twice** the original length. Label this graph **B**. *(1 mark)*

(ii) Draw on a copy of Figure 5 the graph you would expect for a wire of the same material and length as **A**, but of **double** the original diameter. Label this graph **C**. *(1 mark)*

b (i) State **one** piece of evidence from the graph which suggests that the stretching of the wire (by a force of 10 N) is elastic. *(1 mark)*

(ii) Wire **A** has a cross-sectional area of $7.8 \times 10^{-8}\,m^2$ and an original length of 2.00 m.

Calculate the Young modulus of the material of the wire. *(3 marks)*

c Describe metallic bonding on the atomic scale. Include in your description an explanation of how metals such as wire **A** can show elastic behaviour.

In your explanation, you should make clear how the bonding between atoms can account for the large-scale elastic behaviour of the material. *(4 marks)*

OCR Physics B Paper G491 Jan 2011

Section B

1 This question is about taking a self-portrait with a mobile phone camera.

A camera on a mobile phone has a lens of focal length 4.5 mm. It is held 0.5 m from the photographer's face.

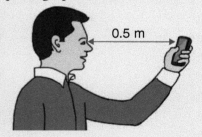

0.5 m

▲ Figure 6

a (i) Show that the power of the lens is about 220 D. *(1 mark)*

(ii) Calculate the distance behind the lens that the image of the photographer will be focused. *(2 marks)*

(iii) Calculate the magnification of the image. *(2 marks)*

The photographer's face has approximate dimensions of 270×225 mm. This fills the picture area of the light-sensitive chip.

b Use your answer to **(a)**(iii) to calculate the dimensions of the light-sensitive chip. *(2 marks)*

c The light-sensitive chip has 1200×1000 pixels. Calculate the resolution of the image of the face and comment on whether the image could resolve an individual eyelash of diameter 0.1 mm. *(3 marks)*

The camera stores three colours for each pixel, each with 256 levels of intensity. The camera memory has 0.9 GB available to store image files.

d (i) Calculate how many images can be stored in the memory. *(3 marks)*

(ii) Suggest how this number could be increased without increasing the size of the memory. Describe the advantages and disadvantages of your suggestion. *(3 marks)*

2 One type of component, called a PTC thermistor, has a resistance which varies with temperature, as shown by the solid line in Figure 7. An 'ordinary' NTC thermistor, as used in sensing circuits, has a resistance which varies as shown by the dashed line.

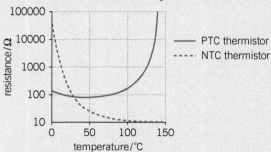

▲ Figure 7

a Compare the behaviour of these two components at different temperatures and suggest, in each case, reasons for the variation of resistance shown by the graph. This question tests your ability to construct and develop a sustained and coherent line of reasoning. *(6 marks)*

b In one sensor application, a chemical reaction vessel which needs to be kept at $50 \pm 2\,°C$ needs to be monitored. Explain why the PTC thermistor is a poorer temperature sensor than the NTC thermistor for this application. *(2 marks)*

c The PTC thermistor is connected to a 6 V battery of negligible internal resistance, as shown in Figure 8

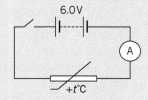

▲ Figure 8

The temperature is 15 °C when the circuit is set up.

(i) Calculate the power dissipated in the PTC thermistor when the switch is closed. *(3 marks)*

(ii) After closing the switch, it is noted that there is an initial change in the current, but that it eventually settles on a fixed value. Explain this observation. *(4 marks)*

Section C

1 This question is about an experiment to determine the focal length of a converging lens.

A student uses a converging lens, a 12 V filament lamp and a ground glass screen. Distances are measured using a pair of metre rules. The basic set up is shown in Figure 9.

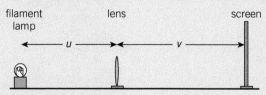

filament lamp   lens   screen

▲ Figure 9

The object distance is varied between −0.3 m and −1.4 m.

a (i) Suggest a possible cause of systematic error in measuring the object distance *u*. *(1 mark)*

(ii) Suggest a possible cause of uncertainty in the measurement of image distance *v*. *(1 mark)*

b Here are two comments students made about the uncertainty in measuring the image distance *v*.

• The actual value of the uncertainty increases as *v* increases.

• The percentage uncertainty in the measurement of *v* remains approximately constant over the range of measurements.

(i) Suggest and explain why the value of the uncertainty in the reading may increase with *v*. *(2 marks)*

(ii) Explain why the percentage uncertainty can remain the same even though the actual value changes. *(2 marks)*

c The student recorded this pair of values $u = 1.000$ m and $v = 0.260$ m

Show that this pair of values leads to a value to a value of the focal length of the lens of about 0.21 m. *(2 marks)*

The uncertainty in the *v* value was estimated at ±10 mm. The uncertainty in the *u* value was estimated at ±1 mm. It was thought that the uncertainty in *u* could be ignored in considering the uncertainty in the final result.

d Using the uncertainty in *v*, calculate the highest and lowest values for the focal length using the data pair in **(c)** to show that the uncertainty in the focal length result is more than ±5 mm. *(4 marks)*

e The student took a range of *u* and *v* readings.

She recalled the equation $\frac{1}{v} = \frac{1}{u} + \frac{1}{f}$ and decided to calculate $\frac{1}{u}$ and $\frac{1}{v}$ values and plot the graph in Figure 10.

Explain why she assumed this would give a straight-line graph. *(2 marks)*

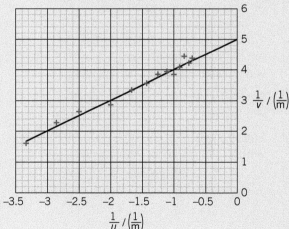

▲ Figure 10

f Use the graph to find the focal length of the lens. *(2 marks)*

g Suggest why the result obtained from the best-fit line in Figure 10 is better than the mean of individual calculations of focal length from *v* and *u* data pairs. *(1 mark)*

Section A — Multiple Choice

1 Here is a list of units:

   A Js^{-1} **B** Jkg^{-1}

   C Jm^{-1} **D** JC^{-1}

  a Which is the equivalent unit for the newton, N?

  b Which is the equivalent unit for the sievert, Sv?

  c Which is the equivalent unit for the watt, W?   (*3 marks*)

2 An alpha particle source of activity 250 Bq emits particles with an energy of 5 MeV.

  What is the dose equivalent delivered by the source into a mass of 12 g of tissue over a period of 48 hours? Assume that the activity of the source remains constant throughout the exposure.

  Quality factor of alpha radiation = 20

  A 0.058 µSv   **B** 1.2 µSv

  C 16 µSv   **D** 58 000 µSv (*1 mark*)

3 An initially uncharged capacitor is charged by a constant current of 1.6 mA over a period of 2.0 s. At the end of this period the potential difference across the capacitor is 8.0 V. What is the capacitance of the capacitor?

  A 0.2 µF   **B** 0.4 µF

  C 200 µF   **D** 400 µF   (*1 mark*)

4 The graph shows how variable y changes with x.

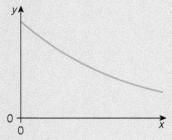

▲ Figure 1

Which pair(s) of variables will produce a graph of a similar shape?

Pair 1: activity of a radioisotope (y) against time (x)

Pair 2: p.d. across a capacitor during discharge (y) against time (x)

Pair 3: Electric field strength of a point charge (y) against distance from charge (x)

  A 1, 2 and 3 are correct

  B Only 1 and 2 are correct

  C Only 2 and 3 are correct

  D Only 1 is correct   (*1 mark*)

5 Figure 2 shows four versions of the same circuit with different component values.

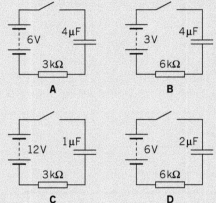

▲ Figure 2

In all four circuits the capacitor is uncharged before the switch is closed.

  a Which circuit (**A**, **B**, **C**, or **D**) has the greatest final charge on the capacitor when the switch is closed?   (*1 mark*)

  b Which circuit (**A**, **B**, **C**, or **D**) takes the least time to charge up the capacitor when the switch is closed?   (*1 mark*)

         OCR Physics B Paper G494 January 2012

6 The half-life of a source is 7.0×10^2 s. The activity of the source at time $t = 0$ s is 1.7 kBq. How many radionuclides are present in the source at this time?

  A 1.1×10^3 **B** 1.7×10^3

  C 1.1×10^6 **D** 1.7×10^6 (*1 mark*)

7 Here are three statements about a mass oscillating in undamped simple harmonic motion between two springs.

  1 The total energy of the oscillator remains constant throughout the oscillation.

  2 The maximum kinetic energy of the oscillator is equal to the total energy of the oscillator.

3 The total energy of the oscillator will double if the amplitude is doubled.

Which of these statements is/are correct?

A 1, 2, and 3 are correct

B Only 1 and 2 are correct

C Only 2 and 3 are correct

D Only 1 is correct (*1 mark*)

8 A simple pendulum oscillates in simple harmonic motion with a time period of 1.2 s. What is the angular frequency, ω, of the oscillation?

A $0.19\,\mathrm{rad\,s^{-1}}$ **B** $1.7\,\mathrm{rad\,s^{-1}}$

C $2.6\,\mathrm{rad\,s^{-1}}$ **D** $5.2\,\mathrm{rad\,s^{-1}}$ (*1 mark*)

9 An object of mass m hangs from a spring and oscillates with time period T. The object is replaced with one of half the mass. What is the new time period of the oscillation?

A $\dfrac{T}{2}$ **B** $\dfrac{T}{\sqrt{2}}$

C $\sqrt{2}\,T$ **D** $2T$ (*1 mark*)

10 The acceleration due to gravity near the surface of the Moon is about $1.7\,\mathrm{m\,s^{-2}}$. The radius of the Moon is about $1.7 \times 10^6\,\mathrm{m}$. What is the mass of the Moon?

A $2.5 \times 10^{22}\,\mathrm{kg}$ **B** $7.3 \times 10^{22}\,\mathrm{kg}$

C $1.0 \times 10^{24}\,\mathrm{kg}$ **D** $1.2 \times 10^{29}\,\mathrm{kg}$

(*1 mark*)

11 The planet Jupiter orbits the Sun in $3.8 \times 10^8\,\mathrm{s}$.

The mean distance between Jupiter and the Sun is $7.8 \times 10^{11}\,\mathrm{m}$.

What is the centripetal acceleration of Jupiter?

A $3.4 \times 10^{-5}\,\mathrm{m\,s^{-2}}$ **B** $6.8 \times 10^{-5}\,\mathrm{m\,s^{-2}}$

C $1.0 \times 10^{-4}\,\mathrm{m\,s^{-2}}$ **D** $2.1 \times 10^{-4}\,\mathrm{m\,s^{-2}}$

(*1 mark*)

12 An electron is accelerated in a laboratory to a relativistic factor of 3.0. What is the velocity of the electron relative to the laboratory?

A $2.0 \times 10^8\,\mathrm{m\,s^{-1}}$ **B** $2.4 \times 10^8\,\mathrm{m\,s^{-1}}$

C $2.7 \times 10^8\,\mathrm{m\,s^{-1}}$ **D** $2.8 \times 10^8\,\mathrm{m\,s^{-1}}$

(*1 mark*)

13 Here is a list of particles emitted in radioactive decays.

A alpha particle   **B** beta particle

C positron   **D** neutrino

a Which particle is a positively charged lepton? (*1 mark*)

b Which particle does not cause ionisation as it passes through matter? (*1 mark*)

c Which particle has a neutral anti-particle? (*1 mark*)

14 The graph in Figure 3 shows how binding energy per nucleon varies with nucleon number.

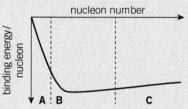

▲ **Figure 3**

Which of the following statements about the graph is/are correct?

1 Energy is released when nuclei in region C split, producing lighter nuclei.

2 The most stable nuclei are at B.

3 The process in which nuclei in region A fuse to produce more massive nuclei is the source of the Sun's energy.

A 1, 2, and 3 are correct

B Only 1 and 2 are correct

C Only 2 and 3 are correct

D Only 1 is correct (*1 mark*)

15 A proton is accelerated to a relativistic factor of 3.0. What is the ratio $\dfrac{\text{kinetic energy of proton}}{\text{rest energy of proton}}$?

A 1.5   **B** 2.0

C 3.0   **D** 4.0 (*1 mark*)

16 Here is an equation representing the beta decay of a neutron.

$$^1_0\mathrm{n} \rightarrow \,^1_1\mathrm{p} + \,^{\ 0}_{-1}\mathrm{e} + \,^0_0\bar{\mathrm{v}}$$

Which of these statements about the decay is/are correct?

1 Mass is conserved.

2 Lepton number is conserved.

3 Momentum is conserved.

A 1, 2, and 3 are correct.

B Only 1 and 2 are correct.

C Only 2 and 3 are correct.

D Only 1 is correct. (*1 mark*)

17 Figure 4 shows a battery of e.m.f. ε and internal resistance r connected to a $6.4\,\Omega$ resistor. The p.d. across the battery terminals is measured with a voltmeter of very high resistance.

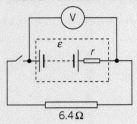

▲ Figure 4

When the switch is open, the voltmeter reads $6.0\,V$. When the switch is closed, the voltmeter reads $4.8\,V$.

Which of the following statements is/are correct?

1 The e.m.f. of the battery is $4.8\,V$.

2 The internal resistance of the battery is $1.6\,\Omega$.

3 The current in the $6.4\,\Omega$ resistor is $0.75\,A$ when the switch is closed.

   A 1, 2, and 3 are correct.

   B Only 1 and 2 are correct.

   C Only 2 and 3 are correct.

   D Only 1 is correct.   *(1 mark)*

18 A $6.4\,\Omega$ resistor is made from wire of conductivity $2.0 \times 10^6\,S\,m^{-1}$. The diameter of the wire is $1.6\,mm$.

What is the length of wire wound into the $6.4\,\Omega$ resistor?

  A $1.6\,m$ **B** $26\,m$

  C $51\,m$ **D** $100\,m$ *(1 mark)*

19 Figure 5 is a displacement–time graph for an object moving with constant acceleration.

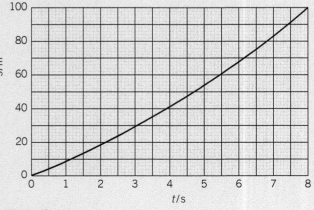

▲ Figure 5

Which of the following statements is/are correct?

1 The mean velocity of the object was more than $10\,m\,s^{-1}$.

2 The acceleration of the object was less than $2\,m\,s^{-2}$.

3 The object started with a velocity of $0\,m\,s^{-2}$.

   A 1, 2, and 3 are correct.

   B Only 1 and 2 are correct.

   C Only 2 and 3 are correct.

   D Only 1 is correct.   *(1 mark)*

20 A rocket of mass $2.5\,kg$ is being launched vertically. It experiences a resultant upwards force of $8.0\,N$. The exhaust gases are ejected with a speed, relative to the rocket, of $250\,m\,s^{-1}$.

What is the rate at which the exhaust gases are ejected? ($g = 9.8\,N\,kg^{-1}$)

  A $32\,g\,s^{-1}$ **B** $66\,g\,s^{-1}$

  C $98\,g\,s^{-1}$ **D** $130\,g\,s^{-1}$ *(1 mark)*

21 A flask of volume $1.5 \times 10^{-3}\,m^3$ contains gas at $294\,K$ at a pressure of $1.0 \times 10^5\,Pa$ and a second flask of $2.0 \times 10^{-3}\,m^3$ contains gas at $330\,K$ at a pressure of $1.5 \times 10^5\,Pa$. The two flasks are connected together so that the gases can mix, and are allowed to reach equilibrium at $294\,K$. Assuming that there is no loss of gas when the flasks are joined, what is the pressure in the flasks? (gas constant $R = 8.3\,J\,mol^{-1}\,K^{-1}$)

  A $1.2 \times 10^5\,Pa$

  B $2.1 \times 10^5\,Pa$

  C $2.4 \times 10^5\,Pa$

  D $2.8 \times 10^5\,Pa$ *(1 mark)*

22 The activation energy ε for a molecule of ethanol to evaporate is $6.5 \times 10^{-20}\,J$.

What is the Boltzmann factor for evaporation of ethanol at a temperature of $50\,°C$? ($k = 1.4 \times 10^{-23}\,J\,K^{-1}$)

  A 4.7×10^{-41}

  B 5.7×10^{-7}

  C 14

  D 93   *(1 mark)*

23 A coil of 250 turns with diameter 15 cm is placed perpendicular to a 0.24 T magnetic field. The coil is flipped through 180° as shown in Figure 6. This flip takes 0.25 s.

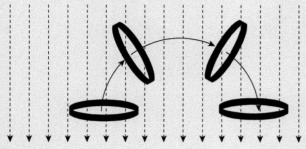

▲ Figure 6

What is the e.m.f. induced in the coil?

A 17 mV   **B** 68 mV

C 42 V   **D** 170 V   (*1 mark*)

24 An electron travelling at $2.3 \times 10^7 \, ms^{-1}$ enters a uniform magnetic field of flux density 0.35 T and travels along a circular path of radius r.

What is r?

($e = 1.6 \times 10^{-19}$ C, $m_e = 9.1 \times 10^{-31}$ kg)

A 46 μm   **B** 0.37 mm

C 19 mm   **D** 2700 m   (*1 mark*)

Section B

25 Two spheres, each of mass 2.5 kg, are placed with their centres 2.0 m apart as shown in Figure 7.

▲ Figure 7

  a Calculate the gravitational force between the spheres.   (*2 marks*)

  b Calculate the energy required to separate the spheres to an infinite distance apart.   (*2 marks*)

26 A 2200 μF capacitor has a charge of 15 V across it.

  a Calculate the charge on the capacitor.   (*2 marks*)

  b Calculate the energy stored on the capacitor.   (*2 marks*)

27 A 470 μF capacitor discharges through a 4 kΩ resistor. The initial p.d. across the capacitor is 6.0 V. Calculate the time it takes for the p.d. across the capacitor to drop to 3.0 V.   (*3 marks*)

28 Cobalt–60 is a radioisotope with a half-life of 5.3 years. Strontium–90 is a radioisotope with a half–life of 28 years.

A cobalt–60 source has twice the activity of a strontium–90 source. Show that the two sources will have approximately the same activity after 6.6 years.   (*4 marks*)

29 The graph shows the motion of a 0.1 kg mass between two springs showing simple harmonic motion.

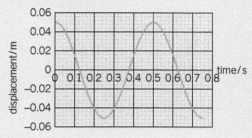

▲ Figure 8

  a Show that the force constant k of the system is about 16 N m^{-1}.   (*2 marks*)

  b Calculate the maximum acceleration of the mass.   (*2 marks*)

30 There is a point along the line joining the centre of the Earth with the centre of the Moon at which the gravitational fields of the two bodies cancel out. At this point the force on a mass due to the Earth is equal and opposite to that due to the Moon. Show that this point is about 3.4×10^8 m from the centre of the Earth.

Mass of Earth = 5.9×10^{24} kg

Mass of Moon = 7.3×10^{22} kg

Distance between the centres of the two bodies = 3.8×10^8 m   (*3 marks*)

31 2.5 moles of nitrogen gas are at a pressure of 8.8×10^5 Pa. The temperature of the gas is 290 K.

  a Calculate the volume of the gas under these conditions, assuming it behaves as an ideal gas.   (*2 marks*)

  b Calculate the root mean square speed of the molecules at this temperature. (molar mass of nitrogen molecules = 28 g mol^{-1})   (*2 marks*)

32 The energy needed for a water molecule to escape into the vapour is about 6.9×10^{-20} J.

The average energy of a particle at 283 K is about 3.9×10^{-21} J.

  a Calculate the Boltzmann factor for the vaporisation of water at 283 K. *(2 marks)*

  b Calculate the temperature that would give a Boltzmann factor for the vaporisation of water of 2.9×10^{-8}. *(2 marks)*

33 The graph shows the variation of flux linkage (mWb-turns) in a coil with time (s).

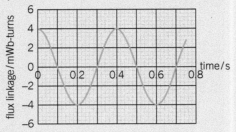

▲ Figure 9

  a Use the graph to estimate the maximum e.m.f. induced in the coil. *(2 marks)*

  b State how the graph would change if the frequency of the variation doubled. What effect would this have on the maximum e.m.f. induced in the coil? *(2 marks)*

34 An electron is accelerated from rest through a p.d. of 1200 V.

  a Calculate the gamma factor of the electron and explain why its value shows that relativistic effects are unimportant at this energy.

   (electron rest energy = 0.51 MeV) *(3 marks)*

  b Calculate the de Broglie wavelength of the accelerated electron.

   (mass of electron = 9.1×10^{-31} kg) *(3 marks)*

35 In a demonstration, a short unmagnetised iron rod is dropped through a vertical copper tube as shown shown in Figure 10.

An identical, magnetised iron rod takes considerably longer to fall through the copper tube. Explain this demonstration using ideas about the induced e.m.f. in the copper tube. *(3 marks)*

OCR Physics B Paper G495 June 2014

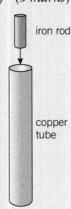

▲ Figure 10

36 The diagram represents the path of an alpha particle near a positively charged nucleus. The nucleus is much more massive than the alpha particle.

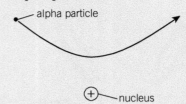

▲ Figure 11

  a State why the curvature of the path of the alpha particle is greatest when the alpha particle is nearest the nucleus. *(1 mark)*

  b How would the path change if the alpha particle had less energy but followed the same initial path? Explain your answer *(2 marks)*

Section C

37 This question is about the energy stored on a capacitor. When the switch in the circuit in Figure 12 is in position A, there is a p.d. of 9.0 V across the capacitor. The switch is moved to position B and the capacitor discharges across the bundle of insulated copper wire of resistance 41 Ω.

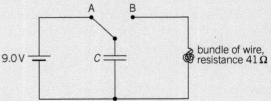

▲ Figure 12

The graph in Figure 13 shows the variation in current with time during this discharge.

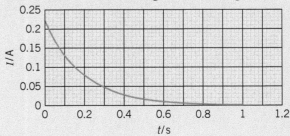

▲ Figure 13

  a Use the graph to show that the charge that flows through the bundle of wire in 1 s is about 40 mC. Explain your method. *(3 marks)*

b Use your answer to **(a)** to find the value of the capacitor. *(2 marks)*

c Use the graph and your answer to **(b)** to show that the capacitor is nearly fully discharged after 5 time constants (5τ). *(3 marks)*

d Show that the energy transferred to the bundle of wire in the first second of the discharge is about 200 mJ. *(2 marks)*

e Calculate the temperature increase of the wire during discharge, assuming that all the energy transferred to the bundle of wire goes to increasing the internal energy of the wire.

Mass of wire = 0.6 g

Specific heat capacity of wire = 420 J kg^{-1} °C^{-1} *(2 marks)*

38 This question is about evidence for the 'hot Big Bang' origin of the Universe.

The graph in Figure 14 shows how the speed of recession of the galaxies v varies with distance d from Earth.

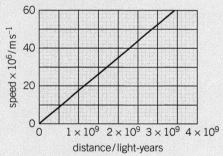

▲ Figure 14

a Describe the observations that show that galaxies are receding from Earth. *(2 marks)*

b Use the graph to determine the Hubble constant, H_0, where $v = H_0 d$.

Include the units in your answer. *(3 marks)*

c Evidence for the hot Big Bang model comes from observation of cosmological redshift and the cosmic microwave background radiation.

Describe these sources of evidence and explain what they suggest about the early Universe. *(6 marks)*

39 This question is about an electric motor, as shown in Figure 15.

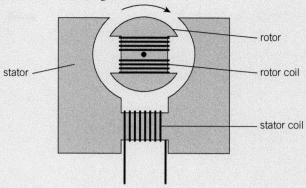

▲ Figure 15

The rotor and stator coils are connected in parallel to the same power supply.

a Current in the stator coil results in magnetic flux in the stator. The stator is made of iron. Explain why this is a good material to use. *(2 marks)*

b The rotor is made from thin sheets of iron, separated by thin sheets of an insulator, instead of solid iron. Explain why. *(3 marks)*

c The output power of the motor can be increased by increasing the voltage of the power supply. Suggest and explain two other modifications to the motor which would result in an increased power output. *(4 marks)*

OCR Physics B Paper 2864 January 2002

Paper 2 Practice questions (A Level)

Section A

1 This question is about the gravitational field of the Earth.

   a (i) Copy Figure 1 and draw eight field lines to show the direction of the Earth's gravitational field.

   ▲ **Figure 1**   (*2 marks*)

   (ii) How does your diagram show that the strength of the gravitational field decreases with distance from the planet?   (*1 mark*)

   b (i) The gravitational field is uniform near the surface of the Earth. Draw a diagram showing the field near the horizontal surface. Include three equipotential lines separated by equal potential differences.   (*3 marks*)

   (ii) State two features of the diagram that show the field is uniform.   (*1 mark*)

   c The gravitational field strength of the Earth is proportional to $-\frac{1}{r^2}$, where r is the distance from the centre of the planet. Use this relationship and the information given at the end of this question to calculate the gravitational field strength of the Earth at a distance from the Earth equal to the Earth-Moon distance.

   Radius of the Earth = 6.4×10^6 m

   Gravitational field strength at the surface of the Earth = 9.8 N kg^{-1}

   Earth-Moon distance = 3.8×10^8 m
                                      (*3 marks*)

2 A type of smoke detector uses an americium–241 alpha source to reveal the presence of smoke in the vicinity of the detector. The presence of smoke decreases the number of alpha particles detected in a given time.

   a Copy and complete the decay equation:
   $^{241}_{95}\text{Am} \rightarrow\ ^{4}_{2}\text{He} +\ ^{...}_{...}\text{Np}$ (*1 mark*)

   b (i) Use the data below to calculate the activity of an alpha source used in a smoke detector.

   Mass of source = 2.8×10^{-10} kg

   Mass of one mole of americium–241 = 0.241 kg

   Avogadro number = 6.0×10^{23} mol^{-1}

   decay constant λ of americium–241 = 4.8×10^{-11} s^{-1} (*3 marks*)

   (ii) Explain, using relevant calculations, why it is reasonable to assume that the activity of the source remains constant over the five-year life of the smoke detector.
   1 year = 3.2×10^7 s.   (*4 marks*)

   c The detector has a plastic casing with vents to allow smoke to enter. Explain why the alpha source does not contribute to the background radiation of the building in which it is housed.   (*2 marks*)

3 This question is about beta decay. The following equation shows beta-minus decay.
   $$^{1}_{0}\text{n} \rightarrow\ ^{1}_{1}\text{p} +\ ^{0}_{-1}\text{e} +\ ^{0}_{0}\overline{v}$$

   a (i) Identify the particle $^{0}_{0}\overline{v}$.   (*1 mark*)

   (ii) Explain how the equation shows that lepton number is conserved in the decay.   (*1 mark*)

   b A student suggests that the essential change in the decay is that a down quark changes into an up quark. Explain this statement.   (*3 marks*)

   c Calculate the energy released in the decay. Assume that $\overline{v}$ has no mass.

   Mass of neutron = 1.6749×10^{-27} kg

   Mass of proton = 1.6726×10^{-27} kg

   Mass of electron = 0.0009×10^{-27} kg
                                      (*2 marks*)

   d (i) Outside the nucleus, free neutrons are unstable with a decay constant λ of about 1.1×10^{-3} s^{-1}. Calculate the half-life of free neutrons.   (*1 mark*)

   (ii) State observational evidence that shows that most neutrons within nuclei are stable.   (*1 mark*)

e Beta-plus decay is shown in the following equation.

$$^1_1p \rightarrow {}^1_0n + {}^0_{+1}e + {}^0_0\nu$$

Use the data given in part **(c)** to explain why a free proton does not decay into neutrons. *(2 marks)*

4 The Boltzmann factor f is given as $f = e^{-\frac{E}{kT}}$. It is a measure of the likelihood of a particle gaining sufficient energy for a physical process to occur.

E is the energy needed for the particular process to take place.

k is the Boltzmann constant $1.38 \times 10^{-23}\,JK^{-1}$.

T is the Kelvin temperature.

a The graph in Figure 2 shows how the Boltzmann factor f varies with temperature.

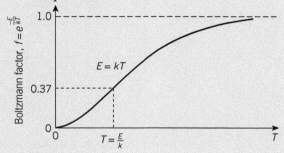

▲ **Figure 2**

Use $f = e^{-\frac{E}{kT}}$ to explain why, as shown in the graph,

(i) the Boltzmann factor f is close to zero at low temperatures, *(1 mark)*

(ii) the Boltzmann factor f is less than 1 even at the highest temperatures *(1 mark)*

(iii) the Boltzmann factor f is about 0.37 when $E = kT$. *(1 mark)*

b Some chemical reactions take place at a more rapid rate in the presence of an enzyme. The enzyme lowers the value of E, the energy needed for the reaction to take place. Enzymes speed up chemical reactions in the body.

The energy, E, required for a particular molecule to decompose is $1.3 \times 10^{-19}\,J$.

(i) Show that the Boltzmann factor f for this process taking place at body temperature, $T = 310\,K$, is about 6×10^{-14}. *(2 marks)*

(ii) The presence of an enzyme lowers the value of E to $6.0 \times 10^{-20}\,J$. Show that the Boltzmann factor f at $310\,K$ is increased by a factor of more than a million in the presence of the enzyme. *(3 marks)*

(iii) Explain why this increase in the value of the Boltzmann factor f increases the rate at which a reaction will take place. *(2 marks)*

OCR Physics B Paper 2863 January 2003

5 This question is about charges in electric fields. Figure 3 shows two conducting parallel plates connected to a power supply.

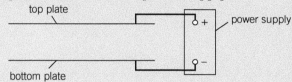

▲ **Figure 3**

a On a copy of Figure 3, sketch five lines to represent the electric field between the plates. *(2 marks)*

b A small metal sphere is placed between the plates as shown in Figure 4.

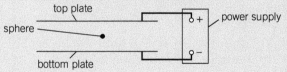

▲ **Figure 4**

Both plates are horizontal and the sphere is charged.

(i) The sphere does not move when the electric field is present.

What sign of charge does the sphere have? Give reasons for your answer. *(3 marks)*

(ii) The magnitude of the charge on the sphere is $3.2 \times 10^{-14}\,C$.

How many electrons had to be removed or added to give the sphere this charge? $e = 1.6 \times 10^{-19}\,C$ *(1 mark)*

(iii) The mass of the sphere is $6.2 \times 10^{-9}\,kg$. The separation of the plates is $14\,mm$.

1 Show that, for the sphere not to move, the electric field strength must be about $2 \times 10^6\,Vm^{-1}$. $g = 9.8\,Nkg^{-1}$ *(3 marks)*

2 Calculate the potential difference across the plates required for the sphere to not move. (*2 marks*)

OCR Physics B Paper 2864 January 2010

Section B

1 This question is about modelling the motion of a simple pendulum undergoing simple harmonic motion.

Figure 5 shows the forces on a simple pendulum of length *L*.

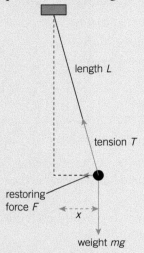

▲ Figure 5

The restoring force *F* is the horizontal component of the tension *T*.

a Show that, for small angles of displacement and assuming that $T = mg$, the acceleration *a* is given by $a = -\dfrac{gx}{L}$. (*3 marks*)

b (i) The acceleration *a* of a simple harmonic oscillator of frequency *f* is given by:
$a = -4\pi^2 f^2 x$ where *x* is the displacement from the equilibrium position.

Use this equation and the equation given in part (a) to show that time period of the pendulum $T = 2\pi\sqrt{\dfrac{L}{g}}$. (*2 marks*)

(ii) Calculate the time period of a pendulum of length 2.5 m. (*1 mark*)

A student uses the equation in (a) to make a simple iterative model of the motion of the 2.5 m pendulum. The acceleration of the pendulum is held constant over each interval of 0.2 s and recalculated at the beginning of each interval.

c (i) Show that, for an initial displacement *x* of the pendulum of 0.040 m, the initial acceleration is about –0.16 m s^{-2}. (*1 mark*)

(ii) Show that the model gives the displacement of the pendulum as about 0.037 m after 0.2 s of uniform acceleration. (*2 marks*)

Further iterations of the calculation produce the graph shown in Figure 6.

▲ Figure 6

d (i) Comment on the graph line between 0.0 and 0.2 s. State and explain what this shows about the velocity of the pendulum bob in this model and why this does not accurately represent the oscillation. (*2 marks*)

(ii) Use the graph to find the time period of the oscillator in the model. Compare the value obtained by this iterative method with the value obtained in (b)(ii). Explain any difference between the values and explain how the iterative model can be adapted to produce a closer match to the value obtained in (b)(ii). (*4 marks*)

2 Figure 7 shows an electric toothbrush being charged. The charging is done with a transformer that has the primary coil in the charger case and the secondary coil inside the toothbrush. There are no electrical contacts joining the toothbrush to the charger case.

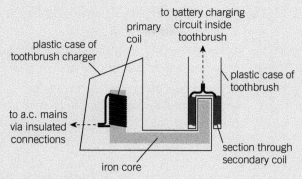

▲ Figure 7

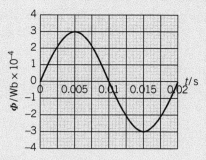

▲ Figure 9

a The transformer is not efficient because the flux in the secondary coil is less than that in the primary coil.

(i) On a copy of Figure 7, draw flux loops through the primary coil to illustrate why flux leakage in the system results in a loss of efficiency. Explain how your additions to the diagram show this.
(3 marks)

(ii) The circuit for a simple mobile phone charger is shown in Figure 8.

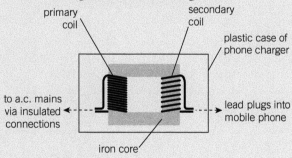

▲ Figure 8

Suggest and explain why this arrangement is suitable for mobile phones but not for electric toothbrushes. *(2 marks)*

b In Figure 8, the maximum value of the flux through the secondary coil (in the toothbrush) is 65% of the maximum value of the flux through the primary coil (in the charger). The secondary coil consists of 20 turns.

(i) The graph in Figure 9 shows how the flux through the secondary coil varies with time. Use the graph to show that the peak output p.d. from that coil is about 2 V. Show your working clearly.
(3 marks)

(ii) Calculate the turns ratio needed in the inefficient charger-toothbrush transformer if an input voltage of 230 V r.m.s is to be stepped down to 2.0 V r.m.s. *(3 marks)*

(iii) The battery inside the toothbrush has a capacity of 9.0 kC. Assuming that it takes 6 hours to charge fully from being completely discharged, calculate the mean current drawn from the secondary coils and the total energy transferred by the charger. *(4 marks)*

3 Figure 10 shows a simple model of a rocket engine. The arrows inside the combustion chamber represent the pressure p acting on all the walls.

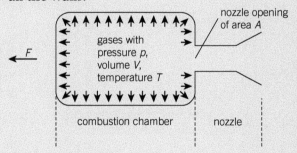

▲ Figure 10

Hydrogen and oxygen gas are pumped into the combustion chamber, where they burn to produce exhaust gases at high pressure and temperature. The hot gases leave through the exhaust at the same rate as they are produced.

a Use Figure 10 to explain why the force F acting on the rocket is given by $F = pA$. *(2 marks)*

b The thrust F provided by the rocket is 1.8×10^6 N. The nozzle opening is a circle of diameter 0.33 m. Show that the pressure inside the combustion chamber is about 20 MPa. *(3 marks)* c   The volume of the combustion chamber is 2.0 m³ and the temperature of the gas is 3500 K.

   (i)   Calculate the number of gas molecules in the combustion chamber.   (3 marks)

   (ii) The molecules each have a mass 3.0 × 10⁻²⁶ kg. Show that the molecules are travelling at about 2000 m s⁻¹.   (3 marks)

   (iii) Calculate the number of gas molecules travelling at this speed that must leave the engine every second to give a thrust of 1.8 × 10⁶ N.   (2 marks)

d   The rocket in Figure 10 is to be used to launch a satellite into Earth orbit. The thrust given above is the initial thrust produced by the engine. Explain why the thrust will not stay at this value as the rocket ascends to its final height.   (3 marks)

4   This question is about oscillations in loaded vehicles.

A small delivery truck can be thought of as a box supported by four springs, one at each wheel (the suspension of the truck). In this question, only the two rear springs will be considered.

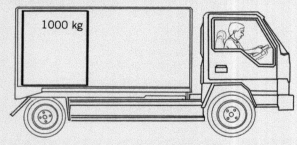

▲ Figure 11

a   The spring constant k in $F = kx$ of each spring is 2.6 × 10⁴ N m⁻¹.

   (i)   Explain why the spring constant of the two rear springs together is 5.2 × 10⁴ N m⁻¹.   (2 marks)

   (ii) Show that the back of the truck will move down about 20 cm when it is loaded with 1 tonne (1000 kg), placed above the rear wheels as in Figure 11.   (2 marks)

The part of the truck body supported by the rear wheels has a mass of 500 kg.

b   Show that the period of oscillation of an unloaded truck is about 0.6 s.   (2 marks)

c   A truck with a certain load oscillates with a time period exactly double that of an unloaded truck. Is it carrying more than the maximum permitted load of 1000 kg? Show your working.   (3 marks)

Speed bumps are put on the road into a large supermarket to slow the traffic (Figure 12).

▲ Figure 12

d   Explain why the truck will oscillate after passing over a speed bump.   (2 marks)

e   If the truck travels at a certain speed over the set of speed bumps, the vertical oscillations can become very large. Explain why this is so.   (2 marks)

f   Free oscillations of the suspension, as shown in the graph below, would be uncomfortable, so friction is used to damp them and reduce the amplitude.

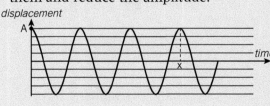

▲ Figure 13

Starting at point A, draw on a copy of Figure 13 a graph showing the displacement-time graph you would get if the oscillation was damped, so that the amplitude at point X was about one-quarter of the starting amplitude.   (3 marks)

OCR Physics B Paper 2865 January 2002

5   This question is about asteroids. These are rocky objects in orbit around the Sun, which are too small to be considered as planets.

a   Table 1 gives some data about two of the largest asteroids, both discovered at the beginning of the nineteenth century.

name	volume / m^3	mass / kg	orbital period / year
Ceres	4.3×10^{17}	8.7×10^{20}	4.6
Vesta	7.8×10^{16}	3.0×10^{20}	3.6

(i) Use the data in Table 1 to calculate the densities of these asteroids.

(2 marks)

(ii) Comment on the composition of the two asteroids Ceres and Vesta.

(1 mark)

b The speed, v, of a satellite in a circular orbit of radius r about a central object is given by

$$v = \sqrt{\frac{GM}{r}}$$

Where G is the universal gravitational constant and M is the mass of the central object.

(i) Use this equation to show that the orbital period, T, of the satellite is given by

$$T = \sqrt{\frac{4\pi^2 r^3}{GM}} \qquad \textit{(3 marks)}$$

(ii) State and explain which asteroid, Ceres or Vesta, has the larger orbital speed around the Sun, assuming that both orbits are circular. *(2 marks)*

c NASA's Shoemaker spacecraft was placed in an orbit of radius 35 000 m around the small asteroid Eros in February 2000.

Use the equation in part **(b)(i)** to show that the spacecraft took nearly a day to orbit the asteroid Eros.

Universal gravitational constant $G = 6.67 \times 10^{-11}\,\mathrm{N\,m^2\,kg^{-2}}$

Mass of Eros $M = 6.69 \times 10^{15}\,\mathrm{kg}$ *(2 marks)*

d After a year in orbit a controlled descent to the surface was planned. At the time Eros was about $3 \times 10^{11}\,\mathrm{m}$ from the Earth.

(i) Calculate the time it took for a control signal to travel from the Earth to the spacecraft. *(1 mark)*

(ii) What problems might this have posed in controlling the descent? *(2 marks)*

e Figure 14 was taken from the Shoemaker spacecraft.

▲ Figure 14

(i) This image, which consists of 398×303 pixels at 8 bits per pixel, was stored in 10 576 bytes of computer memory. Why does this suggest that some kind of data compression was used?

(2 marks)

(ii) Once on the surface the spacecraft continued to function even though it had not been designed to make a landing. Data was transmitted back to Earth at a rate of 10 bits per second. How long would it have taken to transmit the data for the image above?

(2 marks)

OCR Physics B Paper 2865 January 2004

6 This question is about light and astronomy. When an electric current is passed through hydrogen gas in a discharge tube, the hydrogen emits a line spectrum. One of the wavelengths in the visible part of the spectrum is due to the change in the energy levels of the hydrogen atom shown in Figure 15.

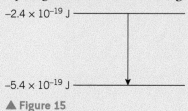

$-2.4 \times 10^{-19}\,\mathrm{J}$ ————————————

$-5.4 \times 10^{-19}\,\mathrm{J}$ ————————————

▲ Figure 15

a Show that photons of light of wavelength 660 nm can be produced by this energy level change.

$h = 6.6 \times 10^{-34}$ J s

$c = 3.0 \times 10^8$ m s^{-1} (*3 marks*)

b Light from a discharge tube containing hydrogen was observed through a diffraction grating with 600 lines per mm.

  (i) Show that the grating spacing d is about 1.7×10^{-6} m.   (*1 mark*)

  (ii) Show that the 660 nm wavelength will be detected at about 23° from the zero order.   (*2 marks*)

c When stars are moving away from us, their spectral lines are red-shifted.

Indicate on a copy of Figure 16 how the position of the lines, originally at 660 nm, would change when viewed through the diffraction grating if the light had been red-shifted. The grey lines indicate the position of these lines if there is no red-shift.

(*2 marks*)

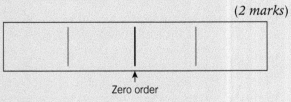

Zero order

▲ **Figure 16**

d The wavelengths of light from the star Regulus were found to be red-shifted by 0.0020%. Use the relationship $\frac{\Delta\lambda}{\lambda} = \frac{v}{c}$ to calculate the velocity of recession v of Regulus relative to Earth.   (*3 marks*)

e Two stars A and B have the same total actual luminosity (brightness). However, star A appears to be emitting 100 times less energy per second than star B, which is at a distance of 30 parsecs from the Earth.

Use the inverse square law to calculate the distance of star A from Earth.

(*3 marks*)

f For distant galaxies, the cosmological redshift z is given by

$$1 + z = \frac{\lambda + \Delta\lambda}{\lambda}$$

One particular spectral line of hydrogen has wavelength 122 nm on Earth. In light from the very distant galaxy *Abell 1835 IR1916* the same spectral line is observed to have a wavelength of 1.34 μm.

  (i) Show that the redshift z of this galaxy is about 10.   (*2 marks*)

  (ii) Use the cosmological redshift equation above to explain why a cosmological redshift of 10 implies that the Universe is now 11 times larger than when the galaxy emitted that light.   (*2 marks*)

Section C

Read the following short article before answering the questions.

The source of the Sun's energy

Thermodynamics developed as a science in the nineteenth century, and it became clear that chemical reactions, such as in a coal fire, could not provide enough energy for the Sun to last more than a few thousand years. Geological evidence at this time suggested that the Earth must
5 be many millions of years old, so another energy source had to be found. The German scientist Helmholtz suggested that gravity could supply the energy as meteors fell into the Sun. However, the mass of meteors needed to provide the 4×10^{26} W that the Sun emits is huge, of the order of 10^{15} kg each second. It seems unlikely that matter is falling into the Sun at a rate equal to the mass of the Moon every year or two.

10 In 1887 Lord Kelvin developed this idea further, and suggested that the Sun began as a huge cloud of gas and dust which began shrinking due to the gravitational attractions between its particles. As the cloud collapsed it got continually hotter. Once reaching the size of the present-day Sun, the core temperature would be millions of degrees, and the pressure of the very hot interior would prevent any further collapse. Under these conditions, the star would take a long
15 time to cool, and once it did, the in-falling matter of the resulting contraction would once again produce heat. Kelvin calculated that the Sun's radius had to shrink only about 50 cm per year to keep the Sun's output fairly steady for tens of millions of years. Geologists still felt that this time was not long enough to agree with the age of the rocks on Earth, but no better explanation could be found at that time.

20 The discovery and study of radioactivity in the first half of the 20th century provided an explanation. Radioactive decay releases large amounts of energy, so it was suggested as a source of the Sun's power. It was eventually realised that nuclear fusion provided the solution. The gravitational collapse that Lord Kelvin had suggested half a century earlier produces extremely high temperatures and pressures. In fact, the temperatures inside stars are so high
25 that atoms are stripped of all their electrons, producing a plasma of positive and negative ions. At these extremes of temperature and pressure, nuclei in this plasma can fuse, releasing energy.

In a small star like our Sun, energy is mostly produced by a series of three nuclear reactions called the proton-proton chain. In this process, protons first combine to form deuterons (hydrogen–2 nuclei). This process is extremely infrequent, happening to a pair of colliding protons
30 only once in about 10^9 years. The deuterons quickly (within about a second) react with further protons to give helium–3 nuclei. Finally, in another process, on a time scale of about a million years, pairs of these helium–3 nuclei combine to give ordinary helium–4 nuclei. The net result of this series of processes is to combine four protons to form a helium–4 nucleus, with the emission of two positrons, two neutrinos, and gamma photons, as shown in Figure 17.

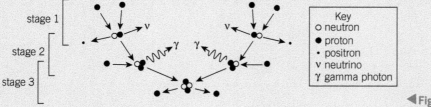

◀ Figure 17

35 The energy liberated in this process is over 24 MeV. The two positrons produced in stage 2 annihilate with two electrons in the Sun's core to produce another 2 MeV, so that over 26 MeV is produced for every four protons converted into helium nuclei. The Sun's core contains about 6×10^{29} kg of hydrogen, and this would allow the production of energy at the current rate of 4×10^{26} W for several billion years. This agrees well with geological estimates of the Earth's age,
40 and confirms the proton-proton chain as a reasonable model of power production in the Sun.

Questions

1 This question is about an early theory of the Sun's energy source (lines 4–8 in the article).

   a A solar water heating panel of area $3\,\text{m}^2$ on a house roof is perpendicular to the solar radiation. Water flows through the panel at a rate of $0.17\,\text{kg s}^{-1}$. The temperature of the water increases by $4\,°C$ when it flows through the panel.

     (i) Show that the solar energy absorbed by the panel in 1 second is about $3000\,\text{J}$. Specific thermal capacity of water, $c = 4200\,\text{J kg}^{-1}\,°C^{-1}$ *(2 marks)*

     (ii) Calculate the solar power per square metre absorbed by the $3\,\text{m}^2$ panel.   *(1 mark)*

     (iii) The solar power per square metre arriving at the outer surface of the Earth's atmosphere is $1400\,\text{W m}^{-2}$.

       Suggest why your answer to part **(ii)** is different from this.   *(1 mark)*

     (iv) At the Earth's distance from the Sun, the energy emitted by the Sun each second passes through the surface of a sphere of area $2.8 \times 10^{23}\,\text{m}^2$, as shown in Figure 18.

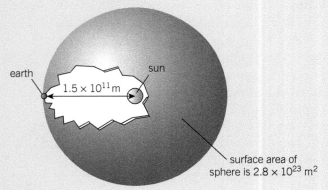

▲ Figure 18

       Use the value given in **(a)(iii)** to show that the total power emitted by the Sun is about $4 \times 10^{26}\,\text{W}$.   *(2 marks)*

   b Figure 19 shows a meteor at point Y before it falls into the Sun.

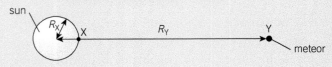

▲ Figure 19

     The gravitational potential difference ΔV_{XY} between points X and Y is given by

$$\Delta V_{XY} = \frac{GM}{R_X} - \frac{GM}{R_Y} \qquad \text{Equation 1}$$

     (i) Explain why, when $R_Y \geq 100$,

$$\Delta V_{XY} \approx \frac{GM}{R_X} \qquad \text{Equation 2}$$
(1 mark)

     (ii) Use Equation 2 given earlier in this question to show that the gravitational potential difference between the surface of the Sun and a distant point is about $2 \times 10^{11}\,\text{J kg}^{-1}$.

         $G = 6.7 \times 10^{-11}\,\text{N m}^2\,\text{kg}^{-2}$
         $M = 2.0 \times 10^{30}\,\text{kg}$
         $R_X = 7.0 \times 10^8\,\text{m}$ *(2 marks)*

     (iii) Explain why the kinetic energy gained when a distant meteoroid of mass $1\,\text{kg}$ falls to the Sun's surface is about $2 \times 10^{11}\,\text{J}$.   *(1 mark)*

     (iv) Calculate the total mass of meteors that would need to fall into the Sun every second to provide the $4 \times 10^{26}\,\text{W}$ that the Sun emits.   *(2 marks)*

2 This question is about the modern theory of the Sun's energy source (lines 26–41 in the article).

   Figure 20 shows stage 1 of the series of nuclear reactions taking place in the Sun's core.

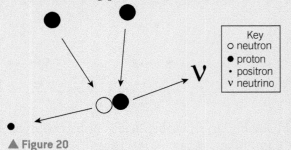

▲ Figure 20

   a Copy and complete the balanced equation for the nuclear reaction shown in Figure 20.

$$^1_1\text{H} + ^1_1\text{H} \rightarrow \underset{\cdots}{\cdots}\text{H} + \underset{\cdots}{\cdots}\underline{\quad} \qquad \textit{(2 marks)}$$

b Stage 2 of the series of nuclear reactions in the Sun's core is

$$^1_1H + ^1_1H \rightarrow ^3_2He + \gamma$$

(i) Use the data in Table 2 to show that the mass of the products of this reaction is about 9×10^{-30} kg less than the mass of the reactants.

nuclear species	mass / u
^1_1H	1.00728
^2_1H	2.01410
^3_2He	3.01605

$u = 1.67 \times 10^{-27}$ kg   (*2 marks*)

(ii) Show that about 8×10^{-13} J of energy is produced in a reaction of this type.
$c = 3.0 \times 10^8$ m s^{-1} (*2 marks*)

c The series of nuclear reactions in the proton-proton chain liberates 4.3×10^{-12} J for every four protons (^1_1H) fused into one helium–4 nucleus.

(i) Show that the energy produced by the fusion of 1 kg of hydrogen is about 6×10^{-14} J.
mass of proton, $m_p = 1.67 \times 10^{-27}$ kg
       (*2 marks*)

(ii) Show that the Sun can produce energy at 4×10^{26} W for several billion years (lines 37–39 in the article), assuming that 2.0×10^{29} kg of hydrogen is available for fusion.
1 year $= 3.2 \times 10^7$ s   (*2 marks*)

OCR Physics B Paper 2865 June 2006

Section A

1   This question is about measuring the acceleration due to gravity using a simple pendulum.
    A student measures the time period of a pendulum for different lengths L.
    The time period is measured by finding the time for ten oscillations to the nearest 0.5 s and dividing the value by 10.

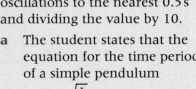

length L

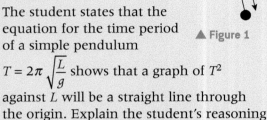

▲ **Figure 1**

a   The student states that the equation for the time period of a simple pendulum
$$T = 2\pi \sqrt{\frac{L}{g}}$$
shows that a graph of T^2 against L will be a straight line through the origin. Explain the student's reasoning and identify the gradient of the line.
        (3 marks)

b   Table 1 shows the results from the experiment. Estimate the uncertainty in the T^2 values, explaining your method.
        (3 marks)

▼ **Table 1**

$L \pm 0.01$ m	$T \pm 0.05$ s	T^2/s^2
1.5	2.45	6.00
1.2	2.20	4.84
0.9	1.90	3.61
0.6	1.55	2.40
0.3	1.10	1.21

c   Plot a graph of L against T^2. Include uncertainty bars on the points. *(4 marks)*

d   Use the gradient of the best fit line to find a value for g. *(2 marks)*

e   Explain how you can estimate the uncertainty in your value for g from the graph. *(2 marks)*

2   This question is about determining the half-life of protactinium, a beta-emitter.

a   Before placing the detector near the protactinium source, the activity of the background radiation is measured. Describe how background count in counts s^{-1} can be determined and explain why a systematic error will arise if the background count is not considered.
        (4 marks)

b   The detector is placed near the prepared source. The number of counts is recorded for each ten-second period over three minutes.

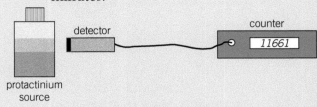

detector   counter

protactinium source

▲ **Figure 2**

Explain how the count rate in counts s^{-1} can be calculated and how it can be corrected for background radiation. *(2 marks)*

c   The relationship between activity A and time t is given by the equation $A = A_0 e^{-\lambda t}$ where A_0 is the initial activity.

(i)   State the difference between activity and count rate. *(1 mark)*

(ii) State why the variation of count rate C with time will follow the relationship $C = C_0 e^{-\lambda t}$ where C_0 is the initial count rate. *(1 mark)*

(iii) This relationship can also be expressed as $\ln C = \ln C_0 - \lambda t$. A graph of $\ln C$ against t for a set of results is given in Figure 3.

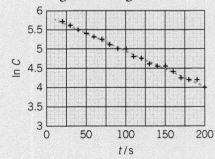

▲ **Figure 3**

Use the graph to find the half-life of the source, making your method clear. *(4 marks)*

3   This question is about determining the internal resistance of a battery.

The circuit is set up as shown in Figure 4. Readings of potential difference and current are taken as the value of resistance R is changed.

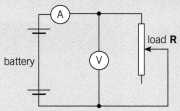

▲ Figure 4

a State why the ammeter must have a low resistance and the voltmeter must have a high resistance. (*2 marks*)

b Table 2 shows the data collected.

▼ Table 2

current / A	p.d. / V
0.11	5.7
0.32	5.2
0.88	3.8
1.32	2.8
1.55	2.3
1.75	1.5

(i) Draw a graph of the data. The p.d. axis must extend to 7.0 V. (*3 marks*)

(ii) Use your graph to find the e.m.f. of the battery. Make your method clear. (*2 marks*)

(iii) Use your graph to find the internal resistance of the battery. Make your method clear. (*3 marks*)

4 This question is about an experiment to measure the strength of a magnetic field.

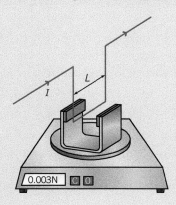

▲ Figure 5

A wire carrying current I passes through a magnetic field of length L. The field is provided by slab magnets either side of the wire. A force F is exerted on the wire. An equal and opposite force is exerted on the magnets. This force is measured by the electronic balance reading in newtons.

a Explain how you could use this apparatus to test whether the field between the two magnets is uniform. (*3 marks*)

The variation of force with current is shown in Figure 6. The resolution of the electronic balance is 0.001 N.

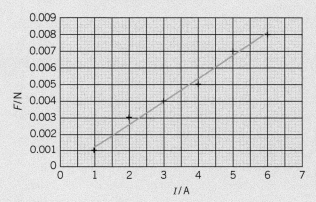

▲ Figure 6

b Use Figure 5 to determine the strength of the field between the two magnets. Length of wire in magnetic field = 28 mm. (*3 marks*)

c Two students discuss the results of the experiment.

• One student suggests that the results for the lower current values are less reliable than larger current values, as the force exerted on the wire at low currents is small and any changes in the force are below the resolution of the balance.

• The other student suggests that the results for higher current values are less reliable because the wire gets hot, changing its resistance.

Give your opinion on these two views. Explain if and why you agree with each statement. (*4 marks*)

5 Students in a class measured the force needed to break 29 samples of copper wire from the same reel.
Their results are shown in the chart in Figure 7.

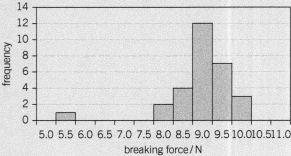

▲ Figure 7

a **(i)** Explain why the single value at 5.5 N needs to be considered to be a possible outlier. *(2 marks)*

(ii) Suggest what may have happened to the wire to give the much smaller breaking force of 5.5 N. *(1 mark)*

b A student calculated the mean value of the remaining 28 values as 9.089 286 N. Use the bar chart to write down the best estimate and uncertainty of the breaking force of the copper, using an appropriate number of significant figures. *(2 marks)*

c The diameter of the wire used is measured at several points along its length with a micrometer. The reading is 0.38 ± 0.005 mm each time.

(i) Calculate the percentage uncertainty in the diameter measurement. Give your answer to an appropriate number of significant figures. *(2 marks)*

(ii) Explain why the uncertainty in the diameter measurement can be ignored when calculating the breaking stress of copper. *(1 mark)*

d Use the answer to **(b)** to calculate the mean breaking stress of copper and to estimate the uncertainty in this value. Cross-sectional area of copper wire = 1.1×10^{-7} m^2 *(2 marks)*

6 a State what is meant by the **resolution** of a measuring instrument. *(1 mark)*

b Two ammeters are shown in Figure 8.

digital ammeter (reading in A)   analogue ammeter

▲ Figure 8

(i) Suggest and explain a value for the resolution of each ammeter in Figure 8. *(4 marks)*

(ii) Calculate the percentage uncertainty in the measurement of the **digital** ammeter when it is used to measure a current of 3 A. *(2 marks)*

c The zero error shown on the analogue ammeter is a systematic error.

(i) State the consequence of ignoring **this** systematic error when taking readings. *(1 mark)*

(ii) Describe one way in which this systematic error could be removed. *(1 mark)*

d The pointer of the **analogue** ammeter moves through an angle of 90° when the current increases from 0 to 50 A. Calculate a value for the current when the angle between the pointer and 0 is 23°. *(1 mark)*

e One application of an ammeter is to monitor a current to check that it does not suddenly increases to a much larger value. Explain why the analogue ammeter may be a better choice than the digital ammeter for this application. *(1 mark)*

f Suggest why an ammeter for use in circuits carrying large currents should have a very low resistance. *(1 mark)*

OCR Physics B Paper G492 June 2010

Section B

1 Figure 9 shows apparatus that can be used to investigate the pressure changes in a gas (air) as it is heated. The air is contained in a glass sphere with attached glass tubing. The glass tubing is joined to a pressure gauge with rubber tubing, which is firmly clamped at each end. The temperature of the air is increased gradually by heating the water bath.

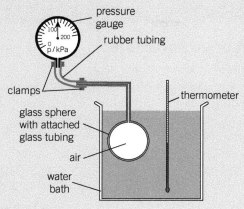

▲ Figure 9

a The pressure gauge is calibrated in kPa from 0 to 200 kPa, with scale divisions every 10 kPa. A student suggests that the uncertainty in pressure readings from the scale shown in Figure 9 is ± 10 kPa. Explain why his estimate is probably too large. *(2 marks)*

b In performing the experiment, the student takes care to heat the water bath slowly, and to wait for a minute, whilst stirring the water bath, before taking each pair of readings of pressure and temperature. Explain why this is good experimental practice. *(2 marks)*

c Alan obtains the data shown in Table 3

▼ Table 3

temperature / °C	13	23	34	41	52	66	76	84	98
pressure / kPa	100	103	107	109	112	117	119	121	126

He then plots the graph shown in Figure 10.

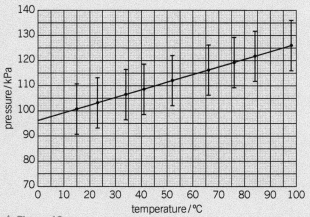

▲ Figure 10

(i) Explain why the uncertainty bars show that the student's estimate of uncertainty of ± 10 kPa is too large. *(2 marks)*

(ii) Calculate the gradient m and the y-axis intercept c of the student's best-fit line. *(3 marks)*

d The fact that the line is straight with a positive intercept at 0 °C suggests it will cut the temperature axis at some temperature θ_{zero} below 0 °C as shown in Figure 11.

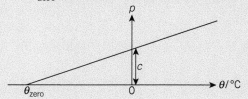

▲ Figure 11

Use this information, together with the values of m and c from part **(c)(ii)**, to find the temperature θ_{zero} of absolute zero in °C. *(3 marks)*

e Experiments using the apparatus shown in Figure 9, with a volume of gas connected to a pressure gauge outside the water bath, often give a value of absolute zero which is at a more negative temperature than −273 °C. Explain why a result of this sort suggests that the gradient of the graph is lower than it should be. *(2 marks)*

f Two other students suggest explanations for the gradient of the student's graph being too small.

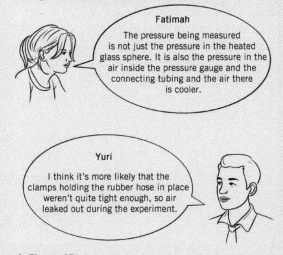

Fatimah

The pressure being measured is not just the pressure in the heated glass sphere. It is also the pressure in the air inside the pressure gauge and the connecting tubing and the air there is cooler.

Yuri

I think it's more likely that the clamps holding the rubber hose in place weren't quite tight enough, so air leaked out during the experiment.

▲ Figure 12

Explain why both suggestions could explain why the gradient of Alan's graph is too small, and suggest how it might be possible to investigate which was responsible for the systematic error. *(6 marks)*

Changing the subject of an equation

An equation shows the relationship between variables. You can change an equation to make any of the variables into the subject of the equation. To do this you do the inverse operation to both sides on an equation to isolate the quantity that you want. For example, dividing is the inverse operation of multiplying. You always add or subtract terms first, and then divide, multiply, or perform other operations, for example, square root.

 Worked example: Make *d* the subject

Change the equation $a = b + \dfrac{c}{d}$ to make *d* the subject.

Step 1: Add or subtract terms to isolate the term involving *d*.

$$a - b = b + \frac{c}{d} - b$$

$$a - b = \frac{c}{d}$$

Step 2: Multiply or divide terms to isolate *d*.

$$a - b = \frac{c}{d}$$

$$(a - b)d = c$$

$$d = \frac{c}{(a - b)}$$

Graphs

The equation of a straight line is $y = mx + c$, where *m* is the gradient and *c* is the *y*-intercept. The gradient and the intercept can be positive or negative. You can deduce information from straight-line graphs. If the line goes through the origin (no intercept) then *y* is *directly proportional* to *x*, and $y = mx$. When you plot a graph you should think about the physical reason why a graph might (or might not) go through (0, 0).

You can test a relationship using a graph. You might collect data relating to the pressure and volume of a gas. Plotting the data produces a curve, but plotting *V* against $\dfrac{1}{p}$ gives a straight line. This shows that pV = constant (equal to the gradient).

> **Hint**
>
> Graphs such as the graph of force against extension for a spring will be a straight line through (0, 0). The equation of the line is $F = kx$, and this is the same as $y = mx + c$ with $y = F$, $m = k$, and $c = 0$.

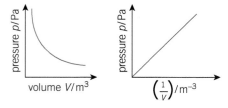

▲ **Figure 1** *The straight line shows that* $p \propto \dfrac{1}{V}$, *so* $p = $ (gradient) $\times \dfrac{1}{V}$

 Worked example: Curve to straight line

You have collected data for two quantities, *t* and *y*, which should be related by the equation $y = \dfrac{k}{t^2}$, where *k* is a constant. State which graph you need to plot to show that this is correct. Explain how the shape of the graph would, or would not, confirm your belief.

Step 1: Change the equation into the form $y = mx$.

If $y = \frac{k}{t^2}$, then it becomes $y = mx$ if $k = m$ and $\frac{1}{t^2} = x$.

So you need to plot y against $\frac{1}{t^2}$.

Step 2: Describe the shape you would get if the relationship was or was not true.

If the graph is a straight line through (0, 0), then this confirms that $y = \frac{k}{t^2}$ and the gradient = k.

If the graph is a curve then y is not equal to $\frac{k}{t^2}$.

Calculations using graphs

You use graphs to display data. When you draw a graph you choose a type of scale, and add numbers vertically and horizontally to the axes. The graph will show the relationship between physical quantities, so you should also include the units as well as the name of the quantity on each of the axes.

You often need to calculate **gradients** and areas from graphs. These may represent important physical quantities. Their units can give an important clue as to what they represent.

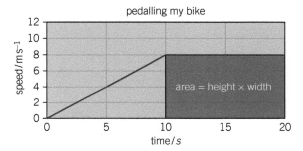

▲ **Figure 2** A bike ride

 Worked example: Gradients and area

Calculate the gradient of the graph in Figure 2 between 0 and 10 s, and the area under the graph between 10 and 20 s.

Step 1: Find the gradient using the equation gradient $= \dfrac{\text{change in } y}{\text{change in } x}$.

$$\text{gradient} = \frac{(8\,\text{m s}^{-1} - 0\,\text{m s}^{-1})}{(10\,\text{s} - 0\,\text{s})} = 0.8\,\text{m s}^{-2}$$

These are the units of acceleration. The gradient of a speed–time graph = acceleration.

Step 2: Substitute values into area = height × width.

Area $= 8\,\text{m s}^{-1} \times 10\,\text{s} = 80\,\text{m}$. These are the units of distance.

The area under a velocity–time graph = distance travelled.

Synoptic link

For worked examples of finding gradients and areas for curved graphs, see Topic 8.1, Graphs of motion.

You need to make sure that the gradient that you are calculating is meaningful. The gradient of a velocity–time graph is equal to acceleration, but resistance is equal to the ratio of the potential difference and the current, not the gradient of a graph of potential

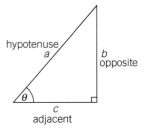

▲ **Figure 3** *You define sin, cos, and tan using the sides of a right-angled triangle*

Hint

You can use Pythagoras' theorem to calculate the length of one side from the other two. In Figure 10, $a^2 = b^2 + c^2$.

In general, $h^2 = x^2 + y^2$

where h = hypotenuse (the longest side of a right-angled triangle), x and y are the other two sides.

Synoptic link

You have used trigonometry when resolving vectors in Topic 8.2, Vectors.

Synoptic link

Small angle approximation is required for Young's double slit experiment in Topic 6.4, Interference and diffraction of light.

Synoptic link

You have seen graphs showing waves in and out of phase in Topic 6.1, Superposition of waves.

difference and current (except in the special case of Ohm's law). The resistance is *not* the change of potential difference with respect to current, but acceleration *is* the change of velocity with respect to time.

Trigonometric (trig) functions

You will use three common trigonometric functions during your physics course. For the triangle in Figure 3

$$\sin\theta = \frac{\text{opp}}{\text{hyp}} = \frac{b}{a}, \cos\theta = \frac{\text{adj}}{\text{hyp}} = \frac{c}{a}, \text{ and } \tan\theta = \frac{\text{opp}}{\text{adj}} = \frac{b}{c}$$

Rearranging these equations means that $b = a\sin\theta$ or $c\tan\theta$, and $c = a\cos\theta$. You will use these ideas when you resolve vectors into two perpendicular components.

Small angle approximation

There may be situations where the angle is very small. Think about what happens to the triangle when this happens.

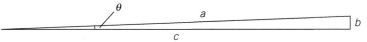

▲ **Figure 11** *When the angle is small the hypotenuse approximately equals the adjacent*

In this situation $a \approx c \Rightarrow \cos\theta = \frac{c}{a} \approx \frac{c}{c} = 1$. Also, $\sin\theta = \frac{b}{a} \approx \frac{b}{c} = \tan\theta$. The triangle in Figure 11 is almost identical to a tiny segment of a circle, where $a = c = $ radius, and $b = $ the length of the arc. The angle θ measured in radians $= \frac{\text{arc length}}{\text{radius}}$, which is also equal to $\frac{b}{a}$ and $\frac{b}{c}$.

So the small angle approximation says that for small angles $\cos\theta = 1$ *and* $\sin\theta = \tan\theta = \theta$ (measured in radians).

Graphs of $\sin\theta$ and $\cos\theta$

The graphs of $y = \sin\theta$ and $y = \cos\theta$ are shown in Figure 12.

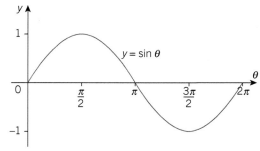

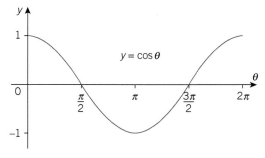

◀ **Figure 12** *Graphs of $y = \sin\theta$ and $y = \cos\theta$ are 90° or $\frac{\pi}{2}$ out of phase*

Logarithms

A **logarithm** is another name for a power (or exponent, or index). You can raise any number to a power, for example, 10^3, 2^3, e^3 ($e = 2.718...$). In all these examples the logarithm (or power) is 3.

The number which is raised to the power is called the **base**. In the example above the bases are 10, 2, and e, respectively. During your physics course you will meet all three of these bases. The logarithm of a number is the power to which you need to raise a base to get that number. For example:

- $10^6 = 1\,000\,000 \Rightarrow \log_{10}(1\,000\,000) = 6$, or $\log_{10}(10^6) = 6$
- $2^6 = 64$. The logarithm to the base 2 of the number 64 is 6. We write this as $\log_2(64) = 6$.
- $e^{-2} = 0.135 \Rightarrow \log_e(0.135) = -2$. This can be written $\ln(0.135) = -2$.

 Worked example: Finding logs

Find $\log_e(2)$ using your calculator. You will need to use the $\log_e$ or ln button. Check that you have the correct answer by finding $e^{(ans)}$.

Step 1: Find $\log_e(2)$ by pressing the ln, then the 2 button.

$$\ln 2 = 0.693 \text{ (3 s.f.)}$$

Step 2: Find $e^{0.693....}$

$$e^{0.693...} = 2$$

Logarithmic scales

When you use any scale (on a graph, on a number line) you can use a **linear scale** or a **logarithmic scale**.

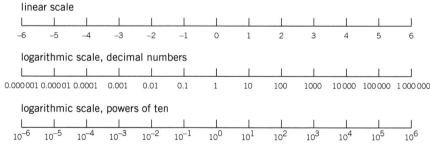
▲ Figure 13 *A logarithmic scale is a 'times' scale*

You move from left to right on a linear scale by *adding* a fixed number. You move from left to right on a logarithmic scale by *multiplying* by a fixed number (usually, but not always, equal to 10). You need to use logarithmic scales when the values vary over many orders of magnitude, and you will meet many logarithmic scales in your physics course: distance, time, resistivity.

▲ **Figure 1** *A fair coin has a 0.5 chance of landing on heads*

Synoptic link

Probability is fundamental to many areas of physics, including quantum mechanics (Chapter 7, Quantum behaviour) and radioactive decay (Chapter 10, Modelling decay).

Hint

Some letters of the Greek alphabet are used for more than one quantity in physics. Here λ is the probability of decay per second, but you have also seen it used to represent wavelength.

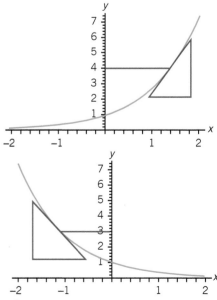

▲ **Figure 2** *In the first graph, if you plot $y = e^x$ then at $y = 4$ the gradient is 4. In the second, if you plot $y = e^{-x}$, then at $y = 3$ the gradient is −3*

Probability

Probability, the chance of an event happening, can be expressed in a variety of ways:

- a ratio (e.g. 1 in 10 or 1:9)
- a percentage (e.g. 10%)
- a fraction (e.g. $\frac{1}{10}$)
- a decimal (e.g. 0.1).

These probabilities all represent the same chance of an event occurring: once in 10 opportunities for it to happen. However, that does not mean that the event *will* happen that often. When you toss a coin the probability of getting a head is 0.5, but you could toss a coin twice and get two tails.

Probability and radioactive decay

You can think of each unstable nucleus as having a fixed probability of decay in one second, for example, $0.01\,\text{s}^{-1}$. The symbol for probability of decay per second, also called the **decay constant**, is lambda, λ. This fixed probability means that the decay of radioactive materials follows an **exponential** curve.

Exponential functions

You may hear the people say that something is growing 'exponentially' when they simply mean that it is increasing very fast. Mathematically, exponential change is not just 'very fast'. Exponential change with respect to time occurs when the rate of growth (or decay) of a quantity is proportional to the quantity at any moment (Figure 2). In an exponential process where x is changing with time:

$$\frac{\Delta x}{\Delta t} \propto \pm x \qquad \text{or} \qquad \frac{\Delta x}{\Delta t} = \pm \text{ constant} \times x$$

where:

- Δ (delta) means 'change in', so $\frac{\Delta x}{\Delta t}$ means change in x divided by change in time, or in other words, 'rate of change of x with time'
- $\propto$ means 'proportional to'
- the sign is positive when the equation shows exponential *growth*, and negative when the equation shows exponential *decay*.

🖩 Worked example: Is radioactive decay exponential?

Suppose you have N unstable nuclei. The probability of decay per second is lambda, λ, for each nucleus. Show that radioactive decay is exponential.

Step 1: Calculate the number of nuclei that will decay in one second.

In one second λN of the nuclei will decay.

Step 2: Calculate the number of nuclei that will decay in a time Δt.

In a time Δt then $\lambda N \Delta t$ will decay.

→

Step 3: Show that radioactive decay is an exponential process.

The change in the number of unstable nuclei, ΔN, will be equal to $\lambda N \Delta t$.

$\Delta N = -\lambda N \Delta t$ (the minus sign shows that the number is decreasing, not increasing)

So $\dfrac{\Delta N}{\Delta t} = -\lambda N$

So $\dfrac{\Delta N}{\Delta t} \propto N$, which is an exponential process.

Equations for exponential processes

You can use mathematical methods (calculus) to find a solution for the equation that links the rate of growth or decay of a quantity to that quantity.

- The equation for exponential *growth* is:
  $X = X_0 e^{at}$ where X is the quantity at time t, X_0 is the quantity at time $t = 0$ seconds, and a is a constant related to the fraction by which X changes in one second.

- The equation for exponential *decay* is:
  $X = X_0 e^{-at}$

You can use these equations to find the value of X at a particular time.

Also, you can take logs of each side of this equation:

- For exponential growth: $\ln X = \ln X_0 + at$
- For exponential decay: $\ln X = \ln X_0 - at$

You can use these equations to find how long it takes for X to reach a certain value.

 Worked example: Using exponential equations

Suppose X is the height of water in a bottle that has a small hole in the bottom, and the height decays exponentially with time. At time $t = 0$, $X = 20\,\text{cm}$, and the constant $a = 0.03\,\text{s}^{-1}$.

a Calculate the height of the water after 10 seconds.

Step 1: $X = X_0 e^{-at}$

Step 2: $X = 20\,\text{cm} \times e^{-(0.03\,\text{s}^{-1} \times 10\,\text{s})}$

$\qquad = 20\,\text{cm} \times e^{-0.3} = 20\,\text{cm} \times 0.74$

$\qquad = 15\,\text{cm}$

b Calculate how long it would take for the level to reach 5 cm.

Step 1: $\ln X = \ln X_0 - at$

$\qquad at = \ln X_0 - \ln X$

$\qquad t = \dfrac{(\ln X_0 - \ln X)}{a}$, but $\ln X_0 - \ln X = \ln \dfrac{X_0}{X}$

Step 2: $t = \dfrac{\ln \dfrac{20\,\text{cm}}{5\,\text{cm}}}{0.03\,\text{s}^{-1}} = \dfrac{1.39}{0.03} = 46\,\text{seconds}$

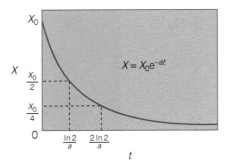

▲ **Figure 3** *Exponential decrease and half-life: X will halve in a time equal to* $\ln \frac{2}{a}$

Synoptic link

Physicists use half-lives to characterise capacitor discharge as well as radioactive decay – both processes are discussed in detail in Chapter 10, Modelling decay.

Hint

The symbol for half-life is $t_{1/2}$, but that does not mean that you need to divide by 0.5. The symbol is the letter t combined with the suffix $\frac{1}{2}$. Whatever the value of the quantity, it will halve in a time equal to the half-life. So after two half-lives you will have one quarter of the initial value, and not zero.

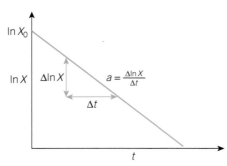

▲ **Figure 4** *A straight-line plot of ln X against t*

Half-life and time constant

In exponential decay **half-life**, $t_{1/2}$, is the time it takes for a quantity to halve, that is, for X to change from X_0 to $\frac{X_0}{2}$.

$$t_{1/2} = \frac{\ln \frac{X_0}{X}}{a} = \frac{\ln \frac{X_0}{X_0/2}}{a}$$

$$t_{1/2} = \ln \frac{2}{a}$$

After one half-life the quantity will halve. After two half-lives it will be equal to one quarter of its initial value, after three half-lives it will be equal to one eighth of its initial value, and so on. In general after n half-lives the fraction remaining will be $\frac{1}{2^n}$ of its initial value (Figure 3).

 Worked example: Using half-life

Use the idea of half-life to calculate the answer to the second part in the previous example.

Step 1: The level decreased from 20 cm to 5 cm. This is one quarter of the original, so it takes a time equal to 2 half-lives.

Step 2: $t_{1/2} = \ln \frac{2}{a}$

$$t_{1/2} = \frac{0.693}{0.03\,\text{s}^{-1}} = 23.1 \text{ seconds}$$

Step 3: One half-life is 23 seconds, so two half-lives are 46 seconds.

In exponential decay the **time constant**, τ, is the time it takes for the quantity to reach $\frac{1}{e}$ of its initial value. The value of e is 2.718, so $\frac{1}{e} = 0.37$, so τ is the time for the value to fall to 37% of its initial value.

$$X = X_0 e^{-at}$$

At time τ, $X = X_0 e^{-1}$

$$a\tau = 1, \text{ so the time constant } \tau = \frac{1}{a}$$

Testing for an exponential relationship

There are three methods that you can use to find out whether a quantity x varies exponentially with time.

1 Find out if x changes by a constant fraction (>1 for growth, or <1 for decay) in the same time interval. Divide the value of x at $t = 0$ by the value of x at $t = 1$, then the value at $t = 1$ by the value at $t = 2$, and so on. In an exponential relationship the ratios will be the same.

2 Plot a graph of the quantity against time and check that the time it takes for x to double (or halve) is constant (see Figure 3).

3 Plot a graph of $\ln X$ against t. Earlier you learnt that $\ln X = \ln X_0 - at$. This equation has the form $y = c - mx$, where $y = \ln X$, $c = \ln X_0$, $m = a$, and $x = t$. So a graph of $\ln X$ against t should be a straight line with a negative gradient equal in value to a. The y intercept will be equal to $\ln X_0$ (Figure 4).

Simple harmonic motion

A **simple harmonic oscillator** moves in simple harmonic motion if the acceleration a is proportional to its displacement x, and in a direction towards the equilibrium position (the position of zero displacement).

In this case:

$a \propto -x$ or $a = -\omega^2 x$, where ω = **angular frequency** (see below).

The solution of this equation is:

$x = A\sin(\omega t + \phi)$, where A is the amplitude of the oscillations and ϕ is the phase angle of the oscillations.

- If $x = 0$ when $t = 0$, then $\phi = 0$ and $x = A\sin(\omega t)$
- If $x = A$ when $t = 0$, then $\phi = \frac{\pi}{2}$ and $x = A\cos(\omega t)$, because

$\sin(\omega t + \frac{\pi}{2}) = \cos(\omega t)$ (Figure 5).

In most physical situations a graph of displacement against time will be $x = A\cos(\omega t)$, because the oscillator will need to be displaced in order to set it in motion.

Angular frequency

The solution to the equation for simple harmonic motion is a trigonometric function. You can link a sine (or cosine) curve with an object moving in a circle with an angular frequency ω.

Angular frequency $\omega = \frac{\theta}{t}$, that is, the angle (in radians) moved per second. An object moving in a circle at a steady speed will move through 2π radians in the periodic time, T. Because $T = \frac{1}{f}$,

$$\omega = \frac{2\pi}{T} = 2\pi f$$

So if the radius of the circle is equal to the amplitude of the oscillation, then the displacement–time graph for a simple harmonic oscillator with a frequency f will be identical to that of an object moving in a circle with an angular frequency $\omega = 2\pi f$.

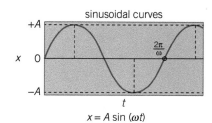

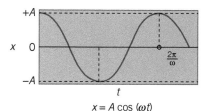

▲ **Figure 5** *Sinusoidal curves*

> ### Synoptic link
> You can review simple harmonic motion in Chapter 11, Modelling oscillations.

> ### Synoptic link
> Look at Figure 3 in Topic 11.1, Introducing simple harmonic oscillators, for an illustration linking circular motion and sine or cosine curves. You can read more about the mathematics of circular motion in Topic 12, Circular motion.

 Worked example: Calculations with angular frequency

A child on a swing completes 5 oscillations of amplitude 50 cm in 10 seconds. At $t = 0$ s, $x = A$.

a Calculate the periodic time and the frequency of the oscillation.

Step 1: Time for one oscillation $= \frac{10}{5} = 2$ s

Step 2: $f = \frac{1}{T} = \frac{1}{2\,\text{s}} = 0.5\,\text{Hz}$

b Calculate the angular frequency of an object moving in a circle that would produce the same displacement–time graph.

Step 1: $\omega = 2\pi f = 2\pi \times 0.5\,\text{Hz} = 3.14\,\text{s}^{-1}$

c Calculate the displacement of the child after 1.8 s.

Step 1: Write down the equation for the displacement of the child as a function of time.

$x = A\cos(\omega t)$

→

Step 2: Substitute the values in and calculate (remember to use radians).

$$x = 0.5\,\text{m} \times \cos(3.14 \times 1.8\,\text{s}) = 0.5\,\text{m} \times 0.81 = 0.4\,\text{m}$$

Synoptic link

Figure 1 of Topic 11.2, Modelling simple harmonic oscillation, shows the relationship between the changing displacement, velocity, and acceleration for simple harmonic oscillators.

Velocity and acceleration in simple harmonic motion

You can find velocity by finding the gradient of a displacement–time graph. Similarly, acceleration is equal to the gradient of the velocity–time graph. In simple harmonic motion, $x = A\cos(\omega t)$ and the shape of the velocity–time graph will be a negative sine curve, whilst the shape of the acceleration–time graph is a negative cosine curve, as illustrated in Figure 1 of Topic 11.2, Modelling simple harmonic oscillation. Table 1 summarises the mathematics.

▼ **Table 1** *Displacement, velocity, and acceleration for simple harmonic oscillators*

	Equation	Maximum value
displacement	$x = A\cos(\omega t)$	A
velocity	$v = -\omega A\sin(\omega t)$	ωA
acceleration	$a = -\omega^2 A\cos(\omega t)$	$\omega^2 A$

Energy in simple harmonic motion

The total energy is the sum of the potential and the kinetic energy:

$$\text{Total energy} = \frac{1}{2}kx^2 + \frac{1}{2}mv^2 = \frac{1}{2}kA^2$$

How energy changes with time

The potential energy of a simple harmonic oscillator depends on the displacement of the oscillator:

$$E = \frac{1}{2}kx^2$$

Since $x = A\cos(\omega t)$, then as a function of time:

$$E = \frac{1}{2}kx^2 = E = \frac{1}{2}k(A\cos(\omega t))^2$$
$$= \frac{1}{2}kA^2\cos^2(\omega t)$$

The kinetic energy of a simple harmonic oscillator depends on the velocity of the oscillator:

$$E = \frac{1}{2}mv^2$$

Since $v = -\omega A\sin(\omega t)$, then as a function of time:

$$E = \frac{1}{2}mv^2 = \frac{1}{2}m(\omega A\sin(\omega t))^2$$
$$= \frac{1}{2}m\omega^2 A^2\sin^2(\omega t)$$

These relationships are illustrated in Figure 3 in Topic 11.4, Resonance.

Synoptic link

See Figure 3 in Topic 11.4, Resonance, for the variation of kinetic and potential energy for an oscillator over time, and Figure 2 in Topic 11.4, Resonance, for the variation of kinetic and potential energy for an oscillator with position. Their sum, the total energy, remains constant.

How energy changes with position

The potential energy of a simple harmonic oscillator depends on the displacement of the oscillator:

$$E = \frac{1}{2}kx^2$$

The kinetic energy of a simple harmonic oscillator depends on the velocity of the oscillator:

$$E = \frac{1}{2}mv^2 = \frac{1}{2}kA^2 - \frac{1}{2}kx^2$$

The variation of kinetic and potential energy with position while their sum remains constant is illustrated in Figure 2 in Topic 11.4, Resonance.

Data, Formulae and Relationships

Data

Values are given to three significant figures, except where more – or fewer – are useful.

Physical constants

speed of light	c	$3.00 \times 10^8\,\mathrm{m\,s^{-1}}$
permittivity of free space	ε_0	$8.85 \times 10^{-12}\,\mathrm{C^2\,N^{-1}\,m^{-2}}\,(\mathrm{F\,m^{-1}})$
electric force constant	$k = \dfrac{1}{4\pi\varepsilon_0}$	$8.98 \times 10^9\,\mathrm{N\,m^2\,C^{-2}}$ $(\approx 9 \times 10^9\,\mathrm{N\,m^2\,C^{-2}})$
permeability of free space	μ_0	$4\pi \times 10^{-7}\,\mathrm{N\,A^{-2}}\ (\text{or } \mathrm{H\,m^{-1}})$
charge on electron	e	$-1.60 \times 10^{-19}\,\mathrm{C}$
mass of electron	m_e	$9.11 \times 10^{-31}\,\mathrm{kg} = 0.000\,55\,\mathrm{u}$
mass of proton	m_p	$1.673 \times 10^{-27}\,\mathrm{kg} = 1.007\,3\,\mathrm{u}$
mass of neutron	m_n	$1.675 \times 10^{-27}\,\mathrm{kg} = 1.008\,7\,\mathrm{u}$
mass of alpha particle	m_α	$6.646 \times 10^{-27}\,\mathrm{kg} = 4.001\,5\,\mathrm{u}$
Avogadro constant	L, N_A	$6.02 \times 10^{23}\,\mathrm{mol^{-1}}$
Planck constant	h	$6.63 \times 10^{-34}\,\mathrm{J\,s}$
Boltzmann constant	k	$1.38 \times 10^{-23}\,\mathrm{J\,K^{-1}}$
molar gas constant	R	$8.31\,\mathrm{J\,mol^{-1}\,K^{-1}}$
gravitational force constant	G	$6.67 \times 10^{-11}\,\mathrm{N\,m^2\,kg^{-2}}$

Other data

standard temperature and pressure (stp)		$273\,\mathrm{K}\ (0\,^\circ\mathrm{C}),\ 1.01 \times 10^5\,\mathrm{Pa}$ (1 atmosphere)
molar volume of a gas at stp	V_m	$2.24 \times 10^{-2}\,\mathrm{m^3}$
gravitational field strength at the Earth's surface in the UK	g	$9.81\,\mathrm{N\,kg^{-1}}$

Conversion factors

unified atomic mass unit	1u	$= 1.661 \times 10^{-27}\,\mathrm{kg}$
	1 day	$= 8.64 \times 10^4\,\mathrm{s}$
	1 year	$\approx 3.16 \times 10^7\,\mathrm{s}$
	1 light year	$\approx 10^{16}\,\mathrm{m}$

Mathematical constants and equations

$e = 2.72$ $\pi = 3.14$ 1 radian $= 57.3°$

$\text{arc} = r\theta$

$\sin\theta \approx \tan\theta \approx \theta$
and $\cos\theta \approx 1$ for small θ

$\ln(x^n) = n\ln x$

$\ln(e^{kx}) = kx$

circumference of
circle $= 2\pi r$

area of circle $= \pi r^2$

surface area of
cylinder $= 2\pi rh$

volume of
cylinder $= \pi r^2 h$

surface area of
sphere $= 4\pi r^2$

volume of sphere $= \frac{4}{3}\pi r^3$

Prefixes

10^{-12}	10^{-9}	10^{-6}	10^{-3}	10^3	10^6	10^9
p	n	μ	m	k	M	G

Formulae and relationships

Imaging and signalling

focal length $\qquad \frac{1}{v} = \frac{1}{u} + \frac{1}{f}$

linear magnification $\qquad m = \frac{v}{u}$

refractive index $\qquad n = \frac{\sin i}{\sin r} = \frac{C_{1st\ medium}}{C_{2nd\ medium}}$

noise limitation on maximum $\quad b = \log_2\left(\frac{V_{total}}{V_{noise}}\right)$
bits per sample

alternatives, N, provided by $\quad N = 2^b,\ b = \log_2 N$
n bits

Electricity

current $\qquad I = \frac{\Delta Q}{\Delta t}$

potential difference $\qquad V = \frac{W}{Q}$

power and energy $\qquad P = IV = I^2 R,\ W = VIt$

e.m.f and potential difference $\quad V = \varepsilon - Ir$

conductors in series and parallel $\quad \frac{1}{G} = \frac{1}{G_1} + \frac{1}{G_2} + \ldots$
$\qquad\qquad\qquad\qquad\qquad G = G_1 + G_2 + \ldots$

resistors in series and parallel $\quad R = R_1 + R_2 + \ldots$
$\qquad\qquad\qquad\qquad\qquad \frac{1}{R} = \frac{1}{R_1} + \frac{1}{R_2} + \ldots$

potential divider $\qquad V_{out} = \frac{R_2}{R_1 + R_2} V_{in}$

conductivity and resistivity $\qquad G = \frac{\sigma A}{L}$
$\qquad\qquad\qquad\qquad\qquad R = \frac{\rho L}{A}$

capacitance $\qquad C = \frac{Q}{V}$

energy stored in a capacitor $\qquad E = \frac{1}{2}QV = \frac{1}{2}CV^2$

discharge of capacitor $\qquad \frac{dQ}{dt} = -\frac{Q}{RC}$
$\qquad\qquad\qquad\qquad\qquad Q = Q_0\,e^{-t/RC}$
$\qquad\qquad\qquad\qquad\qquad \tau = RC$

Materials

Hooke's law $\qquad F = kx$

elastic strain energy $\qquad \frac{1}{2}kx^2$

Young modulus $\qquad E = \frac{stress}{strain},$
$\qquad\qquad stress = \frac{tension}{cross-sectional\ area},$
$\qquad\qquad strain = \frac{extension}{original\ length}$

Gases

kinetic theory of gases $\qquad pV = \frac{1}{3}Nm\overline{c^2}$

ideal gas equation $\qquad pV = nRT = NkT$

Motion and forces

momentum $\qquad p = mv$

impulse $\qquad F\Delta t$

force $\qquad F = \frac{\Delta(mv)}{\Delta t}$

work done $\qquad W = Fx \qquad \Delta E = F\Delta s$

power $\qquad P = Fv, \qquad P = \frac{\Delta E}{t}$

components of a vector in two
perpendicular directions

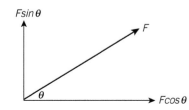

equations for uniformly accelerated motion

$$s = ut + \frac{1}{2}at^2$$
$$v = u + at$$
$$v^2 = u^2 + 2as$$

for circular motion

$$a = \frac{v^2}{r}, F = \frac{mv^2}{r} = mr\omega^2$$

Energy and thermal effects

energy

$$\Delta E = mc\Delta\theta$$

average energy approximation

average energy $\sim kT$

Boltzmann factor

$$e^{-\frac{E}{kT}}$$

Waves

wave formula

$$v = f\lambda$$

frequency and period

$$f = \frac{1}{T}$$

diffraction grating

$$n\lambda = d\sin\theta$$

Oscillations

simple harmonic motion

$$\frac{d^2x}{dt^2} = a = -\left(\frac{k}{m}\right)x = -\omega^2 x$$
$$x = A\cos(\omega t)$$
$$x = A\sin(\omega t)$$
$$\omega = 2\pi f$$

Periodic time

$$T = 2\pi\sqrt{\frac{m}{k}}$$
$$T = 2\pi\sqrt{\frac{L}{g}}$$

total energy

$$E = \frac{1}{2}kA^2 = \frac{1}{2}mv^2 + \frac{1}{2}kx^2$$

Atomic and nuclear physics

radioactive decay

$$\frac{\Delta N}{\Delta t} = -\lambda N$$
$$N = N_0 e^{-\lambda t}$$

half life

$$T_{\frac{1}{2}} = \frac{\ln 2}{\lambda}$$

radioactive dose and risk

absorbed dose = energy deposited per unit mass
effective dose = absorbed dose × quality factor

mass–energy relationship

$$E_{\text{rest}} = mc^2$$

relativistic factor

$$\gamma = \sqrt{\frac{1}{1 - v^2/c^2}}$$

relativistic energy

$$E_{\text{total}} = \gamma E_{\text{rest}}$$

energy–frequency relationship for photons

$$E = hf$$

de Broglie

$$\lambda = \frac{h}{p}$$

Field and potential

for all fields

fields strength $= -\dfrac{dV}{dr} \approx -\dfrac{\Delta V}{\Delta r}$

gravitational fields

$$g = \frac{F}{m}, E_{grav} = -\frac{GmM}{r}$$
$$V_{grav} = -\frac{GM}{r}, F = -\frac{GmM}{r^2}$$

electric fields

$$E = \frac{F}{q} = \frac{V}{d},$$

electrical potential energy $= \dfrac{kQq}{r}$

$$V_{electric} = \frac{kQ}{r}, F = \frac{kQq}{r^2}$$

Electromagnetism

magnetic flux

$$\Phi = BA$$

force on a current carrying conductor

$$F = ILB$$

force on a moving charge

$$F = qvB$$

Induced e.m.f

$$\varepsilon = -\frac{d(N\Phi)}{dt}$$

Glossary

absolute luminosity The brightness that a star would have at a standard distance of 10 parsec.

absolute zero The zero of thermodynamic temperature, $0\,\text{K} = -273.15°\text{C}$.

absorbed dose The number of joules absorbed from ionising radiation per kilogram of tissue, measured in grays (Gy) where $1\,\text{Gy} = 1\,\text{J}\,\text{kg}^{-1}$.

acceleration A vector quantity – the rate of change of velocity, $a = \dfrac{\Delta v}{\Delta t}$.

activation energy A certain amount of energy that is needed to make a given process occur – symbol E.

activity The number of nuclei in a radioactive source decaying per second – unit becquerel, Bq.

alloy An alloy is a material composed of two or more metals, or a mixture of metals and other materials.

alternating current (a.c.) A current, alternating sinusoidally, produced by an alternating p.d. in a circuit.

amorphous At the microscopic scale, an amorphous material has no long range order. The microscopic structure of glass is amorphous – the atoms in glass form strong bonds with one another to make up a rigid structure without any regularity.

amount of substance The number of moles of molecules, atoms, or ions in a certain sample, $n = \dfrac{N}{N_A}$.

ampere (A) The S.I. unit of electrical current. $1\,\text{A} = 1\,\text{C}\,\text{s}^{-1}$

amplitude The greatest displacement of an oscillator from its equilibrium position.

amplitude (wave) The amplitude of a wave at a point is the maximum displacement from some equilibrium value at that point.

angular frequency The angle (in radians) moved per second, given by $\omega = \dfrac{\theta}{t}$.

antinode An antinode is a position of maximum amplitude of oscillation on a standing wave.

anti-particle An anti-particle and its particle counterpart have equal but opposite electric charge (or are both uncharged) and equal rest energy – they also have equal and opposite 'spin'.

antiphase Two oscillations are in antiphase when their phase difference is π radians, or $180°$.

armature The rotating part of an electric motor, which carries the coil.

atomic mass unit One-twelfth of the mass of a carbon–12 atom – unit u, where $1\,\text{u} = 1.661 \times 10^{-27}\,\text{kg}$.

Avogadro constant The number of particles in a mole, 6.02×10^{23} particles mol^{-1}, – symbol N_A.

back e.m.f. An e.m.f. generated by a spinning motor that opposes the p.d. driving the motor and limits the speed of the motor.

background count The activity of background radiation, that is, the radiation detectable from the environment (rocks, air, etc.) that is not related to the experiment.

B-field *see* magnetic flux density.

Big Bang The probable origin of the Universe in a very hot, very dense state from which it has expanded and cooled.

binding energy The binding energy of a nucleus is the amount by which the rest energy of a nucleus is less than the rest energy of its constituent neutrons and protons.

bit A bit is the smallest unit of digital information, represented as a 0 or a 1 corresponding to low voltage or high voltage in a digital circuit.

Boltzmann constant The constant of proportionality k in the ideal gas equation, $pV = NkT$, where k relates temperature, in K, to the energy of a particle.

Boltzmann factor The fraction of particles with extra energy E, compared with the number not having that energy – the exponential term $e^{-\frac{E}{kT}}$.

Boyle's law Pressure p is inversely proportional to volume V, that is, $pV = \text{constant}$.

brittle A brittle material breaks by snapping cleanly. It undergoes little or no plastic deformation before fracture.

byte A byte is a sequence of eight bits coded to represent one of 256 alternatives.

capacitance For a capacitor, the charge separated per volt $C = Q/V$, symbol C, unit farad, F.

capacitor An electrical component used to keep charge separated and usually consisting of a pair of electrical conductors separated by a thin layer of insulator.

chain reaction A reaction in which the products of one reaction go on to start one or more further reactions.

charge A conserved property of some elementary particles (e.g. electron, proton, quarks) which causes them to exert forces upon each other.

Charles' law The volume of an ideal gas increases linearly with temperature at constant pressure – for every °C in temperature, the increase in volume is $3.66 \times 10^{-3} \times$ the volume of the sample at 0°C.

conductance, G Conductance G is the ratio $\dfrac{I}{V}$ for a circuit component, and is measured in siemens (S) for current in A and p.d. in V.

coherence Two sources of waves are coherent if they emit waves with a constant phase difference and have the same frequency. Two waves arriving at a point are said to be coherent if there is a constant phase difference between them as they pass that point.

commutator A switching mechanism that reverses the connections to the coil of an electric motor to maintain its direction of rotation.

component of a vector One of two vectors (in two perpendicular dimensions) into which a vector may be split, e.g. horizontally and vertically.

compression Compressive forces are squashing forces. An object is in compression when two forces act on it in opposite directions to make the object compress (squash) along the line of action of the forces.

conductor (electrical) A material which conducts electricity well on account of having many free charge carriers.

conservation of energy During any interaction, the total energy before the interaction is the same as the total energy after the interaction.

conservation of momentum During any interaction, the total momentum before the interaction is the same as the total momentum after the interaction (see Newton's second law).

cosmic microwave background radiation microwave radiation that permeates the entire Universe, being red-shifted radiation from an early hot stage of the Universe.

coulomb (C) The S.I. unit of electrical charge. 1 C is the charge flowing through a point in 1 s where there is a current of 1 A.

Coulomb's law The force between two charged spheres with charges q and Q is directly proportional to each of the charges, and inversely proportional to the square of the distance r between the centres of the two spheres; $F_{electric} = k\dfrac{qQ}{r^2}$.

covalent bond A strong bond between atoms sharing electrons, e.g. between the atoms in the water (H_2O) molecule.

critical damping Reduction of simple harmonic motion in which the oscillator stops at the equilibrium position without completing a cycle.

current, I The rate at which charge flows through a point in an electrical circuit

current-turns The product of current I and the number of turns N in a solenoid $I \times N$.

damping The action of forces such as friction and air resistance to reduce the amplitude of an oscillator.

decay constant The probability that a nucleus will decay during unit time — symbol λ, for a unit time of 1 s, unit s^{-1}.

density Density is mass per unit volume. The units of density are $kg\,m^{-3}$. Density is often represented by the Greek letter rho, ρ.

differential equation An equation describing how physical quantities change, often with time or position.

diffraction Diffraction is the spreading of waves after passing through a gap or past the edge of an obstacle. The spreading increases if the gap is made narrower or if the wavelength of the waves is increased.

diffusion The slow intermingling of gases or liquids, due purely to the random motion of molecules.

discharge For a charged capacitor connected across its plates, the flow of charge from the negative plate to the positive plate until the potentials are equalised.

dislocation A dislocation is a defect in the regular structure of a crystal or crystalline region of a material. Dislocations in metals are mobile and make metals ductile.

displacement A vector quantity – the distance travelled in a specified direction.

dissipation A thermal energy transfer resulting in an increase in internal energy, often of the surroundings.

distance A scalar quantity – the separation between two points in space, possibly along a curved path, with no reference to direction.

Doppler shift A change in observed wavelength due to the relative motion of source and observer.

dose equivalent The dose equivalent = absorbed dose × quality factor of absorbed radiation. It is measured in sievert (Sv).

drift velocity The mean velocity of charge carriers in a conductor carrying an electrical current.

ductile A ductile material can be easily drawn into a wire (e.g. copper is easier to draw into a wire than tungsten). Metals are ductile because the non-directional metallic bonds allow ions to slide past one another.

dynamo An older term for 'generator' now sometimes reserved for a d.c. generator.

e.m.f., ε The energy per unit charge (in V) given by any source of electrical supply to the charges set in motion.

eddy current A current induced in the core of a transformer.

effective dose The effective dose is the absorbed dose of ionising radiation multiplied by various quality factors that account for the type of radiation absorbed and the tissues that have absorbed it – effective dose = absorbed dose × quality factor, unit sieverts, Sv.

Einstein's first postulate Physical behaviour cannot depend on any 'absolute velocity' — physical laws must take the same form for all observers, no matter what their state of uniform motion in a straight line.

Einstein's second postulate The speed of light c is a universal constant — it has the same value, regardless of the motion of the platform from which it is observed — in effect, the translation between distance and time units is the same for everybody.

elastic deformation When a material deforms elastically it regains its original shape after deformation.

elastic limit The elastic limit is the maximum stress at which an object returns to its original shape after the deforming stress is removed.

electric field A region surrounding an electrical charge where another electrical charge would experience a force.

electric field strength The magnitude and direction of the force on a charge of one coulomb at a given point in an electric field – symbol E, unit NC^{-1} or Vm^{-1}.

electric force constant The constant of proportionality k in Coulomb's law.

electric potential The electrical energy per coulomb of a charge at a position, being the work done to move $+1C$ from infinity to the place in question – unit JC^{-1} or V.

electrical conductivity, σ A constant for electrical conductors given by the equation $G = \dfrac{\sigma A}{L}$.

electrical permittivity of free space A fundamental electrical property of free space, ε_0, related to the electric force constant k by $k = \dfrac{1}{4\pi\varepsilon_0}$.

electrical resistivity, ρ A constant for electrical conductors given by the equation $R = \dfrac{\rho L}{A}$.

electron A fundamental particle, a lepton of charge $-1.6 \times 10^{-19}C$.

electronvolt A practical measure of energy used in atomic and nuclear physics, given by the kinetic energy acquired by an electron falling through a p.d. of 1 V, where 1 electronvolt (eV) = $1.6 \times 10^{-19}J$.

energy A measure of the capacity of a body or a system for doing work.

energy level Confined quantum objects, such as electrons in atoms, exist in discrete quantum states, each with a definite energy — the term 'energy level' refers to the energy of one or more such quantum states (different states can have the same energy).

energy transferred thermally Energy transferred associated with a temperature change, e.g. frictional work can raise the temperature of a moving object.

equilibrium position The position of an oscillator on which no net force is acting.

equipotential surface A continuous surface joining points of the same potential in a three dimensional field — for example, for gravitational or electrical potential.

exponential A process in which the rate of change of a quantity is proportional to the value of the quantity.

Faraday's law of electromagnetic induction The induced e.m.f. $\varepsilon = -$(rate of change of flux linkage) $= -\dfrac{d(\Phi N)}{dt}$.

first law of thermodynamics The change in internal energy ΔU of a system is the sum of the work W done on the system and the energy transferred thermally Q into it, $\Delta U = W + Q$.

flux linkage Flux linkage = number of coils × flux through one coil.

focal length The focal length f of a thin lens is the distance from the centre of the lens to the focal point F.

focal point (focus) The focal point F of a converging lens is the point where light from a very distant object on the axis of the lens is brought to a focus by the lens. This point is also called the focus.

force, F The 'push' or 'pull' acting on an object associated with a change in its momentum: $F = \dfrac{\Delta p}{\Delta t}$ (see Newton's second law).

force constant The ratio force/extension for an elastic specimen.

forced oscillation An oscillation driven by the action of a periodic driving force.

fracture An object fractures when it breaks into two or more pieces when placed under stress.

fracture stress Fracture stress is the stress at which fracture occurs. The fracture stress of a material in tension is sometimes called its tensile strength.

free oscillation An oscillation due to the action of a restoring force without any damping or driving forces.

frequency (for a simple harmonic oscillator) The number of complete oscillations in unit time.

gas constant The constant of proportionality R in the ideal gas equation for one mole of gas, $pV = RT$; $R = N_A k$.

generator An electromagnetic machine that uses rotation, often produced by a turbine, to produce the flux changes needed to induce an e.m.f.

gluon The exchange particle that mediates the strong nuclear force.

grain boundary A grain boundary is the line along which grains meet in a crystalline material.

gravitational field A region surrounding a mass where another mass would experience a force.

gravitational field strength The magnitude and direction of the force on a mass of one kilogram at a given point in a gravitational field – symbol g, unit $N\,kg^{-1}$.

gravitational potential The gravitational energy per kilogram of material at a position, being the work done to move 1 kg from infinity to the place in question – unit $J\,kg^{-1}$.

gravitational potential energy, E_{grav} Potential energy of a mass due to its position in a gravitational field – the gravitational potential energy difference between two points in a uniform field $\Delta E_{grav} = mgh$, where h is the vertical separation between the two points.

hadron A particle composed of two or three quarks – Protons and neutrons are hadrons.

half-life The time required for the number of nuclei in a radioactive sample to fall to half the original value, or the time it takes for the activity of a sample to fall to half the original value – symbol $T_{1/2}$.

half-thickness The thickness of absorber needed to halve the number of photons that can pass through it on average.

hard A material is hard if it is difficult to dent its surface. Many ceramics are very hard.

heavy damping Reduction of simple harmonic motion in which the oscillator returns to the equilibrium position much more slowly than a lightly damped or critically damped oscillator.

Hertzsprung–Russell diagram A logarithmic plot of the luminosity of stars against their surface temperature.

Hubble's law In a plot of the speed of recession of galaxies against their distance, distant galaxies seem to be moving away from us, and their speed is proportional to their distance.

hydrogen bond A weak bond between molecules bearing separated charges, e.g. the negative part of one water molecule and the positive part of a neighbouring one.

ideal gas A gas in which the molecules do not interact at all, and are so tiny that they occupy negligible volume.

impulse The product $F\Delta t$ summed over the whole action of a force on an object resulting in a momentum change Δp = impulse.

in parallel Components joined alongside which share current and have the same p.d.

in series Components joined end to end which share the p.d. and have the same current.

induction The creation of an e.m.f. across a circuit when the magnetic flux in the circuit is changed.

inertia The tendency of any stationary object to remain stationary or of any moving object to continue with the same momentum (see Newton's First Law).

input voltage, V_{in} The p.d. applied to a sensor circuit (frequently a potential divider).

insulator (electrical) A material which conducts electricity poorly on account of having very few charges which are free to move.

intensity The intensity of a wave is the energy per unit time carried by the wave and incident normally per unit area of surface.

interference Interference arises from the superposition of waves on top of one another. When waves overlap, the resultant displacement will be equal to the sum of the individual displacements at that point and at that instant (if the waves superpose linearly). Interference is produced if waves from two coherent sources overlap or if waves from a single source are divided and then reunited.

internal energy The energy U within a system allowing it to do work or transfer energy thermally.

internal resistance, r Resistance within a source of e.m.f. resulting in a drop of terminal p.d. when a current is drawn from the source.

ionic bond A strong bond between ions due to electrostatic attraction, e.g. between Na^+ and Cl^- in sodium chloride.

ionisation Removal of an electron from, or addition of an electron to, an uncharged atom or molecule resulting in formation of an ion.

ionisation energy The energy that must be supplied to a neutral atom or molecule to ionise it.

isotope Isotopes are forms of the same chemical element that have the same proton number but different neutron numbers.

iteration A repeat of a mathematical operation, often used in modelling – for example, iterative calculations are used to model exponential decay.

iterative model A mathematical treatment, often using a computer, whereby small step-wise changes to variables such as displacement and velocity are made at regular time intervals Δt with the assumption that those variables change only at the end of each time interval.

kinematic equations Four equations for motion in a straight line under uniform acceleration (also called suvat equations).

kinetic energy, E_k The energy possessed by an object by virtue of its motion: $E_k = \frac{1}{2}mv^2$.

Kirchhoff's first law At any electrical junction, the total current into the junction = the total current out of the junction.

Kirchhoff's second law Around any electrical circuit, the sum of all e.m.f.s = the sum of all p.d.s.

Lenz's law The direction of an induced e.m.f. is always such as to act against the change that causes the induced e.m.f.

lepton Leptons are a group of fundamental particles that include the electron and the neutrino.

lepton number In any decay, the lepton number (number of leptons) is conserved.

lidar Light detection and ranging (*see* radar).

light clock A pair of mirrors between which pulses of light bounce back and forth.

light damping Reduction of simple harmonic motion in which the maximum displacement of the oscillator is reduced in each oscillation, but the time period of the oscillation is roughly constant.

light-year The distance light travels through a vacuum in one year, approximately 10^{16} m.

lines of flux Lines forming continuous loops from magnetic N-poles to magnetic S-poles – the lines indicate the direction of the vector B (B-field or flux density).

magnetic circuit Analogous to an electrical circuit, an arrangement whereby flux between magnetic poles is carried by a suitable magnetic material between poles.

magnetic field A region surrounding a permanent magnet or electromagnet in which a moving charge would experience a force.

magnetic flux The product of an area A and the component of B-field (flux density) perpendicular to the area A – symbol ϕ, unit weber (Wb), $\phi = BA$.

magnetic flux density A measure of magnetic field strength – symbol B, unit tesla (T) or $Wb\,m^{-2}$.

magnification Linear magnification $= \dfrac{\text{image distance}}{\text{object distance}} = \dfrac{v}{u}$.

malleable A material is malleable if it is easy to hammer or press a sheet of material into a required shape.

mass, m The amount of matter in any object.

mass defect The difference between the mass of a nucleus and the mass of the same number of free protons and neutrons – the mass difference is equivalent to the binding energy of the nucleus.

mass number *see* nucleon number

metallic bond A strong bond beween the ion cores in a solid metal due to a delocalised electron 'gas' circulating in the lattice.

moderator A material (often water) used to slow down neutrons in a fission reactor, as slower neutrons are more likely to be captured by nuclei to prompt further fission.

molar mass The mass of 1 mol of any substance.

mole The unit for amount of substance – 1 mol is the amount of any substance that contains the same number of particles as there are carbon atoms in 12.0 g of carbon–12.

momentum, p A vector quantity: the product mass × velocity.

natural frequency The frequency of a free oscillator.

neutrino An uncharged lepton of insignificant mass – anti-neutrinos are released in beta-minus decay.

neutron An uncharged particle composed of two down quarks and one up quark.

Newton's first law A stationary object will remain stationary, and a moving object will continue moving with the same momentum, unless an external force acts upon it.

Newton's second law When an external force acts on an object, the force is proportional to the rate of change of momentum.

Newton's third law When an object A exerts a force F upon an object B, then the object B exerts a force $(-F)$ on object A.

node A node is a position of minimum amplitude of oscillation on a standing wave.

normal line The normal line is line drawn at ninety degrees to the surface of a transparent material. When light is represented as a ray the angle of incidence at the surface is the angle the ray makes to the normal line.

nuclear fission The splitting of a heavy (high nucleon number) nucleus to form lighter nuclei.

nuclear fusion The fusion of two light (low nucleon number) nuclei to form a heavier nucleus.

nucleon A nucleon is a neutron or a proton.

nucleon number The number of protons and neutrons in each nucleus of an isotope, also called the mass number A.

nucleus The central, relatively massive core of an atom, composed of protons and neutrons.

number density, n The number of mobile charge carriers per unit volume of the conductor, measured in m^{-3}.

ohm (Ω) The S.I. unit of electrical resistance, R:
$$1\,\Omega = \frac{1\,\text{V}}{1\,\text{A}}.$$

output voltage, V_{out} The p.d. measured in a sensor circuit (frequently a potential divider).

parallax The apparent shift in position of near objects against the background of further objects when the observer moves.

parsec A distance at which a star would have a parallax angle from Earth of 1″, where 1 parsec (pc) = 3.09×10^{16} m = 3.26 light-years.

particle accelerator A device using electical fields to accelerate charged particles.

path difference When waves travel from one point to another by two or more routes, the difference in the distance travelled by each wave is the path difference. The importance of a path difference is that it introduces a time delay, so that the phases of the waves differ when they meet. It is the difference in phase that generates interference effects.

Pauli exclusion principle The rule that no two identical fermions can share the same quantum state.

pendulum A mass (bob) on the end of a string that can oscillate back and forth in simple harmonic motion.

permeance For a magnetic circuit, permeance
$$= \frac{\text{flux}}{\text{current-turns}}.$$

phase and phase difference 'Phase' refers to stages in a repeating change, as in 'phases of the Moon'. The phase difference between two objects vibrating at the same frequency is the fraction of a cycle that passes between one object being at maximum displacement in a certain direction and the other object being at maximum displacement in the same direction.

phase angle Phase difference is expressed as a fraction of one cycle, or of 2π radians, or of 360°. This is known as the phase angle. For example, two waves which are half a cycle apart have a phase angle of π radians or 180°.

phasor Phasors are used to represent amplitude and phase in a wave. A phasor is a rotating arrow used to represent a sinusoidally changing quantity.

photons Electromagnetic waves of frequency f are emitted and absorbed in quanta of energy $E = hf$, called photons.

pixel A pixel is a single 'picture element'. In a digital camera, a lens is used to form a real image on a chip with an array of the order of a million very small light-sensitive detectors. Each detector corresponds to one pixel in the final image.

plane wavefronts A plane wavefront is one with zero curvature. Waves from distant sources have plane wavefronts.

plastic deformation When a material deforms plastically it undergoes permanent stretching or distortion before breaking.

polarisation Transverse waves are linearly polarised if they vibrate in one plane only. Unpolarised transverse waves vibrate in a randomly changing plane. Longitudinal waves cannot be polarised.

polycrystalline Polycrystalline materials are composed of tiny crystal grains. Within each grain the material shows an ordered structure but the orientation of each individual grain is random.

positron The anti-particle counterpart of the electron – the charge of a positron is $+1.6 \times 10^{-19}$ C.

potential difference (p.d.), V The energy transfer per unit charge moving between the two points in question.

potential gradient In a radial gravitational field, the change of potential with radius, $\dfrac{dV}{dr}$.

power (of a lens) The power of a lens in dioptres (D) $= \dfrac{1}{f}$, where f is the focal length in metres. The shorter the focal length the more powerful the lens.

pressure law The pressure of an ideal gas increases linearly with temperature at constant volume for every °C in temperature – the increase in pressure is $3.66 \times 10^{-3} \times$ the pressure of the sample at 0°C.

probability A measure of the chance of one of a number of possible things happening.

probability Probability has to do with uncertainty, randomness and quantum effects. Probability is a measure of the chance of one of a number of possible things happening.

projectile An object thrown or fired so that its subsequent motion is affected by a gravitational field.

proton A positively charged particle composed of two up quarks and one down quark.

proton number The number of protons in the nucleus of an atom, Z.

quark A fundamental particle – the building block of protons, neutrons, and other particles.

radar Radio detection and ranging — finding the distance to a remote object by timing an electromagnetic pulse travelling there and back.

radioactive decay The release of particles by the nuclei of atoms of unstable elements, which are thus changed into other elements.

radioactivity The phenomenon whereby unstable nuclei change, emitting ionising radiation.

random walk The random movement of a single molecule during diffusion due to repeated collision with other molecules.

randomness A situation in which an event occurs entirely by chance – a random process is a change or an event with one or more outcomes, including no change, that can occur with a certain probability in a given time interval.

red-shift The shift of spectral lines towards the red (long-wavelength) end of the spectrum, for example, in the light from receding stars.

refractive index The refractive index of a transparent material is the ratio of the speed of light in a vacuum to the speed of light in the material. The refractive index is also given (in Snell's Law) as the ratio of the sine of the angle of incidence to the sine of the angle of refraction as the light enters the transparent material.

$$\text{refractive index} = \frac{\sin \text{ (angle of incidence)}}{\sin \text{ (angle of refraction)}}$$

relative velocity The velocity a moving object appears to have when viewed from another moving object.

resistance, R Resistance R is the ratio p.d. V/current I for a circuit component, and is measured in ohms (Ω) for p.d. in V and current in A.

resolution Splitting up a vector quantity into two components.

resolution (of a digital image) The resolution of an image is the scale of the smallest detail that can be distinguished.

resonance The effect produced when a driving frequency matches the natural frequency of a system, resulting in large-amplitude oscillations.

restoring force A force acting in the direction of the equilibrium position, in simple harmonic motion.

risk When considering the biological effects of radiation, the risk of developing cancer due to the radiation is estimated at between 3% and 5% per sievert.

root mean square speed The square root of the mean square speed, used when quoting molecular speeds.

rotor The rotating part of an electric motor or a generator.

scattering For particles, the deflection of accelerated particles due to interaction with nuclei or with other particles.

second law of thermodynamics Total entropy never decreases.

self-inductance The induction of an e.m.f. ε in the coil of the primary circuit of a transformer by the changing flux induced by the p.d. in that coil.

scalar A quantity (e.g. speed, mass, energy) with magnitude but no direction.

semiconductor A material midway in electrical conductivity and resistivity between conductors and insulators.

sensor circuit A circuit whose electrical properties depend upon environmental variables and which can be used to monitor or measure those variables.

siemen (S) The S.I. unit of electrical conductance, G: $1\,S = \frac{1\,A}{1\,V}$.

simple harmonic oscillator An oscillator in simple harmonic motion, in which the acceleration of the object at any instant is proportional to the displacement of the object from equilibrium at that instant, always directed towards the centre of oscillation.

solenoid A long, current-carrying coil, which acts as an electromagnet.

space–time diagram A representation of the paths of objects moving through space and time, with time shown on the y-axis and distance (in units of x/c) on the x-axis.

special relativity Einstein's theory of motion based on his two postulates.

specific thermal capacity The increase in internal energy required to raise the temperature of 1 kg of a given material by 1 K, $c = \frac{\Delta U}{m\Delta T}$ – symbol c, unit $J\,kg^{-1}\,K^{-1}$.

speed A scalar quantity – the distance per unit time travelled with no account taken of the direction of movement.

standard candle A star of known absolute luminosity that can be used as a comparative measure for other stars.

statistical mechanics The study of how the behaviour of matter and energy is governed by statistics dealing with random distributions.

stator The stationary part of an electric motor or a generator.

stiff A stiff material has a small extension per unit force. The stiffness is indicated by the Young modulus.

strain Strain is the change of length per unit length. Strain is a ratio of two lengths and therefore has no unit. Strain is often represented by the Greek letter epsilon, ε.

stress Stress (tensile and compressive) is the force per unit area acting at right angles to a surface. The units of stress are Pa or $N\,m^{-2}$. Stress is often represented by the Greek letter sigma, σ.

strong interaction A fundamental force that holds nuclei together.

strong nuclear force *see* strong interaction.

superposition When two or more waves meet, their displacements can superpose, that is, be added together at every point. The principle of superposition states that when two or more waves overlap, the resultant displacement at a given instance and position is equal to the sum of the individual displacements at that instance and position.

tension Tensile forces are stretching forces. An object is in tension when two forces act on it in opposite directions to make the object stretch along the line of action of the forces.

tesla A unit for flux density, the force per unit current per unit length of conductor perpendicular to the lines of magnetic flux – units T, where $1\,T = 1\,N\,A^{-1}\,m^{-1} = 1\,Wb\,m^{-2}$.

thermodynamics A study of the relationships between work, internal energy and energy transferred thermally.

threshold frequency The threshold frequency is the minimum frequency of light that will eject photoelectrons from a given surface. The threshold frequency varies with material.

time constant For the discharge circuit of a capacitor, the product RC of the resistance in the circuit and the capacitance – the time taken for the p.d. and charge C on a capacitor to fall to $1/e$ of their original values when discharging through resistance R – symbol τ, unit s.

time period The time for one complete oscillation of a pendulum or other vibrating system

tough A material is tough if it does not break by snapping cleanly, and if it absorbs a lot of energy before fracture when deforming plastically. A tough material is resistant to the propagation of cracks. Toughness is the opposite of brittleness.

transformer An electromagnetic machine that transforms voltages by inducing e.m.f. without physical movement – alternating current produces the necessary flux.

turns ratio The ratio of turns in the primary and secondary coils of a transformer $\dfrac{N_P}{N_S}$.

van der Waals bond A very weak bond between neutral atoms or molecules due to varying charge densities in each, e.g. between argon atoms.

vector A quantity (e.g. velocity, force, momentum) with both magnitude and direction.

velocity A vector quantity – the distance per unit time travelled in a specified direction.

volt (V) The S.I. unit of electrical potential and potential difference, V: $1\,V = \dfrac{1\,J}{1\,C}$

watt (W) The S.I. unit of power, $P = \dfrac{\Delta E}{\Delta t}$. One Watt is equal to one joule per second.

wave-front A wave-front is an imaginary line or surface that moves along with a wave. All points on the wave-front have the same phase. In water waves, the line of the crest of the wave can be thought of as a wave-front.

weber Unit of magnetic flux – if the flux in a simple circuit of one coil changes at a rate of $1\,Wb\,s^{-1}$, an e.m.f. of $1\,V$ will be induced.

work The energy transfer $W = Fs$ associated with a force F moving the point of application a distance s in the direction of the force.

work function The work function Φ is the minimum energy required to eject photoelectrons from a given surface. The work function for a given material is found by multiplying the threshold frequency for that material by the Planck constant.

work, W An energy transfer when a force moves its point of application: W = component of force in the direction of the displacement × displacement = $Fs\cos\theta$.

worldline A line on a space–time diagram representing the path of an object through space and time.

yield stress Yield stress is the stress at which a specimen begins to yield (where plastic deformation begins).

Young modulus The Young modulus is the ratio of stress over strain, $\dfrac{\sigma}{\varepsilon}$. The units of the Young modulus are Pa or $N\,m^{-2}$.

Answers

1.1

1 a curvature = 3.1 D

2 a power = (+) 20 D

   b curvature added = 20 D

3 focal length = 140 m

4 a Note that the wavelength is the same either side of the lens.

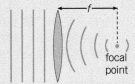

   b The focal length is twice that in **4 (a)**.

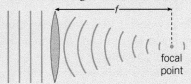

1.2

1 image distance = 53 mm

2 magnification = 1.7

3 Magnification is defined as $\dfrac{\text{image distance}}{\text{object distance}}$. If the image distance is less than the object distance the magnification will be less than one — the image size will be smaller than the object size.

4 a power of lens = +5 D

   b To form an image further from the lens the curvature of waves leaving the lens must be less. Moving the lamp nearer the lens means that the curvature of the waves incident on the lens will be more negative. As the lens adds a constant (positive) curvature to waves passing through it, the (positive) curvature of the waves leaving will be less.

   Alternatively, you can argue using the lens equation. If $\dfrac{1}{u}$ becomes more negative (as it will when u decreases) $\dfrac{1}{v}$ will decrease. If $\dfrac{1}{v}$ decreases, v must increase.

1.3

1 a mean = 111

   b median = 100

   c Replacing the pixel with the median is the best method. A 'noisy' pixel will have a very different value from the surrounding pixels. Using the mean would result in a pixel which still has a different value from the surrounding pixels and so could still show up as noise.

2 number of bits = 12
   number of bytes = 1.5

3 a Resolution is the distance on the object that is represented by one pixel in the image.

   b number of pixels = 1×10^6

   c memory required = 3 Mbyte

1.4

1 wavelength = 3.3 m. Same number of significant figures in the answer as the data provided.

2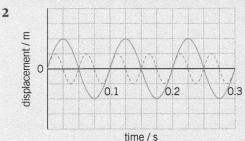

The solid line represents a wave of time period 0.1 s. The dashed line represents a wave of frequency 20 Hz (time period = 0.05 s).

3 Rotate the receiving aerial in a plane at right angles to the direction of travel of the waves. If the radio waves are polarised the received signal will vary in amplitude from maximum to minimum as the aerial is rotated through 90°. If there is no change in amplitude on rotation of the aerial the radio emission is unpolarised.

2.1

1 1016 bits

2 a 64 levels

   b 0.19 V

3 Many possible advantages including: ability to edit, error correction, ease of storage, data compression, digital files are playable on a number of devices. These are all possible because the data is numeric and can be mathematically processed easily and quickly.

4 $\log_2 75 = 6.2$ so 7 bits needed

5 a See Figure 4 on main content page.

   b Quantisation error is the difference between the signal value and the quantisation level that represents the signal value.

2.2

1   256 kHz

2   3.6 ms

3   For accurate recording the sample rate must be greater than twice the highest frequency present in the signal. If the sample rate is lower than this, aliasing can occur producing spurious low frequency signals. If the original signal has a maximum frequency of 20 000 Hz, the sample rate should be greater than twice this, that is, 40 000 Hz.

4   Aliasing could produce lower frequency signals that are within the range of human hearing from frequencies above the range of hearing.

5   a   For example, ease of manipulation can lead to false images and ease of communication can lead to security lapses.

    b   For example, digital images can be altered easily. Techniques such as edge detection and others can be very useful in medical imaging and are of great benefit, but images can also be altered with malicious intent.

3.1

1   a   5 A

    b   60 W

2   $V_1 = 3.4$ V, $V_2 = 2.6$ V and $I_1 = 0.3$ A (one mark each)

3   1.5×10^{18}

4   copper ions: 1.3×10^{20} ions
    chloride ions: 2.6×10^{20} ions

3.2

1   26 V

2   a   As current increases, the wire heats and resistance increases, so $\dfrac{I}{V}$ gets smaller.

    b   If R and G were constant, the I-V graph would be linear with gradient G and the R-V and G-V graphs would be horizontal straight lines. The I-V graph starts linear but drops more and more below that line as V and I increase. This shows that R is increasing and G decreasing, as shown by the G-V and R-V graphs.

3   15.9 °C (3 s.f.)
    if resistance not measured to high precision, $R_\theta - R_0$ will not have 3 s.f.

4   R increases (from a non-zero value) at 0 °C to 100 °C; G decreases (from a non-zero value) over that range (ignore curvature in either)

5   4.1 V

3.3

1   $\rho = \dfrac{RA}{L} = \Omega\ \text{m}^2/\text{m} = \Omega\ \text{m}; \sigma = \dfrac{1}{\rho} = (\Omega\ \text{m})^{-1}$
    $= (\Omega)^{-1}\ (\text{m})^{-1} = \text{S m}^{-1}$

2   1.3 m

3   $G \approx 7 \times 10^{-10}$ S $\Rightarrow I \approx 3 \times 10^{-6}$ A so the experiment should be possible.

3.4

1   Any two points from: silver has more atoms per unit volume, so more free electrons; crystal structure is more closely packed in iron so that electrons are obstructed more; iron has larger ion cores so electrons are obstructed more.

2   $v_B = 10 \times v_A$ because total charge flowing through per second is the same so $(10n_B)\ Av_A e = (n_B)\ Av_B e$.

3   $I = \dfrac{\Delta Q}{\Delta t}$ so $\Delta Q = I\Delta t$ $N = \dfrac{\Delta Q}{e} = \dfrac{I\Delta t}{e}$

    $n = \dfrac{N}{V} = \dfrac{N}{LA} = \dfrac{I\Delta t}{eLA}$ so $L = \dfrac{I\Delta t}{nAe}$

    $v = \dfrac{L}{\Delta t} = \dfrac{\left(\dfrac{I\Delta t}{nAe}\right)}{t} = \dfrac{I}{nAe}$

3.5

1   5.0 V

2   Suitable fixed resistor R in range 5.4 kΩ to 1.0 MΩ, although 200 kΩ to 600 kΩ is better.

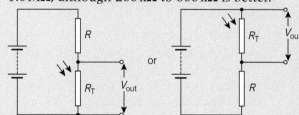

3   The voltmeter has too low a resistance so that when it is in parallel with resistor **A**, the value of parallel resistance drops to 5000 Ω and, as it is $\dfrac{1}{3}$ of the total resistance, it takes $\dfrac{1}{3}$ of the total p.d. = 2.0 V; the same happens when the voltmeter is connected in parallel with resistor **B**.

3.6

1   The p.d. across the terminals $V = \varepsilon - Ir$ and so is less than the 'expected' p.d. ε by the amount Ir, which is the p.d. across the internal resistance. This p.d. seems to have vanished: it is 'lost' inside the cell.

2 3.6 V, 0.41 W

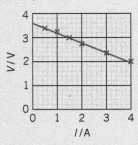

3 The starter motor draws a (very) large current, so the total current through the internal resistance increases. This means the 'lost volts' Ir increase, so that the p.d. across the terminals drops. This p.d. is not enough now to light the car lights to full brightness.

4.1

1 a Steel is a metal, china is a ceramic, rubber and acrylic are polymers.

b Link the properties of the material to its classification. For example, steel is ductile, china is hard and brittle, and the polymers are both tough.

2 For example, keyboard plastic is a tough material, window glass is brittle.

3 Scalpels need to be hard and stiff. Hardness is required to ensure the blade is not blunted. Stiffness is required because the scalpel shouldn't flex.

4 a For example, strong, stiff, hard. (The outer casing of the phone protects the sensitive electronics.)

b The outer sleeve or case needs to be tough to dissipate energy on impact.

c Casing: often steel (metal) with a glass face. Sleeve: often made from a polymer (leather, rubber, or others).

4.2

1 See Figure 5 on main content page.

2 $5500\,\mathrm{N\,m^{-1}}$

3 a The graph is a straight line through the origin.

b $83\,\mathrm{N\,m^{-1}}$

c $0.20\,\mathrm{J}$

4 The student realises that the graph shows proportional behaviour and assumes that this behaviour is maintained up to loads of $35.0\,\mathrm{N}$. Using $F = kx$ and the calculated value of k in **3 (b)** gives an extension of 42.0 cm at a force

of $35.0\,\mathrm{N}$. This may be incorrect as the spring may begin extending non-elastically at a lower force than $35.0\,\mathrm{N}$, making the Hooke's Law equation inapplicable.

4.3

1 a Fracture stress is the stress at which a material breaks (fractures). Yield stress is the stress at which plastic deformation first occurs.

b $10^2\,\mathrm{MN\,m^{-2}}$, $10^2\,\mathrm{N\,mm^{-2}}$

2 a 0.12 or 12%

b 6.3 cm. Assume elastic behaviour.

3 extension (assuming circular cross-section of radius 25 mm) = 7.6 mm

4 a percentage uncertainty in diameter is $\pm 6\%$

b $5.8 \times 10^{10}\,\mathrm{N\,m^{-2}}$

c range: $5.5 - 6.1 \times 10^{10}\,\mathrm{N\,m^{-2}}$

d If the mass and density are known, the volume of the wire can be calculated using the equation volume = mass/density. Measuring the length of the wire allows a calculation of cross-sectional area as cross-sectional area = volume/length. If the mass, density, and length are known to small percentage uncertainties, the percentage uncertainty in cross-sectional area may be smaller than obtained by measuring the diameter of the wire.

e The shape of the curve will be similar to that in Figure 5, Topic 4.3. The yield stress of copper is about 100 MPa. The fracture stress of copper is about 200 MPa.

4.4

1 From Figure 4 you can see that the stiffest metal has a Young Modulus of about 400 GPa. The least stiff metal has a Young Modulus of about 12 GPa (it is difficult to see exactly). The ratio is $\frac{400}{12}$, which is 30 (to 1 s.f.).

2 between 700 and 800

3 a 10.2 kN

b 3.6 mm

c The wire will yield at lower stress. The designers have to include margin for error to cover, for example, the lift being used by eight people of greater average weight than 650 N.

5.1

1 Information required: number of atoms along one side of the image and length represented by one side of the image. The diameter of

one atom can be estimated from the ratio
$\dfrac{\text{length of one side of the image}}{\text{number of atoms along one side}}$

2 The oil patch will be at least one molecule thick so one molecule cannot be longer than the calculated value. If the patch is more than one molecule thick, the length of the molecules will be less than that calculated.

3 order of magnitude = 10^{-9} m (calculated value = 2.1×10^{-9} m)

4 The density of gold is found by measuring the volume of a known mass of gold. To apply this density value to a single atom assumes that there are no spaces between the atoms in a macroscopic specimen of gold.

5.2

1 Amorphous materials are disordered at the microscopic scale. Polycrystalline materials have grains oriented randomly but with an ordered (crystal) structure within the grains.

2 Weakness is a description of breaking stress. Glass (the material) does not have a low breaking stress. Strong materials have a high breaking stress. Brittleness is a description of the fracture process in a material. The fracture process is independent of the strength of the material.

3 Metals may have mobile dislocations. These allow planes of atoms to slip over one another by breaking the bonds one at a time. Plastic deformation can occur at a lower stress if dislocations are present. Alloying atoms can stop the dislocations moving through the metal, making it less plastic. This increases the hardness and brittleness of the material. See Figure 5, Topic 5.2.

4 See explanation under *Stress concentration and crack propagation*, Topic 5.2. The stress is concentrated at the tip of the crack as the area at the tip is small.

5 Dislocations in ceramics are not mobile and so cannot move through the material.

5.3

1 The material may become stiffer because the chains will not unravel so easily.

2 The bonds in metals are strong but non-directional. This allows the ions to slip. Ions in ceramics are held together by directional bonds that lock them into place.

3 Metals deform elastically when the distance between positive ions increases. Up to strains of around 0.1% the ions will be pulled back into position when the deforming force is removed. At strains greater than this the force between the ions grows weaker. Polymer elasticity is explained by the rotation of the bonds in the long chain molecule, allowing it to unravel.

4 Cross-linking joins polymer chains together at points along the chains. This reduces the flexibility of the polymer because the chains cannot unfold to the same degree.

6.1

1 a The two waves are on top of each other.

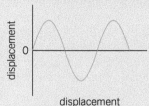

b The two solid lines shown.

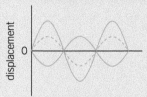

c

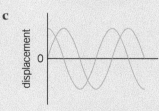

d See dashed line in **1(b)**.

2 a A progressive wave travels along a string or tube. The wave reflects at the end of the string or tube. Waves travelling in opposite directions superpose. At points where the waves always meet in phase an antinode of maximum displacement forms. Where waves always meet in antiphase a node of minimum displacement is formed.

b distance between nodes = 0.25 m

3 2.8 m

4 a 15 GHz

b 0.02 m

c

transmitter   detector   reflector

Place the reflector about 1.5 m from the transmitter. Place the detector between the transmitter and receiver. A minimum signal shows at the point at which the waves from the transmitter and reflector are in antiphase. Measure the position of such a minimum signal. Slowly move the detector towards the reflector, observing the amplitude of the detected signal. Find the next minimum signal position. The distance between one minimum and the next is half a wavelength.

6.2

1 $1.2 \times 10^8 \, \text{m s}^{-1}$. Number of significant figures is that of the least precise value in the data given.

2 27°

3 a

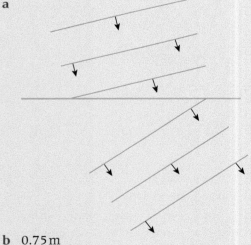

b 0.75 m

4 If the refractive index of the two materials is the same the light will not deviate when it moves from one material to the next. Glycerol and Pyrex are both transparent so the tube will only be detected by the distortion it produces. As there is no deviation of light the observer will not see any distortion and so the tube is invisible.

5 a Refractive index gives a measure of the change of speed when light enters the material.

b Different wavelengths are deviated by different amounts. As the deviation of light in the glass is due to the speed changing on entering the glass it follows that different wavelengths travel at different speeds in glass.

c In this case, refractive index is a measure of the amount of deviation when light enters the glass from air. As different wavelengths are deviated by different amounts the refractive index of the glass is wavelength-dependent.

6 a In this case, refractive index is a measure of the amount of deviation when light enters the glass from air. As different wavelengths are deviated by different amounts the refractive index of the glass is wavelength-dependent.

sin i	sin r +/− 0.04
0	0
0.17	0.09
0.34	0.24
0.5	0.34
0.64	0.44
0.77	0.53
0.87	0.66

b

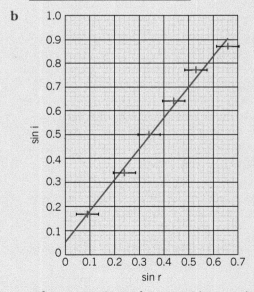

(The uncertainty of +/− 0.04 is a pessimistic estimate based on calculating the range of sin r values when r has an uncertainty of +/− 2°.)

Gradient of graph = 1.3 +/− 0.1

6.3

1 Path difference is the difference in distance between two (or more) sources of waves and a receiver. Phase difference is the difference in the position in the wave cycle of two (or more) waves at a given instant.

2 Coherent waves will show a constant phase difference

3 **a** Path difference = $n\lambda$

    b 0.75 m

    c Path difference has changed by 0.25 m. This represents a change of path difference of one wavelength. Therefore, the wavelength of the sound is 0.25 m.

6.4

1 **a** Diffraction is the spreading of waves when they pass through a gap or around an object. There is no change to speed, wavelength, or frequency on diffraction.

    b 4.8×10^{-7} m

2 **a** 19°

    b 6 orders

3 **a** about 0.03°

    b It is too small to be easily noticed. For example, a first order minimum of 0.09° corresponds to width of about 1.6 mm on a screen 1 m from the slit. This is a very small amount of spreading.

4 **a** Largest value is about 1600 nm, smallest value about 800 nm.

    b The measurement of the gap between the slits contributes most uncertainty. The percentage uncertainty in this measurement is 20% compared to 14% for the fringe separation and less than 1% for the slit-screen distance.

7.1

1 2.0×10^{-24} J, 1.3×10^{-5} eV; 4.0×10^{-19} J, 2.5 eV; 7.9×10^{-16} J, 5.0 keV

2 Use the ideas in the chapter, remembering to describe the phenomenon carefully and then explain it using energy quanta.

3 work function = 4.8×10^{-19} J, threshold frequency = 7.3×10^{14} Hz

4 **a** Energy transfer when an electron of charge e passes through a potential difference $V = eV$. The energy is released as a photon.

    b

wavelength of light / nm	frequency of photon / 10^{14} Hz	striking potential / V	photon energy / 10^{-19} J
470 +/− 30	6.4 +/− 0.4	2.6 +/− 0.2	4.2 +/− 0.4
503 +/− 30	6.0 +/− 0.4	2.5 +/− 0.2	4.0 +/− 0.4
585 +/− 30	5.1 +/− 0.4	2.1 +/− 0.2	3.4 +/− 0.4
620 +/− 30	4.8 +/− 0.4	2.0 +/−0.2	3.2 +/− 0.4

    c Correctly plotted graph using the values given in **4(b)**.

    d gradient from graph gives a value of about 7×10^{-34} J s. Uncertainty can be estimated by considering the variation of gradients possible within the uncertainty bars. These gradients give a wide range of values. The uncertainty can also be estimated by considering the percentage uncertainty in each variable.

7.2

1 rotations per second = 5.0×10^{14}

2 **a** 2.0×10^{-8} s

    b 1.2×10^{7} rotations

3

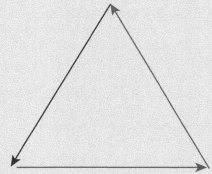

4 **a** All these trip times are the same.

    b Photons from all three paths meet in phase (phasor arrows are all in the same direction).

c Phasor amplitude will be greatest at F, therefore the probability of the arrival of a photon will be greatest at F. As intensity is proportional to probability of arrival, this will also be greatest at F and lower elsewhere where the path lengths are different and the phasors do not meet in phase.

5 a 3×10^{11} mm s^{-1}

b 1 ns

c 5×10^{-11} s

d 2×10^{11} mm s^{-1}

e 1.5

f They have the same trip time because the phasors following route ADEB slow down in the glass but their frequency of rotation remains the same.

7.3

1 1×10^{-10} m

2

Momentum mv	Speed v	Wavelength
6.6×10^{-26} kg m s^{-1}	7.3×10^4 m s^{-1}	10 nm
1.0×10^{-24} kg m s^{-1}	1.1×10^6 m s^{-1}	0.66 nm
1.8×10^{-24} kg m s^{-1}	2.0×10^6 m s^{-1}	0.37 nm

3 The wavelength of an electron of momentum 7×10^{-24} kg m s^{-1} is approximately 9×10^{-11} m. This is far smaller than the wavelength of a photon of equivalent energy. As interference effects only become noticeable when the gap is of the order of magnitude of the wavelength, electrons will require smaller gaps for interference effects to be observed.

4 a A spot is formed when a discrete particle strikes the image in a particular place. The energy of the particle is transferred over a small area. This would not happen with a wave.

b This is a superposition effect. Such effects are usually associated with waves.

c Phasors explore all possible paths. Where the phasor arrows combine to form a large resultant there is a larger probability of an electron being found at that spot. The probability of detection of an electron can be calculated from the square of the resultant phasor.

8.1

1

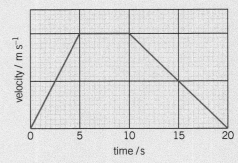

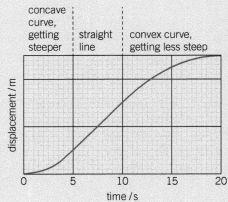

concave curve, getting steeper | straight line | convex curve, getting less steep

2 Answer in range 1950 – 2150. Displacement is not distance travelled as direction changes, so the car is not always heading directly away from the starting place.

3 $a = 3.5$ m s^{-2}
distance = area under curve = 125 m ($\pm$ 8 m)

8.2

1 magnitude of displacement = 21 paces
direction = N 17° E / bearing 017°

2 Diagram has boat pointing upstream at angle θ so that resultant velocity is straight across. $\theta = 24°$.

3 velocity = 240 m s^{-1} in a direction N 54° W / bearing 306°

8.3

1

t / s	v / m s⁻¹	s / m
0	0.00	0.00
0.1	0.34	0.00
0.2	0.68	0.03
0.3	1.02	0.10
0.4	1.36	0.20
0.5	1.70	0.34
0.6	2.04	0.51
0.7	2.38	0.71
0.8	2.72	0.95

2 The increments in velocity are calculated only from the time interval (which is constant) and the previous value of acceleration (which equals zero at terminal velocity), that is, by $a\Delta t$. Therefore, when terminal velocity is reached, $a = 0$, so $a\Delta t = 0$, regardless of the time interval, Δt.

3 On each interval, Δs is 0.8 m downstream. In the diagram, scale = 10 divisions = 2 m.

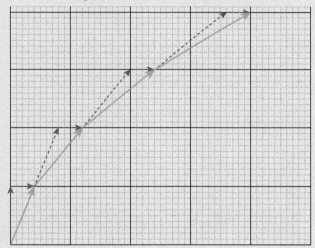

8.4

1 $u = 26.8\,\mathrm{m\,s^{-1}}$ and $s = 55\,\mathrm{m}$ gives $a = -6.5\,\mathrm{m\,s^{-2}}$; $u = 13.4\,\mathrm{m\,s^{-1}}$ and $s = 14\,\mathrm{m}$ gives $a = -6.4\,\mathrm{m\,s^{-2}}$

2 160 m

3 0.86 s

4 0.53 s or 1.9 s: the stone can hit the kite on the way up, or pass the kite and hit it on the way back down again.

9.1

1 a $27\,000\,\mathrm{kg\,m\,s^{-1}}$

b E.g. $98\,\mathrm{kg\,m\,s^{-1}}$, using 70 kg for the mass and $1.4\,\mathrm{m\,s^{-1}}$ for the walking pace.

2 $260\,000\,\mathrm{m\,s^{-1}}$ assuming no other interactions with the atom.

3 $v_{\min}$ car $= 15.9\,\mathrm{m\,s^{-1}}$ for the sum of momenta to be forwards, so yes, the car was going too fast.

9.2

1 $(-)170\,000\,\mathrm{N}$

2 a no momentum before firing, so forwards momentum of bullet = backwards momentum of gun

b gun pushes bullet forwards; bullet pushes gun backwards.

3 By Newton's first law, you carry on going forwards until a force (seatbelt/airbag/windscreen) stops you. The car is being decelerated by a force so, relative to the car, you are still moving forwards.

4 $(-)4000\,\mathrm{N}$ (2 s.f.)

5 $\Delta p = 0.5\,\mathrm{kg} \times 250\,\mathrm{m\,s^{-1}} = 125\,\mathrm{kg\,m\,s^{-1}}$ and $\Delta t = 1\,\mathrm{s}$
$$F = \frac{\Delta p}{\Delta t} = \frac{125\,\mathrm{kg\,m\,s^{-1}}}{1\,\mathrm{s}} = 100\,\mathrm{N}\ (1\ \mathrm{s.f.})$$
At the end of the 3 s, mass lost by the rocket = $0.5\,\mathrm{kg\,s^{-1}} \times 3\,\mathrm{s} = 1.5\,\mathrm{kg}$.

The rocket mass drops from 2.5 kg tp 1.0 kg during the flight.

$a = \dfrac{F}{m}$, so with constant F from the exhaust gases and decreasing m of the rocket, a will increase.

9.3

1 E.g. $530\,000\,\mathrm{J}$

2 0.65 m, or 0.85 m from the lowest point.

3 $\frac{1}{2}mv^2 \times 2m = m^2v^2 = p^2$

4 $3\,\mathrm{m\,s^{-1}}$

9.4

1 Carries on in a straight line until it 'runs out of impetus', and then falls vertically. A real projectile is falling all the time from leaving the mouth of the cannon, and follows a parabolic trajectory.

2 64 m
working to 4 s.f., 45° gives 63.78 m while both 44° and 46° give 63.74 m.

3 $t = 0.79\,\mathrm{s}$ when $s_h = 4\,\mathrm{m}$, which gives $s_v = 1\,\mathrm{m}$ (the net is 1 m above the thrower).

9.5

1 $81\,000\,\mathrm{W}$

2 4.8 J

3 $110\,000\,\mathrm{J}$

10.1

1 11.1 days

2 $1.9 \times 10^{16}\,\text{Bq}$

3 $\lambda = \dfrac{10^6\,\text{Bq}}{10^{15}} = 10^{-9}\,\text{s}^{-1}$

4 a $\dfrac{A}{A_0} = e^{-\lambda t} = e^{-\left(\frac{\ln 2}{0.5}\right)5} = 9.8 \times 10^{-4} \approx 0.001$

   b ~1 000 000

5 a Approximately 500 000

   b Plot log count rate against time

   c 2.28 hours

   d $8.4 \times 10^{-5}\,\text{s}^{-1}$

   e 84

   f About 1.6 million

6 a The decay constant gives the probability of a nucleus decaying in unit time. It is impossible to predict which nuclei will decay in any time interval, so radioactive decay is random. However, knowing the probability of decay of a nucleus and the number of nuclei in a sample allows the number decaying in unit time to be accurately predicted.

   b If few nuclei are present, the random nature of decay becomes more apparent. For example, if one throws six million dice one can predict how many sixes will be thrown to a good level of precision (it will be 'around' a million). The prediction is much less certain when, say, six dice are thrown. The value of the half-life is important because when a large number of half-lives have passed it is extremely likely that the majority of nuclei will have decayed.

10.2

1 $9.4 \times 10^2\,\text{s}^{-1}$

2 1.24 Bq

3 a 25%

   b $\dfrac{A}{A_0} = 0.96\ (96\%)$

4 a $3.2 \times 10^{-5}\,\text{s}^{-1}$

   b 3.1×10^7

   c $6.9 \times 10^{-7}\,\text{J}$

   d $5 \times 10^{-18}\,\text{kg}$

   e 4.0 Bq

10.3

1 a $E = \dfrac{1}{2}CV \times V = \dfrac{1}{2}CV^2$

   b $E = \dfrac{1}{2}Q \times Q/C = \dfrac{1}{2}Q^2/C$

2

Capacitance / F	p.d. / V	Charge / C	Energy / J
2.2×10^{-3}	20	0.044	0.44
1.0×10^{-3}	15	0.015	0.11
2.2×10^{-4}	23	5.1×10^{-3}	0.058
2.2×10^{-6}	54	1.2×10^{-4}	0.0033

3 a $8.46 \times 10^{-3}\,\text{J}$

   b Energy stored at 3.00 V is $2.11(5) \times 10^{-3}\,\text{J}$, so energy released $= 6.35 \times 10^{-3}\,\text{J}$.

4 a From gradient, capacitance = 4700 µF.

   b From area under graph or using $E = \dfrac{1}{2}CV^2$, energy = 0.235 J

   c The amount of charge on the capacitor is proportional to the p.d. across the capacitor, so a capacitor cannot be described as 'fully charged', as a greater p.d. across the capacitor will result in greater charge on the capacitor.

   d If the maximum p.d. is exceeded it is possible that the insulating medium separating the charged surfaces will break down and allow current to flow through the capacitor. This could cause heating and damage the capacitor.

10.4

1 a 2.35 s

   b 6.7×10^{-4} or 0.67%

2

t / s	p.d. / V	Q / C	I / A	ΔQ / C
0	4.5	0.02115	0.0009	0.0045
5	3.542553	0.01665	0.000709	0.003543
10	2.788818	0.013107	0.000558	0.002789
15	2.195453	0.010319	0.000439	0.002195
20	1.728335	0.008123	0.000346	0.001728
25	1.360604	0.006395	0.000272	0.001361
30	1.071114	0.005034	0.000214	0.001071

Note — a spreadsheet was used for this model, hence the precision in the values.

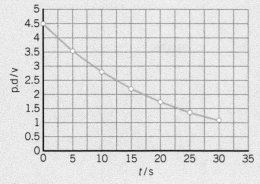

▲ Figure 1

The model could be improved by reducing the time interval between iterations.

3 Ratio $t/RC = 2.3(03)$

4 $700\,\Omega$

5 $4.7 \times 10^{-4}\,F$

11.1

1 $2.27\,ms$

2 a $3.8\,mm$ $(0.000\,38\,m)$

   b $200\,m\,s^{-2}$

3 $0.13\,m$

11.2

1 The displacement curve is a cosine curve with maximum positive displacement at $t = 0$. The acceleration curve is a cosine curve with maximum negative displacement at $t = 0$. See Figure 1 in Topic 11.2.

2 a $2\pi\,m\,s^{-1}$ $(6.3\,m\,s^{-1})$

   b $0.39\,m\,s^{-2}$

3 a $0.74\,Hz$

   b Maximum acceleration is twice that of the first model. Maximum acceleration is proportional to the negative of the amplitude.

   c Maximum velocity is twice that of the first model. Maximum velocity = $2\pi \times$ frequency $\times$ amplitude.

   d It holds the maximum displacement constant for the first iteration. This can be seen from the horizontal line from $0.00\,s$ to $0.05\,s$ in Figure 6, Topic 11.2.

11.3

1 The time period increases by a factor of $\sqrt{2}$.

2 a To increase the time interval measured and hence reduce the percentage uncertainty in the measurement.

b

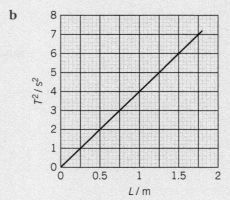

▲ Figure 2

The graph is a straight line through the origin, showing that $T^2 \propto L$. Therefore, $T \propto \sqrt{L}$. The gradient of the graph is $\dfrac{4s^2}{m} = \dfrac{4\pi^2}{g}$.

c

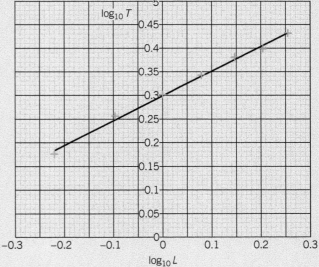

▲ Figure 3

Gradient = 0.5. This shows that $\log_{10} T \propto 0.5\log_{10} L$, therefore $T \propto \sqrt{L}$.

3 a $T = 2\pi\sqrt{\dfrac{L}{g}}$

   b $0.25\,m$

   c The oscillations will be heavily damped, so measuring time intervals beyond a few seconds would be impossible.

11.4

1 $3.1 \times 10^2\,N\,m^{-1}$

2 a $k = \dfrac{F}{x}$. Values substituted and evaluated.

   b $50\,J$

   c $3.2\,Hz$

3 a $5 \times 10^{12}\,\text{Hz}$

   b $5 \times 10^{12}\,\text{Hz}$

   c $49\,\text{N}\,\text{m}^{-1}$

12.1

1 a Frictional force between child and surface of roundabout (assuming child is not holding on).

   b Reaction force between train wheels and the side of the rails.

   c Tension in the arm of the athlete.

2 $4\,\text{m}\,\text{s}^{-2}$

3 a $39\,\text{m}\,\text{s}^{-1}$

   b $2.6\,\text{rad}\,\text{s}^{-1}$

4 Calculated answer = $125\,\text{N}$. The child's weight ($= mg$) is about $200\,\text{N}$, so total force at the bottom of the swing = $325\,\text{N}$.

5 a $F = 8.6 \times 10^{-8}\,\text{N}$, the electrostatic attraction between the proton and electron.

   b Electrostatic force between the particles.

6 The centripetal acceleration of the Earth in orbit $= 0.006\,\text{m}\,\text{s}^{-2}$. This is less than a fifth of that of a body on the surface of the Earth.

12.2

1 $4.4 \times 10^{-8}\,\text{N}$. Forces are equal in magnitude, in opposite directions.

2 Calculated mass = $6.0(4) \times 10^{24}\,\text{kg}$. Calculated value for $G = 6.6(5) \times 10^{-11}\,\text{N}\,\text{m}^2\,\text{kg}^{-2}$.

3 Weight (force due to gravitational field) will decrease with distance from the Moon (dropping to zero at about nine-tenths of the way to the Moon), before increasing as the distance to the Moon decreases. Weight on the Moon will be less than that on Earth. Mass remains constant.

4 i Reading gives your weight.

   ii Reading will decrease because, as the lift is accelerating downwards, the net force on the scales is reduced. (If the lift were to accelerate downwards at $9.8\,\text{m}\,\text{s}^{-2}$ the scales would read zero.

   iii Reading will increase – greater net force on scales.

   iv Reading gives your weight.

5 $7.3\,\text{rad}\,\text{s}^{-1}$

6 $1 : 19\,000$

12.3

1 a $20(.4)\,\text{m}$

   b $200\,\text{J}\,\text{kg}^{-1}$

2 a $50\,\text{kJ}$, $200\,\text{kJ}$, $-50\,\text{kJ}$

   b $50\,\text{J}\,\text{kg}^{-1}$, $200\,\text{J}\,\text{kg}^{-1}$, $-50\,\text{J}\,\text{kg}^{-1}$

   c $50\,\text{J}\,\text{kg}^{-1}$, $200\,\text{J}\,\text{kg}^{-1}$, $-50\,\text{J}\,\text{kg}^{-1}$

   d $150\,\text{J}\,\text{kg}^{-1}$, $100\,\text{J}\,\text{kg}^{-1}$

   e $120\,\text{kJ}$

3 a Calculated value is $9.797\,\text{N}\,\text{kg}^{-1}$, or 9.8 (to 2 s.f.)

   b Hence there is little change in the strength of the field between the surface and $1\,\text{km}$ above the surface, hence the field is uniform.

   c The equipotentials will be equally spaced.

12.4

1 a Calculated value = $-12.5(6) \times 10^8\,\text{J}$

   b $7.4 \times 10^7\,\text{kg}$

2 The spacecraft will be travelling much more quickly through the lower atmosphere on descent. The lower atmosphere is denser so there is more work done against friction, resulting in greater energy transfer to the spacecraft.

3 $8.9 \times 10^{-3}\,\text{m}$

4 a

Radius / 10^6 m	3.4 (surface)	3.9	4.7	**6.1**	**8.6**	14
Potential / 10^7 J kg^{-1}	−1.3	−1.1	−0.9	−0.7	−0.5	−0.3

   b Labelled diagram showing concentric equipotentials with increasing spacing, and radial field lines with arrows towards the planet.

5 Calculated answer using given values is $3.0(04) \times 10^8\,\text{J}\,\text{kg}^{-1}$.

6 Rearrange the equation given in the hint to make v^2 the subject, and then substitute this into the kinetic energy equation.

13.1

1 180 light-seconds (3 light-minutes)

2 The length of time between emission and detection would be too long. The return signal would be too weak.

3

   a Around 800 times.

   b Sirius A is closer to the Earth than other, more luminous stars.

   c The values within the H-R diagram would be too far apart to show any pattern on a linear scale.

4 A suitable logarithmic scale, using A4 paper in landscape orientation, would be 4 cm for each change in the exponent by 10, that is, 10^{-10}, 10^{0}, 10^{10}, 10^{20}, 10^{30}, 10^{40}, 10^{50}.

5

   a 6.06×10^9 m, assuming the asteroid did not move relative to the Earth during the time of measurement.

   b 6.25×10^4 m s^{-1}

13.2

1

   a Hubble's law, the relationship between the distance and the recession velocity of galaxies. Cosmic microwave background radiation.

   b temperature variation and density variation in early Universe

2 8.2×10^{-5} degrees

3

   a 3.2×10^9 years

   b This estimated timescale less than the age of the Earth. The Earth cannot be older than the Universe.

4

   a Calculated value = 44.52 cm

   b Calculated value = 20.3%

5

   a Logarithmic scales, needed to represent a large range of values

   b About 10^{-11} m

   c About 10^5

13.3

1

   a Diagram similar to Figure 2 in Topic 13.3, with worldline passing through (2, 0) and (8, 12).

   b Line at 45° from (0, 1) intercepts worldline at (5, 6). Reflected line intercepts y-axis at $y = 11$ s.

   c 5 light-seconds

2 Calculated answer = $1 + 3.08 \times 10^{-10}$

3

Velocity of object/velocity of light, v/c	Relativistic factor Υ
0.100	1.00
0.200	1.02
0.300	1.05
0.500	**1.15**
0.800	**1.7**
0.900	2.29
0.950	3.20
0.999	**22.4**

4

   a 5.5

   b 2.8×10^8 m s^{-1}

5 This speed corresponds to a gamma factor $\Upsilon = 1.000\,000\,000\,001$, or 1.1×10^{12}.

14.1

1

   a 8 500 000

   b 5.6×10^{30}

2 $4.202\,15 \times 10^{-16}$ m^3

3

   a 7.8×10^4 Pa

   b 1.7×10^5 Pa

14.2

1 230 m s^{-1}

2 8.3×10^{-21} J, 2200 m s^{-1}

3 5 cm = $\sqrt{N_1}$ steps of length L, while 500 cm = $\sqrt{N_2}$ steps of length L, so $\sqrt{N_2} \times L = 100 \times \sqrt{N_1} \times L \rightarrow \sqrt{N_2} = 100 \times \sqrt{N_1} \rightarrow N_2 = 10\,000 \times N_1$, so it should take 10 000 times as long, not 100 times as long.

4 As $\frac{1}{2}mv^2 = \frac{3}{2}kT$, which is the same for both gases, helium has a 7 times smaller m and so a 7 times larger v^2 and a $\sqrt{7}$ times larger v, i.e. 2.6 times larger. The same resonant length of a pipe (the throat and mouth) will have the same wavelength λ fitting into it, but as the velocity v is 2.6 times larger, from $v = f\lambda$ the frequency $f = \frac{v}{\lambda}$ will also be 2.6 times larger, i.e., with a higher pitch (about 2.5 octaves).

14.3

1 $-10\,000\,$J

2 Al, $24\,$J$\,$mol^{-1}K^{-1}; Cu, $25\,$J$\,$mol^{-1}K^{-1}; Fe, $25\,$J$\,$mol^{-1}K^{-1}

3 $0.26\,°$C

4 $V = 430\,$m$^3 \to m = 4.3 \times 10^5\,$kg $\to E = 2.2 \times 10^{10}\,$J $\to P = 130\,$kW. It would need a higher power to compensate for thermal energy transfer to the environment, for example by convection or evaporation.

5 Specific thermal capacity $= 160\,$J$\,$kg^{-1}K^{-1}. Error is due to the efficiency of the energy transfer, with frictional losses into the cardboard tube sides during the fall and by conduction into the top/bottom of the tube when the lead stops falling. Uncertainty due to the resolution of the thermometer results in a 4% percentage uncertainty ($100\% \times \frac{0.1°\text{C}}{2.5°\text{C}}$), so $c = 160 \pm 6\,$J$\,$kg^{-1}K^{-1}.

15.1

1 a $4.2 \times 10^{-21}\,$J

   b $3.8 \times 10^{-23}\,$J

2 $1.0\,$kg $= 36\,$mol $= 2.1 \times 10^{25}$ molecules of lead, so energy per particle $E = 9.3 \times 10^{-21}\,$J. At $40\,$K, $kT = 5.6 \times 10^{-22}\,$J, so $15kT$–$30kT = 8.4 \times 10^{-21}\,$J to $1.7 \times 10^{-20}\,$J. E is in this range, so the process will proceed. At $10\,$K, $kT = 1.4 \times 10^{-22}\,$J so $15kT$–$30kT = 2.1 \times 10^{-21}\,$J to $4.2 \times 10^{-21}\,$J. $E > 30kT$, so the process will not proceed at any reasonable rate.

3 Estimated mean speed $= 400\,$m$\,$s$^{-1} \to E_k = 5.3 \times 10^{-21}\,$J; $kT = 4.2 \times 10^{-21}\,$J, which is similar. The approximation uses a mean speed (not root-mean-square speed). It is difficult to find the mean from the asymmetrical graph. The use of kT is itself only an approximation – kinetic theory would give $\frac{3}{2}kT$.

15.2

1 $kT = 4.1 \times 10^{-21}\,$J, so $5.5 \times 10^{-20}\,$J $= 13kT$. This is just out of the $15kT$–$30kT$ range, but not far, so some conduction would be expected.

2 In each case, f is the ratio of particles in adjacent levels, so $f_A = \frac{1}{2}$, $f_B = \frac{1}{3}$, $f_C = \frac{2}{3}$. A larger f corresponds to a higher temperature (more particles promoted to higher levels) so $T_C > T_A = T_B$.

3 a $n \propto p$, so $n_h = 0.5n_0 \to 0.5 = e^{-\frac{mgh}{kT}} \to \ln(0.5) = -\frac{mgh}{kT} \to -0.69 = -\frac{mgh}{kT} \to h = \frac{0.69kT}{mg}$, where $m = \frac{0.029}{6.0 \times 10^{23}} = 4.8 \times 10^{-26}\,$kg, giving $h = 6000\,$m.

   b The calculation for h includes the term $\frac{T}{g}$, so assuming T and g decrease proportionally to each other, this will have no effect on the answer to (a).

16.1

1 a Flux lines are parallel, straight, and equidistant far from the compass, while near the compass the flux lines curve and vary in separation.

   b The flux lines from the points of the compass are the most distorted. As they shorten and straighten, the compass will align itself with the Earth's magnetic field.

2 a Flux $\Phi = 5.1 \times 10^{-5}\,$Wb

   b Flux linkage $N\Phi = 0.013\,$Wb

   c $0.18\,$V

3 a $2.6\,$V

   b $24\,$A

   c $0.019\,$J

16.2

1 a $16{:}1$ (step-up)

   b $1100\,$A

2 In (b) the plane of each lamination is perpendicular to the flux linkage, allowing the induced e.m.f. to produce current in large loops in each layer. This would cause considerable I^2R joule heating, resulting in energy losses and overheating.

3 $140\,$J

16.3

1 a The e.m.f. is zero at the maxima and minima of the flux linkage graph. The e.m.f. has its extreme values where the flux linkage is zero – maximum where the flux linkage is zero and decreasing, and minimum where flux linkage is zero and increasing.

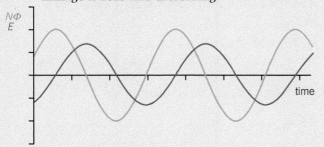

▲ **Figure 4**

b The amplitude of the E–t graph would be double, and the period would halve.

2 a Flux is greatest when the magnet is vertical, and smallest when it is horizontal.

b Assuming that, during a quarter-cycle, the flux changes linearly from 0 to Φ in a time $\Delta t = 0.025\,\text{s}$, then $\dfrac{500\,\Phi}{0.025\,s} = 3.0\,\text{V}$, so $\dfrac{1\,\Phi}{0.025\,s} = 1.5 \times 10^{-4}\,\text{Wb}$.

3 $1000\,\text{V}$ (1 s.f.)

16.4

1 $0.055\,\text{N}$

2 From Figure 4, Topic 16.4, B due to the left-hand wire is coming out of the diagram at the position of the right-hand wire, and this produces a catapult force to the left. Similarly, B due to the right-hand wire is going into the diagram at the position of the left-hand wire, and this produces a catapult force to the right. The wires move toward one another.

3 a $0.13\,\text{N}$

b $0.75\,\text{V}$

17.1

1 a Downwards, equidistant, and parallel in centre, bowing out new the edges.

b Perpendicular to the field lines, equally spaced $\dfrac{1}{5}$ of way down at the centre.

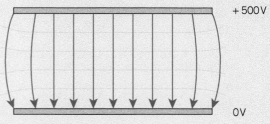

▲ **Figure 5**

c $10\,000\,\text{V}\,\text{m}^{-1}$

2 a $100\,\text{MV}$

b $2\,\text{GJ}$

3 $e = 4.8 \times 10^{-10}\,\text{franklin}$ (1 franklin = $3.3 \times 10^{-10}\,\text{C}$)

17.2

1 a $8.6 \times 10^{-15}\,\text{N}$

b $4.4\,\text{mm}$

2 a The radius of curvature increases as the particle moves along its path, implying that v is decreasing.

b The alpha particle has four times the mass, so identical E_k means it has half the speed. With half the speed and double the charge, qvB is the same and, as $qvB = mv^2/r$ and the alpha particle has four times the mass, the radius of curvature r must be four times greater than that of the proton.

3 a $eV = \dfrac{1}{2}mv^2 \rightarrow v^2 = \dfrac{2eV}{m} \rightarrow v = \sqrt{\dfrac{2eV}{m}}$

b $evB = eE \rightarrow v = \dfrac{E}{B} = \sqrt{\dfrac{2eV}{m}} \rightarrow \dfrac{E^2}{B^2} = \dfrac{2eV}{m}$ and $E = \dfrac{v}{d}$, so $\dfrac{v^2}{d^2B^2} = \dfrac{2eV}{m} \rightarrow \dfrac{e}{m} = \dfrac{V^2}{2Vd^2B^2} = \dfrac{V}{2d^2B^2}$

17.3

1 a $2.5\,\text{N}$

   b $0.38\,\text{J}$

2 a Area under that part of the E–r graph $= \frac{1}{2} \times$
      $(1000\,\text{V} + 500\,\text{V}) \times (0.15\,\text{m} - 0.10\,\text{m}) = 38\,\text{V}$,
      and, from V–r graph, $110\,\text{V} - 72\,\text{V} = 38\,\text{V}$.

   b Gradient of V–r graph at $r = 10\,\text{cm}$ is $\dfrac{210\text{V} - 5\text{V}}{0.20\,\text{m}}$
      $= 1030\,\text{V}\,\text{m}^{-1}$, and the E–r graph shows a little
      over $1000\,\text{V}\,\text{m}^{-1}$ at that distance.

3 a $E = 1.0 \times 10^8\,\text{V}\,\text{m}^{-1}$ vertically upwards

   b $V = 0$

18.1

1 a 1×10^{12}

   b Make assumptions about width of a house
      — e.g., $10\,\text{m}$ — and calculate the ratio of
      lengths.

2 a Greater deflection than original path.

   b Both paths show the greatest deflection
      when the particles are at their closest, as the
      force between particles is proportional to $\frac{1}{r^2}$.
      Deflection is greater for a lower-energy
      particle as it is travelling more slowly, so is
      acted upon by the force for a longer time.

3 $4.1 \times 10^{-12}\,\text{J}$

4 Calculated answer $= 1.23 \times 10^{-14}\,\text{m}$

5 Proton gets closer. Use the equation for
   (electrical) potential energy of two charges in
   the explanation.

18.2

1 a $500\,\text{eV} = 8 \times 10^{-17}\,\text{J}$

   b $1.3 \times 10^7\,\text{m}\,\text{s}^{-1}$

   c The velocity of the electron is not a
      significant fraction of the velocity of light
      ($\Upsilon = 1.001$).

2 a $3.9 \times 10^{-11}\,\text{m}$

   b $6.3 \times 10^{-15}\,\text{m}$

   c $1.2 \times 10^{-16}\,\text{m}$. Relativistic approximation is
      used as $v \approx c$ ($\Upsilon = 20\,000$).

3 a Both particles move through the same
      potential difference and have the same
      magnitude of charge but opposite sign.

   b $v_{\text{electron}}/v_{\text{proton}} = 43$

4 a $(1 + 29)/1 = 30$

   b To 1 s.f., $v = 3 \times 10^8\,\text{m}\,\text{s}^{-1}$. Calculator value $=$
      $2.998 \times 10^8\,\text{m}\,\text{s}^{-1}$.

18.3

1 a neutron

   b $^{216}_{84}\text{Po} \rightarrow\, ^{4}_{2}\text{He} +\, ^{212}_{82}\text{Pb}$

   c 143

2 12

3 Charge of uud $= \left(+\frac{2}{3}\right) + \left(+\frac{2}{3}\right) + \left(-\frac{1}{3}\right) = +1$.
   Charge on udd $= \left(+\frac{2}{3}\right) + \left(\frac{1}{3}\right) + \left(-\frac{1}{3}\right) = 0$
   Change of $-1e$

4 (a) uud; (b) ddu; (c) uuu; (d) ddd. Show how
   the charges on the quarks add up to the required
   charge on the composite particle.

18.4

1 Electrons, positrons, neutrinos, and
   antineutrinos are leptons.

2 They have low mass and no charge so they have
   very little interaction with matter.

3 E.g. Electric charge — all particles are neutral.
   Momentum — particles are created in opposite
   directions.
   Total energy — No new particles are created, so
   the value of $E = mc^2$ does not change.
   Lepton number — No new leptons are created
   or destroyed.

4 Lepton number is not conserved in the equation.
   Correct equation is $^{1}_{0}\text{n} \rightarrow\, ^{1}_{1}\text{p} +\, ^{0}_{-1}\text{e} +\, ^{0}_{0}\overline{\text{v}}$.

5 a It results in one positive charge and one
      equal and opposite negative charge.
      Total charge is zero before and after pair
      production.

   b The two particles move off with equal and
      opposite momenta. Total momentum is zero
      before and after pair production (assuming
      zero velocity collision).

   c A lepton and an anti-lepton are produced.
      Lepton number is zero before and after pair
      production.

6 a $1.6 \times 10^{-13}\,\text{J} = 1\,\text{MeV}$

   b $3.4 \times 10^{-11}\,\text{J} = 210\,\text{MeV}$

18.5

1 Calculator value $= 2.097\,\text{eV}$

2 a $2.2 \times 10^{-18}\,\text{J}$

   b $-3.4\,\text{eV} = -5.4 \times 10^{-19}\,\text{J}$

   c $3.4\,\text{eV} = 5.4 \times 10^{-19}\,\text{J}$

3 Substitute $\text{J} = (\text{kg}\,\text{m}\,\text{s}^{-2}) \times \text{m}$, and $\text{V} = \text{J}\,\text{C}^{-1}$

4 a 4 nm, 2 nm

   b 4.5×10^{-20} J

   c 6.8×10^{13} Hz

   d 4.4×10^{-6} m

   e The difference in kinetic energy calculated in part **(b)** roughly corresponds to the energy of a photon of visible light.

19.1

1 2.1×10^{5} atoms

2 1.2×10^{-7} m

3 a 1.4 m^{-1}

   b 2.1 m, using answer from **(b)**

4 In your answer, include the properties of the radiation that is being used and explain why these properties are important in the application. Amongst other factors, you could consider the half-life required, range, ionising ability, and penetration.

19.2

1 1×10^{8} (calculator value = 1.25×10^{8})

2 9.4×10^{6}

3 14 mSv (activity after 7 days is 1930 Bq, so we can estimate a value by considering the activity to be constant).

19.3

1 $^{208}_{81}$Tl and $^{212}_{84}$Po

2 222 nucleons, 86 protons

3 -1.6×10^{-13} J

4 -1.23×10^{-12} J

19.4

1 -1.23×10^{-12} J

2 Calculated answer = 4.8×10^{-12} J

3 6.5×10^{-13} J, 4.0 MeV

Index

Acknowledgements

Header Photos: CH10-13: Bruce Rolff/Shutterstock; **CH14-15:** IBM Research/Science Photo Library; **CH16-17:** John A Davis/Shutterstock; **CH18-19:** Patrice Loiez, CERN/Science Photo Library;

COVER: DR JUERG ALEAN/SCIENCE PHOTO LIBRARY; **pX:** Rook76/Shutterstock; **pXI:** Esa/Rosetta/Philae/Civa/Science Photo Library; **p7:** Vadim Sadovski/Shutterstock; **p8:** Marques/Shutterstock; **p9:** Roman White/Shutterstock; **p15:** Alexander Mak/Shutterstock; **p16-17:** Giphotostock/Science Photo Library; **p18:** Hikrcn/Shutterstock; **p24:** Johannes Poetzsch; **p25** (R): Andy Crump/Science Photo Library; **p25** (L): Asharkyu/Shutterstock; **p27:** Nasa/Science Photo Library; **p29:** Yanlev/Shutterstock; **p34:** Twin Design/Shutterstock; **p42** (T): Shin Okamoto/Shutterstock; **p42** (B): European Space Agency,J. Huart/Science Photo Library; **p46:** Mino Surkala/Shutterstock; **p55:** Alfred Pasieka/Science Photo Library; **p58:** Andrew Lambert Photography/Science Photo Library; **p59:** Fanfo/Shutterstock; **p63:** Science Photo Library; **p65:** Science Photo Library; **p72-73:** Andrei Seleznev/Shutterstock; **p74:** Trekandshoot/Shutterstock; **p75:** Sebastian Kaulitzki/Shutterstock; **p76:** Tongo51/Shutterstock; **p78:** Bunnyphoto/Shutterstock; **p80** (T): Mycteria/Shutterstock; **p80** (B): Michael Chamberlin/Shutterstock; **p86** (L): Africa Studio/Shutterstock; **p86** (R): Lewis Tse Pui Lung/Shutterstock; **p88** (R): Alan Jeffery/Shutterstock; **p88** (L): Dutourdumonde Photography/Shutterstock; **p92:** Stephan Raats/Shutterstock; **p94:** Philippe Plailly/Science Photo Library; **p96** (L): Science Photo Library; **p96** (C): Power And Syred/Science Photo Library; **p96** (R): Nick Pavlakis/Shutterstock; **p98:** G. Muller, Struers Gmbh/Science Photo Library; **p100:** D. Kucharski K. Kucharska/Shutterstock; **p104** (TL): Bagdan/Shutterstock; **p104** (TR): Antoniomas/Shutterstock; **p104** (C): Boris15/Shutterstock; **p104** (B): Magnetix/Shutterstock; **p105:** Ornl/Science Photo Library; **p108-109:** Richard Peterson/Shutterstock; **p110** (T): Jeng_Niamwhan/Shutterstock; **p110** (B): Nicholas Toh/Shutterstock; **p114:** Andrew Lambert Photography/Science Photo Library; **p117:** Pat_Hastings/Shutterstock; **p119:** Morphart Creation/Shutterstock; **p122:** Mopic/Shutterstock; **p129:** DigitalGlobe; **p130:** Giphotostock/Science Photo Library; **p136:** Marty Pitcairn/Shutterstock; **p137:** David Parker/Science Photo Library; **p138:** Royal Institution Of Great Britain/Science Photo Library; **p144:** A Rose Advances In Biological And Medical Physics 5 211 (1957); **p146:** Science Photo Library; **p149:** OUP; **p150:** Krasowit/Shutterstock; **p152:** Rawpixel/Shutterstock; **p157:** Andrew Lambert Photography/Science Photo Library; **p158:** Clouds Hill Imaging LTD/Science Photo Library; **p160:** American Institute Of Physics/Science Photo Library; **p161:** Molekuul.Be/Shutterstock; **p164-165:** Denis Kuvaev/Shutterstock; **p173:** © XCWeather 2015. All rights reserved.; **p175:** Nicku/Microstock; **p178:** Harvepino/Microstock; **p190:** New York Public Library/Science Photo Library; **p194** (T): 36Clicks/iStockphoto; **p194** (B): Jose Gil/Shutterstock; **p198:** Tonybaggett/iStockphoto; **p203:** Joe Munroe/Science Photo Library; **p207:** Science Photo Library; **p208:** Kristina Postnikova/Shutterstock; **p211:** Pal2Iyawit/Shutterstock; **p213:** 2Happy/Shutterstock; **p216** (T): Library Of Congress/Science Photo Library; **p216** (B): Edstock/iStockphoto; **p217:** Ssuaphoto/iStockphoto; **p220-221:** Bruce Rolff/Shutterstock; **p222:** George Bernard/Science Photo Library; **p231:** Jultud/Shutterstock; **p232:** ZouZou/Shutterstock; **p239**(B): Mariobono/iStockphoto; **p239**(T): Tolga TEZCAN/iStockphoto; **p240:** SHSPhotography/iStockphoto; **p244:** Brian Kinney/Shutterstock; **p255**(B): Library of Congress/Science Photo Library; **p255**(T): Library of Congress/Science Photo Library; **p260**(L): Jim Amos/Science Photo Library; **p260**(R): FelixRenaud/iStockphoto; **p261:** Grzymkiewicz/iStockphoto; **p264:** DeAgostini/Getty Images; **p266**(B): Ria Novosti/Science Photo Library; **p266**(T): Cleanfotos/Shutterstock; **p276:** NASA/Science Photo Library; **p279**(B): NASA/Science Photo Library; **p279**(T): European Space Agency/Science Photo Library; **p283**(B): Meunierd/Shutterstock; **p283**(T): Quayside/Shutterstock; **p284:** Kajano/Shutterstock; **p288:** NASA/ESA/STSCI/S.Beckwith, Hudf Team/ Science Photo Library; **p292:** Harvard College Observatory/Science Photo Library; **p296:** NASA/WMAP Science Team/Science Photo Library; **p297**(B): Dr Jeremy Burgess/Science Photo Library; **p297**(T): NASA/ESA/Stsci/Science Photo Library; **p299:** Sailorr/Shutterstock; **p305**(B): Robert Gendler/Science Photo Library; **p305**(T): NASA/ESA/STSCI/G.Bower & R.Green, Noao/Science Photo Library; **p306:** M J JEE, H FORD/NASA/ESA/STScI/SCIENCE PHOTO LIBRARY; **p310-311:** IBM Research/Science Photo Library; **p312:** Anupan Praneetpholkrang/Shutterstock; **p319**(L): Science Photo Library; **p319**(R): Segre Collection/American Institute of Physics/Science Photo Library; **p325**(B): Science Photo Library; **p325**(T): Harry Willis, Crofton Beam Engines; **p329:** Lefteris Papaulakis/Shutterstock; **p330**(TL): Brian George Jones/Orbiterballoon.com; **p330**(TR): Sheila Terry/Science Photo Library; **p331:** Tatiana Popova/Shutterstock; **p334**(B): Neil langan/Shutterstock; **p334**(T): Pasquale Sorrentino/Science Photo Library; **p338:** Lexaarts/Shutterstock; **p339:** Giideon/Shutterstock; **p344:** Thomas D. Schneider; **p346:** Designua/Shutterstock; **p352**(B): Sergign/Shutterstock; **p352**(T): Giphotostock/Science Photo Library; **p352-353:** John A Davis/Shutterstock; **p357:** JamieWilson/iStockphoto; **p358**(B): Courtesy of Ian Lawrence; **p358**(T): SeDmi/Shutterstock; **p363:** Teun van den Dries/Shutterstock; **p368**(B): Hramovnick/Shutterstock; **p368**(T): Steve Gschmeissner/Science Photo Library; **p375:** Cleanfotos/Shutterstock; **p378**(L): Pavel L Photo and Video/Shutterstock; **p378**(R): Xenotar/iStockphoto; **p379:** Ensuper/Shutterstock; **p383:** Science Photo Library; **p387:** Lawrence Berkeley Laboratory/Science Photo Library; **p388**(B): Emilio Segre Visual Archives/American Institute of Physics/Science Photo Library; **p388**(T): Ted Kinsman/Science Photo Library; **p395:** Geoff Tompkinson/Science Photo Library; **p396:** NASA/Science Photo Library; **p400-401:** Patrice Loiez, CERN/Science Photo Library; **p408:** Maximilien Brice, CERN/Science Photo Library; **p414:** CXC/SAO/F. Seward Et Al/NASA/Science Photo Library; **p417:** Lawrence Berkeley National Laboratory/Science Photo Library; **p420:** Science Photo Library; **p426:** Wellcome Dept. Of Cognitive Neurology/ Science Photo Library; **p427:** Bork/Shutterstock; **p431**(B): Carl Anderson/Science Photo Library; **p431**(T): Science Photo Library; **p443:** AnglianArt/Shutterstock; **p447:** Mikkel Juul Jensen/Science Photo Library; **p450**(B): Argonne National Laboratory/Science Photo Library; **p450**(C): Emilio Segre Visual Archives/American Institute of Physics/Science Photo Library, **p450**(T): EMILIO SEGRE VISUAL ARCHIVES/AMERICAN INSTITUTE OF PHYSICS/SCIENCE PHOTO LIBRARY; **p475:** NASA; **p488:** Bikeriderlondon/Shutterstock;

Artwork by Q2A Media

Although we have made every effort to trace and contact all copyright holders before publication this has not been possible in all cases. If notified, the publisher will rectify any errors or omissions at the earliest opportunity.

Links to third party websites are provided by Oxford in good faith and for information only. Oxford disclaims any responsibility for the materials contained in any third party website referenced in this work.